CAMBRIDGE STUDIES IN MODERN OPTICS

Series Editors

P. L. KNIGHT
Department of Physics, Imperial College of Science, Technology and Medicine
A. MILLER
Department of Physics and Astronomy, University of St Andrews

The physics of laser–atom interactions

TITLES IN PRINT IN THIS SERIES

Optical Holography – Principles, Techniques and Applications
P. Hariharan

Fabry–Perot Interferometers
G. Hernandez

Holographic and Speckle Interferometry (second edition)
R. Jones and C. Wykes

Laser Chemical Processing for Microelectronics
edited by K. G. Ibbs and R. M. Osgood

The Elements of Nonlinear Optics
P. N. Butcher and D. Cotter

Optical Solitons – Theory and Experiment
edited by J. R. Taylor

Particle Field Holography
C. S. Vikram

Ultrafast Fiber Switching Devices and Systems
M. N. Islam

Optical Effects of Ion Implantation
P. D. Townsend, P. J. Chandler and L. Zhang

Diode–Laser Arrays
edited by D. Botez and D. R. Scifres

The Ray and Wave Theory of Lenses
A. Walther

Design Issues in Optical Processing
edited by J. N. Lee

Atom–Field Interactions and Dressed Atoms
G. Compagno, R. Passante and F. Persico

Compact Sources of Ultrashort Pulses
edited by I. Duling

The Physics of Laser–Atom Interactions
D. Suter

The physics of laser–atom interactions

DIETER SUTER

Universität Dortmund

CAMBRIDGE UNIVERSITY PRESS
Cambridge, New York, Melbourne, Madrid, Cape Town, Singapore, São Paulo

Cambridge University Press
The Edinburgh Building, Cambridge CB2 2RU, UK

Published in the United States of America by Cambridge University Press, New York

www.cambridge.org
Information on this title: www.cambridge.org/9780521462396

First published 1997
This digitally printed first paperback version 2005

A catalogue record for this publication is available from the British Library

Library of Congress Cataloguing in Publication data

Suter, Dieter.

The physics of laser–atom interactions / Dieter Suter.

p. cm. – (Cambridge studies in modern optics)

Includes bibliographical references and index.

ISBN 0–521–46239–8

1. Laser manipulation (Nuclear physics) 2. Optical pumping.
3. Laser spectroscopy. I. Title. II. Series: Cambridge studies in modern optics (Unnumbered)

OC689.5.L35S88 1997
539.7–DC21 96–52448
CIP

ISBN-13 978-0-521-46239-6 hardback
ISBN-10 0-521-46239-8 hardback

ISBN-13 978-0-521-01791-6 paperback
ISBN-10 0-521-01791-2 paperback

Contents

Preface *page* xi
Symbols and abbreviations xiii

1 Introduction 1
1.1 Atoms 1
1.1.1 Historical 1
1.1.2 Quantum mechanics 3
1.2 Light 6
1.2.1 The quantum theory of light 6
1.2.2 The classical description 10
1.3 Atom–light interaction 12
1.3.1 General 12
1.3.2 Multilevel atoms 17
1.4 Summary of relevant physical processes 21
1.4.1 Laser spectroscopy 21
1.4.2 Sublevel dynamics 22
1.4.3 Optical properties 25
1.4.4 Magnetic resonance 28
1.4.5 Waves and particles 35

2 Two-level atoms 38
2.1 Quantum mechanical description 38
2.1.1 The Jaynes–Cummings model 38
2.1.2 Summary of results 41
2.2 Semiclassical analysis 46
2.2.1 Model 46
2.2.2 Equation of motion 51
2.2.3 Statics 57

2.2.4 Stationary solution 59
2.3 Dynamics 63
2.3.1 Rabi flopping 63
2.3.2 Free precession 66
2.3.3 Photon echoes 70

3 Three-level effects 74
3.1 Phenomenological introduction 74
3.1.1 Model atoms 74
3.1.2 Coherence transfer 76
3.2 System and Hamiltonian 77
3.2.1 Single-transition operators 77
3.2.2 Irradiation of a single transition 79
3.2.3 Irradiation of two transitions 81
3.3 Three-level dynamics 83
3.3.1 Excitation of a single transition 83
3.3.2 Coherence transfer 85
3.3.3 Three-level echoes 87
3.3.4 Quantum beats 90
3.3.5 Raman excitation 91
3.3.6 Bichromatic excitation 97
3.4 Steady-state effects 99
3.4.1 Coherent population trapping 100
3.4.2 Coherent Raman scattering 105
3.5 Overdamped systems 109
3.5.1 Characteristics 109
3.5.2 Adiabatic limit 112
3.5.3 Optical pumping 113
3.5.4 Light shift and damping 115
3.5.5 Ground state dynamics 117

4 Internal degrees of freedom 119
4.1 Rotational symmetry 119
4.1.1 Motivation 119
4.1.2 Rotations around a single axis 120
4.1.3 Rotations in three dimensions 123
4.1.4 Tensor operators 126
4.1.5 Hamiltonian and Schrödinger equations 128
4.2 Angular momentum 131
4.2.1 Radiation field 131
4.2.2 Atomic angular momentum 134

4.3 Multipole moments 137
4.3.1 Multipole expansion 137
4.3.2 Alignment 139
4.4 Interaction with external fields 140
4.4.1 Electric fields 140
4.4.2 Magnetic interactions 143
4.4.3 Magnetic resonance spectra 146
4.4.4 Larmor precession 149
4.5 Electric dipole transitions 151
4.5.1 Angular momentum exchange 151
4.5.2 Spin–orbit coupling 154
4.5.3 Nuclear spin 158

5 Optical pumping 160
5.1 Principle and overview 160
5.1.1 Phenomenology 160
5.1.2 Historical 161
5.2 Two-level ground states 163
5.2.1 System 163
5.2.2 Longitudinal pumping 165
5.2.3 Relaxation effects 166
5.2.4 Transverse pumping 168
5.2.5 Light shift 171
5.3 Modulated pumping 174
5.3.1 Motivation 174
5.3.2 Equation of motion 177
5.3.3 Rotating frame 178
5.3.4 Polarisation modulation 181
5.4 Multilevel ground states 186
5.4.1 Overview 186
5.4.2 Hyperfine pumping 187
5.4.3 Degenerate multilevel systems 188
5.4.4 The sodium ground state 191
5.4.5 Light shift and damping 196
5.4.6 Diamagnetic ground states 198
5.4.7 Spectral holeburning 200

6 Optically anisotropic vapours 203
6.1 Isotropic atoms 203
6.1.1 The Lorentz–Lorenz model 203
6.1.2 Semiclassical theory 207

6.2 Anisotropic media 209
6.2.1 Introduction 209
6.2.2 System response 211
6.2.3 Magnetooptic effects 215
6.3 Propagation 220
6.3.1 Eigenpolarisations of plane waves 220
6.3.2 Arbitrary polarisation 223
6.3.3 Coherent Raman scattering 226
6.3.4 Transverse effects 228
6.4 Polarisation-selective detection 230
6.4.1 Fundamentals 230
6.4.2 Detection schemes 235
6.4.3 Observables in multilevel ground states 239
6.4.4 The sodium ground state 242

7 Coherent Raman processes 248
7.1 Overview 248
7.1.1 Raman processes 248
7.1.2 Electronic structure of rare earth ions 251
7.1.3 Nuclear spin states 252
7.2 Frequency-domain experiments 256
7.2.1 Spectral holeburning 256
7.2.2 Raman heterodyne spectroscopy 258
7.2.3 Triple resonance 261
7.3 Time-resolved experiments 263
7.3.1 Photon echo modulation 263
7.3.2 Coherent Raman beats 266
7.3.3 Time-domain spectroscopy 269
7.3.4 Examples 279

8 Sublevel dynamics 280
8.1 Experimental arrangement 280
8.1.1 General considerations 280
8.1.2 Setup 282
8.1.3 Historical overview 284
8.1.4 Phenomenology 285
8.2 Spin nutation 286
8.2.1 Signal 286
8.2.2 Experimental control 288
8.3 Free induction decay 291
8.3.1 Theory 291

8.3.2 Experimental control 294
8.4 Spin echoes 296
8.4.1 Introduction 296
8.4.2 Mechanism 298
8.4.3 Control parameters 300
8.5 Modulated excitation 303
8.5.1 Laboratory-frame detection 303
8.5.2 Phase-sensitive detection 305
8.5.3 Frequency-domain experiments 307
8.6 Time-domain spectroscopy 308
8.6.1 Example 308
8.6.2 Microscopic analysis 310
8.6.3 Possible extensions 312

9 Two-dimensional spectroscopy 314
9.1 Fundamentals 314
9.1.1 Motivation and principle 314
9.1.2 Theoretical analysis 316
9.1.3 Coherence transfer echoes 321
9.1.4 Possible applications 322
9.2 Coherence transfer 324
9.2.1 Introduction 324
9.2.2 Example 325
9.2.3 System and Hamiltonian 328
9.2.4 Light-induced dynamics 330
9.2.5 Signal 332
9.3 "Forbidden" multipoles 336
9.3.1 Observables 336
9.3.2 Rotations 339
9.3.3 Separation of multipole orders 346
9.3.4 Coherence transfer echoes 349

10 Nonlinear dynamics 354
10.1 Overview 354
10.1.1 Resonant vapours as optically nonlinear media 354
10.1.2 Wave mixing 357
10.1.3 Coupled absorption 359
10.2 Nonlinear propagation: self-focusing 361
10.2.1 Light-induced waveguides 361
10.2.2 Self-focusing 366
10.2.3 Experimental observation 369

10.2.4 Other structures 372
10.3 Temporal instabilities 374
10.3.1 Feedback 374
10.3.2 Evolution 376
10.3.3 Limit cycles 380
10.3.4 Chaos 382

11 Mechanical effects of light 385
11.1 Light-induced forces 385
11.1.1 Momentum conservation 386
11.1.2 Optical potential 388
11.2 Spontaneous forces 389
11.2.1 Scattering force 389
11.2.2 Doppler cooling 392
11.2.3 Velocity diffusion 395
11.2.4 Doppler limit 397
11.3 Stimulated forces 399
11.3.1 Gradient force 399
11.3.2 Applications 402
11.3.3 Rectified dipole force 406
11.4 Forces on multilevel atoms 409
11.4.1 Multilevel effects 409
11.4.2 Magnetooptic traps 412
11.4.3 Sisyphus cooling 415
11.4.4 Stimulated magnetooptic force 417
11.4.5 Raman transitions 419

References 423
Index 449

Preface

Light interacting with material substances is one of the prerequisites for life on our planet. More recently, it has become important for many technological applications, from CD players and optical communication to gravitational-wave astronomy. Physicists have therefore always tried to improve their understanding of the observed effects. The ultimate goal of such a development is always a microscopic description of the relevant processes. For a long time, this description was identical with a perturbation analysis of the material system in the external fields. More than a hundred years ago, such a microscopic theory was developed in terms of oscillating dipoles. After the development of quantum mechanics, these dipoles were replaced by quantum mechanical two-level systems, and this is still the most frequently used description.

However, the physical situation has changed qualitatively in the last decades. The development of intense, narrowband or pulsed lasers as tunable light sources has provided not only a new tool that allows much more detailed investigation, but also the observation of qualitatively new phenomena. These effects can no longer be analysed in the form of a perturbation expansion. One consequence is that the actual number of quantum mechanical states involved in the interaction becomes relevant. It is therefore not surprising that many newly discovered effects are associated with the details of the level structure of the medium used in the experiment. Two popular examples are the discovery of sub-Doppler laser cooling and the development of magnetooptical traps, which rely on the presence of angular momentum substates.

Chapters 1 and 2 may be considered as a qualitative overview of the subjects relevant in this context and an introduction to the phenomena that can be analysed within the two-level model. Chapter 3 presents a mathematical analysis of effects that are incompatible with the two-level model, whereas Chapter 4 discusses the physical quantities associated with the more complicated level structures. Chapter 5 summarises optical pumping, the oldest technique that allows atomic systems to be driven far from thermal equilibrium. As shown in Chapter 6, such optically pumped vapours are optically anisotropic. Chapter 7 discusses coherent Raman processes, using atomic ions in a crystal matrix as the optical medium. Chapter 8 contains a summary of transient effects in the ground state sublevel system of atomic vapours driven by polarised light and magnetic fields, using laser light to probe the microscopic structure of the vapour. This can be done in much greater detail by using two-dimensional spectroscopy, a technique discussed in Chapter 9. As Chapter 10 shows, the system can be driven far from equilibrium by the interaction with laser light. As a result, the system can show spontaneous symmetry breaking and spontaneous structure formation. The final section summarises the recent development in the field of laser cooling and trapping, where the consideration of multilevel effects has significantly enhanced the possible experimental tools.

I am grateful to many colleagues who have helped me during the course of this work. In particular I should like to mention my former students Tilo Blasberg and Harald Klepel as well as my former supervisor Jürgen Mlynek. Among the people who helped to improve this manuscipt, I should like to mention specifically Rudi Grimm and Scott Holmstrom.

Symbols and abbreviations

Symbol	Explanation
a, $a^\dagger$	annihilation (or lowering), creation (or raising) operator
$\vec{B} = (B_x, B_y, B_z)$	magnetic induction
B_0	amplitude of static magnetic induction
β	Bohr magneton, $9.27 \cdot 10^{-24}$ J/T
c	velocity of light in vacuum, $2.9979 \cdot 10^8$ ms^{-1}
d	dipole moment
FID	free induction decay
γ_{eff}	damping rate
γ_F	gyromagnetic ratio of total angular momentum
γ_J	gyromagnetic ratio of total electronic angular momentum
γ_S	gyromagnetic ratio of electron spin
Γ_1	(optical) spontaneous emission rate
Γ_2	(optical) dephasing rate
Δ	optical detuning
$\overline{\Delta}$	normalised optical detuning
e	charge quantum, $1.60218 \cdot 10^{-19}$ C
ϵ_0	dielectric constant, $0.88542 \cdot 10^{-11}$ As/Vm
$\vec{E}$	electric field vector
$\mathscr{E}$	energy
EFG	electric field gradient
$\vec{F} = (F_x, F_y, F_z)$	total (nuclear + electronic) angular momentum operator
$\mathscr{F}$	Fourier transform operator
h, $\hbar$	Planck's constant, $\hbar = 1.05459 \cdot 10^{-34}$Js

$\vec{H} = (H_x, H_y, H_z)$	magnetic field
$\mathcal{H}$	Hamiltonian
$\vec{I} = (I_x, I_y, I_z)$	nuclear spin angular momentum operator
$\vec{J} = (J_x, J_y, J_z)$	total electronic angular momentum
$\vec{k}$	wave vector
k_B	Boltzmann constant, $1.38066 \cdot 10^{-23}$ JK^{-1}
$\vec{L} = (L_x, L_y, L_z)$	electron orbital angular momentum
l	sample length
λ	wavelength
$\vec{M}$	angular momentum of radiation field
$\vec{m} = (m_x, m_y, m_z)$	magnetisation
m_e	electron mass, $9.1094 \cdot 10^{-31}$ kg
μ_0	magnetic permeability, $4\pi \cdot 10^{-7}$VsA^{-1}m^{-1}
$\vec{\mu}_e$	electric dipole moment
$\vec{\mu}_m$	magnetic dipole moment
$P_R(z,\alpha)$	object rotation by α around z
$\vec{P}$	optical polarisation
$\vec{p}$	linear momentum
$R(z,\alpha)$	coordinate rotation by α around z
$R_E(\alpha, \beta, \gamma)$	Euler rotation
$\vec{r} = (x, y, z)$	position
ρ	density operator
$\dot{\rho}$	derivative of density operator
rf	radio time frequency
$\vec{S} = (S_x, S_y, S_z)$	electron spin angular momentum operator
$\sigma_x, \sigma_y, \sigma_z, \sigma_+, \sigma_-$	Pauli spin operators
$\sigma_i = \|i\rangle\langle i\|$	level shift operator
t	time
ω_0	atomic resonance frequency
ω_x	Rabi frequency
$\vec{\Omega}$	effective magnetic field
Ω_L	Larmor frequency
$\vec{\nabla}$	Nabla operator

1
Introduction

The interaction between matter and radiation has fascinated physicists for a long time. On the material side, the most detailed investigations of these processes concentrate on atoms, the basic constituents of matter. The radiation that is involved in these processes is primarily light, i.e., radiation whose wavelength is in the range of a few tenths of a micron to a few microns. Under today's laboratory conditions, this radiation is generally produced by a laser. This introduction outlines our picture of these constituents and presents some of the concepts and models that we will use throughout this book.

1.1 Atoms

1.1.1 Historical

Early models: atoms as building blocks

The term "atom" was coined by the Greek philosopher Democritus of Abdera (460–370 B.C.), who tried to reconcile change with eternal existence. His solution to this dilemma was that matter was not indefinitely divisible, but consisted of structureless building blocks that he called atoms. According to Democritus and other proponents of this idea, the diverse aspects of matter, as we know it, are a result of different arrangements of the same building blocks in empty space (Melsen 1957; Simonyi 1990). The most important opponent of this theory was Aristotle (384–322 B.C.), and his great influence is probably the main reason that the atomic hypothesis was not widely accepted, but lay dormant for two thousand years. It reappeared only in the eighteenth century, when the emerging experimental science found convincing evidence that matter does indeed consist of elementary building blocks. Chemists discovered that elements react in constant proportions with each other, and that these proportions are related by fractions of small integer numbers. Aristotle's teachings could not explain these experimental findings, but the atomic hypothesis

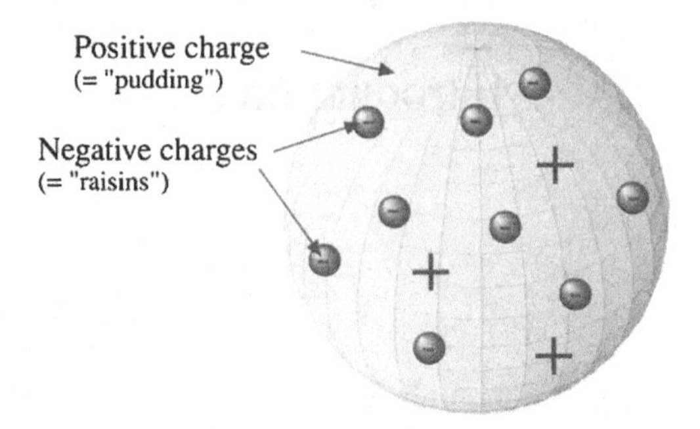

Figure 1.1. Sir Joseph Thomson's model of the atom as a "pudding with raisins."

gave a convincing explanation. During the nineteenth century, most types of atoms were discovered and classified in the periodic system of the elements. The atomic hypothesis also gained recognition as it could explain diverse findings, like the variation of the pressure of gases with temperature or Brownian motion. Although nobody had yet seen an atom, the atomic hypothesis was already quite well established by the end of the nineteenth century.

Also during the nineteenth century, pieces of evidence emerged that indicated that atoms were not the unchanging and structureless ultimate constituents of matter that Democritus had postulated. The discovery of radioactivity showed that they were not eternal, but subject to decay and change. In lightning, cathode rays, and in electrolysis, particles showed up that apparently were smaller constituents of the atom. It soon turned out that these negatively charged particles, which were called electrons, carried only a small part of the atomic mass. Thus, most of the mass had to be concentrated in the remaining part. To make the atom electrically neutral, this remaining part had to carry a positive electric charge. A model for the atomic structure that was quite popular at this time was Sir Joseph Thomson's (1856–1940) "pudding with raisins" that considered the electron as the "raisins" in the much larger, positively charged "pudding" (see Figure 1.1).

Internal structure

It therefore came as a big surprise when the scattering experiments which Ernest Rutherford (1871–1937) performed in the years 1911–1913, showed that the positive charge was concentrated in a region many orders of magnitude smaller than the volume of the whole atom. Although a model of electrons orbiting the positively charged nucleus could explain the apparent size of the atoms, it was in direct contradiction to the newly established field of

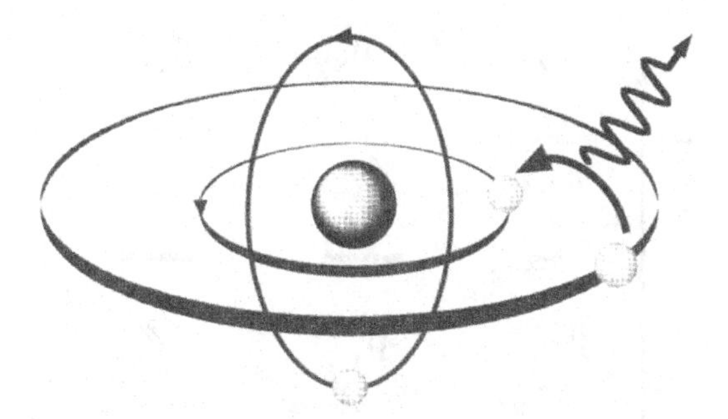

Figure 1.2. Bohr's model of the atom.

electrodynamics: Accelerated charges were known to radiate. Electrons orbiting a nucleus should therefore radiate and lose energy. This would cause them to fall into the nucleus on a short timescale. This prediction of classical electrodynamics was evidently in contradiction to the experimental fact of the stability of atoms and constituted one of the famous inconsistencies that were later on resolved by the quantum theory. Another important hint came from spectroscopy, where Joseph von Fraunhofer (1787–1826) had discovered dark lines in the solar spectrum and Michael Faraday (1791–1867) had shown that magnetic fields could influence the optical properties of various substances.

Niels Bohr's (1885–1962) model of the atom (see Figure 1.2) tackled these problems by postulating stationary states in which the electron did not evolve in time. Energy exchange through absorption or emission of light would be associated with discrete jumps of the electron between different stationary states. A few years later, justification for these assumptions was provided by the quantum mechanics of Erwin Schrödinger (1887–1961) and Werner Karl Heisenberg (1901–1976).

1.1.2 Quantum mechanics

Energy levels

The quantum mechanical picture, still relevant today, tells us that the electrons do not orbit the nucleus in planetlike trajectories, but in spatially extended "orbitals." Depending on the energy of the system, the atom can be in one of an infinite number of stationary states, which are represented mathematically as the eigenstates of the Hamiltonian of the system (Weissbluth 1978).

Figure 1.3 shows the usual representation of the lowest of these orbitals for the hydrogen atom. The lowest state is spherically symmetric, whereas higher lying states have lower symmetry. This description implies that the atom has

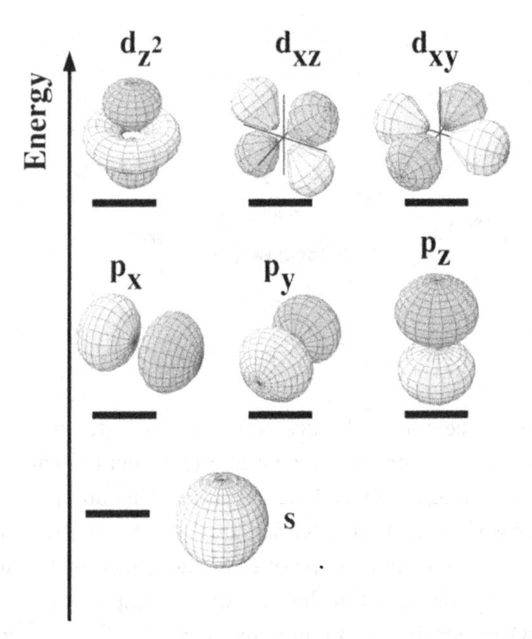

Figure 1.3. Hydrogenlike orbitals.

only a single electron, but even for atoms with a large number of electrons, it has turned out to be a useful model.

Since the main subject of this book is not the structure of the atom, but its interaction with radiation, we do not describe the atom in detail. In particular, we will not have to consider the complete set of energy levels. The most important ones are the energetically highest of the occupied orbitals and the lowest unoccupied orbital, since absorption and emission of light are often accompanied by transitions between these two levels.

Sublevels

As Figure 1.3 indicates, the energy level structure of atoms and molecules consists in most cases of multiplets of degenerate or near-degenerate states. We refer to these states as substates or sublevels. Although their energy differs very little, other physical properties, in particular the angular momentum, exhibit significant variations between them. If a measurement is performed on such a system without distinguishing the various sublevels, the result is a weighted average of the contributions from all states. This is the typical situation in many experiments that probe material properties with electromagnetic fields. Since the interaction between the material and the probing radiation is strongly enhanced when the photon energy is close to an allowed electronic transition, distinguishing the different states is in most cases achieved through their energy. This method cannot easily distinguish between states that are energetically degenerate, however.

Historically, the energy level structure of atomic systems has been investigated primarily with optical spectroscopy. From this point of view, the energy differences are usually classified into electronic transitions, fine structure, hyperfine structure, and Zeeman multiplets. Whereas the fine structure and in many cases the hyperfine structure of these systems can be resolved by optical spectroscopy, the Zeeman level splitting is considerably smaller than the natural linewidth of the optical transitions unless the applied magnetic field is significantly stronger than that of the earth.

In most experimental situations, one does not deal with individual atoms, but with a large number, typically $\geq 10^9$. The observable properties depend then on the population of the various levels, i.e., on the number of atoms occupying a particular eigenstate of the Hamiltonian. According to equilibrium statistical mechanics, these populations depend only on the energy of the states and the temperature of the ensemble. For degenerate states, they are identical, and a measurement performed on such a system averages over all degenerate sublevels with equal weights.

As long as the system is in internal equilibrium, the influence of the sublevel structure on the macroscopically observable properties is small and does not depend on the details of the experimental situation. If the interaction with external fields drives the system far from internal equilibrium, however, the distribution of the sublevel populations may become nonthermal, as in the right-hand side of Figure 1.4. The population of each sublevel, and therefore the weighting coefficients in the averaging process, thus depend on the details of the experimental parameters and usually also on the history of the system. In addition, the external fields may put the atoms not into a single eigenstate of the Hamiltonian, but into superposition states of two or more eigenstates. These superposition states can have physical properties qualitatively different from either of the constituent states, e.g., a nonvanishing electric dipole moment. Without a detailed knowledge of the internal state of the microscopic system, it is difficult to make predictions about its macroscopic properties.

This situation represents the main theme of this book, which concentrates on methods for obtaining detailed and precise information about the stationary properties and on the dynamics of the internal degrees of freedom of atomic multilevel systems. In a somewhat different context, the investigation of the sublevel structure is also the subject of magnetic resonance spectroscopy,

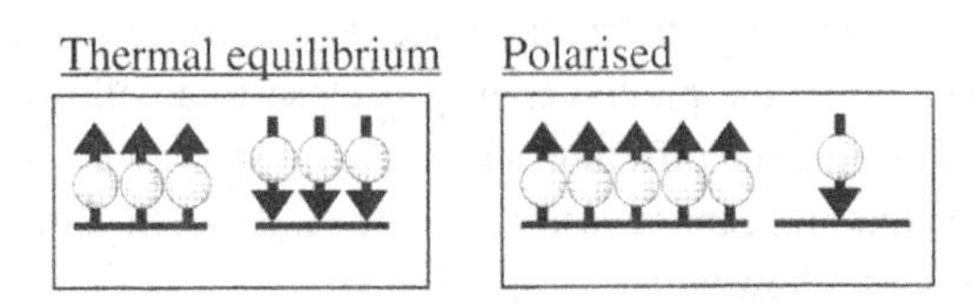

Figure 1.4. Change of overall properties by polarisation of sublevels.

where radio frequency fields are used to induce and observe transitions between different substates. This field, which was the first to introduce coherent methods to spectroscopy, has had a large impact on laser spectroscopy and we will use many of its results in the discussion of optical experiments.

1.2 Light

1.2.1 The quantum theory of light

Introduction

The interaction of light with matter, in particular blackbody radiation and the photoelectric effect, were among the major experimental discoveries that initiated the development of quantum mechanics. The new theory quickly allowed a better description of the material world, in particular atoms. Almost in parallel, P. A. M. Dirac (1902–1984), Heisenberg and Wolfgang Pauli (1900–1958) formulated in 1928 a quantum mechanical theory of light, which was later extended by Shinichiro Tomonaga (1906–1979), Julian Schwinger (1918–1994) and Richard Feynman (1918–1988) (Dyson 1949) and others. This theory, known as quantum electrodynamics, or QED, is today among those physical theories that have been most thoroughly tested experimentally. In all respects, these tests have confirmed the predictions of the theory and today it has an extraordinarily good status – both experimentally and theoretically.

Nevertheless, most theoretical descriptions of light and its interaction with matter describe the radiation field classically, using Maxwell's equations (James Clerk Maxwell, 1831–1879), as we will for most of this book. In a few cases, however, particularly when the conservation laws for energy, momentum and angular momentum are involved, the quantum nature of the light is important. For this reason, we include here a brief, qualitative outline of the main features of the quantum mechanical description of light.

Modes

Even the quantum mechanical theory uses Maxwell's equations to describe the propagation of the light. Quantum mechanical aspects are important only

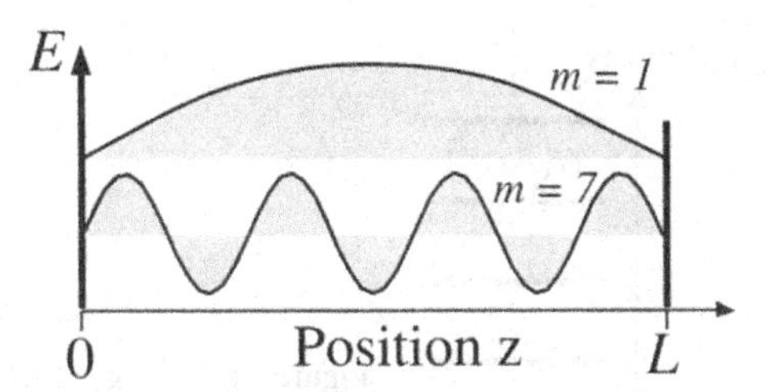

Figure 1.5. One-dimensional resonator. The two curves represent the field between two conducting surfaces for two different modes.

for the interaction of light with matter and when statistical features become important. The usual formulation of quantum electrodynamics expands the radiation field in the eigenmodes of optical resonators with fixed frequency, polarisation, and field distribution. Although the selection of modes is in principle arbitrary, the most popular expansion uses the eigenmodes of rectangular boxes with perfectly conducting walls.

Figure 1.5 illustrates this expansion for the one-dimensional case: The mirrors at positions 0, L impose the boundary conditions that the field vanish at those positions. The eigenmodes have field distributions that vary with $\sin(m\pi z/L)$, where m is the mode number, z the spatial variable, and L the separation of the two mirrors. The spatiotemporal variation of the field is, in complex notation,

$$E_m = \sin\left(m\,\pi\frac{z}{L}\right)e^{i\omega_m t} \qquad (m = 1, 2, \ldots) \tag{1.1}$$

where the angular frequency is

$$\omega_m = m\frac{\pi\,c}{L} \tag{1.2}$$

The dynamics of each mode of the field is governed by the Hamiltonian

$$\mathcal{H}_m = \frac{1}{2}(\omega_m^2 q^2 + p^2) \tag{1.3}$$

which is equivalent to that of a harmonic oscillator with unit mass. Here and through most of the book, we use frequency units for the energy, which correspond to units in which $\hbar = 1$. This not only allows a more compact notation, but also emphasises that energy differences will always be measured in

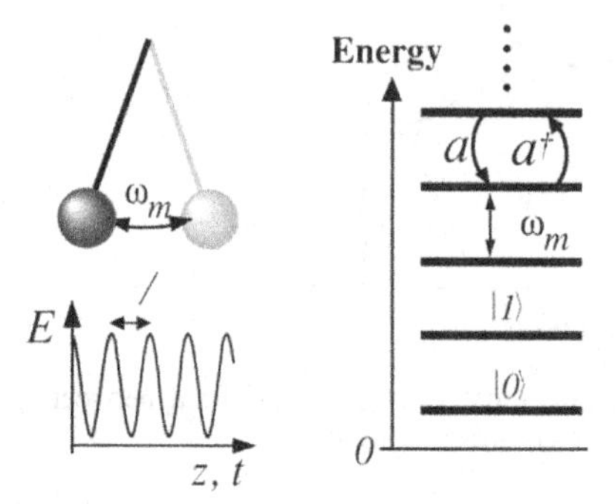

Figure 1.6. Energy levels of the harmonic oscillator, which may represent a pendulum or a single mode of the radiation field.

the form of frequencies. In the harmonic oscillator case, the variables p and q correspond to momentum and position of the harmonic oscillator and in the electromagnetic case to the electric and magnetic components of the field. In the mechanical harmonic oscillator, the energy oscillates between kinetic and potential energy. In the case of the radiation field, it oscillates between the electric and the magnetic fields.

The eigenvalues of the harmonic oscillator

$$\mathcal{E}_{m,n} = \left(n + \frac{1}{2}\right) \omega_m \qquad (n = 0, 1, 2, \ldots) \tag{1.4}$$

increase in units of the oscillator frequency ω_m. The corresponding eigenstates are commonly written as $|n\rangle$, as indicated in Figure 1.6.

In contrast to the classical oscillator, the energy of the lowest level is not zero but $\omega_m/2$. This zero-point energy is an important distinction between the quantum mechanical and the classical systems. It cannot be extracted from the system but nevertheless has observable consequences. It is responsible, e.g., for spontaneous emission (Weisskopf and Wigner 1930), the Lamb shift (Lamb and Retherford 1947), the electron g factor, and the shot noise in the detection of light (Walls 1979).

Photons

The operators p and q are closely related to the electric and magnetic fields in the mode

$$E(z,t) = 2\,q\,\sqrt{\frac{\hbar\omega_m}{\epsilon_0 V}}\,\sin\!\left(m\,\pi\,\frac{z}{L}\right) \quad H(z,t) = 2\,p\,\sqrt{\frac{\hbar\!-\!\omega_m}{\mu_0 V}}\,\cos\!\left(m\,\pi\,\frac{z}{L}\right) \tag{1.5}$$

where V is the resonator volume. In the quantum mechanical analysis, one often uses linear combinations of the operators p and q

$$a = \frac{1}{\sqrt{2\hbar\omega_m}}(\omega_m q + \mathrm{i}p) \qquad a^t = \frac{1}{\sqrt{2\hbar\omega_m}}(\omega_m q - \mathrm{i}p) \tag{1.6}$$

which act on the eigenstates of the Hamiltonian as

$$a\,|n\rangle = \sqrt{n}\,|n-1\rangle \qquad a^\dagger\,|n\rangle = \sqrt{n+1}\,|n+1\rangle \tag{1.7}$$

Their effect on the eigenstate |n⟩ is thus a decrease or increase of the energy by an amount ω_m. These excitations of the field mode may be taken as the constituents of the radiation field and are known as photons. The operators $a^\dagger$ and a create and destroy photons, usually during the interaction with atoms and are referred to as creation and annihilation (or raising and lowering) operators. The relevant properties of the photons include, apart from their energy $\mathscr{E}_\phi = \hbar\omega$, a linear momentum $p_\phi = \hbar k = \mathscr{E}_\phi/c$ and an angular momentum $S_\phi = \hbar$.

Each eigenstate of the Hamiltonian corresponds to a definite number of photons in the field and is therefore known as a number state. The operator

$$a^\dagger\, a\,|n\rangle = n\,|n\rangle \tag{1.8}$$

counts the number of photons in each state; it is known as number operator and may be used to rewrite the Hamiltonian as $\mathscr{H}_m = \omega_m(a^\dagger a + 1/2)$.

Field states

Although the number states $|n\rangle$ are useful to describe some properties of the isolated field, they have an important drawback when it comes to describing the interaction with matter. As an evaluation of the field operator $q = a + a^\dagger$ shows, the expectation value of the field vanishes for number states. This is a consequence of the Heisenberg uncertainty relation: Since the eigenstates of the Hamiltonian have a definite amplitude, their phase, which is the conjugate variable of the amplitude, must be completely uncertain. This contrasts with the classical description of a field mode: For a complete description, we need to specify two parameters, e.g., amplitude and phase, or cosine and sine components of the field. The state of the field is thus a point in a two-dimensional coordinate system, as shown in Figure 1.7.

Quantum mechanically, such a state would violate the uncertainty relation, since the two variables correspond to operators that do not commute with each other. For the number state, the uncertainty relation is satisfied by the complete phase uncertainty. This behaviour makes such a field state unsuitable for

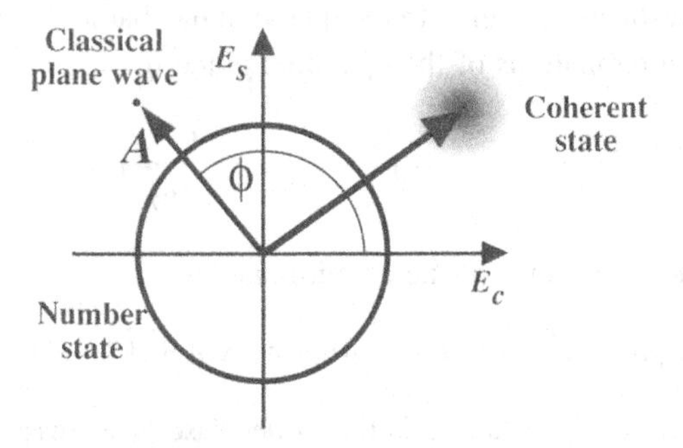

Figure 1.7. Probability distribution for the field of a number state. The two coordinate axes are the cosine and sine components; the polar coordinates A and ϕ represent amplitude and phase of the field.

describing phase-sensitive experiments, such as interference phenomena, and differs markedly from what we expect for a laser field.

The output of a laser is best approximated by a coherent state (Glauber 1963b; Glauber 1963a), which is the closest quantum mechanical approximation to a classical field. It can be expanded in the basis of the number states as

$$|\alpha\rangle = \mathrm{e}^{-|\alpha|^2/2} \sum_n \frac{\alpha^n}{\sqrt{n!}} |n\rangle \tag{1.9}$$

The complex parameter α characterises the state completely; it represents the mean excitation of the field. Coherent states are states with minimum uncertainty, i.e., the uncertainty is the minimum permitted by the Heisenberg relation. In addition, the uncertainty is distributed equally between conjugate variables like amplitude and phase.

As Figure 1.8 shows, the coherent state has a nonvanishing mean field, in contrast to the number state, where the probabilities at positive and negative field are equal, thus cancelling each other. The width of the field distribution is independent of the mean excitation. The relative uncertainty decreases correspondingly with the inverse square root of the mean photon number $1/\sqrt{\bar{n}}$. For large excitations, it is often sufficient to use a classical description that approximates the field with its mean. For most of this book, this approximation will be sufficient and we describe the field classically.

1.2.2 The classical description

Formalism

The corresponding classical description (Born and Wolf 1986) uses the Maxwell equations to describe the electromagnetic field. Here, we introduce

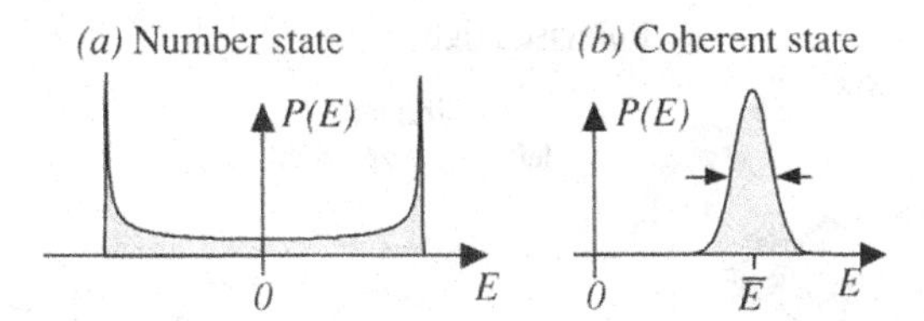

Figure 1.8. Probability distribution for one component of the electric field for a number state (left) vs. a coherent state (right).

the formalism and the basic assumptions by writing down the relevant equations. The basic variables are the electric and magnetic fields $\vec{E}$ and $\vec{H}$. In most cases, we will have to deal with plane waves, where the fields are

$$\vec{E} = \mathrm{Re}[\{E_x, E_y, 0\}\, \mathrm{e}^{\mathrm{i}(\omega t - k_z z)}] \qquad \vec{H} = \mathrm{Re}[\{H_x, H_y, 0\}\, \mathrm{e}^{\mathrm{i}(\omega t - k_z z)}] \tag{1.10}$$

The z axis is along the direction of propagation of the wave, and k_z is the component of the wavevector in this direction. We take the material properties into account through the displacement fields $\vec{D}$ and $\vec{B}$. For homogeneous and isotropic media, they are related to the external fields through the dielectric susceptibility χ as

$$\vec{D} = \epsilon_0(1 + \chi)\,\vec{E} \qquad \vec{B} = \mu_0\,\vec{H} \tag{1.11}$$

where we use SI units and consider only nonmagnetic materials ($\mu = 1$).

To describe the propagation of the radiation field, we use Maxwell's equations. In dielectric media, where currents can be neglected, they read

$$\vec{\nabla} \times \vec{H} = \frac{\partial}{\partial t}\vec{D} \qquad \vec{\nabla} \times \vec{E} = -\frac{\partial}{\partial t}\vec{B}$$

$$\vec{\nabla} \cdot \vec{D} = \rho_\mathrm{e} \qquad \vec{\nabla} \cdot \vec{B} = 0 \tag{1.12}$$

where ρ_e is the charge density. Inserting the fields of equation (1.11) into Maxwell's equations and using the dielectric constant $\epsilon = 1 + \chi$ yields the refractive index n as

$$\frac{\omega}{k_z} = \frac{c}{n} = \frac{c}{\sqrt{\epsilon}}. \tag{1.13}$$

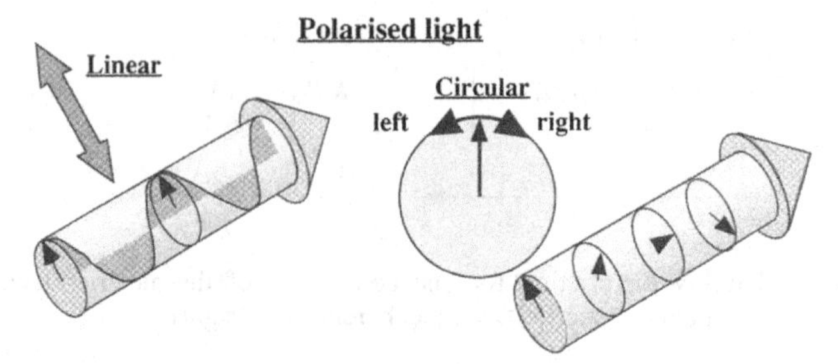

Figure 1.9. Linear and circular polarisation of the light.

Polarisation

The space of solutions to Maxwell's equations is spanned by two linearly independent waves that are degenerate in an isotropic medium, which is the case considered here. Depending on the problem at hand, it may be useful to choose the basis as two orthogonal linear polarisations or as opposite circular polarisations.

In the case of linear polarisation (left-hand side of Figure 1.9), the two components E_x and E_y have the same phase, whereas circular polarisation (right-hand side of Figure 1.9) is described by a wave of the form

$$\vec{E} = \mathrm{Re}[E_0\{1, \pm \mathrm{i}, 0\}\ \mathrm{e}^{\mathrm{i}(\omega t - k_z z)}] \tag{1.14}$$

In this case, the field vector rotates around the direction of propagation as a function of time or, if we observe the field at a given time, it rotates along the direction of propagation. Since these solutions are all degenerate, the polarisation of the light does not change during propagation. Modification of the polarisation always requires an anisotropic medium. We discuss these effects for atomic systems in Chapter 4.

1.3 Atom–light interaction

1.3.1 General

Phenomenological versus microscopic theories

As in all scientific theories, the description of the interaction between radiation and atoms can be discussed on different levels. As two important layers, we distinguish between a phenomenological and a microscopic description of the phenomena. The phenomenological level may be considered as a relation between the incident and the transmitted laser field that depends on various

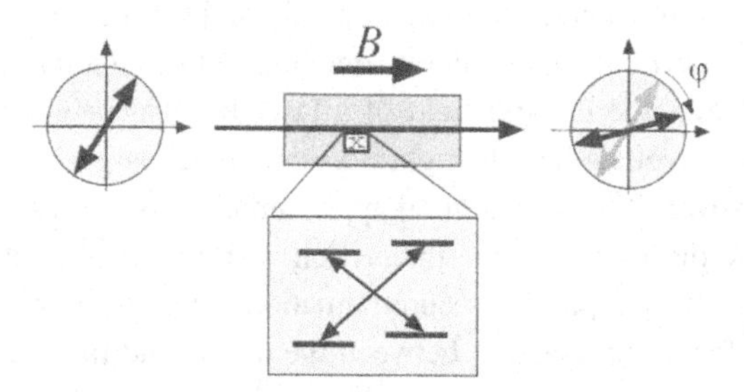

Figure 1.10. Phenomenological vs. microscopic analysis of Faraday rotation.

experimental and material parameters. A good example of such a relation is the Faraday rotation of the polarisation of light, which has the form

$$\Delta\varphi = B\, C\, l \tag{1.15}$$

where $\Delta\varphi$ is the rotation angle, B the magnetic induction, l the length of the sample, and C a material-dependent constant, which contains all the information specific for the medium that generates the Faraday rotation.

This phenomenological level, represented by the upper part of Figure 1.10, is always a necessary step for the description of experimental results; it is not the main subject of this book, however. Instead, we devote most of the space to the calculation of the material parameters that appear in the phenomenological equations, like C in equation (1.15). The ultimate goal of such a microscopic theory is the derivation of these parameters from first principles, using only the experimentally controllable parameters for the calculation. In the case of the Faraday rotation, the microscopic analysis, represented by the lower part of Figure 1.10, calculates the constant C from the known shift of energy levels inside the medium.

It turns out that such a program is surprisingly realistic for atomic media, within certain limitations. In particular, we largely restrict the discussion to atoms with a single valence electron and disregard the contribution of the filled shells to the optical properties of the system. Describing these systems as single electron atoms allows us to use analytical approximations to the true evolution of the system. These approximations can often provide more insight into the relevant physics than an exact solution, which may not exist in analytical form. In addition, we use some parameters, like the lifetime of electronically excited states, which can be calculated in principle, but whose derivation falls outside the scope of this book.

Furthermore, we usually disregard spatial variations of the laser fields and other inhomogeneous effects. These limitations restrict the accuracy of the results, but they do not change them qualitatively and can be controlled experimentally to any desired degree, albeit sometimes at considerable expense.

Like every wave, the electric field of a laser beam depends on time and space. Since the wavelength of the light is usually much larger than the atomic dimension, however, it is often a good approximation to disregard the spatial variation for the discussion of the interaction with an atom as the prototype for all atoms in the sample. This approximation leads to the electric dipole approximation for the interaction between the atoms and the radiation field. Some effects like phase matching or self-focusing require the consideration of spatially inhomogeneous media. In those cases it is sufficient to consider the spatial variation of the interaction at the phenomenological level. On the microscopic level, we have to take it into account only for dealing with light-induced mechanical effects on the atoms, which we discuss in Chapter 11.

Compared to macroscopic systems, atoms interact only very weakly with external fields. The static electric dipole moment vanishes and the induced dipole moment is less than 10^{-28} cm. Such small dipole moments become observable only through the resonant enhancement close to atomic transition frequencies. As oscillators, atoms have quality factors that are very large compared to macroscopic systems. For transitions in the visible part of the spectrum, numerical values for the quality factors range from about 10^8 to more than 10^{15}. For continuous lasers, it is therefore often a good approximation that the interaction between the laser field and the atom occurs exclusively through that resonance whose frequency is closest to the laser frequency. This is a useful approximation when relatively weak and narrowband radiation is used to drive and probe the systems.

Quantum mechanical description

A fully quantum mechanical description of the interaction between atoms and radiation is far from trivial. There is an important model, however, that allows us to derive some important results analytically. This is the Jaynes–Cummings model (Jaynes and Cummings 1963). Analytical solutions are obtained by reducing the radiation field to a single mode and approximating the atom with a quantum mechanical two-level system. We include a brief discussion of this model in Chapter 2, but throughout the remainder of this book, we use a semiclassical analysis with a classical description of the field.

One aspect that complicates a fully quantum mechanical analysis is that in such a discussion, there is no unique distinction between matter and field: Photons are shared between matter and field modes, and correlations between them can make it impossible to split the system into two parts that can be discussed independently.

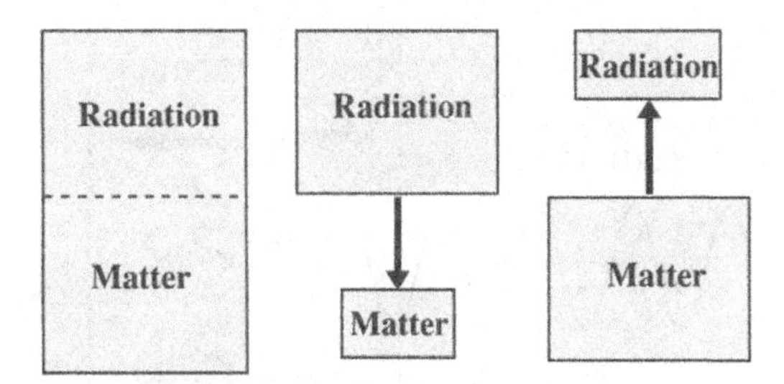

Figure 1.11. Schematic representation of the interaction between matter and radiation. Quantum mechanically, they form part of one complete system with only a somewhat arbitrary separation line (left).

Figure 1.11 analyses the situation schematically: The box at the left-hand part of the figure represents the quantum mechanical description of the physical system, which includes the matter as well as the radiation field. A separation into subsystems, as indicated by the dashed line, is usually not trivial. This is one of the reasons we use here a semiclassical treatment.

Semiclassical description

Figure 1.12 outlines the idea of the semiclassical analysis: A classical electromagnetic wave interacts with a quantum mechanical system represented by two energy levels. The quantum mechanical system includes only the atom. Such a treatment does not cover quantum effects of the radiation field. We refer readers who are interested in those effects to specialised books on that subject (Heitler 1953; Loudon 1983; Meystre and Sargent 1990; Cohen-Tannoudji, Dupont-Roc and Grynberg 1992). In some cases, quantum effects can be added to the results in a perturbative way (Courty et al. 1991; Hilico et al. 1992a).

On the other hand, this reduction of the electromagnetic field to a classical wave allows us to consider separately two aspects of the interaction between atoms and light, which are represented as independent effects in Figure 1.11. The central part symbolises the modification of the microscopic dynamics of the atoms by the radiation field, which appears as a parameter in the Hamiltonian of the atomic system. The right-hand side of Figure 1.11 represents the reverse effect, the modification of the light by the material system. To describe this process, we solve the dynamics of the microscopic system and use the expectation value of the optical polarisation as a material parameter in Maxwell's equations.

This discussion of the interaction as two independent one-way processes is also reflected in the structure of the book: Chapters 2 through 5 and 11 discuss primarily the effect of the radiation field on the microscopic system. The

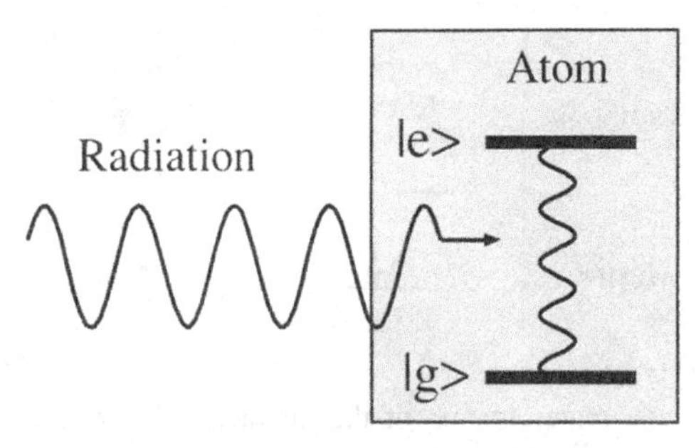

Figure 1.12. Semiclassical model of atom–radiation interaction.

reverse effect is the subject of Chapter 6, whereas Chapters 7 to 10 include both types of effects. In most cases, this separation can be approximated experimentally to a large degree, but we also discuss examples of systems where the breakdown of this approximation leads to qualitatively new phenomena.

Degrees of freedom

Another separation that is convenient but not always exact is the distinction between internal degrees of freedom and external degrees of freedom of the atoms. The internal degrees are associated with the quantum mechanical energy level structure in the centre of mass coordinates and include primarily electronic energy, electric dipole moment, and the angular momentum of the atom. The external degrees of freedom are position and momentum. The interaction with radiation can affect all degrees of freedom.

Figure 1.13 shows a simple example: An atom that is initially at rest absorbs a photon. Energy E, momentum p and angular momentum J are conserved quantities, so the atom must change its state to accommodate the contribution of the photon. Immediately after the absorption, it is therefore in an excited state with energy close to $\hbar\omega$, and its momentum has increased by $\hbar k$ in the propagation direction of the photon. This example shows that the interaction with radiation affects both types of atomic degrees of freedom. To keep things simple, however, we discuss the two types separately. Our primary interest is with the internal degrees of freedom, assuming that the atoms are at rest. The effect of the radiation on the external degrees of freedom is dealt with in Chapter 11.

The two-level model for the atomic system is probably the most frequently used description of the interaction between light and matter, in particular atoms. It is closely related to the classical Lorentz–Lorenz theory of dispersion, where the material part is a collection of dipoles instead of quantum me-

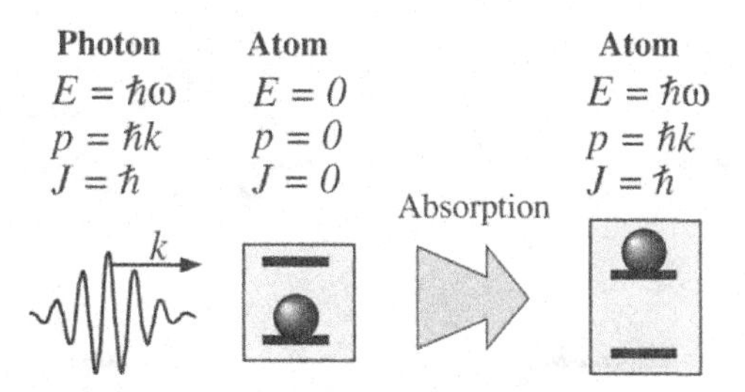

Figure 1.13. The absorption of a photon changes internal degrees of freedom, like the energy E, as well as external ones, like the momentum p.

chanical two-level systems. The semiclassical version is due to Felix Bloch (Bloch 1946), who describes the interaction of a magnetic dipole with a magnetic field, whereas Feynman, Vernon and Hellwarth (Feynman, Vernon and Helwarth 1957) realised that the evolution of all quantum mechanical two-level systems is identical to that of a spin 1/2.

Two-level atoms are prototypical model systems as their dynamics can be solved analytically and because their evolution has an appealing geometrical representation in three-dimensional space (Feynman, Vernon and Helwarth 1957). Real atoms have an infinite number of energy levels. Nevertheless, the success of the Lorentz–Lorenz theory of dispersion and the two-level models in explaining many aspects of the interaction of light with material systems shows that it is often possible to neglect most of the energy levels in the discussion and to consider the material subsystem as a collection of two-level systems – one for every transition. This is a suitable approximation whenever the system under investigation is not too far from thermal equilibrium.

1.3.2 Multilevel atoms

Limitations of the two-level model

The two-level model has been extremely successful, but, like every model, it has its limitations. A simple and straightforward example of a failure occurs whenever the behaviour of the physical system is anisotropic. A two-level atom can interact with the optical field through an induced dipole moment that oscillates at an optical frequency. In three-dimensional space, however, a dipole moment has three components; the two-level model, which can describe only a single component of the electric dipole moment, cannot describe a situation in three-dimensional space. Experimental evidence shows that atomic media may show strongly anisotropic behaviour. Examples include

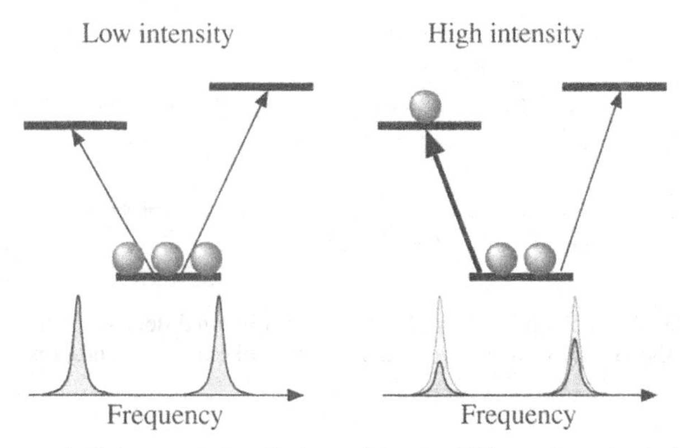

Figure 1.14. Two coupled optical transitions at different laser intensities.

media that pass light of one polarisation unattenuated but absorb light of an orthogonal polarisation or systems whose refractive index varies with the direction of polarisation. On a microscopic level, effects of this type are associated with the creation of order in the atomic system. The most widely known of these processes is optical pumping. Atomic states that are fully isotropic, i.e., invariant under rotations in three-dimensional space, do exist and are relevant for many experimental situations. A transition between two such states, however, cannot couple to the electromagnetic field and the corresponding two-level system is therefore of no practical interest.

Some aspects of multilevel systems can be described approximately in the form of a superposition of two-level systems. This possibility exists particularly in those cases where the light interacting with the system does not drive it too far from equilibrium. As the system moves away from its thermal equilibrium position, however, the deviations between the predictions of the two-level model and the experimentally observed behaviour increase.

The difference between the high- and low-power case can be appreciated, e.g., by considering the energy level scheme of Figure 1.14. At low light intensity, the different states are populated according to Boltzmann's law. If the absorption of light by this system is measured, the two transitions can be observed as resonance lines in the spectrum. From the spectrum, one can determine the energy differences, but it is not possible to distinguish the situation depicted here from a mixture of two types of two-level atoms whose transition frequencies match the observed ones.

This situation changes, if the laser intensity is increased. If one of the two transitions is irradiated with a strong laser field, indicated by the thick arrow

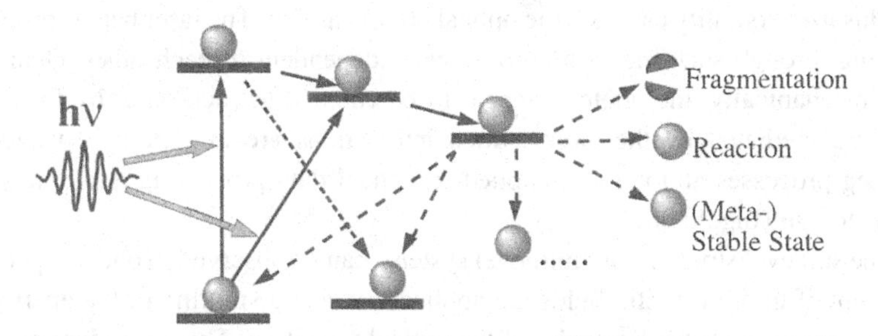

Figure 1.15. Some aspects of multilevel systems: different excitation pathways, and escape to other environments.

in the right-hand side of Figure 1.14, it moves part of the population into the electronically excited state and the population of the ground state decreases. This decrease of the ground-state population reduces the population difference for both transitions and hence reduces the signal strength for both absorption lines. The modification of the height of the second resonance line proves that the two optical transitions share a common energy level and excludes the possibility that the two resonance lines belong to different atomic species.

With few exceptions, the low-power approximation was quite safe until lasers became widely available as powerful sources of narrowband light. Since then, the situation has changed dramatically, as lasers make it possible to drive resonant systems very far from equilibrium. At the same time, laser media must consist of systems with more than two states; they are therefore among the most thoroughly studied multilevel systems.

Multilevel effects

Figure 1.15 illustrates some aspects of multilevel systems that are lost in the two-level approximation. Possibly the largest difference between true two- and multilevel systems is the occurrence of irreversibility. A two-level atom that absorbs a photon can only fall back into the ground state. For multilevel atoms, the situation is different, as shown schematically in Figure 1.15: The atom does not always fall back into the state from which it absorbed the photon, but may fall into different excited states, metastable states, or, in the case of molecules, it may undergo fragmentation or chemical reactions. Such processes are particularly important when the absorbed photon should not be reemitted, as in the context of photosynthesis or the perception of light – in biological as well as artificial systems.

This irreversibility causes large optical nonlinearities. The laser beams propagating through such media are no longer independent of each other. Quantum mechanically, the photons appear to be coupled to each other by forces that are mediated by the atoms. Such interactions are important in wave-mixing processes and may be applied to optical data processing and optical data storage (Ingold et al. 1992).

The sublevel structure of multilevel systems can be observed in optical spectroscopy if high magnetic fields are applied, causing a splitting of the energy levels that exceeds the linewidth of the optical transition. This procedure cannot provide high-resolution spectra, since the resulting linewidth is that of the optical transition, many orders of magnitude larger than that of the sublevel transition. An important part of this book is the discussion of experimental methods that rely on the interaction of the atomic systems with optical radiation to obtain spectroscopic information about the multiplet structure with a resolution independent of the spectral properties of the light. Although the laser frequency, or more precisely the difference between the laser frequency and the electronic transition frequency of the atoms, is an important parameter, it does not influence the resolution of the measurements that we discuss. Experimental techniques with these possibilities are necessarily nonlinear, i.e., they involve the absorption and emission of more than one photon.

The interest in the optical properties of atomic multilevel systems has been considerably enhanced in recent years in the context of laser cooling and trapping of neutral atoms (Chu 1991). The processes involved in these experiments were originally described within the two-level approximation for the atomic system, but it was soon realised that the details of the level structure of the atoms used in these experiments were of considerable importance and made it possible to achieve temperatures far below the limit predicted for two-level atoms.

Experimental investigations of two-level systems require the measurement of only a few variables of the system to get a complete description. In multilevel systems, the complexity of the systems increases dramatically with the number of levels involved, in particular when the system is not in a pure state. In this case, a complete description requires the knowledge of its density operator. For a system with n levels, the density operator contains n^2 elements, which are independent variables. A complete characterisation therefore requires a large number of independent measurements, which quickly exceeds the possibility of conventional experiments. In Chapter 9, we discuss a possible extension to experimental techniques that provides significantly more information, thus allowing complete characterisation, even for complicated systems.

1.4 Summary of relevant physical processes

1.4.1 Laser spectroscopy

Historical

Spectroscopy is a technique that measures the response of material systems to radiation as a function of frequency. Through the relation $\mathscr{E} = h\nu$ between energy E and frequency ν, observed resonances provide information about energy level spacings in the material. Spectroscopy with visible light was traditionally performed with incoherent light sources such as light bulbs or discharge lamps. The power within the desired frequency band available from these light sources was relatively low and hence the number of photons absorbed by the system within its relaxation time was much smaller than the number of atoms. The system was therefore always very close to thermal equilibrium. The process of interaction between the radiation field and the atomic systems was usually described with perturbation theory, which is useful when the rate of absorption is small compared to the spontaneous emission rate.

With the introduction of coherent radiation sources like lasers, the situation changed qualitatively. Although the total output power of the new radiation sources is often lower than that of the thermal sources, it is concentrated within a narrow spatial direction and a narrow frequency range or in a short time window. The available light intensity and the coherence properties of the radiation field have improved in such a way that many experiments not feasible before have now become routine. Important examples include the generation of ultrashort pulses to obtain a very high time resolution (Zewail 1988) or spectroscopy with individual atoms (Diedrich and Walther 1987) and molecules (Moerner and Kador 1989).

On the theoretical side, many concepts and descriptions of physical processes have had to be revised. The high spectral intensity of the laser light causes a nonlinear response of the system to the optical field. The high spectral purity of a cw laser leads to additional phenomena like velocity-selective excitation (Pinard, Aminoff and Laloe 1979). With incoherent light sources, the spectral width of the light was generally much broader than the inherent linewidth of the optical transition and the process of absorption could be described with rate equations for the populations. With narrowband lasers, the situation is reversed; in many cases, the optical coherences have to be taken into account and the dynamics must be formulated in terms of the density operator (Brewer 1977b). Stimulated emission, for example, which could safely be neglected when experiments were performed with discharge lamps, must now be taken into account.

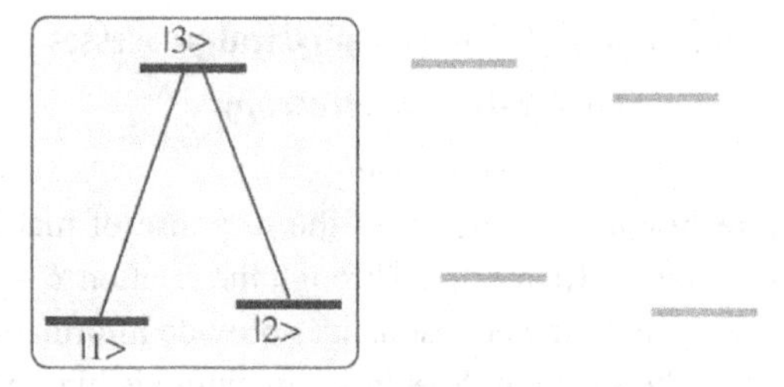

Figure 1.16. Quantum mechanical three-level systems as part of a larger system.

Generic level schemes

The discussion of physical phenomena must always navigate between the Scylla of a narrow scope and the Charybdis of formulations that are too abstract. We try to minimise these problems in the following way: The initial Chapters (2 to 6), which introduce the relevant phenomena, use abstract level schemes chosen to minimise the mathematical difficulties and not associated with real atoms. Only after the introduction of the necessary tools does the discussion focus on real atomic systems.

Apart from the two-level scheme, outlined above, we frequently use three-level schemes.

An example of a generic three-level system, known as a Λ-type system, is shown in Figure 1.16. Although systems of this type do not occur in nature, they represent parts of real level systems, as indicated by the additional levels. The optical transitions indicated in the figure may couple to different fields with coupling strengths that depend on the polarisation of the light. The third transition between states $|1\rangle$ and $|2\rangle$ is usually not an allowed optical transition but is often a magnetic dipole transition.

1.4.2 Sublevel dynamics

In the systems of interest here, the near-degenerate states have the same electronic structure but different angular momentum. An exchange of population between them, the generation and modification of coherence, and other processes that we will discuss later, modify the atomic angular momentum. Loosely speaking, we may associate angular momentum with the "shape" of an atom and angular momentum dynamics with rotations or shape changes. Our analysis of these systems centres on two parts: We discuss methods to influence the sublevel dynamics and we present techniques that make it possible to observe the order between angular momentum substates through the optical properties of the system. Since optical radiation does not couple to transitions between angular momentum sublevels, these techniques must rely on higher order processes. We distinguish three aspects of these experiments, the creation of order within the angular momentum multiplets, the dynamics of the ordered state, and the detection of the order.

Optical pumping

The generation of order within the angular momentum multiplets can be summarised under the term "optical pumping," which we use for all processes where the absorption and emission of photons increases the order within a manifold of near-degenerate states (Bernheim 1965; Kastler 1967; Happer 1972; Balling 1975). Alfred Kastler was the first to suggest that optical excitation should allow the creation of nonthermal populations in multilevel systems (Kastler 1950). The first experimental observations (Brossel, Kastler and Winter 1952; Hawkins 1955; Dehmelt 1957b) showed that under a wide range of experimental parameters like polarisation, intensity, and frequency of the light, nonthermal populations of ground state sublevels could be excited.

Optical pumping is one of the few experiments where conventional light sources could drive the system far from its thermal equilibrium. The pumping rates that could be achieved were relatively low, but appreciable population differences in the sample were possible if the relaxation that counteracts the optical pumping could be kept slow. If most of the atoms are in the electronic ground state, the radiative lifetime of these states is essentially infinite. Relaxation occurs primarily through two types of processes: Zeeman precession in magnetic fields and collisions with walls. Magnetic field relaxation can be reduced by compensation coils and by enclosing the vapour cells with multiple layers of μ-metal shielding. The addition of inert buffer gas reduces the mean free path of the atoms sufficiently that they remain in the vapour phase without hitting a wall for much longer times than in high vacuum. In addition, wall coatings reduce the relaxation efficiency of wall collisions.

Optical pumping relies on the transfer of angular momentum between photons and atoms: Absorption and emission processes change the angular momentum state of the atom. The nature of this change depends on the polarisation of the photons and on their direction of propagation. In the case of circularly polarised light, the absorption of a photon changes the angular momentum component parallel to the direction of propagation by $\pm\hbar$, depending on the sense of polarisation. When the photon absorbed by the atom is reemitted after the excited state lifetime, it carries away one quantum of angular momentum whose direction depends on the polarisation of the fluorescence and on the direction of propagation. If the radiation incident on the atom differs from the reemitted radiation, the difference in angular momentum remains on the atom, thereby polarising the spin system.

Since the optical pumping process is a consequence of the conservation of angular momentum during absorption, it primarily changes the electronic angular momentum. In most atoms, however, the coupling between the electron and the nuclear spin (the hyperfine interaction) is strong enough to cause a fast exchange of polarisation between the two reservoirs, so that the nuclear spin is equally polarised. Since this mixing occurs in the excited state as well as in the ground state, the nuclear spin system can also acquire polarisation, even if the ground state of the atom is diamagnetic (Lehmann 1964). This possibility of orienting nuclear spins has turned optical pumping into an attractive tool also in nuclear physics: It allows one to polarise atomic nuclei for use as a target in accelerator experiments (Anderson 1979) or to orient nuclei of short-lived isotopes for the study of nuclear momenta (Huber et al. 1978).

Light shift

Light interacting with atoms not only creates order in the system, it also changes the dynamics of the order already present in the angular momentum multiplets (Arditi and Carver 1961; Barrat and Cohen-Tannoudji 1961b; Barrat and Cohen-Tannoudji 1961a; Barrat, Cohen-Tannoudji and Ribaud 1961; Cohen-Tannoudji 1961; Cohen-Tannoudji 1962; Pancharatnam 1966; Dupont-Roc et al. 1967; Happer 1970; Cohen-Tannoudji and Dupont-Roc 1972). Two types of effects of the radiation field on the atomic dynamics can be distinguished: a dissipative damping effect whose strength has an absorptive dependence on the detuning of the laser from the electronic transition frequency, and the so-called light shift, which preserves the order in the system. As first suggested by Barrat and Cohen-Tannoudji (Barrat and Cohen-Tannoudji 1961b; Barrat and Cohen-Tannoudji 1961a; Barrat, Cohen-Tannoudji and Ribaud 1961) and confirmed experimentally by Arditi and Carver (Arditi and Carver 1961) and Cohen-Tannoudji and colleagues (Cohen-Tannoudji 1961; Cohen-Tannoudji 1962; Dupont-Roc et al. 1967; Cohen-Tannoudji and Dupont-Roc 1972), strong off-resonant radiation causes an apparent shift of the energy levels associated with the optical transition. This shift is proportional to the intensity of the light and has a dispersionlike dependence on the optical detuning.

When light-shift effects were first discovered, the low spectral intensity of the available light sources limited them to a few Hertz (Cohen-Tannoudji 1961). In contrast, even the light of cw lasers can cause shifts of the order of kHz to MHz and can therefore strongly affect the dynamics of the sublevel coherences (Rosatzin, Suter and Mlynek 1990; Boden, Dämmig and Mitschke 1992). Under appropriate experimental conditions, these shifts have the same effect on the spin dynamics as magnetic fields, allowing not only a shift of selected energy levels, but also a change in the orientation of the quantisation axis of the effective Hamiltonian (Rosatzin, Suter and Mlynek 1990; Suter and Marty 1993b). The light-shift effect thus represents an interesting experimental tool, provided its properties are well understood and can be suitably controlled.

1.4.3 Optical properties

Linear effects

Apart from the effect of the laser field on the atomic dynamics, it is also necessary to consider the effect of the polarised atomic system on the laser field. The interest in this process is again twofold: Knowledge of the process allows one to use the interaction to modify the properties of the light, for example to generate nonclassical states of light (Slusher et al. 1985; DeOliveira, Dalton and Knight 1987) and, second, the modification of the properties of light passing through an atomic medium can yield information about the microscopic degrees of freedom of the atomic system. In many of the experiments described here, the light behind the sample contains a frequency component that differs from the original frequency by an amount in the radio frequency range. These experiments may therefore be considered as examples of Raman scattering.

The first optical pumping experiments were monitored through the polarisation of the fluorescence (Kastler 1967; Alzetta et al. 1976; Haroche 1976; Dodd and Series 1978), but it was soon realised that the transmitted laser beam contained similar information (Dehmelt 1957a; Pancharatnam 1968). We consider primarily this latter type of experiment, where the microscopic order modifies a coherent radiation field propagating through the medium. The most prominent aspect of this interaction is the reduction of the absorptivity of the sample for the light that drives the optical pumping. This reduction is accompanied by an increased absorptivity for the orthogonal polarisation; it should therefore be considered not a reduced absorption, but an induced dichroism. As required by the Kramers–Kronig relations, this dichroism is associated with a birefringence, which can also be measured experimentally. If light propagates through a polarised atomic medium, its polarisation therefore changes in a way that depends not only on the parameters of the atomic system, but also on external parameters like the laser frequency.

These measurements can access different types of information about the systems being studied: They may contain information about the populations and coherences present in the system, as well as information about energy level differences. In this context it is important to realise that the coherence properties of the optical radiation field do not limit the spectroscopic resolution of these experiments. This becomes evident when one considers the early experiments with discharge lamps, which could already measure energy level splittings of less than one Hertz – many orders of magnitude below the linewidth of the radiation used in the experiments. This can be understood qualitatively by considering that no net absorption of photons occurs during the detection process. Instead, the magnetic interaction leads primarily to a modulation of the polarisation of the probe beam, which can be measured with the same absolute frequency resolution as in purely magnetic resonance experiments.

Depending on the actual level system, different components of the atomic density operator contribute to the resulting signal. Since the emission and absorption processes are always associated with a transfer of angular momentum, group theory is extremely helpful in analysing the possible contributions. Apart from the effects on the polarisation of the light, order within the atomic system can also cause a lateral displacement of the laser beam. This effect is required by angular momentum conservation, since the laser beam modifies the transverse components of the internal angular momentum of the atomic system while these do not affect the polarisation of the light. Instead, they change the external angular momentum of the laser beam, i.e., its lateral position.

Nonlinear effects

In linear optics, the photons that pass through the sample are independent of each other. The frequency of the radiation field behind the sample is always the same as the frequency of the radiation entering the sample. Until the introduction of the laser, deviations from this behaviour became noticeable only under relatively rare circumstances. With the availability of intense monochromatic radiation sources, the situation has changed completely. At high intensities, the optical properties of the medium change and the light no longer interacts with the same medium as when it is close to equilibrium. In this situation, the photons are no longer independent, as in the vacuum, but they "feel" each other's presence. The atoms can thus mediate virtual forces between photons and couple laser beams to each other.

These effects are generally referred to as wave-mixing processes. In a typical case, the medium can convert two incident photons into a single photon at twice the frequency of the individual photons. The first experiment that demonstrated doubling of an optical frequency (Franken et al. 1961) used one

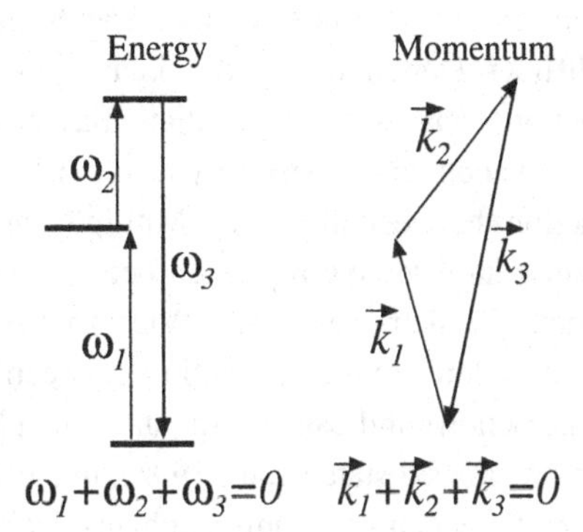

Figure 1.17. Conservation of energy (left) and momentum (right) during wave-mixing processes.

of the earliest lasers. Today, there is a large variety of frequency conversion experiments (Shen 1984), including frequency doubling (Gheri, Saavedra and Walls 1993), tripling (Bloom et al. 1975b; Bloom, Young and Harris 1975a) or down-conversion (Giordmaine and Miller 1965). In all these cases, the new frequency differs from the old frequency by an amount that is of the same order of magnitude as the original frequency.

All these processes can be explained classically, using Maxwell's equations with the relevant material properties. The conservation of energy and momentum, however, which are among the most important aspects of wave-mixing processes, can best be appreciated in the photon picture, as shown schematically in Figure 1.17. The left-hand side of the figure shows how energy conservation requires that the frequency ω_3 of the outgoing photon equal the sum of the frequencies of the two incident photons. The phase-matching condition of nonlinear optics becomes in the photon picture the requirement that the wave vector of the outgoing photon equal the sum of the wavevector of the two incident photons, as required by momentum conservation. Many similar processes have been investigated, but most important for atomic media are four wave-mixing experiments (Bloom, Liao and Economou 1978; Bloch and Ducloy 1981; Köster, Mlynek and Lange 1985; Slusher et al. 1985; Pinard, Grancelement and Grynberg 1986; Slusher et al. 1987; Bai and Kachru 1991; Ducloy and Bloch 1994).

Material-induced interactions between photons can change not only their frequency, but also their propagation. An important aspect of these interactions is the modification of the refractive index by the light. Probably the best known effect resulting from such a modification is self-focusing (Grischkowsky 1970; Bjorkholm and Ashkin 1974; Tam and Happer 1977; Boshier and Sandle 1982; Yabuzaki, Hori and Kitano 1982; McCord, Ballagh and Cooper 1988; Suter and Blasberg 1993). Here, the modified refractive index acts as a waveguide for the laser beam, counteracting its usual tendency to diffract. Depending on the polarisation and frequency of the light, the interactions that the atoms mediate may be attractive or repulsive.

The converse possibility also exists: The light can mediate effective interactions between the atoms. Multiple absorption and reemission processes, e.g., cause an effective repulsion between atoms that limits the density in optical traps (Walker, Sesko and Wieman 1990). Similarly, polarised light can mediate a long range interaction between atomic spins (similar to an exchange interaction) and cause a spontaneous phase transition of the spin polarisation to an ordered state (Suter 1993), in close analogy to ferromagnetism. Such interactions can be strongly enhanced when the atomic medium is placed in an optical resonator, where the photons undergo multiple passages through the medium (Mitschke et al. 1986; Boden, Dämmig and Mitschke 1992; Häger and Kaiser 1992; Möller and Lange 1992). These systems have been studied extensively as model systems for nonlinear dynamics.

1.4.4 Magnetic resonance

Principle

The angular momentum that distinguishes the atomic substates is associated with a magnetic dipole moment. States with different angular momentum interact differently with the magnetic field. Static magnetic fields separate the energy of angular momentum states, whereas external radio frequency (rf) fields can induce transitions between them. These transitions have been investigated for many years in the field of magnetic resonance spectroscopy, which measures the interaction of atomic and molecular systems with external magnetic fields. In most cases, the experimentalist tries to isolate the magnetic interaction from other interactions by exciting transitions between states that differ only in their angular momentum, while all other quantum numbers remain unchanged, i.e., between angular momentum substates within one electronic state.

Figure 1.18 illustrates the principle of magnetic resonance spectroscopy: The material is placed in a strong magnetic field, which lifts the degeneracy of the angular momentum substates, and a radio frequency field applied to the coil induces transitions between these levels. From the point of view of optics, magnetic resonance investigates transitions between angular momentum substates.

The gyromagnetic ratio measures the strength of the coupling between the angular momentum and external magnetic fields. Its size depends on the type of angular momentum; for orbital and spin angular momentum of electrons, it is of the same order of magnitude and exceeds that of nuclear spins by

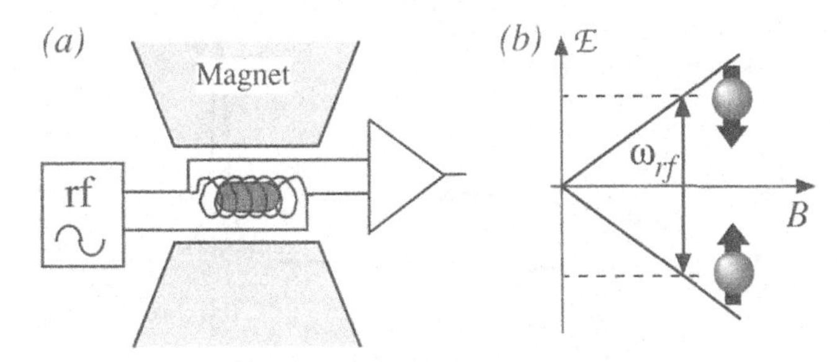

Figure 1.18. (a) Schematic setup of magnetic resonance spectroscopy. (b) An alternating magnetic field induces transitions when the Zeeman splitting matches the resonance condition.

roughly three orders of magnitude. If direct excitation of the magnetic resonance transitions is used, the different orders of magnitude result in different frequency ranges and therefore different requirements on the experimental apparatus. Accordingly, there is a clear distinction between the fields of electron paramagnetic resonance (EPR), investigating mainly electron spin transitions, and nuclear magnetic resonance (NMR), dealing with nuclear spins.

Evolution of the technique

Among the first experiments probing transitions between different spin states were those performed with atomic beams (Rabi et al. 1938). In those experiments, an inhomogeneous magnetic field separated atoms with different spin states in space and a second field gradient of opposite sign recombined them. Between the two inhomogeneous regions, a resonant radio frequency field induced transitions between spin states in a region of homogeneous field. The atoms whose spin state changed during their passage through the rf field would no longer recombine on the detector, thereby lowering the count rate. If the frequency of the rf field was scanned, a drop in the detected count rate indicated a resonance. In this experiment, every atom participated in the resonance process and very few were needed for the detection of a resonance.

Several years later magnetic resonance transitions were measured in bulk material (Zavoisky 1945; Bloch, Hansen and Packard 1946; Purcell, Torrey and Pound 1946). Instead of a few atoms, these experiments were done on samples containing some 10^{24} atoms, indicating that their sensitivity was considerably lower than that of the beam experiments. The difference can be traced to two major causes: In the bulk, only the small fraction of spins that corresponds to the population difference between the two stationary spin states

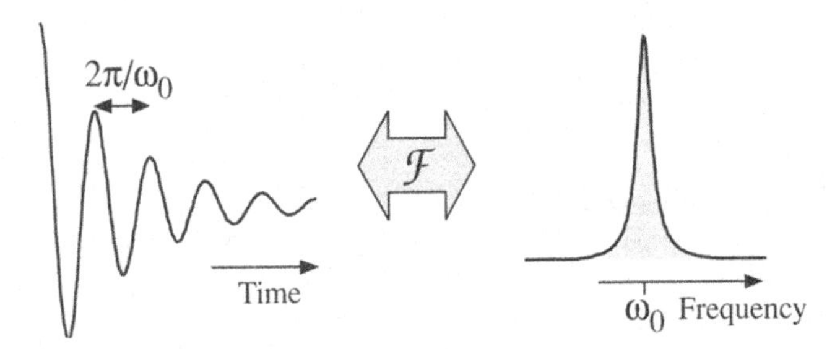

Figure 1.19. Conversion of the time-domain response into a frequency spectrum by Fourier transformation.

participates in the experiment, and, second, the resonance is detected through the rf photons absorbed by the resonating spins; these photons are much harder to detect than the atoms observed in Rabi's beam experiment.

In these early experiments, the sample was irradiated with a continuous, monochromatic field, and the strength of the static magnetic field was adjusted such that the energy difference between the Zeeman states became equal to the photon energy, as shown in Figure 1.18b. These slow-passage or continuous-wave (cw) experiments, which study the system as a function of frequency, were to a large degree superseded by experiments where a time-dependent perturbation excites the system under study and the response of the system is measured as a function of time (Ernst 1966; Ernst and Anderson 1966). The time-domain experiment also measures the response to a perturbation as a function of frequency, but it does the measurement for all frequency components simultaneously. The response to the different frequency components is separated through Fourier transformation of the time-dependent system response (Ernst 1966; Ernst and Anderson 1966).

Figure 1.19 shows the simplest case, where a single frequency component is present. The two experiments contain the same information and can be converted into each other (Lowe and Norberg 1957), but the time-domain experiment provides higher sensitivity and allows extensions that can measure quantities not directly accessible with the frequency-domain method (Ernst, Bodenhausen and Wokaun 1987).

The primary motivation for magnetic resonance experiments is not the desire to measure the magnetic interactions more accurately, but the possibility of using the magnetic resonance spectra as a probe for other degrees of freedom of the material that hosts the spins. This possibility has made magnetic resonance an indispensable tool in many areas of science, most notably chemistry, medicine, and solid state physics.

Sensitivity

Today, the largest remaining weakness of the technique is its low sensitivity compared, e.g., to optical experiments. In the area where magnetic resonance has become most popular, nuclear magnetic resonance (NMR) of liquids, the minimum number of atoms that can be detected is of the order of 10^{18}. This may be compared to optical spectroscopy, where experiments with individual atoms (Hurst et al. 1979; Goy et al. 1983; Grangier, Roger and Aspect 1986; Diedrich and Walther 1987; Walther 1988b; Walther 1988a; Dehmelt 1990) and molecules (Moerner and Kador 1989; Ambrose and Moerner 1991; Basche et al. 1992; Wild et al. 1992b; Güttler et al. 1993a; Güttler et al. 1993b; Pirotta et al. 1993; Sepiol et al. 1993) have become rather popular recently.

Several reasons contribute to this low sensitivity of magnetic resonance spectroscopy: The small size of the interaction energy leads to small thermal population differences between the energy levels of a magnetic dipole transition, to small detector efficiency, and to high thermal noise levels. Whereas population differences in thermal equilibrium are close to unity for optical transitions ($\nu \approx 10^{14}$ Hz), those associated with radio frequency (rf) transitions ($\nu \approx 10^{8}$ Hz) are of the order of $e^{-h\nu/k_{B}T} \sim 10^{-5}$. Similarly, optical photons can be detected with an efficiency near unity whereas a large number of rf photons are required for a signal that exceeds the noise level. Optical methods for exciting and detecting magnetic resonance can often transfer some of the sensitivity from optical spectroscopy to magnetic resonance. This possibility was strikingly demonstrated by optically detected magnetic resonance of single molecular spins in solids (Köhler et al. 1993; Wrachtrup et al. 1993).

This sensitivity enhancement relies on different processes (Suter 1992b). Two important steps are an increase of the population difference between the different magnetic sublevels and improved efficiency of the detection process. Among other methods, optical pumping of the system allows an increase in the population difference (Kastler 1967; Cates et al. 1990; Raftery et al. 1991; Raftery et al. 1992). Like the population difference between ground and electronically excited states, the population difference between levels differing only with respect to their spin state can then reach values near unity.

Conversely, the population difference and coherence between the magnetic substates can change the optical properties of the system. It is therefore possible to detect the magnetisation optically, with a much higher sensitivity than if radio-frequency photons are detected (Bitter 1949; Brossel and Bitter 1952; Dehmelt 1958; Kastler 1967; Pancharatnam 1968; Mlynek et al. 1979; Suter and Mlynek 1991b). This gain in sensitivity can be understood qualitatively as an amplification of the signal energy: If a magnetic dipole transition is detected, a spin can radiate a radio-frequency photon while making a transition. With optical detection, the same transition is detected by optical photons, whose energies are typically six orders of magnitude higher.

Instead of transferring the angular momentum to an optical transition to increase the sensitivity, it is also possible to excite a resonance line that corresponds to a transition between states that differ in their electronic, vibrational or rotational states as well as in their magnetic quantum numbers. The transition frequency is then the sum of the optical and magnetic energy. This technique has been applied successfully in the far infrared region of the spectrum (Davies 1981). Since the width of the observed resonance lines is the same as that of the purely optical or vibrational transitions, this technique is not likely to provide sufficient resolution in the visible part of the spectrum, where the broadening mechanisms of the optical transitions, such as Doppler broadening and spontaneous emission from the excited state, are much larger than in the far infrared.

Apart from the gain in sensitivity, the use of optical radiation also provides the option to perform magnetic resonance spectroscopy of electronically or vibrationally excited states (Geschwind, Collins and Schawlow 1959; Franken 1961; Breiland, Harris and Pines 1973; Bos, Buma and Schmidt 1985; Bitto and Huber 1990; Bitto, Levinger and Huber 1993). Since the thermal population of these states vanishes, the atoms or molecules that are to be studied must be brought into the excited state before magnetic resonance can be performed. If the excitation with light is possible, it is often advantageous to use selective excitation of the magnetic substates to obtain a polarised system. This possibility represents an important advantage since the populations that can be achieved are often smaller than in the ground state so that sensitivity again becomes an important issue. The fluorescence that these atoms emit is often polarised and permits a direct measurement of the excited state magnetisation. The high polarisation, small background, and good detection efficiency of this type of experiment has allowed measurement of magnetic resonance spectra of individual molecules (Köhler et al. 1993; Wrachtrup et al. 1993). The resolution that can be achieved in such a system is of course limited by the lifetime of the excited state.

Other attractive features of optical excitation in magnetic resonance spectroscopy include the possibility of obtaining additional information, which is not accessible by conventional methods (Blasberg and Suter 1993) or of using the shorter wavelength of optical radiation to localise the region from which signals are obtained, e.g., to the surface of a solid (Lukac and Hahn 1988) or to the interface between a solid and a vapour (Suter, Aebersold and Mlynek 1991).

Comparison

The history of laser spectroscopy and magnetic resonance shows some similarities and analogies between the two fields; there are also many differences, either of technological or of fundamental physical nature. The common features allow one to take advantage of the experience gained in one domain and use it in the other domain; the differences limit the applicability of these analogies. They may, however, help in gaining a deeper understanding of the phenomena by distinguishing them from related phenomena. Probably the most prominent example of a fruitful analogy between the two fields is the realisation that the Bloch equations, originally derived for a spin 1/2 (Bloch 1946), describe every two-level system (Feynman, Vernon and Helwarth 1957).

Other developments in laser spectroscopy followed the path laid out by the field of nuclear magnetic resonance during the fifties and sixties. Although the transition probabilities in magnetic resonance are small, the number of photons is large and the spontaneous emission rate is negligibly small. Under typical experimental conditions, the evolution of nuclear spin systems is therefore much faster than the decay processes. In contrast to the description of earlier optical experiments, the theoretical analysis of magnetic resonance experiments treats the system as an ensemble of identical spins interacting primarily with the applied field, whereas relaxation processes are considered only as perturbations. Only the development of sufficiently powerful and monochromatic lasers moved optical experiments into a parameter range where processes become relevant that are typical for magnetic resonance experiments. Examples of these experiments include coherent transients and echoes (Abella, Kurnit and Hartmann 1966; Brewer and Shoemaker 1971; Brewer 1977b; Allen and Eberly 1987; Golub, Bai and Mossberg 1988).

A major distinction between magnetic resonance and coherent optics is the ratio between the wavelength of the radiation used and the typical dimensions of the sample. For magnetic resonance spectroscopy, the sample size d is always small compared to the wavelength λ, $d << \lambda$, whereas the opposite is usually true in optics, $d >> \lambda$. One consequence is that in optics, the radiation field is not homogeneous throughout the sample. The whole concept of coherence is therefore different in optics compared to magnetic resonance spectroscopy. In the radio frequency range, two spins that are oriented in the same direction in space at the same instant contribute to the overall signal with the same amplitude and phase. If an ensemble of spins is excited with a radio-frequency pulse, they evolve coherently, always pointing in the same (time-dependent) direction.

In optical spectroscopy, however, atoms at different locations "see" fields with different phases and their induced electric dipole moments point in different directions. The induced polarisation of these atoms may still lead to a coherent radiation field, if their phase differs just by the amount that the field accumulates when it travels from one to the other. Averaged over the whole sample, this results in a preferred direction in which the emitted radiation interferes constructively and in which the sample emits most of the radiative power. If a single laser beam is used for the preparation of the sample, this direction usually coincides with the direction of propagation of the laser beam. In the general case, where multiple laser beams with different directions of propagation interact with the medium, the direction of propagation of the signal is determined by conservation of the linear momentum contributed by the different photons.

When optical methods are used to study sublevel transitions, the sensitivity does not depend on the strength of the magnetic field. It is therefore possible to use relatively low field strengths. Under these conditions, the couplings between different angular momenta often exceed the coupling to the magnetic field and the individual angular momentum operators are no longer constants of the motion. Only the total angular momentum is a "good," quantum number for the eigenstates of the Hamiltonian. As a consequence, the distinction of the various types of angular momenta is less important than in magnetic resonance spectroscopy, where a clear distinction exists between nuclear magnetic resonance (NMR) and electron spin resonance (EPR). In the examples discussed later on, the angular momentum to which the magnetic field couples usually includes contributions from different sources of angular momentum.

Other differences between optical and magnetic resonance spectroscopy can be traced to the following properties:

Transition strengths. Nuclear spin systems in high magnetic fields exhibit a very limited set of transition strengths: In weakly coupled systems, a transition is either allowed with an amplitude that depends only on the gyromagnetic ratio of the spin, or it is forbidden. In optical systems, on the other hand, transition strengths can differ by many orders of magnitude.

Relaxation mechanisms. Spontaneous emission is completely negligible in almost every spin system, but it plays a dominant role in optical spectroscopy.

Optical pumping. Because of the different transition strengths and relaxation mechanisms, the phenomenon of optical pumping has no direct analogue in magnetic resonance spectroscopy. Nevertheless, a large number of experiments use the exchange of spin polarisation between different reservoirs, as in the Overhauser effect or Hartmann–Hahn cross-polarisation (Hartmann and Hahn 1962; Pines, Gibby and Waugh 1973).

Polarised light. The polarisation of the radiation has a profound influence on the dynamics of multilevel atomic systems. Although experiments with circularly polarised microwave fields have been performed in magnetic resonance (Schweiger and Guenthard 1981), they are not widely used today.

1.4.5 Waves and particles

Quantum mechanics keeps reminding us that the border between waves and particles is not a well defined line. In the field of optics, the debate about the nature of light – wave or a stream of particles – has lasted for centuries and only quantum electrodynamics has united the views. Today, we have become used to talking about waves when we discuss the propagation of light and about particles when it comes to the interaction with matter. Nevertheless, much of the research effort in the field of quantum optics is directed towards proving that the particle nature of light and the wave properties of the atoms are also important.

Light-induced forces on atoms

The most important aspect of the particlelike nature of light is probably that absorption and emission of radiative energy occur in units of $\hbar\omega$. The same holds true for the linear and angular momentum of the photons, which must be conserved during absorption and emission. Although we discuss the conservation of angular momentum primarily in the context of optical pumping, the exchange of linear momentum between atoms and photons can be observed as light-induced forces on the atoms, as indicated in Figure 1.20.

The possibility of using light for controlling the motion of atoms was realised early and even observed experimentally as early as 1933 (Frisch 1933). Since the momentum transfer from a visible photon to an atom is orders of magnitude smaller than the momentum of a thermal atom, however, this possibility was not explored further. The widespread use of lasers changed the situation dramatically: The repeated absorption of photons from an intense laser beam can induce a force more than a million times as large as the earth's gravity. The realisation that these forces might be useful for reducing the atomic velocity of atoms (Hänsch and Schawlow 1975; Wineland and Dehmelt

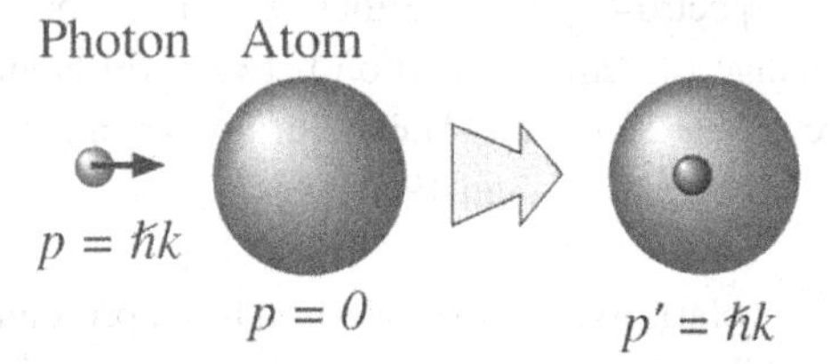

Figure 1.20. Conservation of linear momentum during absorption of a photon by an atom.

1975), thus cooling them to temperatures in the milli- and micro-Kelvin range (Chu et al. 1985; Aspect et al. 1988; Cohen-Tannoudji 1992b) has opened a whole new field in quantum electronics.

Besides the cooling, it is possible to store the atoms in traps whose walls consist of the potential energy from the interaction with electromagnetic fields (Phillips, Prodan and Metcalf 1985; Cornell, Monroe and Wieman 1991; Sesko, Walker and Wieman 1991; Aspect 1992; Emile et al. 1992; Lindquist, Stephens and Wieman 1992; Phillips 1992; Ketterle et al. 1993). Such experiments primarily use atoms, but they are also applicable to molecules (Berns et al. 1989), macroscopic particles (Ashkin 1970b; Ashkin 1970a; Ashkin and Dziedzik 1971) and even to living cells (Ashkin, Dziedzic and Yamane 1987). Apart from cooling and confining atoms, optically induced forces can also deflect atomic beams, acting like a mirror (Cook and Hill 1982; Moskowitz, Gould and Pritchard 1985; Balykin et al. 1988; Freyberger and Herkommer 1994).

Atoms as waves

Atoms are usually considered as solid particles like balls that are localised in space and move with well defined velocities. Such a classical analysis cannot be complete, as it would violate Heisenberg's uncertainty relation. According to quantum mechanics, atoms should also have a wavelike nature, characterised by the de Broglie wavelength $\lambda = h/p$. This wavelike behaviour was first verified in 1929 by Estermann and Stern (Estermann and Stern 1930), who observed diffraction of molecular beams from single crystalline surfaces. For many years, these findings were not followed by other experiments, but during the last decade, a whole new field has sprung up that uses just this effect (Adams, Sigel and Mlynek 1994).

Figure 1.21 illustrates such an experiment, performed at the Massachusetts Institute of Technology by Dave Pritchard and colleagues (Martin et al. 1988).

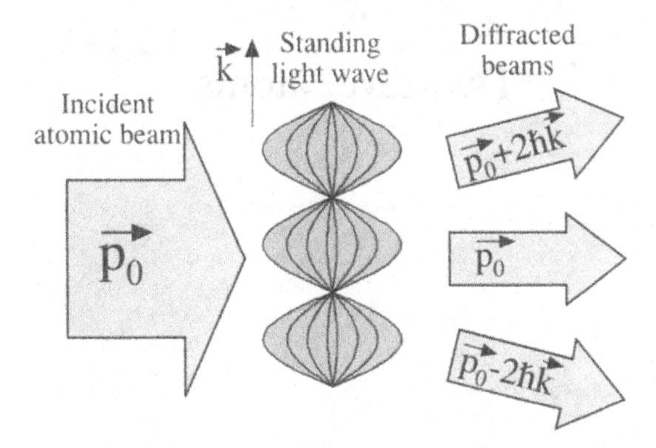

Figure 1.21. Diffraction of an atomic beam off a grating defined by two counter-propagating laser beams.

They used two counter-propagating laser beams that produced a standing wave. When the frequency of the laser was close to the resonance frequency of sodium atoms passing through the standing wave, the light acted as a grating, splitting the incident atomic beam into two partial beams with different momentum components in the direction of the standing wave. For atoms that behave like waves, it is possible to build not only beam splitters (Pfau et al. 1993; Grimm, Söding and Ovchinnikov 1994), but also lenses (Pearson et al. 1980; Sleator et al. 1992) and mirrors (Cook and Hill 1982; Balykin et al. 1988). Lenses, mirrors, and beam splitters for atoms may be made of light, but also of matter: Microfabricated structures can diffract atom waves if the structures are of a dimension comparable to the de Broglie wavelength of the atoms (Carnal and Mlynek 1991). Experiments of this type work best with light atoms at low temperatures, where the de Broglie wavelength is long. The field of atom optics is therefore closely connected to laser cooling: At the lowest temperatures reached so far, the de Broglie wavelength of the atoms can exceed one micrometer.

When atoms pass through a beam splitter, the partial waves in the two beams are coherent with each other. This must be interpreted in such a way that the atoms do not pass through one or the other slit, but through both arms simultaneously. This is best demonstrated by interferometers for atoms (Carnal and Mlynek 1991; Kasevich and Chu 1991; Keith et al. 1991). Since the separation between the two arms can be a macroscopic distance (Kasevich and Chu 1991), atoms in such an interferometer may be considered as Schrödinger cats.

2
Two-level atoms

The simplest and most frequently used model for describing the interaction between atoms and light describes the atom as a quantum mechanical two-level system interacting with a classical electromagnetic wave. This description is closely analogous to the classical model. Although atoms with only two energy levels do not exist in nature, they are a convenient fiction and often a good approximation to reality if the frequency of the radiation field is close to an atomic transition. We include here a brief review of its most important properties.

2.1 Quantum mechanical description

2.1.1 The Jaynes–Cummings model

Outline

Classically, the interaction between an atom and the radiation field is determined by the scalar product $-\vec{E}\cdot\vec{d}$ of the electric field $\vec{E}$ with the atomic dipole moment $\vec{d}$. The correspondence principle allows us to translate this interaction into a quantum mechanical Hamiltonian by replacing the classical variables with the corresponding quantum mechanical operators. As outlined in Chapter 1, the radiation field has an infinite number of modes. The Hamiltonian therefore contains an infinite sum over all modes, each with an infinite number of states. The complete form of this Hamiltonian is therefore far from trivial.

There is, however, a simplified form, which allows the derivation of analytical results and yet contains many of the significant features. It neglects all but a single mode of the radiation field (Jaynes and Cummings 1963). This model, known as the Jaynes–Cummings model, was proposed originally as a purely theoretical tool for studying fundamental aspects of matter–radiation interaction. Several groups have since built experimental systems capable of enhancing the coupling with one field mode and suppressing all other modes

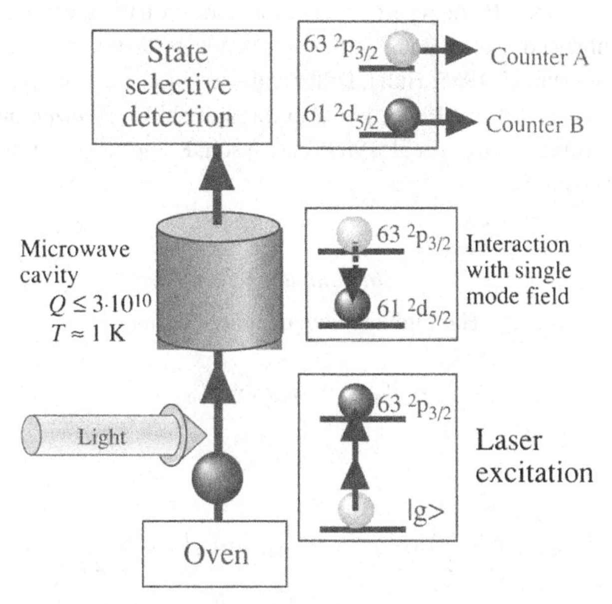

Figure 2.1. The micromaser allows an experimental realisation of the Jaynes–Cummings model.

strongly enough that the dynamics of the system is essentially equivalent to that of a single mode field (Goy et al. 1983; Walther 1988a; Haroche and Kleppner 1989; Hinds 1991; Haroche 1992; Walther 1992).

Figure 2.1 shows schematically one possibility for reducing the radiation field to a single spatial mode: If the interaction region is completely enclosed with conducting walls, the resulting resonator supports only field modes with a discrete frequency spectrum. Using a suitable geometry, it is then possible to arrange that only one mode is close enough to the atomic resonance frequency and all other modes can be neglected (Walther 1988a). For practical reasons, the transition frequencies must lie in the microwave part of the electromagnetic spectrum. The atoms that interact with the field are therefore prepared in high lying Rydberg states before they enter the resonator (Haroche and Raimond 1985). The result of the interaction is measured by a state-selective detector after they have passed through the resonator.

In the optical part of the spectrum, it is not possible to build structures that are of the dimension of a wavelength and completely enclose the atom with conducting walls, although quantum well structures in semiconductors are approaching this goal. Nevertheless, it is possible to modify the mode structure in such a way that the interaction of the atom with one mode of the field is much stronger than with the rest of the radiation field. Such effects were observed first in Fabry–Perot type optical resonators, in particular through inhibited or enhanced spontaneous emission (Kleppner 1981; Goy et al. 1983; Gabrielse and Dehmelt 1985; Hulet, Hilfer and Kleppner 1985; Jhe et al. 1987; DeMartini and Jacobovitz 1988), but also through shifts (Heinzen and Feld 1987) and splittings (Raizen et al. 1989; Thompson, Rempe and Kimble 1992) of optical transitions.

Hamiltonian

The Jaynes–Cummings Hamiltonian contains three terms

$$\mathcal{H}_{\text{tot}} = \mathcal{H}_{\text{atom}} + \mathcal{H}_{\text{field}} + \mathcal{H}_{\text{int}} \tag{2.1}$$

where

$$\mathcal{H}_{\text{atom}} = -\frac{1}{2}\hbar\omega_0\sigma_z \qquad \mathcal{H}_{\text{field}} = \hbar\omega_{\text{L}}\left(a^\dagger a + \frac{1}{2}\right) \tag{2.2}$$

represent the free atom and the isolated field. The operator σ_z is the Pauli spin operator acting on the two-level atom, ω_0 the atomic resonance frequency, and ω_{L} the frequency of the radiation field.

Figure 2.2 shows the two components of the model. The interaction between them is the electric dipole interaction. Using the correspondence principle, we write the interaction Hamiltonian as

$$\mathcal{H}_{\text{int}} = \hbar\omega_1(a + a^\dagger)(\sigma_+ + \sigma_-) \approx \hbar\omega_1(a\sigma_+ + a^\dagger\sigma_-) \tag{2.3}$$

with ω_1 describing the coupling strength, which is determined by the atomic dipole moment, the mode volume and the position of the atom in the mode. The operators $\sigma_\pm$ are Pauli spin matrices that describe transitions between the ground and excited states of the two-level atom. The first form of the Hamiltonian is obtained directly from the classical interaction term through the correspondence principle, replacing the atomic dipole moment by the operator $\sigma_x = (\sigma_+ + \sigma_-)$ and the electric field by $E = a + a^\dagger$. The approximate form is obtained by neglecting the terms involving $a^\dagger\sigma_+$ and $a\sigma_-$. The first of these terms describes the transition of the atom from the ground to the excited state while simultaneously creating a photon, the second the reverse process. These processes violate energy conservation and therefore play a role only in very high order. The approximate form corresponds in the semiclassical analysis

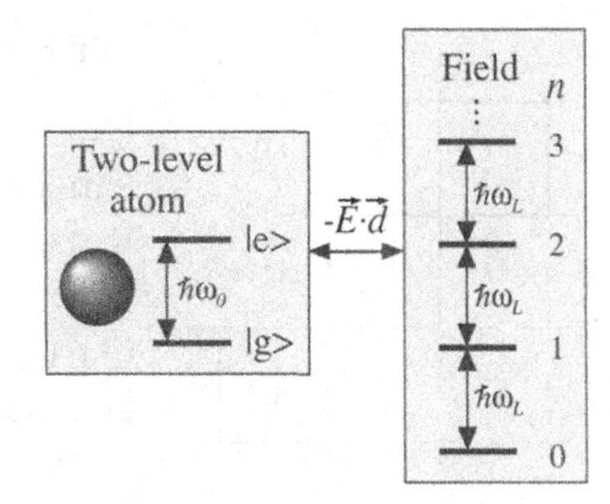

Figure 2.2. The components of the Jaynes–Cummings model are a two-level atom and a single mode of the radiation field.

to the "rotating wave approximation," which we discuss in detail in Section 2.2.

2.1.2 Summary of results

Structure of Hamiltonian

To evaluate the Hamiltonian, we use a basis that consists of the product states $|g, n\rangle$ and $|e, n\rangle$ of the individual subsystems. They correspond to states where the atom is in its ground state $|g\rangle$ or excited state $|e\rangle$, and the field contains n photons. The Hamiltonian acts on these states as

$$
\begin{aligned}
\mathcal{H}_{\text{atom}} |g;n\rangle &= -\frac{1}{2}\hbar\omega_0 |g;n\rangle; \\
\mathcal{H}_{\text{atom}} |e;n\rangle &= \frac{1}{2}\hbar\omega_0 |e;n\rangle; \\
\mathcal{H}_{\text{field}} |g;n\rangle &= \hbar\omega_L\left(n + \frac{1}{2}\right) |g;n\rangle; \\
\mathcal{H}_{\text{field}} |e;n\rangle &= \hbar\omega_L\left(n + \frac{1}{2}\right) |e;n\rangle \\
\mathcal{H}_{\text{int}} |g;n\rangle &= \hbar\omega_1 \sqrt{n}\, |e;n-1\rangle \\
\mathcal{H}_{\text{int}} |e;n\rangle &= \hbar\omega_1\sqrt{n+1}\, |g;n+1\rangle
\end{aligned}
\tag{2.4}
$$

The operators for the partial systems are diagonal in this basis, while the coupling operator contains off-diagonal elements between pairs of states having almost the same energy. The total Hamiltonian has a block-diagonal structure consisting of 2×2 blocks, as shown in Figure 2.3.

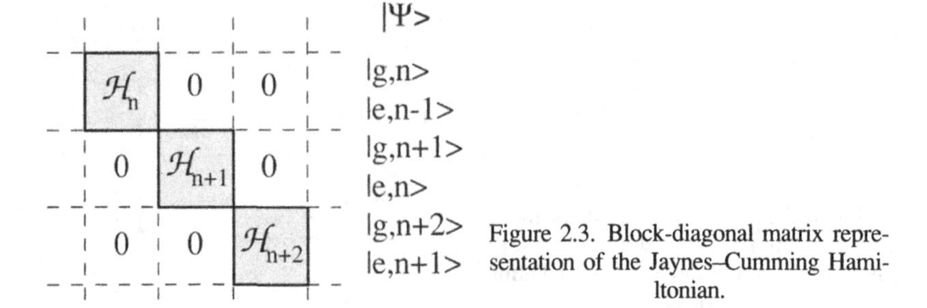

Figure 2.3. Block-diagonal matrix representation of the Jaynes–Cumming Hamiltonian.

The left-hand side of the figure indicates the structure of the Hamiltonian, with the shaded part containing the nonvanishing elements. The basis is indicated to the right of the Hamiltonian. The individual blocks along the diagonal take the form

$$\mathcal{H}_n = \hbar \begin{pmatrix} \left(n + \frac{1}{2}\right)\omega_{\mathrm{L}} - \frac{\omega_0}{2} & \omega_1 \sqrt{n} \\ \omega_1 \sqrt{n} & \left(n - \frac{1}{2}\right)\omega_{\mathrm{L}} + \frac{\omega_0}{2} \end{pmatrix} \tag{2.5}$$

The upper of the two states of this partial system corresponds to n photons in the field; the atom is in the electronic ground state. The second state has almost the same energy, but the photon has been absorbed: The field now contains $n - 1$ photons, whereas the atom is in the excited state. In the absence of an interaction, the energy difference between these two states is the frequency difference between the atomic transition frequency ω_0 and the laser frequency ω_{L}. The off-diagonal elements of the Hamiltonian, which describe absorption and emission of a photon, couple these two states to each other, but not to additional states. They thus represent a closed two-level system that can be diagonalised analytically.

Eigenvalues and eigenvectors

The eigenvalues of this two-level system are

$$\mathcal{E}_{n;\alpha,\beta} = \hbar\left(n\ \omega_{\mathrm{L}} \mp \frac{\Omega}{2}\right) \tag{2.6}$$

where α, β are indices that label the eigenstates within the two-level system.

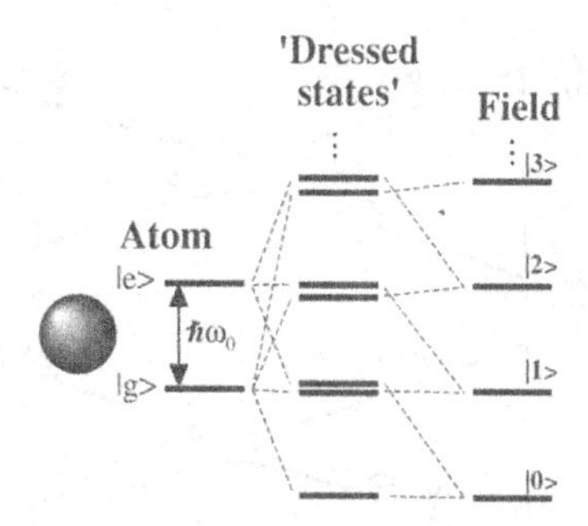

Figure 2.4. Energy level scheme for the isolated subsystems (left, right) and the coupled Jaynes–Cummings system (centre).

The energy difference,

$$\Omega = \sqrt{(\omega_0 - \omega_L)^2 + 4n\omega_1{}^2} \tag{2.7}$$

is the generalised Rabi frequency. The eigenstates are

$$\begin{aligned} |n;\alpha\rangle &= -\sin\theta_n|g;n\rangle + \cos\theta_n|e;n-1\rangle \\ |n;\beta\rangle &= \cos\theta_n|g;n\rangle + \sin\theta_n|e;n-1\rangle \end{aligned} \tag{2.8}$$

with the mixing angle

$$\tan(2\theta_n) = \frac{2\sqrt{n}\ \omega_1}{(\omega_L - \omega_0)} \tag{2.9}$$

determined by the ratio of coupling strength $2\sqrt{n}\ \omega_1$ to frequency mismatch $\omega_L - \omega_0$. These eigenstates of the coupled systems are superpositions between two states that are identical apart from one photon being either in the field or absorbed by the atom. This process is referred to as the "dressing" of the atom by the photons and the states are known as "dressed states."

As Figure 2.4 indicates, each pair of dressed states includes contributions from two field states and both atomic states. The splitting Ω within each pair of dressed states increases with the square root of the photon number, i.e., with the field strength.

Figure 2.5 shows the energy of the dressed states as a function of the laser frequency detuning. For large detuning, the eigenstates of the total Hamiltonian are identical with the product states. In the figure, the curves representing their energy are therefore labelled with the product states. At resonance, the coupling term causes an avoided crossing (anticrossing) of two

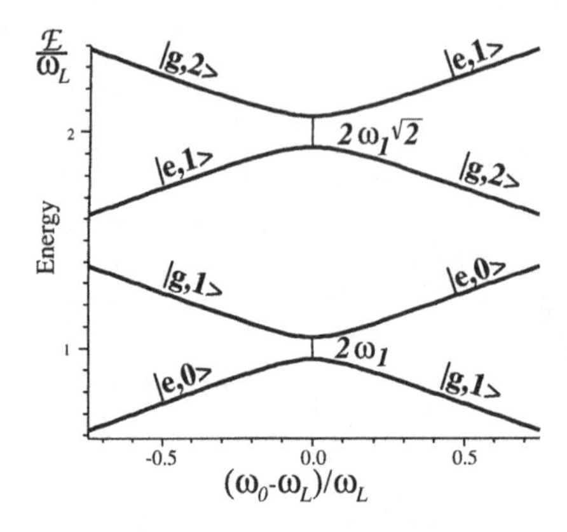

Figure 2.5. Energy of dressed states as a function of the atomic transition frequency.

states. The separation between the two states at resonance is proportional to the square root of the number of photons, i.e., proportional to the absolute value of the field strength. An interesting aspect of this is that even the product state $|e;0\rangle$ between the excited atom $|e\rangle$ and the vacuum state $|0\rangle$ of the field is not an eigenstate of the coupled Hamiltonian. The usual interpretation of this is that even the empty cavity couples to the atom. This also explains the occurrence of spontaneous emission. The only product state that is also an eigenstate of the coupled Hamiltonian is the state $|g,0\rangle$, which is not included in the drawing.

Dynamics: the resonant case

The Hamiltonian not only determines the energy of the eigenstates, but also generates the time evolution of the system. We consider the case where an atom in the excited state couples to the field mode occupied initially by $n-1$ photons and assume that the field frequency is exactly resonant ($\theta = \frac{11}{4}$) with the atomic transition frequency. We expand the initial state in the eigenbasis of the Hamiltonian as

$$\Psi(0) = |e;n-1\rangle = \frac{1}{\sqrt{2}}\left(|n;\alpha\rangle + |n;\beta\rangle\right) \tag{2.10}$$

The Schrödinger equation provides the time evolution

$$\Psi(t) = \frac{1}{\sqrt{2}}\left(|n;\alpha\rangle e^{-i\mathscr{E}_{n;\alpha}t} + |n;\beta\rangle e^{-i\mathscr{E}_{n;\beta}t}\right) \tag{2.11}$$

where

$$\mathscr{E}_{n;\alpha,\beta} = \hbar\ (n\ \omega_{\mathrm{L}} \mp \omega_1 \sqrt{n}) \tag{2.12}$$

Transforming this result back to the product basis, we find

$$\begin{aligned}
\Psi(t) & \\
&= \frac{1}{2}\ [(-|\mathrm{g};n\rangle + |\mathrm{e};n-1\rangle)\mathrm{e}^{-\mathrm{i}\mathscr{E}_{n;\alpha}t} + (|\mathrm{g};n\rangle + |\mathrm{e};n-1\rangle)\mathrm{e}^{-\mathrm{i}\mathscr{E}_{n;\beta}t}] \\
&= \frac{1}{2}\ [|\mathrm{g};n\rangle\ (-\mathrm{e}^{-\mathrm{i}\mathscr{E}_{n;\alpha}t} + \mathrm{e}^{-\mathrm{i}\mathscr{E}_{n;\beta}t}) + |\mathrm{e};n-1\rangle(\mathrm{e}^{-\mathrm{i}\mathscr{E}_{n;\alpha}t} + \mathrm{e}^{-\mathrm{i}\mathscr{E}_{n;\beta}t})] \\
&= [|\mathrm{g};n\rangle\ (-\mathrm{i})\ \mathrm{e}^{-\mathrm{i}\hbar n\omega_{\mathrm{L}}t}\sin\ (\omega_1\sqrt{n}\ t) + |\mathrm{e};n-1\rangle\ \mathrm{e}^{-\mathrm{i}\hbar n\omega_{\mathrm{L}}t}\cos\ (\omega_1\sqrt{n}\ t)]
\end{aligned} \tag{2.13}$$

The probability of finding the atom in the excited state is then

$$|\langle\Psi(t)|\mathrm{e};n-1\rangle|^2 = \frac{1}{2}(1 + \cos\ (2\ \omega_1\sqrt{n}\ t)) \tag{2.14}$$

This result describes an oscillatory exchange of one quantum of energy between the atom and the field. This behaviour is the quantum mechanical version of the Rabi oscillation. As we see in the following chapter, the semiclassical analysis yields a closely related result. Quantum mechanical features enter the Jaynes–Cummings model only when the occupation of the field states is taken into account.

A well known example of nonclassical behaviour occurs when the field is initially not in a number state, but in a coherent state. The atom then "sees" not a well defined field strength, but a distribution of fields that is determined by the occupation probabilities for the various states.

Figure 2.6 illustrates this effect: For a coherent state, the field strength does not have a sharp value, but a distribution whose width is proportional to the square root of the mean photon number. An atom that enters such a field therefore evolves in a qualitatively different way from an atom that enters a mode in a number state. Its evolution contains all the Rabi frequencies represented in the left-hand side of Figure 2.6, and the oscillatory motion ceases on a timescale inversely proportional to the width of the field distribution. This is shown in the right-hand side, where the system quickly reaches a thermal equilibrium, where half of the atoms are in the ground state and the other half in the excited state. Since the Rabi frequencies form a discrete set, however, the signal reappears after a time inversely proportional to the spacing between the

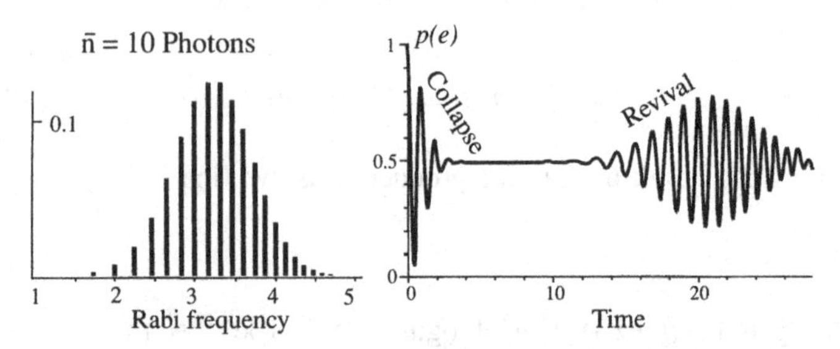

Figure 2.6. Cummings collapse and revival. The left-hand side represents the occupation probabilities for the different states; the right-hand side shows the probability of finding the atom in the excited state.

Rabi frequencies (Rempe, Walther and Klein 1987). These two phenomena are known in the literature as Cummings-collapse and revival and represent an important example of nonclassical behaviour of radiatively driven atoms.

2.2 Semiclassical analysis

2.2.1 Model

Outline

For the semiclassical description, we start with a two-level model for the atomic system, connected by an electric dipole-allowed optical transition, and driven by a classical electromagnetic wave (Brewer 1977a; Allen and Eberly 1987). This model is a direct quantum-mechanical analogue of the classical Lorentz–Lorenz theory of dispersion (Lorentz 1880; Lorenz 1881; Born and Wolf 1986), which describes the resonant interaction of light with a collection of dipoles. There, the oscillating electrical field drives the dipoles, and the accelerated charges of the oscillating dipole act in turn as the source of a secondary wave. The two waves propagate together and the interference between them causes damping, i.e., absorption, as well as a phase shift, i.e., dispersion. Physically, the semiclassical model describes the situation in a very similar way: An electromagnetic wave excites atoms acting as electric dipoles. The main difference is that they are described quantum mechanically, often as two-level systems.

Experimentally used samples contain a large number of atoms. In most cases, the theoretical description does not distinguish these atoms, but writes the Hamiltonian for an average representative of the ensemble and uses a density operator that represents the ensemble average of the whole system. As long as the atoms of the sample are sufficiently isolated from each other and their environment, this approximation is quite safe. The most important devi-

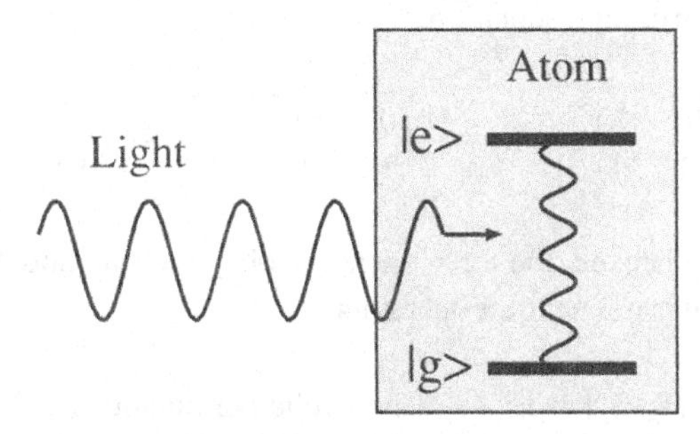

Figure 2.7. Summary of the model system: The atom is a quantum mechanical two-level system driven by a classical electromagnetic wave.

ations from this idealisation are relaxation effects and inhomogeneities. Relaxation effects are due to interactions between the various subsystems and imperfect isolation from the environment, mainly electromagnetic fields and translational degrees of freedom. The most important inhomogeneous effect in this context is Doppler broadening due to the thermal velocity distribution of the atoms.

As in the quantum mechanical case, we label the two energy levels of the atoms with $|g\rangle$ (ground state) and $|e\rangle$ (excited state; see Figure 2.7). The two states are connected by an allowed electric dipole transition.

The transition connecting the two states should be electric-dipole allowed, so that the states are of opposite parity. With an appropriate choice of their phases, the dipole operator $\hat{d}$ has the matrix elements

$$\langle e|\hat{d}|g\rangle = \mu_e \qquad \langle g|\hat{d}|e\rangle = \mu_e \qquad \langle e|\hat{d}|e\rangle = \langle g|\hat{d}|g\rangle = 0 \tag{2.15}$$

where μ_e is the real-valued matrix element of the electric dipole operator.

Operator base

The system has three degrees of freedom, i.e., its operators can be expanded in a basis set of three operators in addition to the unity operator. For this basis, we choose the three spin operators S_x, S_y, and S_z. Using units where $\hbar = 1$, the matrix elements of these operators are

$$\begin{array}{lll} \langle g|S_x|g\rangle = 0 & \langle g|S_x|e\rangle = 1/2 & \langle e|S_x|e\rangle = 0 \\ \langle g|S_y|g\rangle = 0 & \langle g|S_y|e\rangle = -i/2 & \langle e|S_y|e\rangle = 0 \\ \langle g|S_z|g\rangle = 1/2 & \langle g|S_z|e\rangle = 0 & \langle e|S_z|e\rangle = -1/2 \end{array} \tag{2.16}$$

or, written in matrix representation

$$S_x = \frac{1}{2}\begin{pmatrix} 0 & 1 \\ 1 & 0 \end{pmatrix} \qquad S_y = \frac{\mathrm{i}}{2}\begin{pmatrix} 0 & -1 \\ 1 & 0 \end{pmatrix} \qquad S_z = \frac{1}{2}\begin{pmatrix} 1 & 0 \\ 0 & -1 \end{pmatrix} \tag{2.17}$$

As can easily be checked, the three operators obey the standard commutation relations of angular momentum operators

$$[S_x, S_y] = \mathrm{i}\, S_z \qquad \text{and cyclic permutations} \tag{2.18}$$

Two-level systems of this type were analysed first by Felix Bloch (Bloch 1946) in the context of nuclear magnetic resonance spectroscopy. He considered a collection of spins $S = 1/2$ and derived the relevant equations of motion, which became known as the Bloch equations. Roughly ten years later, Feynman, Vernon and Helwarth (FVH) showed that every two-level system obeys the same equations of motion and can therefore be analysed in terms of the Bloch equations (Feynman, Vernon and Helwarth 1957). It has therefore become customary to consider the system as a pseudo-spin 1/2.

In this picture, the ground state is the $+1/2$ state of a pseudo-spin 1/2, whereas the excited state corresponds to the $-1/2$ state. The energy difference between the ground and excited states is an apparent Zeeman splitting. The z component of the pseudo-spin describes the population difference between the two states; the x- and y-components describe the coherence between the two levels. In the case of an optical two-level system, the quantum mechanical coherence between the two states corresponds physically to an oscillating electric dipole moment. This pseudo-spin is extremely helpful for visualising the dynamics of the system. Its evolution is identical to that of a classical magnetic dipole coupled to an angular momentum vector.

Hamiltonian

The Hamiltonian for the atom interacting with a monochromatic electric field $E = \frac{1}{2}\,\{E_0\, \mathrm{e}^{\mathrm{i}\omega_\mathrm{L} t} + \text{c.c.}\}$ can be written as

$$\mathcal{H} = \mathcal{H}_\text{atom} + \mathcal{H}_\text{int} = -\omega_0\, S_z - 2\, \omega_x \cos(\omega_\mathrm{L} t)\, S_x \tag{2.19}$$

where ω_0 represents the electronic excitation energy, ω_L the laser frequency, and the coupling constant

$$\omega_x = E_0\, \mu_e \tag{2.20}$$

the interaction between the atomic dipole moment μ_e and the electric field E_0. In this form of the Hamiltonian, as well as in the subsequent calculation, we express energies as frequencies to obtain a direct relationship with the experiment, where energy differences always appear as frequencies.

In the Hamiltonian of equation (2.19), the electric field of the optical radiation does not appear as a vector wave, but as a scalar. This is an inherent difficulty of the two-level model which can only be removed by abandoning the two-level approximation and considering a more complete level structure of the atom. The form of the Hamiltonian that we use here sets the origin of the energy axis not to the lowest state or to the ionisation limit, but halfway between the two states. This normalisation helps to emphasise the symmetry of the system. We may use the FVH analogy to relate this Hamiltonian to a spin-1/2 system in a magnetic field: The electronic energy ω_0 corresponds to the Zeeman splitting arising from a magnetic field oriented along the z axis. The coupling to the laser field is mathematically equivalent to the interaction of the spin 1/2 with an oscillating magnetic field along the x axis.

Density operator

In many of the experiments of interest here, the physical system consists of an ensemble of atoms that cannot be described by a quantum mechanical wave function. A complete description of these systems can be obtained in terms of the density operator ρ (Fano 1957). Like every quantum mechanical operator, it can be expanded in a suitable set of basis operators. In the case of the two-level system, the preferred basis set consists of the spin operators S_x, S_y, and S_z defined above. Writing the expansion coefficients as s_x, s_y, and s_z, the density operator becomes

$$\rho' = \frac{1}{2}\mathbb{1} + s_x\, S_x + s_y\, S_y + s_z\, S_z \tag{2.21}$$

The unity operator $\mathbb{1}$ does not depend on time and does not contribute to any observable quantity; as an abbreviation it is therefore often simpler to use the traceless part of the density operator

$$\rho = s_x\, S_x + s_y\, S_y + s_z\, S_z \tag{2.22}$$

often referred to as the reduced density operator. The coefficients s_x, s_y, and s_z are not only expansion coefficients of the density operator in the chosen

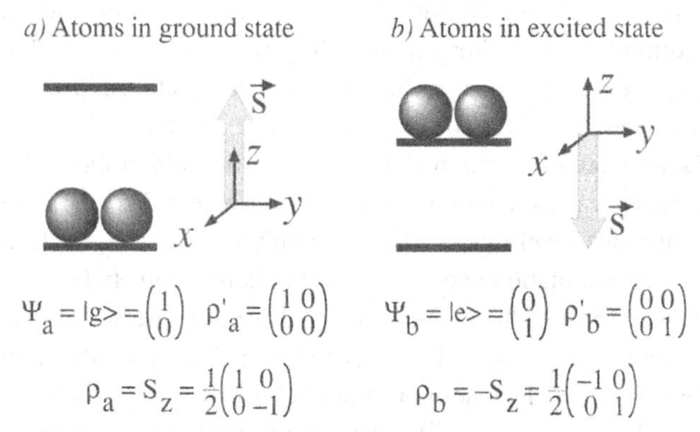

Figure 2.8. Comparison of wavefunctions and density operator representation for atoms in ground and excited state.

basis, but also the components of the pseudo-spin that describes the motion of the system in the FVH picture. Its z component contains the population information: For $S_z = 1$, all atoms are in the ground state $|g\rangle$, and the energy of the system reaches its minimum, $\mathrm{tr}\{\mathcal{H}\rho\} = -\omega_0/2$.

Figure 2.8 illustrates the orientation of the pseudo-spin $\vec{s}$ together with the wavefunction and density operator representation for atoms in the electronic ground state and the electronically excited state. Since we have eliminated the component proportional to the unity operator, the density operators ρ_a, ρ_b are traceless.

The transverse components of the pseudo-spin $\vec{s}$ indicate an induced optical dipole moment of the atoms – they represent the quantum mechanical analogue of the oscillating dipoles in the Lorentz–Lorenz theory. As Figure 2.9 shows, they indicate a coherent superposition of the two levels.

The orientation of the pseudo-spin vector in the transverse plane corresponds to the phase of the oscillating dipole moment, in close analogy to the complex notation in the description of the electromagnetic field. The main difference between the classical and the quantum mechanical electric dipole is the maximum size of the induced dipole moment. In the classical case, the induced dipole moment can become arbitrarily large if the driving field is large enough. In the quantum mechanical case, however, the dipole moment cannot exceed a maximum size. Mathematically, this limitation derives from the condition for the trace of the density operator,

$$\mathrm{tr}\{\rho'^2\} = \frac{1}{2}(1 + s_x{}^2 + s_y{}^2 + s_z{}^2) \leq 1 \text{ or } s_x{}^2 + s_y{}^2 + s_z{}^2 \leq 1$$

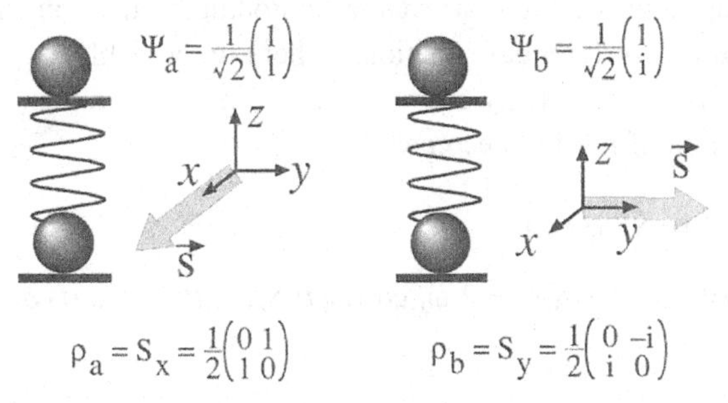

Figure 2.9. The transverse components of the pseudo-spin vector indicate a coherent superposition of the two quantum states. The orientation of the pseudo-spin indicates the phase of the optical polarisation.

In the FVH picture, this implies that the length of the spin vector is limited to the unit sphere; for a mixed state ($\mathrm{tr}\{\rho'^2\} < 1$), it can be smaller, but never larger.

This difference can also be traced to the energy level structure: Even in quantum mechanics, a harmonic oscillator can reach arbitrarily large oscillation amplitudes, since its energy level structure extends upwards to infinity. The two-level atom, however, has a level structure that is bounded from both below and above. Every atom can therefore absorb at most one photon, whereas a harmonic oscillator can absorb an arbitrary amount of energy. This also limits the amplitude of the oscillation. For small oscillation amplitudes, when the system is mostly in the ground state (i.e., $s_z^2 \sim 1$ and therefore $s_x^2 + s_y^2 \ll 1$), the behaviour of the two-level atom is identical to that of the harmonic oscillator or a classical dipole. This is the regime in which perturbation theory can be used. The result of a perturbation analysis of the two-level system is therefore identical to that of the harmonic oscillator or the classical Lorentz–Lorenz model (Heitler 1953).

2.2.2 Equation of motion

Laboratory frame

The equation of motion for the density operator

$$\dot{\rho}(t) = -\mathrm{i}[\mathcal{H}(t), \rho(t)] \tag{2.23}$$

has no closed solution when the Hamiltonian is time-dependent. For a discussion of the equation of motion, we expand the density operator in the usual spin-1/2 operator base, using time-dependent coefficients

$$\rho(t) = s_x(t)\, S_x + s_y(t)\, S_y + s_z(t)\, S_z \tag{2.24}$$

This expansion allows us to convert the Schrödinger equation into a linear system of ordinary differential equations, where the variables are the expansion coefficients of the density operator. Using the Hamiltonian of equation (19), the equation of motion becomes

$$\dot{\rho} = \dot{s}_x(t)\, S_x + \dot{s}_y(t)\, S_y + \dot{s}_z(t)\, S_z$$
$$= -\mathrm{i}[\mathcal{H}, \rho(t)] = -\mathrm{i}[-\omega_0 S_z - 2\,\omega_x \cos(\omega_{\mathrm{L}} t)\, S_x,\, s_x(t)\, S_x + s_y(t)\, S_y + s_z(t)\, S_z] \tag{2.25}$$

The equation of motion for the density operator coefficients can be written in matrix form as

$$\frac{\mathrm{d}}{\mathrm{d}t}\begin{pmatrix} s_x \\ s_y \\ s_z \end{pmatrix} = \begin{pmatrix} 0 & \omega_0 & 0 \\ -\omega_0 & 0 & 2\omega_x \cos(\omega_{\mathrm{L}} t) \\ 0 & -2\omega_x \cos(\omega_{\mathrm{L}} t) & 0 \end{pmatrix}\begin{pmatrix} s_x \\ s_y \\ s_z \end{pmatrix} \tag{2.26}$$

This equation is completely equivalent to the equations of motion for a classical angular momentum. The three expansion coefficients are the components of a vector in three-dimensional space and the matrix describes the effect of a constant (ω_0) and a time-dependent ($2\,\omega_x \cos(\omega_{\mathrm{L}} t)$) torque on this vector. The time dependence of these matrix elements makes the task of finding a solution nontrivial. The only straightforward solution arises if the interaction term vanishes, corresponding to the free atom. In this case, the solution is

$$\begin{aligned} s_x(t) &= s_x(0)\cos(\omega_0 t) + s_y(0)\sin(\omega_0 t) \\ s_y(t) &= s_y(0)\cos(\omega_0 t) - s_x(0)\sin(\omega_0 t) \\ s_z(t) &= s_z(0) \end{aligned} \tag{2.27}$$

corresponding to a precession of the spin vector around the z axis, i.e., around the virtual magnetic field, as shown in Figure 2.10.

For the optical polarisation of the system, $2\,\mu_e\, \mathrm{tr}\{\rho(t)\, S_x\} = \mu_e\, s_x(t)$, this precession corresponds to an oscillation at frequency ω_0, as the right-hand side of Figure 2.10 shows. This oscillation is the quantum mechanical analogue of the oscillating dipole in the Lorentz–Lorenz theory.

Rotating frame

The precession of the pseudo-spin occurs at the atomic resonance frequency $\omega_0/2\pi$, which is, for the visible or near infrared part of the spectrum, of the order of 10^{14}–10^{15} Hz. This precession frequency is not directly observable,

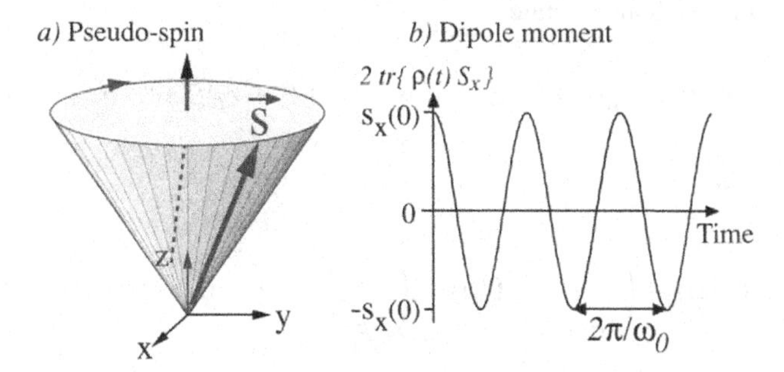

Figure 2.10. **(a)** Free evolution of the pseudo-spin. **(b)** Corresponding time dependence of the dipole moment.

and masks slower processes in the system, those in which we are interested. It is therefore convenient to remove this fast motion from the description of the dynamics. In addition, this transformation removes the time dependence of the Hamiltonian and allows us to obtain analytical solutions to the equation of motion. Mathematically, this can be achieved by a unitary transformation which we describe below. Physically, this transformation corresponds to a rotation of the coordinate system around the z axis. Since a rotating coordinate system is not an inertial reference frame, we expect that the equation of motion and therefore the Hamiltonian describing the system in the rotating frame should be different from those in the laboratory-fixed system.

To find the equation of motion in the rotating reference frame, we rewrite the Hamiltonian in a new basis. Instead of the usual basis $|\alpha\rangle$, $|\beta\rangle$, we use a time-dependent basis

$$|\alpha^{\mathrm{r}}\rangle = |\alpha\rangle\, \mathrm{e}^{\mathrm{i}\omega_{\mathrm{L}}t/2} \qquad |\beta^{\mathrm{r}}\rangle = |\beta\rangle\, \mathrm{e}^{-\mathrm{i}\omega_{\mathrm{L}}t/2} \tag{2.28}$$

In operator notation, this basis transformation appears as

$$\begin{pmatrix} |\alpha^{\mathrm{r}}\rangle \\ |\beta^{\mathrm{r}}\rangle \end{pmatrix} = \begin{pmatrix} \mathrm{e}^{\mathrm{i}\omega_{\mathrm{L}}t/2} & 0 \\ 0 & \mathrm{e}^{-\mathrm{i}\omega_{\mathrm{L}}t/2} \end{pmatrix} \begin{pmatrix} |\alpha\rangle \\ |\beta\rangle \end{pmatrix} = U(t) \begin{pmatrix} |\alpha\rangle \\ |\beta\rangle \end{pmatrix} \tag{2.29}$$

The transformation operator can be written as

$$U(t) = \mathrm{e}^{\mathrm{i}\omega_{\mathrm{L}}tS_z} \tag{2.30}$$

which represents a rotation around the z axis at the frequency ω_{L} of the laser field. This interaction representation is therefore referred to as the "rotating frame."

A quantum mechanical state

$$\Psi(t) = c_\alpha|\alpha\rangle + c_\beta|\beta\rangle = c_\alpha|\alpha^r\rangle e^{-i\omega_L t/2} + c_\beta|\beta^r\rangle e^{i\omega_L t/2} \tag{2.31}$$

has the representations

$$\Psi^{lab}(t) = \begin{pmatrix} c_\alpha \\ c_\beta \end{pmatrix} \qquad \Psi^{r}(t) = \begin{pmatrix} c_\alpha e^{-i\omega_L t/2} \\ c_\beta e^{i\omega_L t/2} \end{pmatrix} = U^{-1}\, \Psi^{lab}(t) \tag{2.32}$$

The upper index indicates the basis states used for that representation. The transformation of the wavefunction therefore uses the inverse of the operator that transforms the basis states. The transformation property of the wavefunction implies that the density operator transforms like

$$\rho^r = (|\Psi\rangle\langle\Psi|)^r = U^{-1}|\Psi\rangle\langle\Psi|U = U^{-1}\rho U \tag{2.33}$$

The same transformation property must hold for all observables, since the expectation values must be independent of the basis transformation.

Time dependence

To find the equation of motion in the rotating frame, we consider the time derivative of a state

$$\frac{d}{dt}\Psi^r(t) = \frac{d}{dt}(U^{-1}(t)\Psi^{lab}(t)) = \dot{U}^{-1}(t)\Psi^{lab}(t) + U^{-1}(t)\frac{d}{dt}\Psi^{lab}(t) \tag{2.34}$$

Using the Schrödinger equation for the laboratory frame representation and expressing the state in the rotating frame, we find

$$\frac{d}{dt}\Psi^r(t) = (\dot{U}^{-1}(t)\, U(t) - U^{-1}(t)\, i\, \mathcal{H}^{lab}\, U(t))\Psi^r(t) \tag{2.35}$$

If the Hamiltonian is to remain the generator of the time evolution, we must require that its interaction representation be

$$\mathcal{H}^r = U^{-1}\mathcal{H}^{lab}U + i\,\dot{U}^{-1}U \tag{2.36}$$

This is the general form of the transformation of the Hamiltonian: In addition to the usual operator transformation, $U^{-1}\mathcal{H}^{lab}U$, it contains a "correction term" $i\,\dot{U}^{-1}U$ that is nonzero whenever the transformation operator U is time

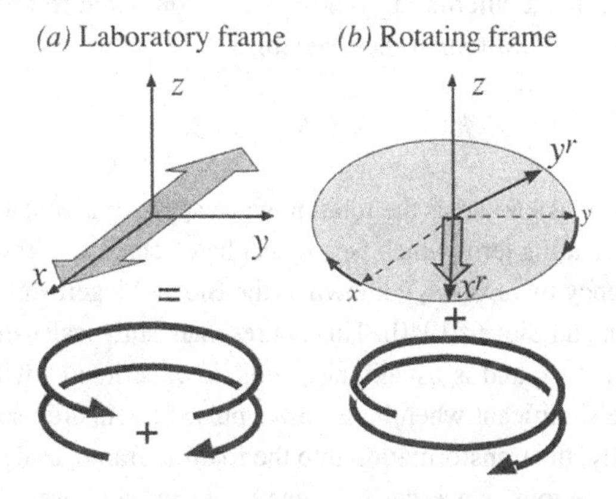

Figure 2.11. The interaction between the two-level atom and the laser field appears as an oscillatory field in the laboratory frame, i.e., as a superposition of two counter-rotating circularly polarised fields. In the rotating frame, the co-rotating component appears static, whereas the counter-rotating component rotates at twice the laser frequency.

dependent. It accounts for the fact that the rotating frame is not an inertial frame and can be compared to a Coriolis-force term in mechanics. It does not depend on the Hamiltonian, but only on the transformation operator U. If we apply this general form to the Hamiltonian of equation (19), we find

$$\begin{aligned} \mathcal{H}^{\mathrm{r}} &= U^{-1}(t)\, \mathcal{H}_{\mathrm{atom}}\, U(t) + U^{-1}(t)\, \mathcal{H}_{\mathrm{int}}\, U(t) + \mathrm{i}\dot{U}^{-1}U(t) \\ &= -\Delta\omega_0\, S_z - \omega_x\, S_x - \omega_x\, S_x \cos(2\omega_{\mathrm{L}}t) - \omega_x\, S_y \sin(2\omega_{\mathrm{L}}t) \end{aligned} \quad (2.37)$$

with

$$\Delta\omega_0 = \omega_0 - \omega_{\mathrm{L}} \quad (2.38)$$

The interaction energy with the static virtual magnetic field is thus reduced by the energy of one photon. The interaction term consists of two parts, a time-independent contribution, and a contribution rotating around the z axis at twice the laser frequency. This separation has a straightforward physical interpretation: The two contributions represent the circularly polarised components of the linearly polarised field.

The component which rotates in the same direction as the rotating coordinate system in the laboratory frame becomes time independent in the rotating frame, while the other component rotates now at twice the frequency (see Figure 2.11). Perturbation theory shows that this component can be ignored in lowest order, i.e., if the alternating field is small compared to the static field. The rotating frame Hamiltonian then becomes

$$\mathcal{H}^{\mathrm{r}} \approx -\Delta\omega_0 \, S_z - \omega_x \, S_x \tag{2.39}$$

This approximation is known as the rotating wave approximation. To second order, the counter-rotating term which we neglect here causes a shift of the atomic resonance frequency by $\omega_x^2/(4\omega_L)$, known as the Bloch–Siegert shift in magnetic resonance (Bloch and Siegert 1940). This nonresonant shift scales inversely with the laser frequency ω_L and is usually negligible in experiments with cw lasers, but may become significant when high-power pulsed lasers are used.

Mathematically, the transformation into the rotating frame, as defined above, is somewhat ambiguous, since the frequency ω_L can be chosen with positive or negative sign. Physically, however, the only meaningful choice is such that the remaining longitudinal component $\Delta\omega_0$ is minimised. This implies that the reference frame rotates with respect to the laboratory frame in the same direction and approximately the same angular velocity as the pseudo-spin under the internal Hamiltonian. In the rotating frame, the pseudo-spin still undergoes precession, but at a frequency which is reduced from ω_0 to $\Delta\omega_0$ by the laser frequency ω_L. It can therefore become very small or even negative.

Phase shifts

So far, we have treated the electromagnetic field as a monochromatic wave, but actual experiments often use time-dependent fields. Here we are interested primarily in fields that modulate either the amplitude or the phase of a monochromatic carrier wave. Amplitude modulations do not require any additions to the formalism, but phase variations require some comment.

In the laboratory frame Hamiltonian, we can introduce phase shifts as

$$\mathcal{H}_{\mathrm{int}} = -2 \, \omega_x \cos(\omega_L t + \phi) \, S_x \tag{2.40}$$

Transforming this operator to the rotating frame, we obtain

$$\begin{aligned} \mathcal{H}_{\mathrm{int}}{}^{\mathrm{r}} = U^{-1}(t) \, \mathcal{H}_{\mathrm{int}} \, U(t) &= -2 \, \omega_x \cos(\omega_L t + \phi) \, \mathrm{e}^{\mathrm{i}\omega_L t S_z} \, S_x \, \mathrm{e}^{-\mathrm{i}\omega_L t S_z} \\ &\approx -\omega_x \, (\cos\phi \, S_x + \sin\phi \, S_y) \end{aligned} \tag{2.41}$$

A shift of the phase of the optical field corresponds in the rotating frame to a rotation of the transverse field by the phase angle. This allows the experimenter to choose in which direction the field should point in the rotating frame.

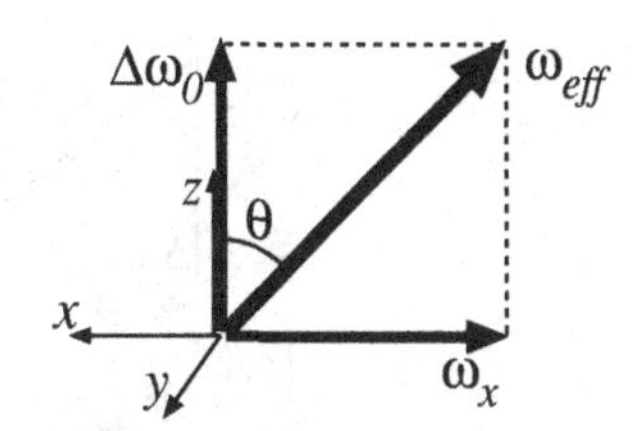

Figure 2.12. Direction of effective field.

Experimentally, the phase of the laser field can be controlled through modulators and through path length differences.

2.2.3 Statics

Geometrical analysis

The Feynman–Vernon–Helwarth picture of the two-level system provides the eigenvalues and eigenvectors of this problem through geometrical considerations, circumventing the need for diagonalising the Hamiltonian.

As Figure 2.12 shows, the rotating frame Hamiltonian can be interpreted as representing the interaction of a spin-1/2 particle with two magnetic field components along the x and z axis. In frequency units, the z component is the frequency difference $\Delta\omega_0 = \omega_0 - \omega_{\mathrm{L}}$, and the x component ω_x arises from the interaction between the atomic dipole moment and the laser field. The effective field ω_{eff}, the vector sum of the two components, lies in the xz plane, at an angle

$$\theta = \tan^{-1}\frac{\omega_x}{\Delta\omega_0} \tag{2.42}$$

from the z axis. The strength of this field is

$$\omega_{\mathrm{eff}} = \sqrt{\Delta\omega_0{}^2 + \omega_x{}^2} \tag{2.43}$$

as can readily be verified in Figure 2.12.

To transform the system into its eigenbase, we rotate the quantisation axis into the direction of ω_{eff},

$$\mathcal{H}^{\mathrm{d}} = \mathrm{e}^{\mathrm{i}\theta S_y}\,\mathcal{H}^{\mathrm{r}}\,\mathrm{e}^{-\mathrm{i}\theta S_y} = -\,\omega_{\mathrm{eff}}\,S_z \tag{2.44}$$

The two exponential operators rotate the quantisation axis of the Hamiltonian by an angle θ from the z axis to the direction of the effective field. The upper index d indicates that the Hamiltonian is diagonal in this basis.

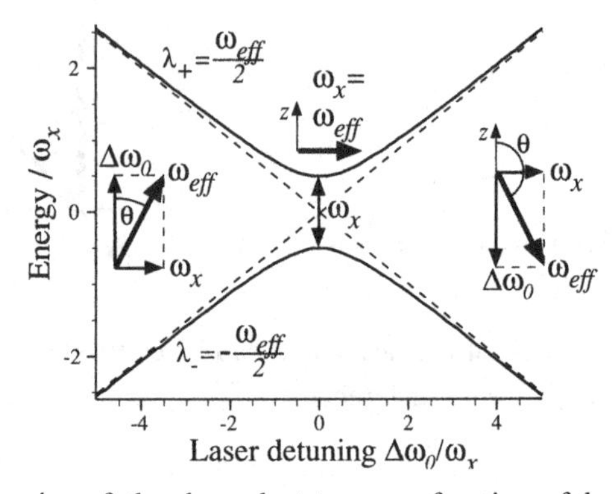

Figure 2.13. Energies of the dressed states as a function of laser detuning. The dashed lines indicate the energies for vanishing coupling strength. The insets indicate the direction of the effective field in the rotating frame for different laser detunings.

Eigenvalues and eigenvectors

In this basis, the Hamiltonian is diagonal and we find its eigenvalues as

$$\lambda_{\pm} = \pm\frac{1}{2}\,\omega_{\text{eff}} = \pm\frac{1}{2}\sqrt{\omega_x{}^2 + \Delta\omega_0{}^2} \tag{2.45}$$

The eigenvectors are parallel and antiparallel to the direction of the effective field. In the basis of the rotating frame, their representation is

$$\xi_+ = \begin{pmatrix} \cos\frac{\theta}{2} \\ \sin\frac{\theta}{2} \end{pmatrix} \qquad \xi_- = \begin{pmatrix} -\sin\frac{\theta}{2} \\ \cos\frac{\theta}{2} \end{pmatrix} \tag{2.46}$$

These states are superpositions between ground and excited states. In a quantum mechanical picture, the photon interacting with the atom is partly bound to the atom, but not completely absorbed. This is often summarised in the phrase that the atom is "dressed" with the radiation field, as we have discussed for the Jaynes–Cummings model. The energy of these states depends not only on the internal Hamiltonian of the atom, but also on the frequency and strength of the laser field.

Figure 2.13 shows the dependence of the energies of the dressed states on the laser frequency. The strongest effect of the laser field occurs on resonance, where the two levels would be degenerate in the absence of the laser field, as

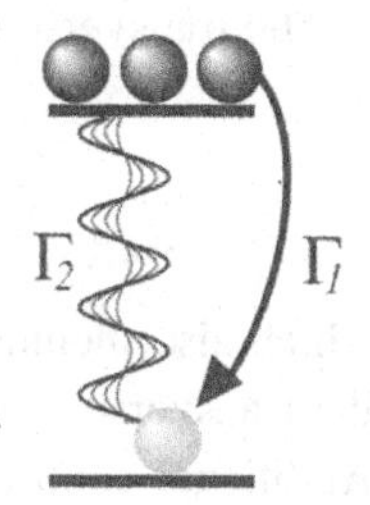

Figure 2.14. Relaxation rates Γ_1, Γ_2 for the two-level atom: Γ_1 describes the population transfer, Γ_2 the relaxation of the coherence.

indicated by the dashed lines. The insets show the direction of the effective field for negative, zero, and positive laser detuning. The double arrow marks the anticrossing due to the coupling with the field. At this point, the effective field consists only of ω_x and the mixing angle θ becomes $\pi/2$. The eigenstates are then linear combinations of the basis states with equal weights. For increasing laser detuning, the dressed states approximate the uncoupled states. The difference between the uncoupled states and the dressed states is parametrised by the mixing angle θ, which increases from 0 to π as the detuning increases from left to right. The dependence of the dressed states on the laser detuning is closely analogous to that encountered in the context of the Jaynes–Cummings model discussed in Section 2.1.

2.2.4 Stationary solution

Relaxation

To calculate the stationary state of the system, we need to take relaxation processes into account. Without discussing the physical origin of the relaxation, we describe it by the two phenomenological parameters Γ_1 and Γ_2 (see Figure 2.14).

Γ_1 is the inverse of the excited state lifetime and describes the transfer of population from the excited state to the ground state

$$\dot{\rho}_{11} = -\dot{\rho}_{22} = \Gamma_1\,\rho_{22}$$

or, in terms of the pseudo-spin variables

$$\dot{s}_z = \dot{\rho}_{11} - \dot{\rho}_{22} = 2\Gamma_1\,\rho_{22} = \Gamma_1(1 - s_z) \tag{2.47}$$

In free atoms, this process is driven by spontaneous emission, but interactions with the environment may also cause relaxation effects like energy transfer to other atoms or, in solids, to phonons.

The transverse components decay at the rate Γ_2

$$\dot{s}_x = -\Gamma_2 \, s_x \qquad \dot{s}_y = -\Gamma_2 \, s_y \tag{2.48}$$

which also includes contributions from spontaneous emission. In free atoms, the transverse relaxation rate is half the excited state decay rate, $\Gamma_2 = \Gamma_1/2$. Additional contributions to the decay of transverse coherence include collisions, which can cause fast dephasing processes, or inhomogeneous effects, like the Doppler shift.

To obtain the complete equation of motion, we expand the Schrödinger equation as

$$\dot{\rho} = -\mathrm{i}[\mathcal{H}, \rho] = -\mathrm{i}[-\Delta\omega_0 \, S_z - \omega_x \, S_x, \, s_x(t) \, S_x + s_y(t) \, S_y + s_z(t) \, S_z] \tag{2.49}$$

Adding the relaxation contribution to the Schrödinger part of the equation of motion, we obtain differential equations for the density operator coefficients s_x, s_y, s_z

$$\begin{aligned} \dot{s}_x &= \Delta\omega_0 \, s_y - \Gamma_2 \, s_x \\ \dot{s}_y &= -\Delta\omega_0 \, s_x + \omega_x \, s_z - \Gamma_2 \, s_y \\ \dot{s}_z &= -\omega_x \, s_y + \Gamma_1(1 - s_z) \end{aligned} \tag{2.50}$$

This equation of motion for a quantum mechanical two-level system was derived by Felix Bloch (Bloch 1946) for the motion of a spin 1/2 in a magnetic field. In optical spectroscopy, they are known as the optical Bloch equations (Allen and Eberly 1987).

The relaxation rates define the natural frequency scale of the system. It is therefore helpful to rewrite the equations of motion in terms of scaled frequencies $\delta = \Delta\omega_0/\Gamma_2$ and $\kappa = \omega_x/\Gamma_2$, which are now dimensionless parameters:

$$\begin{aligned} \dot{s}_x &= \Gamma_2(\delta \, s_y - s_x) \\ \dot{s}_y &= \Gamma_2(-\delta \, s_x + \kappa \, s_z - s_y) \\ \dot{s}_z &= \Gamma_2\left(-\kappa \, s_y - \frac{\Gamma_1}{\Gamma_2}(s_z - 1)\right). \end{aligned} \tag{2.51}$$

Stationary state

The equation of motion (2.51) has the stationary solution

$$(s_x, s_y, s_z)_\infty = \frac{1}{1 + \delta^2 + \kappa^2\Gamma_2/\Gamma_1}(\delta\kappa, \kappa, 1 + \delta^2) \tag{2.52}$$

If we use the dimensional quantities, the same expression reads

$$(s_x, s_y, s_z)_\infty = \frac{1}{\Gamma_2^2 + \Delta\omega_0^2 + \omega_x^2\Gamma_2/\Gamma_1}(\Delta\omega_0\, \omega_x, \omega_x\, \Gamma_2, \Gamma_2^2 + \Delta\omega_0^2) \tag{2.53}$$

In free atoms, the transverse components decay at a rate $\Gamma_2 = \Gamma_1/2$, half the spontaneous emission rate Γ_1. For this case, the expression becomes

$$(s_x, s_y, s_z)_{\infty,f} = \frac{1}{\Gamma_1^2 + 4\Delta\omega_0^2 + 2\omega_x^2}(4\Delta\omega_0\, \omega_x, 2\omega_x\, \Gamma_1, \Gamma_1^2 + 4\Delta\omega_0^2) \tag{2.54}$$

A frequently used alternative form introduces the saturation parameter

$$s = \frac{2\omega_x^2}{\Gamma_1^2 + 4\Delta\omega_0^2} \tag{2.55}$$

to write the stationary state as

$$\begin{pmatrix} s_x \\ s_y \\ s_z \end{pmatrix}_\infty = \frac{s}{1+s}\begin{pmatrix} 2\Delta\omega_0/\omega_x \\ \Gamma_1/\omega_x \\ 1/s \end{pmatrix} \tag{2.56}$$

The dimensionless saturation parameter is a measure of the number of atoms in the excited state. At $s = 1$, the longitudinal component s_z has decreased to 1/2, indicating that one quarter of the atoms are in the excited state. For free atoms, this saturation intensity is reached when

$$\kappa^2 = 2(1 + \delta^2) \tag{2.57}$$

or

$$\omega_x^2 = \frac{1}{2}\Gamma_1^2 + 2\Delta\omega_0^2 \tag{2.58}$$

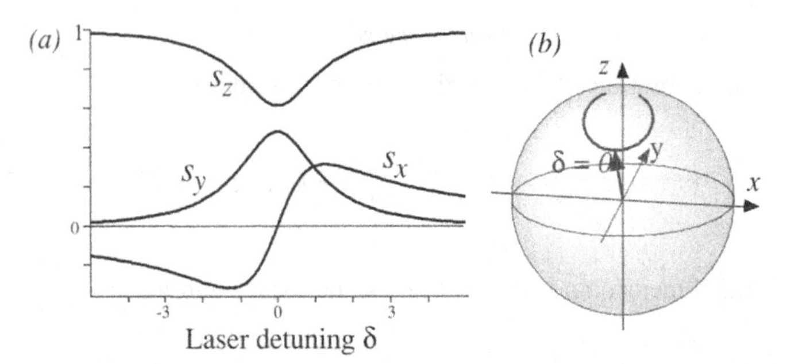

Figure 2.15. Stationary solution of the equations of motion for the two-level system as a function of normalised laser detuning δ. (a) The three components as a function of the laser detuning. (b) The stationary state on resonance ($\delta = 0$, arrow) and as a function of detuning (solid line) on the unit sphere.

At even higher intensity, the population difference s_z decreases even further, but it approaches zero only at infinite intensity. For all finite intensities, the stationary population of the excited state remains somewhat smaller than that of the ground state.

Another interesting quantity is the rate r at which the atom scatters photons. We consider here only the spontaneous emission as the excited state population times the longitudinal relaxation rate Γ_1

$$r = \frac{1 - s_{z\infty}}{2}\Gamma_1 = \frac{\kappa^2\Gamma_2}{2(1 + \delta^2 + \kappa^2\Gamma_2/\Gamma_1)} = \frac{\omega_x^2\Gamma_2}{2(\Gamma_2^2 + \Delta\omega_0^2 + \omega_x^2\Gamma_2/\Gamma_1)} \tag{2.59}$$

Energy conservation requires that this also be the rate at which energy is removed from the laser beam, i.e., the absorption of the atoms. The dependence on the laser frequency detuning $\Delta\omega_0$ shows the usual Lorentzian lineshape with width Γ_2. The third term in the denominator represents the effect of power broadening. With increasing intensity, the line broadens while the scattering rate saturates at $\Gamma_1/2$.

Frequency dependence

Figure 2.15 shows the dependence of the three components of the pseudo-spin on the laser detuning. The left-hand side of the figure shows the behaviour of the three components individually. The right-hand side of the figure shows them in the unit sphere in the form of a three-dimensional vector on resonance (arrow) and as a function of the laser frequency (curve). The transverse components describe absorption and dispersion of the medium, as discussed in detail in Chapter 6. Here, we only note that in the low intensity limit ($s \to 0$), these results are identical to those of the classical Lorentz–Lorenz theory of dispersion (Lorentz 1880; Lorenz 1881), which is discussed in Chapter 6. The z component, which represents the population difference, tends towards unity for large detunings, indicating that the atoms are predominantly in their ground state. Close to resonance, the population of the excited state increases with the laser intensity.

2.3 Dynamics

2.3.1 Rabi flopping

Undamped precession

The general solution of the equation of motion (2.50) has been given by Torrey (Torrey 1949). Since this general solution is too complicated for analytical use, we consider here only some limiting cases which are of interest under our conditions. For vanishing relaxation rates $\Gamma_{1,2}$, the motion of the spin vector corresponds to a precession of the magnetisation around the total field ω_{eff}. The resulting evolution may be reduced to the case of free precession by writing it as

$$
\begin{aligned}
s_x'(t) &= s_x'(0)\cos(\omega_{\text{eff}}t) + s_y'(0)\sin(\omega_{\text{eff}}t) \\
s_y'(t) &= s_y'(0)\cos(\omega_{\text{eff}}t) - s_x'(0)\sin(\omega_{\text{eff}}t) \\
s_z'(t) &= s_z'(0)
\end{aligned}
\tag{2.60}
$$

The coefficients s_x', s_y' and s_z' refer to a tilted coordinate system whose z' axis is parallel to the effective field:

$$
s_x' = s_x\cos\theta - s_z\sin\theta \qquad s_y' = s_y \qquad s_z' = s_x\sin\theta + s_z\cos\theta \tag{2.61}
$$

Figure 2.16 summarises some common situations. In the absence of a laser field, $\omega_x = 0$, the pseudo-spins precess around the z axis. This is the case of free precession, which is represented by the left-hand column of Figure 2.16.

Another important case is that of on-resonance irradiation, $\Delta\omega_0 = 0$ and $\omega_x \neq 0$, shown on the right-hand side of Figure 2.16; in this case, the precession occurs around the x axis. The angle between the pseudo-spin and the rotation axis depends on the initial condition. If the system is initially in the ground state, the precession occurs in the yz plane. The atoms oscillate then between the ground and excited states: After half a period, $\tau_{1/2} = \pi/\omega_x$, all atoms are in the electronically excited state characterised by $s_z = -1$. This is the situation of Rabi flopping, where the laser field exchanges the ground and

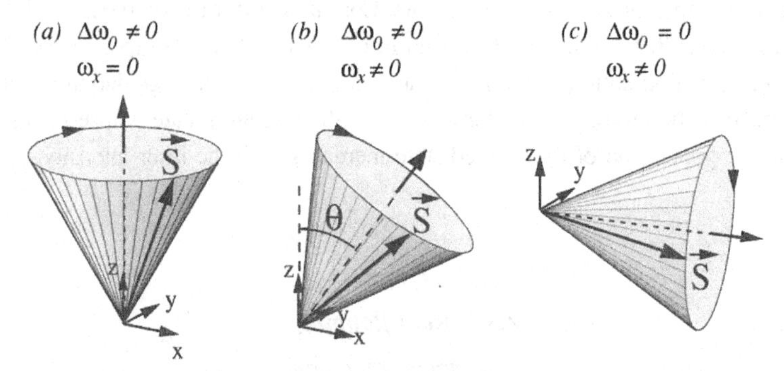

Figure 2.16. Evolution of a spin 1/2 in the rotating frame for the case of free precession (a), off-resonance irradiation (b), and on-resonance irradiation (c).

excited state populations periodically. The general case of off-resonance irradiation, where the effective field lies in the *xz* plane, is shown in the centre.

Evolution with damping

A simple analytic solution of the equations of motion is obtained in the case of isotropic relaxation, $\Gamma_1 = \Gamma_2 = \gamma$. The eigenvalues and eigenvectors of the homogeneous part of eq. (2.51) are then

$$\begin{aligned} \xi_1 &= (\kappa, 0, \delta) \qquad && \lambda_1 = -\gamma \\ \xi_{2,3} &= (\delta, 0, -\kappa) \pm \mathrm{i}(0, \omega_{\mathrm{eff}}/\gamma, 0) \qquad && \lambda_{2,3} = -\gamma \pm \mathrm{i}\omega_{\mathrm{eff}} \end{aligned} \tag{2.62}$$

The first eigenvector ξ_1 is parallel to the direction of the effective field; it represents the component of the density operator that commutes with the Hamiltonian and its evolution is an exponential decay at the rate γ. The other eigenvectors span the plane perpendicular to ξ_1. They describe a rotation perpendicular to the effective field at the frequency ω_{eff}, damped with the same rate γ.

The motion of the system can be calculated by decomposing it into the eigenvectors.

$$(s_x, s_y, s_z)(t) = (s_x, s_y, s_z)_\infty + \sum_{j=1}^{3} c_j\, \xi_j\, \mathrm{e}^{\lambda_j t} \tag{2.63}$$

where the expansion coefficients c_j must be determined from the initial condition. We find that the pseudo-spin vector moves on the surface of a cone whose tip is located at the equilibrium position and whose symmetry axis is

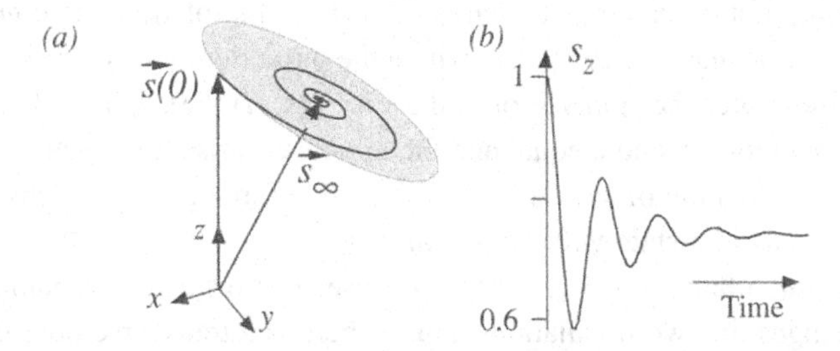

Figure 2.17. (**a**) Nutation of the pseudo-spin upon application of a near-resonant laser field, as described by equation 2.62. The trajectory was calculated for the parameters $\kappa = 3$, $\delta = 5$. (**b**) Evolution of the z components as a function of time.

the direction of the effective field. The opening angle is determined by the initial conditions. The most common situation is the case, where the system is initially in thermal equilibrium, corresponding to a pseudo-spin vector $\vec{s}(0) = \{0, 0, 1\}$. In this case, the cone degenerates to a plane and the tip of the spin vector moves on a spiral, as shown in part (a) of Figure 2.17.

In magnetic resonance, in the context of which they were first discussed, these equations of motion describe the evolution of an angular momentum. In analogy to the corresponding motion of a rotating top, this evolution is referred to as "nutation." In an actual experiment, it is not possible to observe the motion of the spin vector directly. Instead, one usually sees one component, e.g., the z component which describes the population difference between ground and excited state. For the individual component, the nutation appears as a damped oscillation, as shown in part (b) of Figure 2.17. This effect was first observed in molecules like SF_6 (Hocker and Tang 1968) and $^{13}CH_3F$ (Brewer and Shoemaker 1971).

Pulses

In many cases the laser is not used to drive the system to an equilibrium position, but only to excite it into a definite state. This application normally uses a laser exactly matching the atomic resonance frequency ($\Delta\omega_0 = 0$). Neglecting relaxation, the equation of motion then becomes

$$\dot{s}_x = 0 \qquad \dot{s}_y = \omega_x\, s_z \qquad \dot{s}_z = -\,\omega_x\, s_y \tag{2.64}$$

If the system is initially in thermal equilibrium ($s_x(0) = 0$, $s_y(0) = 0$, $s_z(0) = 1$), it evolves as

$$(s_x, s_y, s_z)(\tau) = (0, \sin(\omega_x\tau), \cos(\omega_x\tau)) \tag{2.65}$$

The resulting motion can thus be parametrised by the rotation angle $\omega_x \tau$. An important case is that of a $\pi/2$ pulse, where the pulse duration is $\tau = \pi/(2\ \omega_x)$. Such a pulse rotates the spin vector to the y axis ($s_x(\tau) = 0$, $s_y(\tau) = 1$, $s_z(\tau) = 0$): The populations become equal and the optical polarisation reaches a maximum. The application of such a $\pi/2$ pulse is therefore the optimal choice for preparing atoms in a state with large coherences.

Another important case is that of a π pulse: For resonant irradiation, the laser exchanges the two populations and, if there is a transverse polarisation, the y component is also inverted: $\{s_x(0),\ s_y(0),\ s_z(0)\} \rightarrow \{s_x(0),\ -s_y(0),\ -s_z(0)\}$. These pulses are the ideal tool for inverting the optical phase of a coherence, required, e.g., in photon echo experiments. They also allow putting all atoms into the excited state, which is not possible by continuous irradiation.

2.3.2 Free precession

Evolution

Another interesting case is that of free precession, which occurs when an optical polarisation has been established in the system by the application of a resonant laser pulse (Brewer and Shoemaker 1972; Brewer and Genack 1976; Genack, Macfarlane and Brewer 1976; Brewer 1977b; Brewer 1977a; DeVoe and Brewer 1978; DeVoe and Brewer 1979; DeVoe et al. 1979; Macfarlane, Shelby and Shoemaker 1979; DeVoe and Brewer 1983). After the end of the pulse, the system evolves freely under the internal atomic Hamiltonian. Under these conditions, the longitudinal and transverse components of the pseudo-spin decouple: The z component relaxes exponentially towards its equilibrium value of 1, while the transverse components perform a spiralling motion towards the origin:

$$s_z(t) = 1 + (s_z(0) - 1)e^{-\Gamma_1 t}$$

$$s_x(t) = s_{xy}(0)\cos(\Delta\omega_0 t + \alpha)\ e^{-\Gamma_2 t} \qquad s_y(t) = -s_{xy}(0)\sin(\Delta\omega_0 t + \alpha)\ e^{-\Gamma_2 t} \tag{2.66}$$

where s_{xy} denotes the length of the transverse part of the pseudo-spin and

$$\alpha = \tan^{-1}\frac{s_y(0)}{s_x(0)} \tag{2.67}$$

is the angle between the initial direction of the transverse polarisation vector and the x axis. These two parameters depend on the initial condition estab-

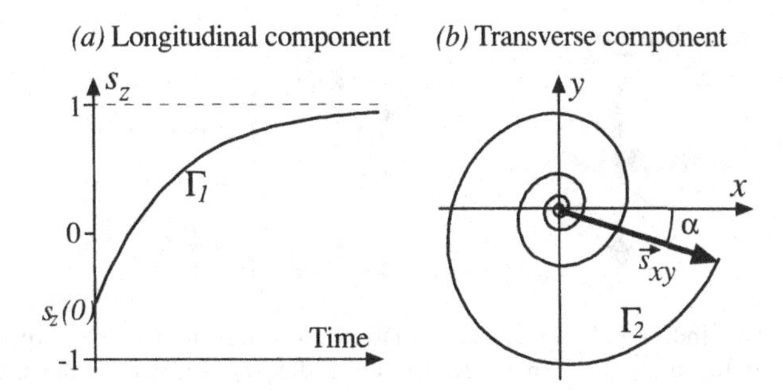

Figure 2.18. Evolution of longitudinal (a) and transverse (b) components of the pseudo-spin during free precession.

lished by the preparation pulse and can be calculated from the equations of motion discussed under the subject of Rabi flopping. Figure 2.18 graphically represents the evolution of the pseudo-spin. The left-hand side shows the relaxation of the longitudinal part towards thermal equilibrium; the right-hand side represents the damped precession of the transverse component in the *xy* plane. Physically, the transverse components $s_x(t)$, $s_y(t)$ represent the optical polarisation of the atoms and this precession corresponds to the oscillation of the atomic dipole.

Coherent emission

In a macroscopic analysis, the optical polarisation appears as a source term in the Maxwell equations. This freely precessing polarisation radiates an exponentially decaying pulse of light with duration $1/\Gamma_2$ and optical frequency ω_0. The amplitude of the field is directly proportional to the optical polarisation. Within the two-level model, the electric field appears again as a scalar. This description therefore gives no details of the spatial distribution of the emitted radiation. In an ensemble of atoms, every individual atom radiates as a dipole and therefore emits primarily in the plane perpendicular to the axis of the dipole. If the sample volume that has been prepared by the laser pulse is comparable to or bigger than an optical wavelength, interference between the wave packets emitted by the different atoms must be taken into account. Since the orientation of the individual dipoles differs only by a phase due to the propagation delay of the exciting laser pulse, the ensemble emits the radiation coherently in the direction of propagation of the laser beam, as sketched in Figure 2.19.

We describe this coherent emission process by a simple one-dimensional model: If the medium with number density N was excited by a resonant ($\omega_L =$

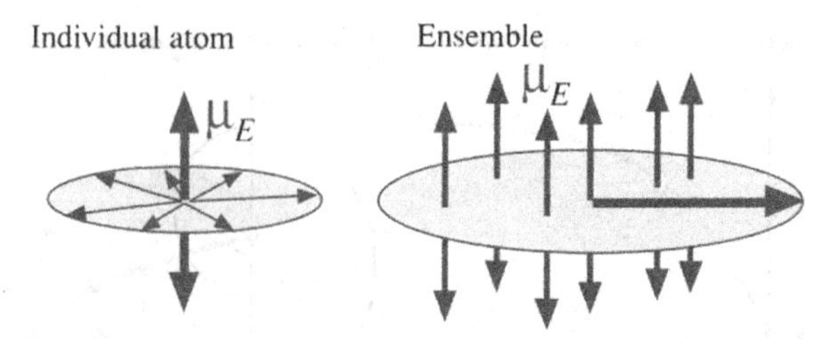

Figure 2.19. An individual atom emits radiation uniformly in a plane perpendicular to the axis of the dipole; in an ensemble, the radiation emitted by the individual dipoles interferes, causing the ensemble to emit radiation coherently in a direction determined by phase matching.

ω_0) plane wave propagating along the z axis, the optical polarisation depends on space and time as

$$P(t,\, z) = 2\ \mu_e\, N\, \mathrm{tr}\{\rho\, S_x\} = \mu_e\, N\, s_x(t,\, z) \tag{2.68}$$

This polarisation generates an optical wave according to the one-dimensional wave equation

$$\left[\frac{\partial^2}{\partial z^2} - \frac{1}{c^2}\frac{\partial^2}{\partial t^2}\right] E(t,z) = \frac{4\pi}{c^2}\frac{\partial^2}{\partial t^2} P(t,z) \tag{2.69}$$

The optical polarisation $s_x(t,z) = s_x(t)\ exp[-ikz]$ that appears in this source term refers to the laboratory frame. To find the laboratory-frame expression, we reverse the transformation to the rotating frame,

$$\begin{aligned} \rho^{\mathrm{lab}} &= U\, \rho'(t)\, U^{-1} \\ &= [S_x\, s_{xy}(0)\cos(\omega_0 t + \alpha) - S_y\, s_{xy}(0)\sin(\omega_0 t + \alpha)]\mathrm{e}^{-\Gamma_2 t} \\ &\quad + S_z\,[1 + (s_z(0) - 1)\mathrm{e}^{-\Gamma_1 t}] \end{aligned} \tag{2.70}$$

In the laboratory frame, the polarisation thus oscillates at the atomic resonance frequency ω_0. The usual description of the optical waves separates the oscillation at the laser frequency ω_L from other time-dependent effects through the "slowly varying amplitude approximation" (Born and Wolf 1986). In the present case, this notation reduces the oscillation frequency by ω_L to $\Delta\omega_0 = \omega_0 - \omega_L$, and we find

$$s_x(t) = s_{xy}(0)\cos(\Delta\omega_0 t + \alpha)\ \mathrm{e}^{-\Gamma_2 t} \tag{2.71}$$

which is just the rotating frame component. We may thus insert the results of the calculation in the rotating frame and use them as the "slowly varying amplitude." At the same time, we switch to complex notation, which provides the in- and out-of-phase component of the field simultaneously.

With this notation, we find

$$E_{\mathrm{FID}}(t, z) = \frac{2\,\pi}{c^2}\,k\,z\,\mu_e\,N\,s_{xy}(0)\,\exp[\mathrm{i}(\Delta\omega_0 t + \alpha - kz)]\,\mathrm{e}^{-\Gamma_2 t} \tag{2.72}$$

i.e., a wave whose field strength increases linearly along the z axis. Energy and intensity thus increase quadratically with z as well as with the atomic polarisation. This is only possible as long as the emitted wave is too weak to change the polarisation, i.e., as long as absorption and emission stimulated by the emitted wave can be neglected. The subscript of the electric field originates from the designation of this signal as an optical free induction decay (FID). The term originates from magnetic resonance, where this type of signal was observed before the development of laser technology allowed its observation in the optical domain.

Heterodyne detection

The derivation of the FID signal involved an approximation of low intensity of the emitted wave. If this approximation holds, the light emitted by the atoms is often too weak for direct detection. It is then advantageous to enhance the signal by superimposing it with a laser beam. This laser beam has the function of a local oscillator, in close analogy to heterodyne detection schemes used in the radio frequency domain. Experimentally, one uses a laser beam that is derived from the same laser as the excitation pulse. To distinguish the signal from the local oscillator, one shifts either the frequency of the local oscillator field with respect to that of the excitation pulse, or the precession frequency of the medium, e.g., by applying a Stark pulse (Brewer and Shoemaker 1971; Brewer and Shoemaker 1972).

If signal and local oscillator fields are superimposed on a photodetector, the resulting signal is proportional to the square of the sum of the two fields:

$$\begin{aligned} S_{\mathrm{tot}}(t) &= |E_{\mathrm{LO}}\,\mathrm{e}^{\mathrm{i}(\omega_L t+\theta)} + E_{\mathrm{FID}}\,\mathrm{e}^{\mathrm{i}\omega_0 t}|^2 \\ &= |E_{\mathrm{LO}}|^2 + |E_{\mathrm{FID}}|^2 + 2|E_{\mathrm{LO}}E_{\mathrm{FID}}|\cos(\Delta\omega_0 t + \phi) \end{aligned} \tag{2.73}$$

Since the local oscillator is strong compared to the FID signal, $|E_{\mathrm{LO}}| >> |E_{\mathrm{FID}}|$, the second term in equation (73) can be neglected. Apart from a constant offset $|E_{\mathrm{LO}}|^2$, the observed signal becomes linear in the emitted field.

Using the expression (72) for E_{FID}, we write the linear part of the signal as

$$S_1(t, z) \,\alpha\, |E_{\mathrm{LO}}|\, s_{xy}(0) \cos(\Delta\omega_0 t + \alpha - \phi)\, \mathrm{e}^{-\Gamma_2 t} \qquad (2.74)$$

It directly reflects the motion of one component of the pseudo-spin vector in the rotating coordinate system. The orientation of this component in the rotating coordinate system can be chosen experimentally by the phase ϕ of the local oscillator field. A heterodyne experiment thus allows a direct measurement of quantities that are static in the rotating reference frame. It is therefore useful to consider the rotating frame as the coordinate system in which the heterodyne experiment is actually performed.

The free induction decay yields not only the atomic resonance frequencies of the system, but also the decay rate Γ_2, which contains information about collisions and other relaxation mechanisms. The information thus obtained is therefore equivalent to an absorption spectrum. Mathematically, the free induction decay signal and the complex index of refraction of the system form a Fourier transform pair, as shown in Figure 2.20.

The left-hand side of the figure shows the free induction decay signal as a function of time. The beat frequency is the difference between the laser frequency and the atomic resonance frequency, whereas the decay is, in a free atom, due to spontaneous emission. Fourier transformation of this time domain signal transforms it into a single resonance line at the precession frequency $\Delta\omega_0$ in the rotating frame. The width of the resonance line (in angular frequency units) is the decay rate.

2.3.3 Photon echoes

Inhomogeneous decay

Free induction decays are never observed from a single atom, but always from an ensemble of atoms that are not exactly equivalent. The observed signal S_{tot} is therefore a sum over the atoms in the ensemble

$$S_{\mathrm{tot}} = \sum_i s_x^{(i)}(t) = a_0 \sum_i \cos(\Delta\omega_0^{(i)} t)\, \mathrm{e}^{-\Gamma_2 t} \qquad (2.75)$$

where we have assumed that the atoms of the ensemble have been prepared identically in a state with the polarisation along the x axis, and that the local oscillator phase selects the x component of the rotating-frame polarisation for observation.

The resonance frequencies $\Delta\omega_0^{(i)}$ of the individual atoms are not exactly identical, as various physical mechanisms may cause differences. In vapours, the

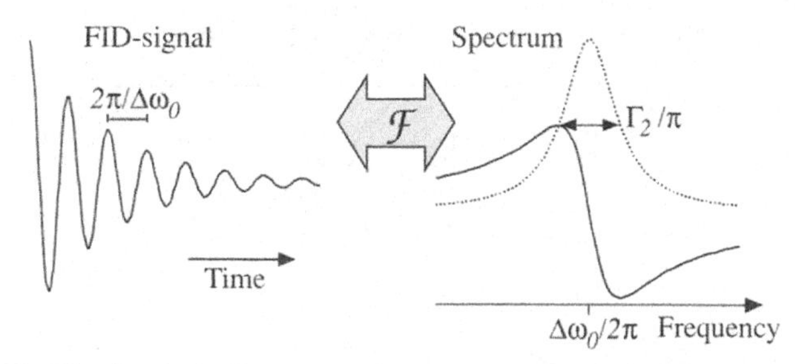

Figure 2.20. The free induction decay signal (FID) and the complex index of refraction form a Fourier transform pair.

most important interaction that causes this so-called inhomogeneous spread of the resonance frequencies is the Doppler shift due to atomic motion. In solids, the interaction of the atoms with their environment shifts the atomic resonance frequency to a value that depends on the position of the atom within the crystal. Crystal imperfections and interactions with neighbouring atoms in different quantum states can shift their resonance frequencies by many GHz. As noted above, the free induction decay signal is the Fourier transform of the frequency distribution. It decays therefore with a time constant that is equal to the inverse width of the frequency distribution.

For a microscopic analysis consider a system of atoms described by their pseudo-spins $S^{(i)}$. If the system is initially in a state

$$\rho(0) = \sum_i S_x^{(i)} \tag{2.76}$$

free precession under the inhomogeneous internal Hamiltonian brings it to a state

$$\rho(T) = \sum_i (\cos(\Delta\omega_0^{(i)}T)\, S_x^{(i)} - \sin(\Delta\omega_0^{(i)}T)\, S_y^{(i)})\, \mathrm{e}^{-\Gamma_2 T} \tag{2.77}$$

where the phases $\Delta\omega_0^{(i)}\, T$ have an essentially random distribution in the xy plane if the duration T exceeds the inverse width of the frequency distribution.

Figure 2.21 illustrates the dephasing process after the excitation with a laser pulse. The central trace represents the spreading of two pseudo-spin vectors during the free precession in the rotating frame. The phases $\Delta\omega_0^{(i)}t$ of the individual atoms evolve linearly in time with different slopes, as represented by the bottom trace. The signal, which is represented after the excitation pulse in the top trace, is the Fourier transform of the spectral distribution of the resonance frequencies.

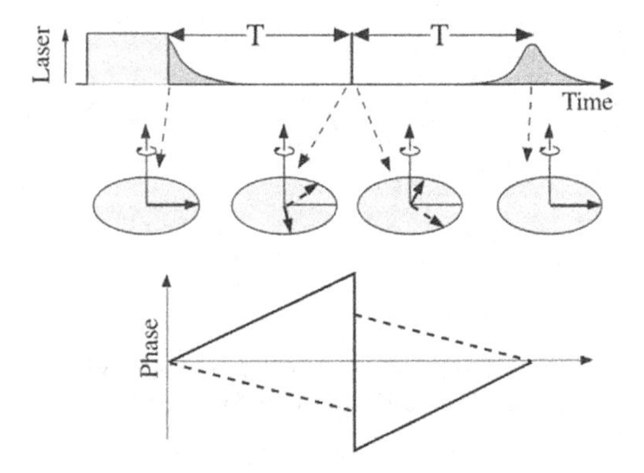

Figure 2.21. Excitation of a photon echo. The first laser pulse prepares coherence along the x axis, as represented by the arrow in the second row. Differences between the resonance frequencies of the individual atoms cause the different contributions to precess out of phase. A second laser pulse inverts the phase of the optical polarisation. The bottom row shows the evolution of the phase of one atom that is faster (full line) and one that is slower (dashed line) during the experiment. The second laser pulse inverts both phases and when both come back to zero, an echo forms.

Refocusing

The decrease of the total signal arises from destructive interference of the polarisations of the individual atoms spread in the xy plane. The problem is therefore often compared to the decay of order in a system of atoms originally stored in a small region of space and subsequently allowed to expand into a larger volume: A Maxwell demon could bring them back into the original ordered state by simultaneously inverting their velocities. In the present case, this would correspond to inverting the atomic resonance frequencies $\Delta\omega_0^{(i)}$. If the frequency distribution is due to Doppler broadening, the frequency shift of the individual atoms is directly proportional to their velocity, and the reversal of the frequency is identical to reversing the velocity and therefore equally impossible.

There is an alternative possibility in the present case, however. It is possible to invert the phase of the atomic polarisation from $\exp(\mathrm{i}\Delta\omega_0^{(i)}T)$ to $\exp(-\mathrm{i}\Delta\omega_0^{(i)}T)$ by rotating the pseudo-spins by an angle π around the x axis:

$$\begin{aligned}\rho(T_+) &= \mathrm{e}^{-i\pi S_x}\sum_i\left(\cos(\Delta\omega_0^{(i)}T)\,S_x^{(i)} - \sin(\Delta\omega_0^{(i)}T)\,S_y^{(i)}\right)\\ &\quad\times\mathrm{e}^{-\Gamma_2 T}\,\mathrm{e}^{i\pi S_x}\\ &= \sum_i\left(\cos(\Delta\omega_0^{(i)}T)\,S_x^{(i)} + \sin(\Delta\omega_0^{(i)}T)\,S_y^{(i)}\right)\mathrm{e}^{-\Gamma_2 T} \qquad (2.78)\\ &= \sum_i\left(\cos(-\Delta\omega_0^{(i)}T)\,S_x^{(i)} - \sin(-\Delta\omega_0^{(i)}T)\,S_y^{(i)}\right)\mathrm{e}^{-\Gamma_2 T}\end{aligned}$$

The state after the pulse is therefore equal to that before the pulse, except for the sign of the precession frequency $\Delta\omega_0^{(i)}$. The pulse has inverted the phase and therefore the evolution of the atoms before the pulse. As discussed in Section 2.3.1, a resonant laser pulse ($\Delta\omega_0 = 0$) with field strength ω_x and duration $\tau = \pi/\omega_x$ (a "π pulse") achieves just such a rotation. After the pulse, the atoms continue to accumulate phase under their internal Hamiltonian. As long as the additional degrees of freedom of these atoms stay invariant, the phase grows at the same frequency as before the pulse. The density operator evolves therefore as

$$\rho(T+t) = \sum_i \left[\cos(\Delta\omega_0^{(i)}(t-T))\, S_x^{(i)} - \sin(\Delta\omega_0^{(i)}(t-T))\, S_y^{(i)}\right] \mathrm{e}^{-\Gamma_2(t+T)} \quad (2.79)$$

The time t measures the duration of the free precession period after the refocusing pulse. When this time becomes equal to the duration of the first free precession period, $t = T$, all the phase factors acquired by the atoms vanish, independent of their precession frequency. At this time, the signal contributions of all atoms add coherently, as they did immediately after the preparation pulse. Coherent emission by the medium returns to a maximum for a duration that is again the inverse of the spectral width. This second coherent emission period is centred around the time $t = T$. The corresponding signal, which appears without an immediately preceding excitation, appears like an echo and is therefore commonly referred to as "photon echo."

Photon echoes have been observed in atomic (Kachru et al. 1979; Kachru, Mossberg and Hartmann 1980; Yodh et al. 1984; Ghosh et al. 1985) and molecular (Brewer and Shoemaker 1971; Chebotayev et al. 1978; Zewail 1980) vapours as well as in inhomogeneously broadened solid materials, in particular crystals of rare earth (Chen, Chiang and Hartmann 1979; Macfarlane, Shelby and Shoemaker 1979; Erickson 1991) and transition metal ions (Kurnit, Abella and Hartmann 1964; Abella, Kurnit and Hartmann 1966). Probably the most important information that can be obtained from photon echo experiments are homogeneous decay rates: The decay of the echo intensity as a function of the delay between the two pulses occurs, according to equation (2.79), at the homogeneous decay rate Γ_2.

3

Three-level effects

After the introduction of the two-level model, we add a third level to the quantum mechanical system to discuss some of the effects that cannot be described with only two states. The main feature is the exchange of order between states and transitions, which may proceed through populations or coherences. We concentrate our attention on the latter possibility.

3.1 Phenomenological introduction

3.1.1 Model atoms

In the preceding section, we discussed the interaction between light and two-level atoms – probably one of the most popular physical models. The basis of this popularity is its intuitively simple interpretation combined with the possibility of explaining a wide range of physical phenomena. An interesting aspect of the two-level system is that, although its dynamics are formally equivalent to those of a classical angular momentum, it can explain many aspects of quantum mechanics. Once these aspects are understood, it is tempting to look further into the behaviour of real systems, trying to find patterns inconsistent with the predictions of the two-level model.

This section discusses aspects of the interaction between matter and radiation that are incompatible with the two-level model. We do not consider specific atomic systems or attempt a complete analysis of the dynamics that a three-level system can exhibit. Instead we select a number of phenomena that play an important role in the discussion of those physical systems that form the subject of the subsequent sections. For most of those processes it is sufficient to discuss them in the context of a three-level system.

Three-level atoms

Figure 3.1 shows the relevant model system consisting of three energy levels connected by three transitions of which the two that are indicated by the ar-

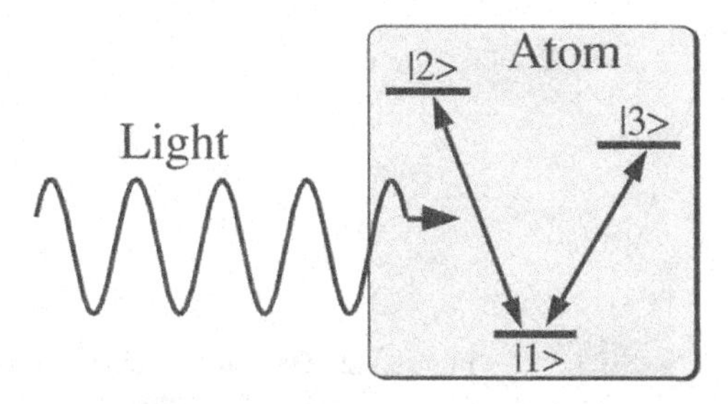

Figure 3.1. The relevant model atom includes three energy levels and three transitions, of which two couple to the laser field.

rows are electric dipole transitions. The third transition may also be an optical transition, but as long as we assume that the states have definite parity, two of them must have the same parity and the transition between them cannot be an electric dipole transition. In actual physical systems discussed in later sections, the third transition is often a magnetic dipole transition. Here, we do not yet discuss the actual physics, but start with the formal mathematical analysis of the possible processes.

One important property that distinguishes three- or multilevel systems from two-level systems is the coupling between laser fields with different frequencies. A mixture of different two-level atoms may be resonant with more than one monochromatic field. Each laser field, however, is then resonant with only one type of atom and even if the intensity is high enough to push the atoms away from equilibrium, one laser field does not affect the others and there is no coupling between the different beams. In three-level systems, on the other hand, two waves with different frequencies can interact with the same atom. If they are strong enough to push the atomic system away from equilibrium, one optical wave can influence the other – the medium "mixes" the waves.

The atomic medium allows different mechanisms to couple different laser fields, which we may classify into the broad groups of "coherent" and "incoherent" mechanisms. In the case of a coherent coupling, the atoms couple phase and amplitude information between the two beams, whereas only intensity information is mediated in the incoherent case.

A typical example of an incoherent mechanism is optical pumping, where one strong pump wave (thick arrow in Figure 3.2) moves population away from the transition to which it couples and thus increases the number of atoms that can interact with the probe beam. Coherent interactions, on the other hand, exchange coherence between different transitions (in contrast to populations). This type of process includes coherent Raman scattering and other related processes discussed below.

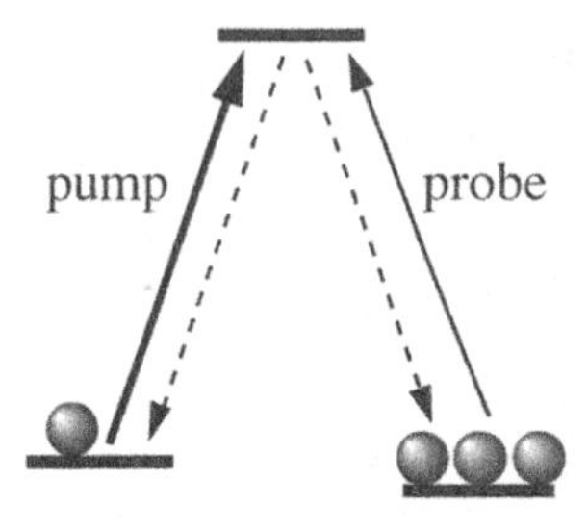

Figure 3.2. Optical pumping in three-level systems.

3.1.2 Coherence transfer

A coupling between matter and radiation always involves coherent superpositions between different eigenstates of the unperturbed atomic Hamiltonian. In density operator notation, they correspond to off-diagonal elements ρ_{rs} or coherences. Systems in thermal equilibrium have no coherences; in most cases, they are generated by the interaction with a laser field. From the discussion of two-level atoms, we also expect that in three-level systems light that couples to one of the optically allowed transitions converts population difference between the two levels into coherence in that transition. In addition to these processes, the presence of several transitions in the same system allows an additional type of process that involves transfer of coherence from one transition into coherence of another transition. This type of process has no classical analogue. There is no possibility of representing it in the three-dimensional space used in the case of the two-level system for a geometric representation of its evolution. This is probably the main reason why such effects have been studied less thoroughly than the transfer of populations and other effects that can be explained in terms of two-level models.

In terms of state vectors, coherence represents a definite phase relation between two states: A coherence implies that for an ensemble of atoms in a superposition state

$$\Psi_{12} = c_1\,\Psi_1 + c_2\,\Psi_2 \tag{3.1}$$

the product $\langle c_1{}^* c_2\rangle$ has a nonvanishing average. This is typically the case when the superposition state was prepared by a laser pulse.

If another laser pulse couples to the transition between level Ψ_2 and a third level, as shown in the left-hand side of Figure 3.3, it transforms the state Ψ_{12} into

$$\Psi_{123} = c_1\,\Psi_1 + c_2{}'\,\Psi_2 + c_3\,\Psi_3 \tag{3.2}$$

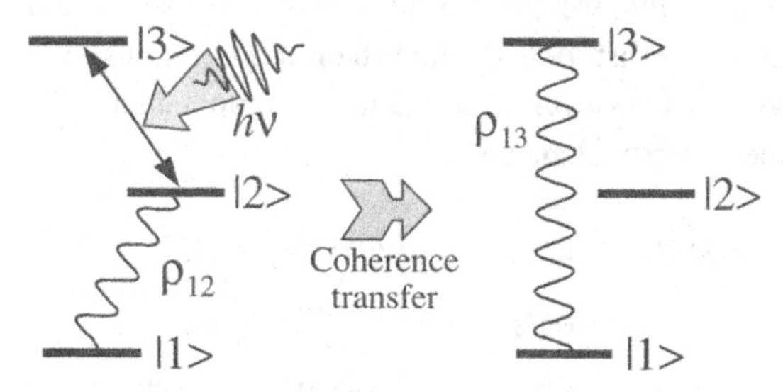

Figure 3.3. Transfer of coherence from transition $|1\rangle \leftrightarrow |2\rangle$ to transition $|1\rangle \leftrightarrow |3\rangle$.

If we examine this state, we find that the laser pulse has created coherence in the transition 23, to which it coupled; this agrees with the behaviour that we expect from the analysis of the two-level model. If we examine the product $\langle c_1{}^* c_3 \rangle$, however, we also find a nonvanishing coherence in transition $|1\rangle \leftrightarrow |3\rangle$. This result may come as a surprise, since it implies that a coherence exists in a transition to which no electromagnetic field was ever coupled. As more detailed analysis will show, the most appropriate interpretation of this situation is to state that the pulse applied to transition $|2\rangle \leftrightarrow |3\rangle$ transferred coherence from transition $|1\rangle \leftrightarrow |2\rangle$ into transition $|1\rangle \leftrightarrow |3\rangle$.

This basic phenomenon of coherence transfer can explain a large number of physical processes. The brief analysis given here should already motivate the possibility that three-level systems can induce wave mixing, as the fields coupling to one transition may modify the polarisation in a neighbouring transition. The situation described above, where two independent fields generated a coherence in the third transition, may also be interpreted as the absorption of a photon from one field, accompanied by a stimulated emission into the other field – an example of a Raman process.

3.2 System and Hamiltonian

3.2.1 Single-transition operators

To prepare the mathematical description of these processes, we need to define the operator basis in which we expand the interaction Hamiltonian and the density operator of the system. Commonly used are the irreducible tensor operators. They form an ideal basis for discussing anisotropic interactions and implementing the rotational symmetry of free space. The three-level systems discussed here cannot, however, include a complete basis for the description of rotations, so spherical tensor operators cannot be used.

The operator basis that we use instead is known as "single-transition operators" (Wokaun and Ernst 1977), which facilitate the transfer of the results from the analysis of two-level atoms to more complicated systems. We define them by their matrix elements

$$\begin{aligned} \langle r|S_x^{(rs)}|s\rangle &= 1/2 & \langle s|S_x^{(rs)}|r\rangle &= 1/2 \\ \langle r|S_y^{(rs)}|s\rangle &= -\mathrm{i}/2 & \langle s|S_y^{(rs)}|r\rangle &= \mathrm{i}/2 \\ \langle r|S_z^{(rs)}|r\rangle &= 1/2 & \langle s|S_z^{(rs)}|s\rangle &= -1/2 \end{aligned} \tag{3.3}$$

and zero otherwise. This definition is motivated by the analogy of one transition with a two-level model: Every three-level system contains three two-level systems and three corresponding sets of single-transition operators associated with one particular transition $|r\rangle \leftrightarrow |s\rangle$. For each set, the usual angular momentum commutation relations are valid,

$$[S_x^{(rs)}, S_y^{(rs)}] = \mathrm{i}\, S_z^{(rs)} \qquad \text{and cyclic permutations} \tag{3.4}$$

The fact that the three transitions are not independent of each other, but share a common energy level, is reflected mathematically in the commutation relations between operators belonging to different transitions. In arbitrary systems, the commutators vanish if the operators share no common state. In a three-level system consisting of the states $|r\rangle$, $|s\rangle$, $|t\rangle$, the following commutation relations hold:

$$\begin{aligned} &[S_x^{(rt)}, S_x^{(st)}] = [S_y^{(rt)}, S_y^{(st)}] = \mathrm{i}/2\, S_y^{(rs)} \\ &[S_z^{(rt)}, S_z^{(st)}] = 0 \qquad\qquad [S_x^{(rt)}, S_y^{(st)}] = \mathrm{i}/2\, S_x^{(rs)} \\ &[S_x^{(rt)}, S_z^{(st)}] = -\mathrm{i}/2\, S_y^{(rt)} \qquad [S_y^{(rt)}, S_z^{(st)}] = \mathrm{i}/2\, S_x^{(rt)} \end{aligned} \tag{3.5}$$

This system of operators, together with the unity operator, is overcomplete, since the z components are connected by the conditions

$$S_z^{(rs)} + S_z^{(st)} + S_z^{(tr)} = 0 \tag{3.6}$$

For actual calculations it may be necessary to eliminate these dependencies by choosing a reduced set of operators. One possibility is to replace the z components of the single-transition operators by level-shift operators $\sigma_i = |i\rangle\langle i|$. This set of operators, together with the x and y components of the single-transition operators, is exactly complete. The expectation values of the level-shift operators σ_i are the populations of the corresponding levels, whereas the expectation values of the single transition operators $S_z^{(rs)} = (1/2)\,(\sigma_r - \sigma_s)$ is the population difference across the relevant transition. The x and y components of the single-transition operators describe coherences in the transitions. Physically, these coherences correspond to the components of the (electric or magnetic) atomic dipole moment perpendicular to the quantisation axis.

3.2.2 Irradiation of a single transition

In the most general case, the three-level system can be coupled to three different fields. We start the discussion with the case of the excitation of a single transition and continue to the case of two transitions coupled to the laser field. We do not discuss explicitly the case of three coupled transitions, which does not add qualitatively different processes to the dynamics of the system.

Hamiltonian

Starting from an isolated atom, the unperturbed Hamiltonian may be written in a basis of shift operators σ_i as

$$\mathcal{H}_0 = \sum_{i=1}^{3} \mathcal{E}_i\, \sigma_i \tag{3.7}$$

where the $\mathcal{E}_i$ indicate the energies of levels 1, . . . , 3. The term that describes the coupling between the atom and the laser field is

$$\mathcal{H}_{\text{int}} = -2\; \omega_{12}\; S_x^{(12)} \cos(\Omega_{12} t) \tag{3.8}$$

in close analogy to the two-level case. Ω_{12} indicates the frequency of the laser field and ω_{12} the coupling strength, which is proportional to the dipole matrix element and the laser field amplitude. Here, we have assumed that the phase of the basis states is chosen in such a way that the matrix elements of the electric dipole moment are real and positive.

As in the case of the two-level atom, we would like to transform this Hamiltonian into a time-independent form. For the three-level system, there is no longer a unique rotating coordinate system, but we can select among different possible transformations that eliminate the time-dependent terms. The most useful transformation depends on the actual system parameters, such as coupling strengths and the frequency (or frequencies) of the laser field. Finding

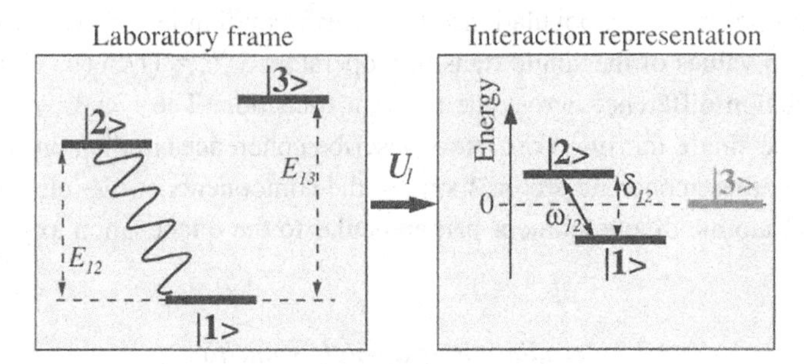

Figure 3.4. Comparison of energy level schemes in laboratory frame (left) and interaction representation (right).

a solution to the equations of motion depends in many cases on finding a suitable interaction representation where the Hamiltonian can be diagonalised.

Interaction representation

For the present case of a field interacting with a single transition, the most useful transformation is

$$U_1 = \exp\left[\mathrm{i}\, t\left\{\Omega_{12}\, S_z^{(12)} - \frac{\mathscr{E}_1 + \mathscr{E}_2}{2}(\sigma_1 + \sigma_2) - \mathscr{E}_3\, \sigma_3\right\}\right] \tag{3.9}$$

which transforms the coupling Hamiltonian into

$$\mathscr{H}^{\mathrm{r}} = U_1^{-1}(t)\,(\mathscr{H}_0 + \mathscr{H}_{\mathrm{int}})\, U_1(t) + \mathrm{i}\, \dot{U}_1^{-1}(t)\, U_1(t) \approx -\delta_{12}\, S_z^{(12)} - \omega_{12}\, S_x^{(12)} \tag{3.10}$$

where $\delta_{12} = (\mathscr{E}_2 - \mathscr{E}_1) - \omega_{\mathrm{L}}$ is the reduced level splitting. In this form, the nonresonant time-dependent terms have been eliminated, in close analogy to the treatment of the two-level system in Section 2.2.2. As shown in Figure 3.4, the two coupled energy levels $|1\rangle$ and $|2\rangle$ are split by the laser detuning δ_{12} and coupled by the interaction ω_{12}, while the energy of the uncoupled third level is arbitrarily set to zero.

This choice of the interaction representation is not unique. In particular, the energy of level $|3\rangle$ can be chosen freely. This system looks much like a two-level system, as the third level is not coupled to the two other levels. As we shall see below, however, there are important cases in which the initial condition includes a superposition between level $|3\rangle$ and level $|1\rangle$ or $|2\rangle$. In those cases, the third level has important effects on the behaviour of the system, even if the radiation field does not couple directly to it.

3.2.3 Irradiation of two transitions

Monochromatic laser field

We next consider the case when the laser radiation couples to both optical transitions. The relevant interaction Hamiltonian is

$$\mathcal{H}_{\text{int}} = -2\,(\omega_{12}\,S_x^{(12)} + \omega_{13}\,S_x^{(13)})\cos(\Omega\,t) \tag{3.11}$$

in an obvious extension of the case of a single transition. The coupling strengths ω_{12}, ω_{13} may differ, but as we deal with a single optical field, their ratio ω_{12}/ω_{13} is equal to the ratio of the dipole matrix elements for the two transitions.

A useful interaction representation for this system is defined by the unitary operator

$$U_{\text{mc}} = \exp[\mathrm{i}\,t\,(-\Omega\,(\sigma_2 + \sigma_3) - \mathcal{E}_1\,\mathbb{1})] \tag{3.12}$$

which transforms the interaction Hamiltonian into

$$\mathcal{H}_{\text{int}}^{\text{r}} = U_{\text{mc}}^{-1}(t)\,\mathcal{H}_{\text{int}}\,U_{\text{mc}}(t) \approx -\omega_{12}\,S_x^{(12)} - \omega_{13}\,S_x^{(13)} \tag{3.13}$$

if we neglect, as usual, nonresonant terms that appear at frequency 2Ω in this reference frame. The total Hamiltonian in the interaction representation is

$$\begin{aligned}\mathcal{H}^{\text{r}} &= U_{\text{mc}}^{-1}(t)\,\mathcal{H}\,U_{\text{mc}}(t) + \mathrm{i}\,\dot{U}_{\text{mc}}^{-1}(t)\,U_{\text{mc}}(t)\\ &= \delta_{12}\,\sigma_2 + \delta_{13}\,\sigma_3 - \omega_{12}\,S_x^{(12)} - \omega_{13}\,S_x^{(13)}\end{aligned} \tag{3.14}$$

with

$$\delta_{12} = \mathcal{E}_2 - \mathcal{E}_1 - \Omega \qquad \delta_{13} = \mathcal{E}_3 - \mathcal{E}_1 - \Omega \tag{3.15}$$

representing the reduced level splittings arising from the mismatch between the atomic resonance frequency and the laser frequency.

As Figure 3.5 shows, this transformation sets the origin of the energy scale to the position of level $|1\rangle$.

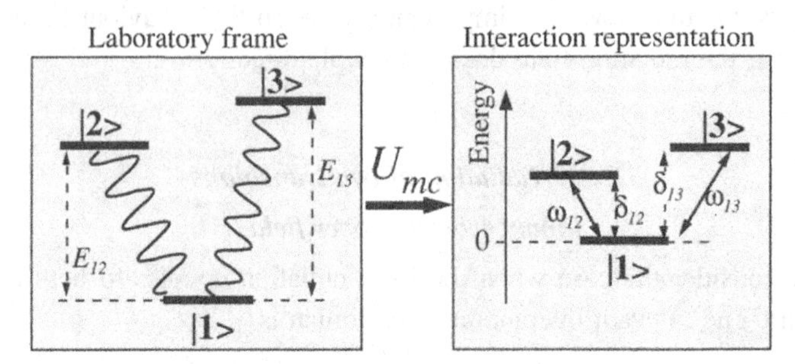

Figure 3.5. Comparison of energy level schemes in laboratory frame (left) and interaction representation (right).

Bichromatic field

If the two optical transitions couple to two different fields, the interaction Hamiltonian becomes

$$\mathcal{H}_{\text{int}} = -2\ \omega_{12}\, S_x^{(12)} \cos(\Omega_{12}\, t) - 2\ \omega_{13}\, S_x^{(13)} \cos(\Omega_{13}\, t) \tag{3.16}$$

where Ω_{12} and Ω_{13} are the frequencies of the two fields and we assume that each field interacts only with one transition. To remove the time dependence, we use the transformation operator

$$U_{\text{bc}} = \exp[-\mathrm{i}\ t(\Omega_{12}\, \sigma_2 + \Omega_{13}\, \sigma_3 + \mathcal{E}_1\, \mathbb{1})] \tag{3.17}$$

which is closely related to the unperturbed Hamiltonian. The reference frame defined by this transformation is not associated with a physical rotation. Nevertheless, it is useful to consider it as rotating at different velocities for the two optical transitions – for each transition at the relevant laser frequency.

The transformed interaction Hamiltonian is formally identical to that of the monochromatic case, defined in equation (3.14). The only difference is the value of the level splittings, which become

$$\delta_{12} = \mathcal{E}_2 - \mathcal{E}_1 - \Omega_{12} \qquad \delta_{13} = \mathcal{E}_3 - \mathcal{E}_1 - \Omega_{13} \tag{3.18}$$

For each transition, the apparent splitting is reduced by the energy of the photons that couple to that transition.

The Hamiltonians discussed here all use real matrix elements to describe the coupling between atom and radiation. This choice can always be realised

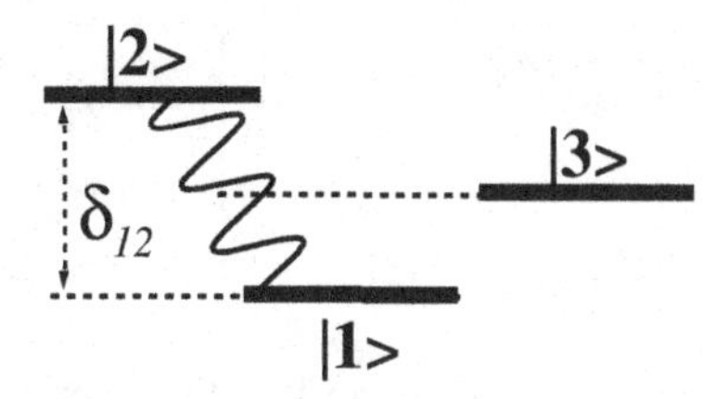

Figure 3.6. Level scheme for the excitation of a single transition.

by a suitable choice of the phases of the atomic eigenstates as well as by the choice of the phase of the optical field. Different optical phases, which may appear in a single experiment, change the phase of the matrix elements, making them complex. The general form of the interaction Hamiltonian then becomes

$$\mathcal{H}^{\mathrm{r}}_{\mathrm{int}} = -\omega_{12}\,(S_x^{(12)}\cos\phi_{12} + S_y^{(12)}\sin\phi_{12}) - \omega_{13}(S_x^{(13)}\cos\phi_{13} + S_y^{(13)}\sin\phi_{13}) \tag{3.19}$$

For arbitrary phases of the optical pulse, the effective field in the interaction representation is then no longer oriented along the x axis, but at an angle ϕ from the axis.

3.3 Three-level dynamics

Using the Hamiltonians of the preceding section, we now discuss the dynamics that they generate in the systems. For all situations considered in the following, we calculate the dynamics in the interaction representation, where we can use the time-independent form of the Hamiltonians derived above.

3.3.1 Excitation of a single transition

We first discuss the case of a three-level system in which only a single transition is coupled to the radiation field.

As Figure 3.6 shows, the optical field couples states $|1\rangle$ and $|2\rangle$, while state $|3\rangle$ is uncoupled. We assume that all three levels are nondegenerate and use the Hamiltonian of equation (3.10)

$$\mathcal{H}^{\mathrm{r}} = -\delta_{12}\,S_z^{(12)} - \omega_{12}\,S_x^{(12)} \tag{3.20}$$

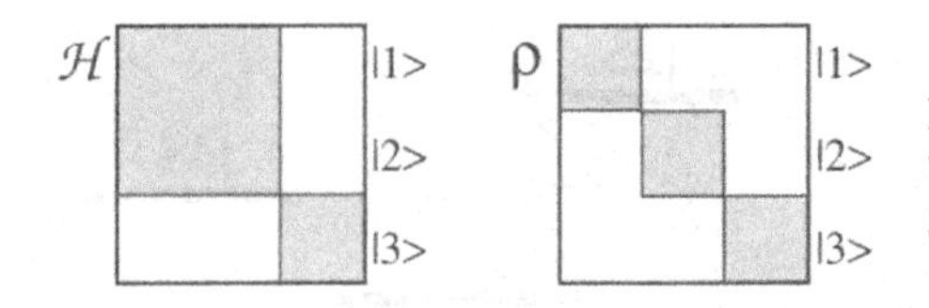

Figure 3.7. Block structure of Hamiltonian and density operator for single transition irradiation and diagonal initial condition.

The first term describes the energy difference between states $|1\rangle$ and $|2\rangle$, which arises from the laser detuning δ_{12} from the optical resonance frequency. The second term describes the coupling to the laser field.

For a description of the dynamics of such a system, we consider the evolution of different parts of the density operator. The linearity of the Schrödinger equation allows us to use the results for calculating the evolution for arbitrary initial conditions.

Diagonal initial condition

In the simplest case, the initial condition is diagonal, i.e., the coherences vanish. We may then write the initial density operator as

$$\rho_0 = z_{12}\, S_z^{(12)} - z_3\,(\sigma_1 + \sigma_2 - 2\,\sigma_3) \tag{3.21}$$

where z_{12} and z_3 are expansion coefficients for the corresponding terms in the density operator. This form separates the density operator into a term proportional to $S_z^{(12)}$, which describes the population difference in a two-level system, and an additional term, which commutes with the Hamiltonian. The second term is time-independent and, since we wish to discuss evolution here, we may drop it from the discussion without loss of generality. The density operator and the Hamiltonian thus contain no term that refers to level $|3\rangle$. This implies that for irradiation of a single transition and a diagonal initial condition, we can reduce the three-level system to the two-level case.

This possibility of neglecting level $|3\rangle$ in the analysis of the evolution of the three-level system is associated with the block structure of density operator and Hamiltonian that is sketched in Figure 3.7: The total Hilbert space can be decomposed into two subspaces, one describing the two-level system $|1\rangle$, $|2\rangle$, and the second subspace containing only level $|3\rangle$. These two subspaces are independent of each other and their dynamics may be discussed independently.

This possibility of separating the total Hilbert space into two independent subspaces implies that as long as the initial condition contains only popula-

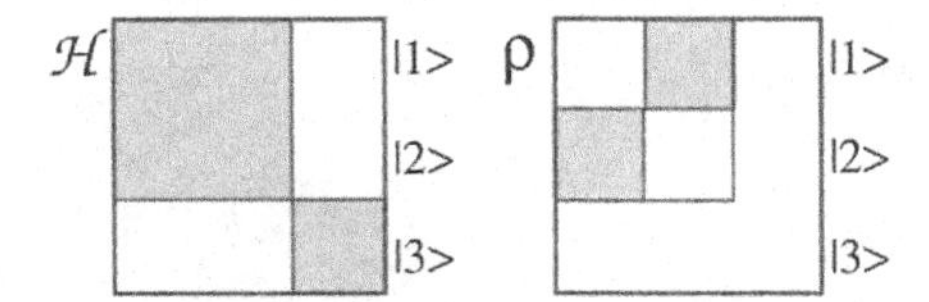

Figure 3.8. Block structure of Hamiltonian and density operator if it contains only coherence in transition $|1\rangle\leftrightarrow|2\rangle$.

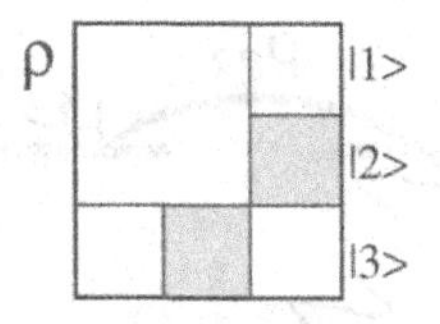

Figure 3.9. Coherence in transition $|2\rangle\leftrightarrow|3\rangle$ changes the block structure of the density operator.

tions (no coherences), and the external field couples only to a single transition, there are no effects specific for three-level systems. This finding may be interpreted as a demonstration that the levels do not "know" about each other as long as no coherence exists between them.

3.3.2 Coherence transfer

Interesting effects arise when the initial density operator includes off-diagonal elements, i.e., coherences among the three states. Again, we may consider the coherences in the three transitions separately. If coherence exists only in the irradiated transition $|1\rangle\leftrightarrow|2\rangle$, we find again that the resulting evolution is confined to the two levels $|1\rangle$ and $|2\rangle$, since the third state is not coupled to the others: Neither the Hamiltonian nor the density operator contains off-diagonal elements connecting level $|3\rangle$ to the levels $|1\rangle$ and $|2\rangle$.

The block structure of Hamiltonian and density operator, represented in Figure 3.8, shows that in this case as well, it is possible to split the Hilbert space into the same uncoupled subspaces.

Coherence in uncoupled transition

The situation changes when a coherence is present in one of the two transitions not coupled to the radiation field, e.g., in transition $|2\rangle\leftrightarrow|3\rangle$.

As Figure 3.9 shows, this coherence changes the block structure of the density operator. It is no longer possible to decompose the Hilbert space into two independent subspaces, since the coherence does not commute with the interaction term of the Hamiltonian. With an arbitrary choice of phase, the relevant part of the initial density operator is

$$\rho_0 = x_{23}(0)\, S_x^{(23)} \tag{3.22}$$

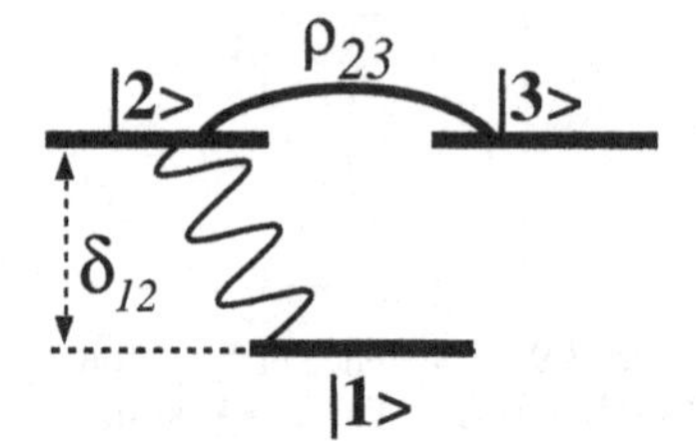

Figure 3.10. Level scheme after shift of level $|3\rangle$.

where we disregard all terms discussed above, i.e., diagonal terms and coherence in the transition being irradiated.

Before we describe the effect of the irradiation on the coherence in the adjacent transition, we eliminate effects not due to the irradiation. This refers primarily to the energy difference between states $|2\rangle$ and $|3\rangle$: It causes the coherence ρ_{23} to precess under the atomic Hamiltonian. The phase it acquires during this evolution is not of particular interest in this context, and we can easily eliminate it by choosing yet another interaction representation. Quite generally, we may shift the energy of any state by a suitable amount if it is not coupled to other states through off-diagonal elements in the Hamiltonian.

Formally, the shift can be performed by another unitary transformation, using the operator

$$U_s = \exp\left[-\mathrm{i}\, t\, \frac{\delta_{12}}{2}\, \sigma_3\right] \tag{3.23}$$

which commutes with the Hamiltonian. After this shift, the Hamiltonian is

$$\mathcal{H}^{r'} = -\delta_{12}\, S_z^{(12)} - \omega_{12}\, S_x^{(12)} + \frac{\delta_{12}}{2}\, \sigma_3 \tag{3.24}$$

In this representation, the energy of level $|3\rangle$ is now the same as that of level $|2\rangle$, as indicated in Figure 3.10.

Evolution

For vanishing laser detuning, $\delta_{12} = 0$, this Hamiltonian and the initial condition of equation (3.22) result in an evolution

$$\rho_{\mathrm{res}}(t) = x_{23}\, [S_x^{(23)} \cos(\omega_{12}/2\ t) - S_y^{(13)} \sin(\omega_{12}/2\ t)] \tag{3.25}$$

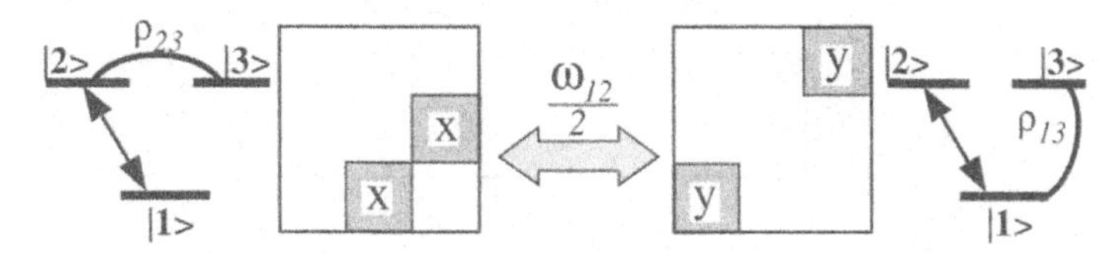

Figure 3.11. Coherence transfer induced by the laser field.

The laser pulse thus causes an oscillatory exchange of order between transitions $|2\rangle\leftrightarrow|3\rangle$ and $|1\rangle\leftrightarrow|3\rangle$. After a quarter period, the system has coherence only in transition $|1\rangle\leftrightarrow|3\rangle$. The phase and amplitude information associated with x_{23} that was created in transition $|2\rangle\leftrightarrow|3\rangle$ has now been transferred completely to the adjacent transition. As equation (3.25) shows, this transfer is in fact an oscillatory exchange. The frequency of this exchange is half the Rabi frequency ω_{12}, because only one of the two levels is coupled to the field. See Figure 3.11.

If the laser is not exactly resonant with transition $|1\rangle\leftrightarrow|2\rangle$, there is no complete transfer of the coherence. As in the two-level case, the exchange frequency increases to an effective Rabi frequency $\omega_e = \sqrt{\delta_{12}^2 + \omega_{12}^2}$, but the amplitude of the coherence in transition $|2\rangle\leftrightarrow|3\rangle$ diminishes. The system evolves as

$$\rho(t) = x_{23}\,[S_x^{(23)} \cos(\omega_e/2\; t) - (\sin\theta\, S_y^{(13)} - \cos\theta\, S_y^{(23)}) \sin(\omega_e/2\; t)] \quad (3.26)$$

where the angle $\theta = \tan^{-1}(\omega_{12}/\delta_{12})$ parametrises the laser detuning. It represents the angle between the z_{12} axis and the direction of the effective field for the transition $|1\rangle\leftrightarrow|2\rangle$.

3.3.3 Three-level echoes

An interesting experiment that clearly shows the effect of coherence transfer is the tri-level echo (Schenzle, Grossman and Brewer 1976; Mossberg et al. 1977), which involves all three transitions of a three-level system.

Figure 3.12 illustrates the sequence of events. A first laser pulse, ideally a $\pi/2$ pulse, excites coherence in transition $|1\rangle\leftrightarrow|2\rangle$. If the system is initially in thermal equilibrium, it evolves like a two-level system and the third level is not involved. After the pulse, the coherence precesses freely under the atomic Hamiltonian. Inhomogeneous interactions, such as Doppler broadening, cause dephasing, which is observed as an optical free induction decay. A second

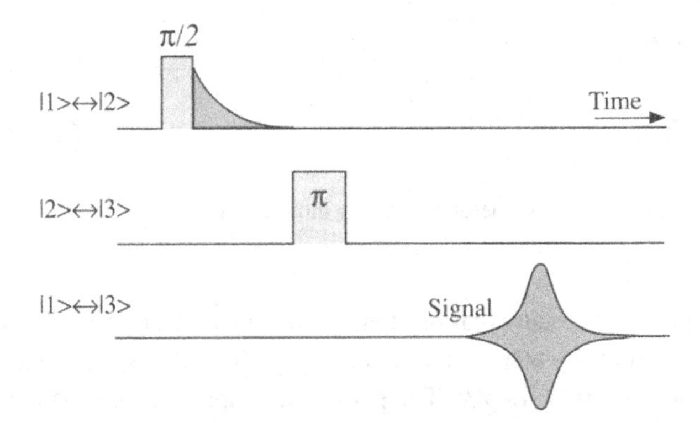

Figure 3.12. Survey of three-level echoes. A $\pi/2$ excitation pulse creates coherence in transition $|1\rangle\leftrightarrow|2\rangle$, where it can be observed as an optical FID. A second pulse, applied to transition $|2\rangle\leftrightarrow|3\rangle$, transfers the coherence to the third transition, where it refocuses and appears as an echo.

laser pulse, ideally a π pulse, couples to transition $|2\rangle\leftrightarrow|3\rangle$ and transfers the coherence to the third transition $|1\rangle\leftrightarrow|3\rangle$. If this transition is also inhomogeneously broadened, the coherence may refocus there. In this case, an echo can be observed in this transition. This experiment is a direct demonstration of the coherence transfer processes discussed above.

Formal analysis

For a more formal analysis, we assume that the first pulse of duration $\tau = \pi/(2\omega_{12})$ acts on transition $|1\rangle\leftrightarrow|2\rangle$ and converts an initial population difference z_{12} of this transition into coherence. The density operator after the pulse is then

$$\rho(0+) = z_{12}\, S_y^{(12)} \tag{3.27}$$

After the pulse, this coherence precesses under the inhomogeneous Hamiltonian

$$\mathcal{H}_0^{\mathrm{r}} = \delta_{12}\,\sigma_2 + \delta_{13}\,\sigma_3 \tag{3.28}$$

as

$$\begin{aligned}\rho(t) &= z_{12} \int \mathrm{d}\delta_{12}\, p(\delta 12)\, \mathrm{e}^{-\mathrm{i}\mathcal{H}_0^{\mathrm{r}}t}\, S_y^{(12)}\, \mathrm{e}^{\mathrm{i}\mathcal{H}_0^{\mathrm{r}}t} \\ &= z_{12} \int \mathrm{d}\delta_{12}\, p(\delta 12)\, [S_y^{(12)} \cos(\delta_{12}t) + S_x^{(12)} \sin(\delta_{12}t)]\end{aligned} \tag{3.29}$$

where δ_{12} represents the resonance frequency of transition $|1\rangle\leftrightarrow|2\rangle$ and $p(\delta_{12})$ the relative frequency of atoms with this resonance frequency. The integral runs over the inhomogeneously broadened resonance line. The different frequencies contributing to this integral cause a dephasing and a signal cancellation after a time that is inversely proportional to the inhomogeneous linewidth, as indicated by the shaded shape behind the excitation pulse in Figure 3.12.

After this free precession period of duration T, a second laser pulse of duration $\tau_2 = \pi/(\omega_{23})$ is applied to transition $|2\rangle\leftrightarrow|3\rangle$. It converts the density operator into

$$\begin{aligned}\rho(T+) &= \mathrm{e}^{-\mathrm{i}\pi S_y^{(23)}}\,\rho(T)\,\mathrm{e}^{\mathrm{i}\pi S_y^{(23)}} \\ &= z_{12}\int \mathrm{d}\delta_{12}\, p(\delta 12)\,[S_y^{(13)}\cos(\delta_{12}T) + S_x^{(13)}\sin(\delta_{12}T)]\end{aligned} \tag{3.30}$$

It thus appears as if the coherence is created in transition $|1\rangle\leftrightarrow|3\rangle$ instead of $|1\rangle\leftrightarrow|2\rangle$. The phase information $\delta_{12}T$ associated with it, however, "remembers" where it originated. Apparently, the laser pulse has transferred this information to the connected transition.

Echo formation

The system continues to evolve with the coherence now in transition $|1\rangle\leftrightarrow|3\rangle$. At a time t after the π pulse, the density operator is

$$\begin{aligned}\rho(T+t) &= z_{12}\int \mathrm{d}\delta_{12}\, p(\delta_{12}, \delta_{13}) \\ &\quad\times [S_y^{13}\cos(\delta_{12}T + \delta_{13}t) + S_x^{(13)}\sin(\delta_{12}T + \delta_{13}t)]\end{aligned} \tag{3.31}$$

The probability distribution $p(\delta_{12},\delta_{13})$ now depends on two variables, δ_{12} and δ_{13}, and the total phase acquired by the coherence depends on the same two frequencies. δ_{12} is the frequency at which the coherence evolves between the two pulses, δ_{13} the precession frequency for the second transition.

The average of the coherence over the inhomogeneous line can assume a nonvanishing mean if the two frequencies are correlated to each other, e.g., as

$$\delta_{12} - \overline{\delta_{12}} = -c_\delta\,(\delta_{13} - \overline{\delta_{13}}) \tag{3.32}$$

for all atoms. The constant c_ω is a proportionality constant that indicates how strongly the inhomogeneity of the two transitions are correlated. The phase scatter

$$\begin{aligned}(\delta_{12} - \overline{\delta_{12}})T + (\delta_{13} - \overline{\delta_{13}})t &= -c_\delta (\delta_{13} - \overline{\delta_{13}})T + (\delta_{13} - \overline{\delta_{13}})t \\ &= (\delta_{13} - \overline{\delta_{13}})(-c_\delta T + t)\end{aligned} \tag{3.33}$$

vanishes at $t = c_\delta T$ for all atoms. At this time, all coherences come back into phase to interfere constructively, thus giving rise to an echo. Such a proportionality exists, e.g., in Doppler broadened systems, where the inhomogeneity is proportional to the transition frequency and the atomic velocity. The proportionality constant is then $c_\omega = \pm\overline{\delta_{12}}/\overline{\delta_{13}}$ and the echo occurs at $t = (\delta_{12}/\delta_{13})\, T$.

For this simple analysis, we have assumed that all three transitions are optically allowed transitions. For states with definite parity, this is not possible, as we discuss later. At least two of the three states must have the same parity, and the transition between them cannot be an electric dipole transition. Even in such systems, the effect can be observed. The first observations of this type were reported in a nuclear magnetic resonance (NMR) experiment (Maudsley, Wokaun and Ernst 1978), where all three states had equal parity and the radiation field coupled to magnetic dipole transitions. In the first optical experiment of this type (Mossberg et al. 1977), a two-step process was used to prepare the initial coherence, $S_y^{(12)}$ in our notation, in a transition that is not an electric dipole transition. This excitation process used two laser pulses that coupled to allowed optical transitions; the first prepared a coherence in an adjacent transition, while the second pulse transferred it to the forbidden transition. This is another interesting aspect of this type of experiment: The coherence-transfer process can prepare coherence in a "forbidden" transition and another coherence transfer process can make it observable. Chapter 9, on two-dimensional spectroscopy, provides more details about this possibility.

3.3.4 Quantum beats

According to the usual interpretation of quantum mechanics, interference effects occur whenever two possible events cannot be distinguished experimentally. An example of such a situation is the emission of a fluorescence photon from an atom that has several closely spaced excited states. If the experiment does not distinguish between the two excited states, during either excitation or detection, the observer cannot know from which of the excited states the atom emitted the fluorescence photon. We expect to observe the superposition of the two possibilities. The interference between the two possi-

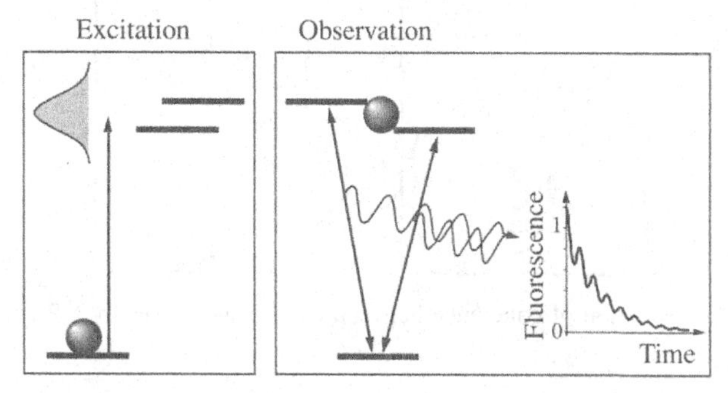

Figure 3.13. Quantum beat excitation (left) and observation as beats in the fluorescence (right).

ble decay channels leads to a time-dependent enhancement and inhibition of the fluorescence, which is known as quantum beats (Aleksandrov 1964; Dodd, Kaul and Warrington 1964; Dodd, Sandle and Zissermann 1967; Haroche, Paisner and Schawlow 1973; Haroche 1976; Dodd and Series 1978).

Figure 3.13 summarises the relevant steps of a quantum beat experiment: A laser pulse, which couples to both transitions from the ground state to two excited states brings the atom in a superposition of these two states. The details of this preparation step are discussed in the following section. The fluorescence of the system in a specific spatial direction is then detected as a function of time, usually after passing through a polariser. The beat signal is always superimposed on an exponentially decaying background which arises from the decay of the excited state. The oscillation frequency measures the splitting between the two excited states and can be Fourier transformed to obtain a spectrum of the excited state sublevels.

Quantum beats appear to be one of the possibilities through which to distinguish between neoclassical theories and quantum electrodynamics (Chow, Scully and Stoner 1975). They allow spectroscopic experiments in electronically excited states with high resolution and have found application primarily in molecular systems (Bitto and Huber 1990; Hack and Huber 1991).

3.3.5 Raman excitation

Principle

If a laser field couples to two transitions, it excites coherence in both of them. If the two transitions share a common energy level, they do not interact independently with the atoms, but the simultaneous presence of the two fields has an additional effect: As just discussed, irradiation of one transition in a

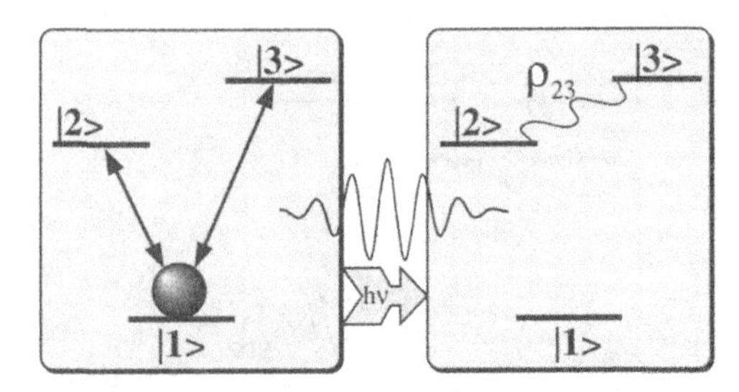

Figure 3.14. Excitation of coherence between two excited states by a Raman pulse.

three-level system can transfer coherence from a neighbouring transition into a third. We expect, therefore, that simultaneous irradiation of two adjacent transitions creates coherence in all three transitions, even in the one that is not irradiated. The effect occurs in V-type systems and Λ-type three-level system, as well as in ladder-type systems. In Λ-type systems, the effect is usually viewed as a Raman process: Absorption of a photon in one transition and simultaneous emission of a photon from an adjacent transition creates a coherence in the third transition. In a V-type system, it corresponds more to the simultaneous absorption of two photons in the two optical transitions. Nevertheless, the semiclassical description is identical for both cases and we refer to both as Raman excitation.

Figure 3.14 illustrates the process for a V-type three-level system. A laser pulse that is resonant with both optical transitions acts on a ground-state atom and excites coherence between the two excited states. This situation occurs, e.g., in the case of quantum beats (Haroche 1976; Hack and Huber 1991) or Raman beat experiments (Brewer and Hahn 1973), where the simultaneous irradiation of two optical transitions $|1\rangle\leftrightarrow|2\rangle$ and $|1\rangle\leftrightarrow|3\rangle$ excites a coherent superposition of the states $|2\rangle$ and $|3\rangle$. In the quantum beat experiment, this coherence causes the modulation of the fluorescence. In the coherent Raman beat experiment, discussed in detail below, the coherences give rise to coherent Raman scattering of a probe laser beam.

System and Hamiltonian

For a description of the process, we use the V-type three-level system of Figure 3.14 and assume that the system is initially in its ground state $|1\rangle$. For resonant irradiation of both transitions, the Hamiltonian in the interaction representation is, according to Section 3.2.3,

$$\mathcal{H}^{\mathrm{r}} = -\omega_{12}\, S_x^{(12)} - \omega_{13}\, S_x^{(13)} \tag{3.34}$$

This operator can be diagonalised with the unitary transformation

$$U = \exp(\mathrm{i}\ \theta\ S_y^{(23)})\ \exp(-\mathrm{i}\ \pi/2\ S_y^{(12)}) \tag{3.35}$$

where the mixing angle

$$\theta = 2\ \tan^{-1}\left(-\ \frac{\omega_{13}}{\omega_{12}}\right) \tag{3.36}$$

is determined by the ratio of the Rabi frequencies for the two transitions. With this transformation, the Hamiltonian assumes the form

$$\mathcal{H}^{\mathrm{rd}} = U^{-1}\ \mathcal{H}^{\mathrm{r}}\ U = -\ \omega_{\mathrm{e}}\ S_z^{(12)} \tag{3.37}$$

where the splitting $\omega_{\mathrm{e}} = \sqrt{\omega_{12}^2 + \omega_{13}^2}$ between the highest and lowest energy is determined by the two Rabi frequencies.

We assume that this Hamiltonian acts on a system that is initially in its electronic ground state,

$$\rho(0) = \sigma_1 \tag{3.38}$$

Evolution

We do not discuss the evolution of the system in general, but concentrate our attention on those features specific for the three-level system. This includes primarily the excitation of coherence in transition $|2\rangle \leftrightarrow |3\rangle$. In our usual notation, the real part of this density operator term is

$$x_{23}(t) = \mathrm{tr}\{\rho(t)\ S_x^{(23)}\}\ /\ \mathrm{tr}\{S_x^{(23)2}\} = 2\ \mathrm{tr}\{\rho(t)\ S_x^{(23)}\} \tag{3.39}$$

and a corresponding expression for y_{23}. The evolution is most easily described in the eigenbase of the Hamiltonian, where the initial density operator is

$$\rho_{\mathrm{EB}}(0) = U^{-1}\sigma_1 U = -S_x^{(12)} + \frac{1}{2}\ (\sigma_1 + \sigma_2) \tag{3.40}$$

It evolves under the Hamiltonian $\mathcal{H}^{\mathrm{rd}}$ as

$$\begin{aligned} \rho_{\mathrm{EB}}(t) &= \exp[-\mathrm{i}\ \mathcal{H}^{\mathrm{rd}}\ t]\ \rho_{\mathrm{EB}}(0)\ \exp[\mathrm{i}\ \mathcal{H}^{\mathrm{rd}}\ t] \\ &= -S_x^{(12)}\ \cos(\omega_{\mathrm{e}}t) + S_y^{(12)}\ \sin(\omega_{\mathrm{e}}t) + \frac{1}{2}\ (\sigma_1 + \sigma_2) \end{aligned} \tag{3.41}$$

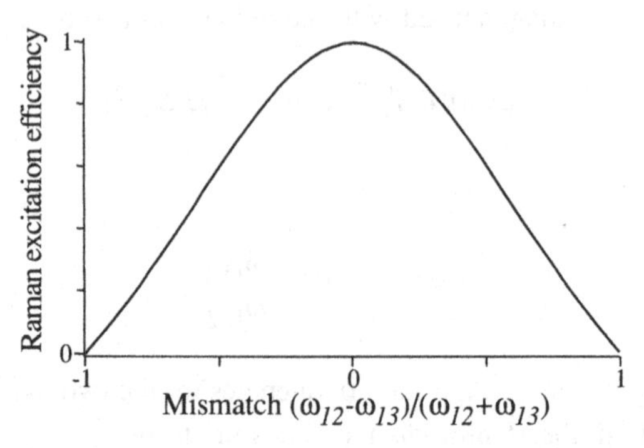

Figure 3.15. Efficiency of the Raman excitation process as a function of mismatch. The curve shows the amplitude of the coherence x_{23} as a function of the difference $\omega_{12} - \omega_{13}$ between the coupling strengths of the two fields for constant sum $\omega_{12} + \omega_{13}$.

This is the evolution of the density operator in the eigenbase of the Hamiltonian. To find the evolution of the coherence in transition $|2\rangle \leftrightarrow |3\rangle$, we have to transform back to the original representation

$$\begin{aligned} \rho(t) &= U\, \rho_{\mathrm{EB}}(t)\, U^{-1} \\ &= U \left[-S_x^{(12)} \cos(\omega_{\mathrm{e}} t) + S_y^{(12)} \sin(\omega_{\mathrm{e}} t) + \frac{1}{2}(\sigma_1 + \sigma_2) \right] U^{-1} \end{aligned} \tag{3.42}$$

Among the various terms, we are specifically interested in the coherence in transition $|2\rangle \leftrightarrow |3\rangle$

$$\begin{aligned} x_{23}(t) &= 2\, \mathrm{tr}\{\rho(t)\, S_x^{(23)}\} = -\sin\theta \sin^2\!\left(\frac{\omega_{\mathrm{e}} t}{2}\right) \\ &= -\frac{2\omega_{12}\, \omega_{13}}{\omega_{\mathrm{e}}{}^2} (1 - \cos(\omega_{\mathrm{e}} t)) \\ y_{23}(t) &= 0 \end{aligned} \tag{3.43}$$

The coherence in the uncoupled transition $|2\rangle \leftrightarrow |3\rangle$ is thus excited in an oscillatory manner. The signature of the Raman transition appears in the initial time behaviour: It starts quadratic in time, indicating the two-photon nature of the process, in contrast to direct excitation, where the coherence appears linear in time. The Rabi frequency for this process, ω_{e}, is determined mainly by the stronger of the two transitions.

The amplitude of this coherence depends, however, on the difference between the two coupling strengths. Figure 3.15 plots the amplitude of the coherence as a function of the relative difference between the two coupling

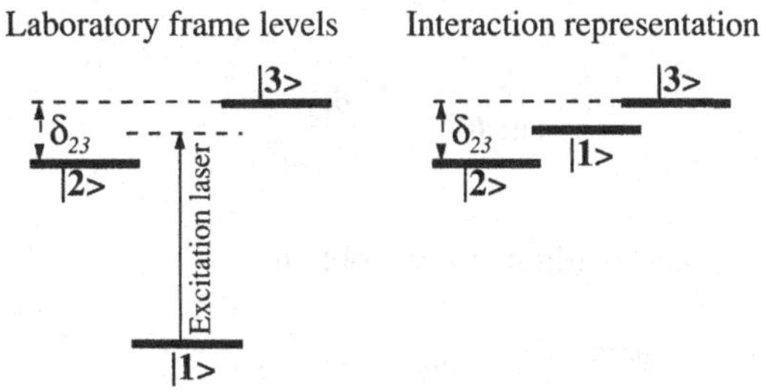

Figure 3.16. Energy level scheme with a splitting between levels $|2\rangle$ and $|3\rangle$ in the laboratory frame (left) and in the interaction representation (right).

strengths $\omega_{12} - \omega_{13}$. For equal coupling strengths, $\omega_{12} = \omega_{13}$, where $\theta = \pi/2$, the amplitude of x_{23} reaches its maximum value of unity. It falls off for mismatched amplitudes and vanishes when either of the two coupling strengths falls to zero. Efficient Raman excitation does not therefore primarily require strong fields, but matched coupling strengths.

Off-resonance effects

So far we have assumed that the laser frequency is resonant with both optical transitions. With a single laser field, this is possible only if the two excited levels $|2\rangle$ and $|3\rangle$ are degenerate. The excitation of coherence between two degenerate states should be considered as a limiting case of primarily pedagogical interest; in practice, the more interesting case arises when states $|2\rangle$ and $|3\rangle$ are nondegenerate. This is a necessary condition for the observation of quantum beat or Raman beat phenomena, since the beat frequency is the difference of the two energies. On the other hand, any difference between their energies results in a decrease of the excitation efficiency.

For a discussion of these effects, we assume that a monochromatic laser field couples to both transitions and that the laser frequency matches the energy separation between the ground state and the average of the two excited states, as shown in Figure 3.16.

In addition, we assume matched fields, $\omega_{12} = \omega_{13} = \omega_1$. The Hamiltonian for this system is in the interaction representation

$$\mathcal{H} = -\delta_{23}\, S_z^{(23)} - \omega_1 (S_x^{(12)} + S_x^{(13)}) \tag{3.44}$$

which can be diagonalised by a transformation with

$$U = \exp(-\mathrm{i}\pi/2\, S_y^{(23)})\, \exp(\mathrm{i}\,\theta\, S_y^{(13)})\, \exp(-\mathrm{i}\pi/2\, S_y^{(12)}) \tag{3.45}$$

The mixing angle θ is

$$\tan\theta/2 = \frac{\delta_{23}}{\sqrt{2}\,\omega_1} \tag{3.46}$$

For the diagonalised Hamiltonian, we obtain

$$\mathcal{H}^{\text{dia}} = U^{-1}\,\mathcal{H}\,U = -\omega_{\text{eo}}\,S_z^{(12)} \tag{3.47}$$

where the effective frequency $\omega_{\text{eo}} = \sqrt{\delta_{23}^2 + 2\omega_1^2}$ increases with the detuning δ_{23} as well as with the coupling strength ω_1. The coherence between the two excited states evolves as

$$\begin{aligned} x_{23}(t) = \frac{1+\cos\theta}{8}\,[(\cos\theta - 3)\cos(t\omega_{\text{eo}}) \\ + 4(1-\cos\theta)\cos(t\omega_{\text{eo}}/2) - 1 + 3\cos\theta] \end{aligned} \tag{3.48}$$

$$y_{23}(t) = \frac{1+\cos\theta}{2}\left[\sin\!\left(\frac{\theta}{2}\right)\sin(t\omega_{\text{eo}}) - 2\sin\!\left(\frac{\theta}{2}\right)\sin(t\omega_{\text{eo}}/2)\right] \tag{3.49}$$

In this case, the time dependence of the excitation includes terms at the effective Rabi frequency ω_{eo} as well as at half of it.

Without discussing the details of this time dependence, we try to get an idea of the overall behaviour by looking at the common prefactor of the two terms

$$\frac{1+\cos\theta}{4} = \frac{\omega_1^2}{\delta_{23}^2 + 2\,\omega_1^2} \tag{3.50}$$

Like an absorption line, it depends on the sublevel splitting, falling off to half its maximum when the splitting δ_{23} exceeds the coupling strength ω_1. The quadrature component y_{23} has a dispersive dependence on the sublevel splitting but the same width.

This analysis shows that excitation of the coherence with a monochromatic field becomes inefficient when the splitting between the two states exceeds the optical Rabi frequency (Shelby and Macfarlane 1984). For continuous lasers, this limits the possible excitation bandwidth to relatively small splittings. In the rare earth crystals frequently used for this type of experiment, the Rabi frequencies and therefore the excitation bandwidth is typically less than one MHz. This limitation to small splittings between energy levels can be circumvented if the laser is simultaneously resonant with both optical transitions;

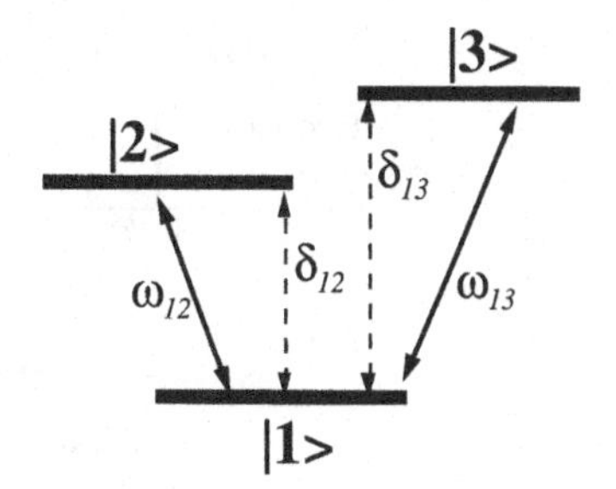

Figure 3.17. Graphical representation of the energy level scheme considered for the bichromatic irradiation.

the laser field should then contain two frequency components, matching the two optical resonance frequencies.

3.3.6 Bichromatic excitation

A bichromatic field whose frequency components are resonant with the two optical transition frequencies can decrease the mismatch between atomic energy differences and photon energy. As shown in Section 3.2.3, the Hamiltonian in the interaction representation can become completely resonant if the two frequencies are set to the two optical transitions. It is then possible to remove any effects of the level splitting during the excitation process.

System

For a description of bichromatic excitation, we start with the interaction-frame Hamiltonian

$$\mathcal{H}^{\mathrm{r}} = \delta_{12}\,\sigma_2 + \delta_{13}\,\sigma_3 - \omega_{12}\,S_x^{(12)} - \omega_{13}\,S_x^{(13)} \tag{3.51}$$

This form of the Hamiltonian implies that each frequency component interacts only with one optical transition. This is a good approximation for those systems that require the use of a bichromatic laser field. The frequency separation of the two transitions exceeds the Rabi frequency and the interaction with the nonresonant frequency component is too weak significantly to influence the dynamics of the system.

Figure 3.17 gives a graphical representation of the energy differences δ_{12}, δ_{13} and coupling strengths ω_{12}, ω_{13} between the energy levels. The frequency mismatches δ_{12} and δ_{13} can be controlled independently through the frequencies of the two components of the laser field. If both frequencies are resonant, $\delta_{12} = \delta_{13} = 0$, we return effectively to the case of vanishing level split-

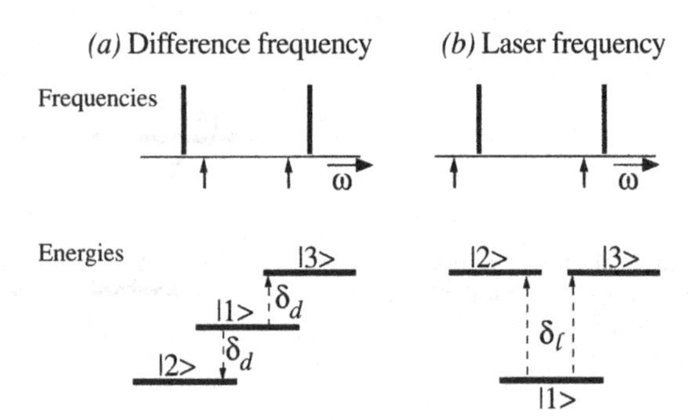

Figure 3.18. Two cases of frequency mismatch. The upper part of the figure shows the frequencies in the laboratory frame. The vertical lines represent the atomic resonance frequencies, the arrows the two frequencies of the laser field. The lower part shows the effective energies in the interaction representation.

ting, where the excitation efficiency depends only on the ratio of the two coupling strengths. Nonvanishing detunings are best discussed in analogy to the case of monochromatic excitation: We distinguish "difference frequency mismatch," where $\delta_d = \delta_{12} = -\delta_{13}$, from "laser detuning," where $\delta_l = \delta_{12} = \delta_{13}$.

In the case of difference frequency mismatch, shown in the left-hand side of Figure 3.18, the average of the two laser frequencies, represented as arrows in the upper part of the figure, coincides with the average of the resonance frequencies, which are shown as vertical lines. In the case of laser frequency detuning, represented in the right-hand side, the two laser frequencies are shifted to the same side of the resonances. The lower part of Figure 3.18 shows the resulting energy level schemes in the interaction representation.

If only difference frequency detuning is present, the Hamiltonian becomes

$$\mathcal{H}^{\mathrm{r}} = -\delta_d\, S_z^{(23)} - \omega_{12}\, S_x^{(12)} - \omega_{13}\, S_x^{(13)} \tag{3.52}$$

This is formally identical to the case of monochromatic irradiation. Therefore, we do not discuss this case here, but consider only the case of laser detuning.

Laser detuning

In an inhomogeneously broadened system, laser detuning as represented in the right-hand side of Figure 3.18, is always present for most atoms. The overall response of the atomic medium is in those cases a weighted average over the resonance line. The Hamiltonian is, in this case,

$$\mathcal{H}^{\mathrm{r}} = \delta_l(\sigma_2 + \sigma_3) - \omega_{12}\, S_x^{(12)} - \omega_{13}\, S_x^{(13)} \tag{3.53}$$

This operator can be diagonalised with the unitary operator

$$U = \exp(-\mathrm{i}\ \theta_1\, S_y^{(23)})\ \exp(-\mathrm{i}\ \theta_2\, S_y^{(12)}) \tag{3.54}$$

where the mixing angles are

$$\tan\frac{\theta_1}{2} = \frac{\omega_{13}}{\omega_{12}} \qquad \tan(\theta_2) = \frac{\omega_e}{\delta_1} \tag{3.55}$$

and ω_e represents, as before, the effective frequency $\omega_e = \sqrt{\omega_{12}^2 + \omega_{13}^2}$. The two mixing angles depend on the ratio of the coupling strengths and the ratio of the field strength to the laser detuning. The diagonalised Hamiltonian is

$$\mathcal{H}^{\mathrm{dia}} = U^{-1}\, \mathcal{H}^{\mathrm{r}}\, U = \frac{\delta}{2}(1 + \sigma_3) - \omega_{e\delta}\, S_z^{(12)} \tag{3.56}$$

where the level splitting is $\omega_{e\delta} = \sqrt{\omega_e^2 + \delta_1^2}$. On resonance, the second mixing angle becomes $\theta_2 = \pi/2$ and we recover the case treated in equation (3.35).

As before, we restrict the attention to the coherence in transition $|2\rangle \leftrightarrow |3\rangle$. If the system is initially in thermal equilibrium ($\rho_{11} = 1$), the coherence evolves as

$$\begin{aligned} x_{23}(t) &= \frac{2\omega_{12}\,\omega_{13}}{\omega_{e\delta}^2}\ \sin^2\!\left(\frac{\omega_{e\delta}\, t}{2}\right) = \sin(\theta_1)\ \sin^2(\theta_2)\ \sin^2\!\left(\frac{\omega_{e\delta}\, t}{2}\right) \\ y_{23}(t) &= 0 \end{aligned} \tag{3.57}$$

The mixing angles θ_1 and θ_2 determine the amplitude. The coherence is maximised for $\theta_1 = \theta_2 = \pi/2$, i.e., for matched coupling strengths $\omega_{12} = \omega_{13}$ and on-resonance irradiation, $\delta_1 = 0$. With increasing laser detuning, the amplitude falls off like a Lorentzian whose width is ω_e.

3.4 Steady-state effects

Time-dependent effects like those that formed the subject of the preceding section are manifestations of a freely precessing system in an ordered state. Various mechanisms contribute to the damping of such a motion and a decrease of the order. The effects described here can therefore be observed only for a finite time, while the system evolves towards a new stationary state. Even these stationary states may exhibit closely related effects, provided the damping effects are balanced by an interaction with external fields that reestab-

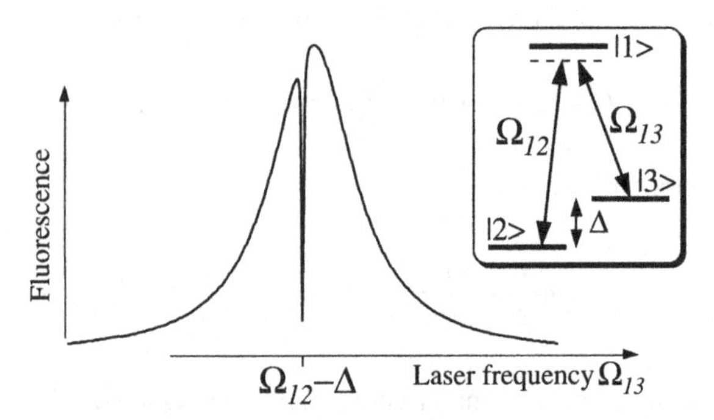

Figure 3.19. A sharp decrease in the fluorescence intensity indicates that the two laser frequencies match the condition $\Omega_{12} - \Omega_{13} = \Delta$.

lish the order that the damping mechanisms destroy. In the following, we discuss some effects of this type, again selecting those that are specific to systems with three or more states.

3.4.1 Coherent population trapping

Coherent population trapping is closely related to optical pumping: In both cases, the interaction with a laser field brings the system into a state that does not absorb light although the atoms are in the electronic ground state(s). The difference between optical pumping and coherent population trapping is that, in the latter case, the atoms are not in a single ground state, but in a superposition of two states that may have different energies. Such a state can be nonabsorbing if the transition probabilities from the two states to the excited state interfere destructively. Under these conditions, the system cannot absorb light and appears as a "dark resonance" (Arimondo and Orriols 1976; Gray, Whitley and Stroud 1978).

Dark resonances can be observed if two laser fields drive a three-level system like the one shown in Figure 3.19. If one of the laser frequencies remains fixed close to one of the optical transition frequencies and the second laser scans across the other transition, the observed fluorescence shows a narrow dip when the difference $\omega_{12} - \omega_{13}$ between the two laser frequencies matches the transition frequency between the two ground-state levels.

Hamiltonian

Consider a Λ-type three-level system that is driven by two laser fields. We can describe it by the Hamiltonian of equation (14) derived for V-type sys-

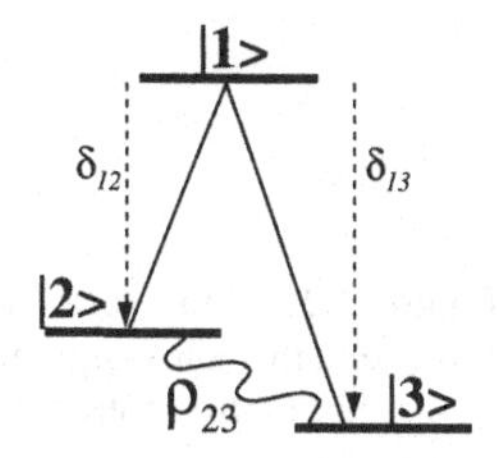

Figure 3.20. Λ-type three-level system for the analysis of coherent population trapping.

tems, if we invert the sign of the energies. The Hamiltonian in the interaction representation is the usual three-level form with different laser fields

$$\mathcal{H}^{r} = -\delta_{12}\,\sigma_2 - \delta_{13}\,\sigma_3 - \omega_{12}\,S_x^{(12)} - \omega_{13}\,S_x^{(13)} \tag{3.58}$$

thus making level $|1\rangle$ the energetically highest state.

Coherent population trapping occurs when the two laser detunings are equal, $\delta = \delta_{12} = \delta_{13}$. See Figure 3.20. In this case, the Hamiltonian reduces to

$$\mathcal{H}^{r} = -\delta\,(\sigma_2 + \sigma_3) - \omega_{12}\,S_x^{(12)} - \omega_{13}\,S_x^{(13)} \tag{3.59}$$

In the representation in Figure 3.21, the levels $|2\rangle$ and $|3\rangle$ are degenerate. The first term of the Hamiltonian commutes not only with all diagonal terms, but also with the ground-state coherence $S_x^{(23)}$ and $S_y^{(23)}$ and has therefore no effect on their evolution – a necessary condition for coherent population trapping.

Trapping state

We now search the "dark state," i.e., a density operator with no population in the excited state and therefore no optical coherences. We make the ansatz

$$\rho^{r}_{\mathrm{cpt}} = z_{23}\,S_z^{(23)} + x_{23}\,S_x^{(23)} + \frac{1}{2}\,(\sigma_{22} + \sigma_{33}) \tag{3.60}$$

This state contains only two free parameters, z_{23} and x_{23}. The state must be stationary, so we can determine the coefficients by requiring that the commutator between the Hamiltonian and the density operator vanish,

$$[\mathcal{H}^{r}_{\mathrm{cpt}},\rho^{r}_{\mathrm{cpt}}] = \omega_{12}\,(z_{23}\,[S_x^{(12)}, S_z^{(23)}] + x_{23}\,[S_x^{(12)}, S_x^{(23)}] + \frac{1}{2}\,[S_x^{(12)}, \sigma_2 + \sigma_3])$$
$$+\ \omega_{13}\,(z_{23}\,[S_x^{(13)}, S_z^{(23)}] + x_{23}\,[S_x^{(13)}, S_x^{(23)}] + \frac{1}{2}\,[S_x^{(13)}, \sigma_2 + \sigma_3]) = 0 \tag{3.61}$$

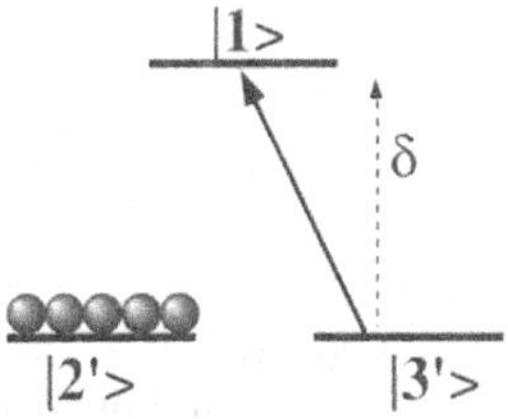

Figure 3.21. Transformed representation of the three-level scheme showing the analogy between coherent population trapping and optical pumping.

The solution is

$$x_{23} = -2\,\frac{\omega_{12}\,\omega_{13}}{\omega_{12}^2 + \omega_{13}^2} \qquad z_{23} = -\frac{\omega_{12}^2 - \omega_{13}^2}{\omega_{12}^2 + \omega_{13}^2} \tag{3.62}$$

The trapping state is determined by the optical Rabi frequencies ω_{12} and ω_{13} alone, while the common detuning δ of the two laser beams vanishes. When this condition is satisfied, the system absorbs light as long as the density operator has components orthogonal to the trapping state. Every atom going through an absorption/emission cycle, however, has a certain chance of falling into the trapping state. Once it arrives there, it stops absorbing light and the "dark resonance" appears.

An inspection of equation (3.62) shows that the trapping state is purely diagonal, i.e., contains only populations, when one of the two coupling strengths vanishes; at that point, the whole population assembles in a single state – the one not coupled to the laser field. This is the situation of optical pumping, which we discuss in Chapter 5. Another special case arises when the two coupling strengths match, $\omega_{12} = \omega_{13}$. In this case, the longitudinal term z_{23} vanishes, i.e., both ground-state sublevels are equally populated. At the same time, the off-diagonal term x_{23} reaches a maximum, indicating a definite phase relation between the two levels which results in destructive interference of the two absorption probabilities.

Connection to optical pumping

We can make the connection to optical pumping explicit and complete by a basis transformation for this system, using the unitary transformation

$$U = \mathrm{e}^{-\mathrm{i}\theta S_y^{(23)}} \qquad \text{with} \qquad \theta = \tan^{-1}\frac{x_{23}}{z_{23}} \tag{3.63}$$

The density operator then transforms into

$$\rho_{2lev}^{(23)} = U^{-1}\,\rho_{cpt}^{r}\,U = \sigma_2 \tag{3.64}$$

In this basis, the density operator is diagonal and only level $|2'\rangle$ is populated. The Hamiltonian, on the other hand, becomes

$$\mathcal{H}_{2lev}^{(23)} = U\,\mathcal{H}_{cpt}^{r}\,U^{-1} = \delta\,(\sigma_2 + \sigma_3) - \omega_e\,S_x^{(13)} \tag{3.65}$$

where the effective coupling strength is $\omega_e = \sqrt{\omega_{12}^2 + \omega_{13}^2}$.

In this representation, the field couples only the unpopulated level $|3'\rangle$ to the excited state, whereas level $|2'\rangle$ is uncoupled. This is the situation of optical pumping. Schemes of this type are used, e.g., for laser cooling, where atoms are accumulated into an uncoupled state if their velocity component in the direction of the laser beams vanishes (Aspect et al. 1988; Kasevich et al. 1991; Shang et al. 1991).

Amplification without inversion

The same scheme that "hides" population in the lower levels from creating absorption can also be used to allow amplification through stimulated emission, even if the population of the upper level is smaller than the sum of the populations in the two lower levels (Veer et al. 1993). In terms of the three-level system of equation (3.60), amplification "without inversion" can be obtained in the sense that the excited state $|1\rangle$ has a nonzero population (say $\rho_{11} = 0.1$), while level $|3'\rangle$ is empty ($\rho_{33} = 0$), and level $|2'\rangle$ contains the largest amount of population ($\rho_{22} = 0.9$). See Figure 3.22.

Formally, there is no inversion in the system, as the total ground-state population is larger than the excited state population, $\rho_{22} + \rho_{33} > \rho_{11}$. On the other hand, the system can show amplification, since stimulated emission from the excited state exceeds absorption from level $|3'\rangle$.

Various other schemes have been suggested to obtain gain without inversion. In the simplest case, it is sufficient to drive a two-level atomic system with a strong, monochromatic laser field (Wu et al. 1977). More elaborate schemes use irradiation with multiple fields, where the resulting gain is best described as a wave-mixing process (Boyd et al. 1981), or multilevel systems with a coherent excitation in the ground state or in an excited state (Fill et al. 1990; Imamoglu et al. 1991; Kocharovskaya 1992; Scully 1992; Zhu 1992). This coherent excitation can be generated either by optical fields or by radio-frequency fields applied to transitions between angular momentum states (Wei and Manson 1994).

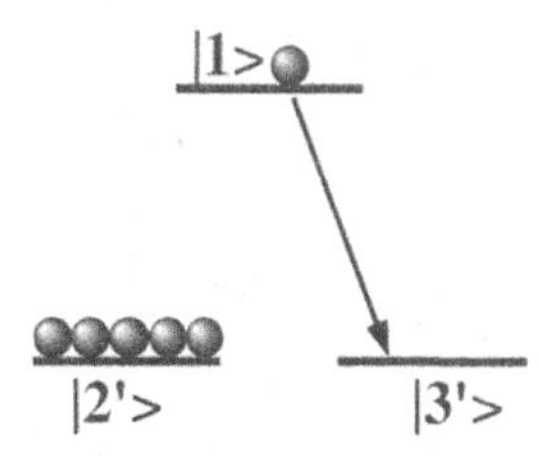

Figure 3.22. Amplification without an overall inver-

As the basis transformation in Equation (3.63) shows, the question of whether a system shows inversion or not is often somewhat ambiguous. This is particularly the case for steady-state amplification, which requires some energy flow into and out of the system. Such systems are necessarily open and allow no unique definition of stationary states and "populations." Correspondingly, the definition of the term "inversion" is somewhat arbitrary.

Dispersion without absorption

Their large nonlinearities make atomic vapours an attractive tool for nonlinear optical experiments. For many experiments, the dispersive nonlinearities are most interesting, whereas absorption represents an unwanted loss mechanism. The ideal system for such experiments would have large dispersion, but no absorption. The two are not independent of each other, however: the Kramers–Kronig relations establish a definite relationship between the two. In a two-level atom, absorption can be reduced only if the dispersion is reduced correspondingly. In multilevel systems, one has more degrees of freedom that allow the creation of situations in which the absorption is decreased or eliminated for certain laser frequencies without a significant decrease of the dispersive properties. To explain how such situations can be established, we discuss here the simplest possible model.

Figure 3.23 shows a system with two transitions: If one of them has inverted populations, its absorption, as well as its dispersion, is inverted. The medium provides gain. At the centre frequency between the two transitions, the contributions to the absorption cancel, whereas the dispersions add. Such a system should allow experiments with large optical nonlinearities without absorption losses. In reality, it is difficult to find two atomic species with almost identical resonance frequencies and invert one of them selectively. Similar schemes can be implemented, however, with a single atomic species if its energy level scheme has an appropriate structure (Harris, Field and Imamoglu 1990; Hakuta, Marmet and Stoicheff 1991; Scully 1991; Hakuta, Marmet and Stoicheff 1992; Scully 1992).

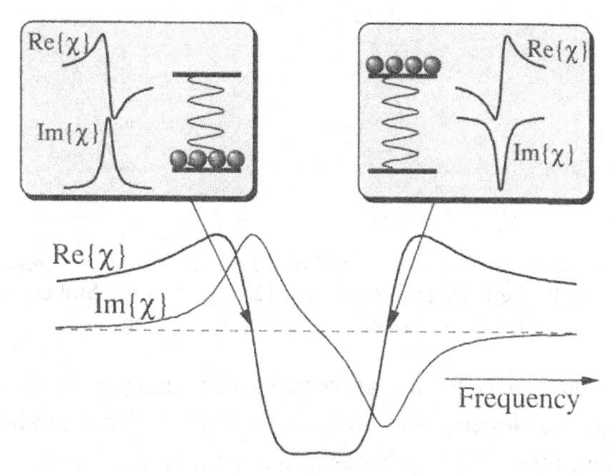

Figure 3.23. Schematic representation of a simple system that displays large dispersion without absorption. It consists of two transitions with different frequencies. The transition in the right-hand box is inverted and therefore exhibits gain instead of absorption. The arrows indicate the resonance frequencies for the two transitions. The curves at the bottom show the wavelength dependence of the absorption, $\mathrm{Im}\{\chi\}$, and dispersion, $\mathrm{Re}\{\chi\}$, of the combined system. At the centre, the dispersion is large, whereas the absorption vanishes.

3.4.2 Coherent Raman scattering

Phenomenological introduction

Raman scattering is a nonlinear optical process in which the scattered light contains, apart from elastically scattered photons, inelastically scattered photons that indicate a resonant energy transfer to the material. In spontaneous Raman scattering, this exchange of energy occurs with a vibrational transition in a molecule. If Raman scattering occurs in a medium that is in thermal equilibrium, most molecules are in their electronic and vibrational ground state. In a scattering process, these molecules can only increase their energy while simultaneously decreasing the photon energy. In spontaneous Raman scattering, most of the scattered photons are thus converted to lower energies (Stokes-shifted).

The left-hand side of Figure 3.24 illustrates this with the frequency spectrum of the scattered light, where the largest peak is at the laser frequency Ω_L. Most of the inelastically scattered light is Stokes-shifted to a frequency $\Omega_S = \Omega_L - \Omega_R$. In the adjacent level scheme, the thick arrow indicates the incident laser field, which couples to the ground-state molecules; the thin arrow represents the scattered Raman field. The reverse process, which increases the photon energy to $\Omega_{AS} = \Omega_L + \Omega_R$ and simultaneously de-excites molecules, can only occur for the small number of molecules initially in the vibrationally excited state. In both cases, the scattering process does not involve real absorp-

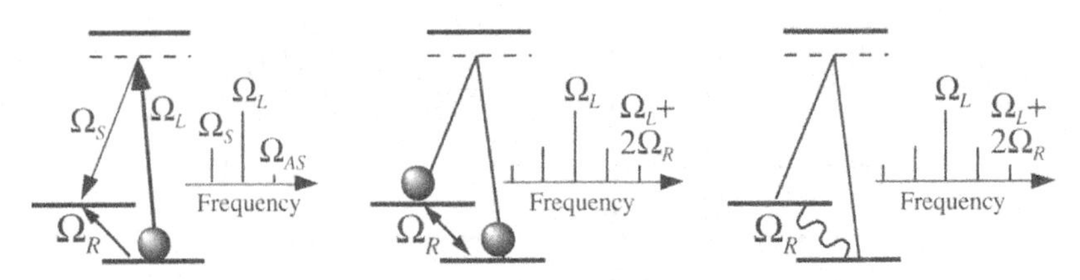

Figure 3.24. Comparison of different types of Raman processes. Ω_R refers to the Raman frequency, Ω_L to the laser frequency, and Ω_S (Ω_{AS}) to the Stokes- (anti-Stokes-) shifted frequencies of the scattered light.

tion of photons but proceeds through virtual states, indicated by the dashed line in the level scheme. Nevertheless, the presence of a real state, indicated by the full line, enhances the coupling between the medium and the radiation field.

In the case of stimulated Raman scattering (Hellwarth 1963), the use of lasers increases the intensity of the exciting wave sufficiently that a significant fraction of the molecules populates the energetically higher state. This allows the anti-Stokes process to become important. The Stokes and anti-Stokes components in the scattered light thus have comparable intensities. In addition, the shifted photons can again undergo Raman processes, thus giving rise to multiple scattering processes that lead to intensity in higher order sidebands, as indicated by the central part of Figure 3.24. Under conditions of stimulated Raman scattering, several coherent optical waves propagate through the material, coupled by the material excitation (Shen 1984).

The same situation can also be established without using the high intensity laser irradiation required to drive Raman scattering into the stimulated regime by a technique known as coherent Raman scattering (Giordmaine and Kaiser 1966). For this purpose, it is necessary to prepare the coherent material excitation before the actual Raman process, as shown in part (c) of Figure 3.24. Again, the Stokes and anti-Stokes components of the scattered light have the same intensity, and multiple scattering processes are possible. In contrast to stimulated Raman scattering, however, the scattering process is now linear in the amplitude of the incident light. The price that must be paid for this enhanced coupling efficiency is the preparation of a coherent excitation, which can be achieved through the Raman excitation scheme discussed above and also by direct irradiation with radio frequency or infrared radiation.

Scattering process

Coherent Raman processes occur with coherent superpositions of electronic ground states as well as in electronically excited states. Formally, the two cases can be treated identically, but to make the discussion more specific, we treat explicitly only the V-type three-level system of Figure 3.25.

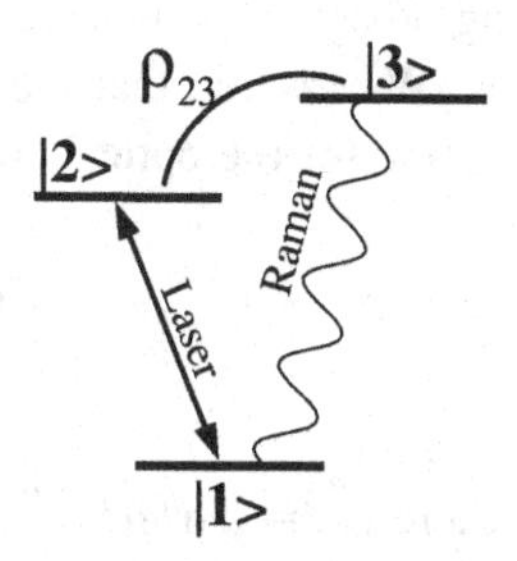

Figure 3.25. V-type level system as a model of coherent Raman scattering.

In this figure, the coherent excitation that provides the coupling between the laser and the Raman field is between the two excited state sublevels $|2\rangle$ and $|3\rangle$. The relevant part of the initial density operator is

$$\rho_0 = x_{23}\, S_x^{(23)}$$

which can be created, e.g., by the Raman excitation processes described above. Examples of actual experiments will be discussed in Chapter 7. We assume that the incident laser field is monochromatic and couples only with transition $|1\rangle \leftrightarrow |2\rangle$ and is resonant with this transition ($\delta_{12} = 0$). In the interaction representation, the Hamiltonian is then

$$\mathcal{H}^{\mathrm{r}} = -\omega_{12}\, S_x^{(12)} \tag{3.66}$$

As usual, the coupling strength ω_{12} is proportional to the product of the field strength and the electric dipole moment.

According to the commutation relations of equation (3.5), this Hamiltonian causes a time evolution

$$\frac{\mathrm{d}}{\mathrm{d}t}\rho_0 = -\frac{1}{2}\omega_{12}\, x_{23}\, S_y^{(13)} \tag{3.67}$$

i.e., a transfer of coherence from the sublevel transition $|2\rangle \leftrightarrow |3\rangle$ to the optical transition $|1\rangle \leftrightarrow |3\rangle$. The incident laser field excites an optical coherence in the transition with which it is not resonant. The resulting optical polarisation is the source term for the Raman wave.

Stationary solution

Since we are interested in the steady-state behaviour, we need to find the equilibrium density operator. The stationary state of the system is determined by the balance of the excitation process described by equation (3.67) and damping processes, which we take into account on a purely phenomenological basis: We assume that the optical coherence decay rate is γ. The equation of motion for the optical coherence in the Raman transition is then

$$\frac{\mathrm{d}}{\mathrm{d}t} y_{13} = -\frac{1}{2} \omega_{12} x_{23} - \gamma y_{13} \tag{3.68}$$

where $y_{13} = 2\,\mathrm{tr}\{\rho S_y^{(13)}\}$ is the expansion coefficient of the optical coherence in transition $|1\rangle \leftrightarrow |3\rangle$. The stationary optical polarisation is

$$y_{13\infty} = -\frac{1}{2} \frac{\omega_{12}}{\gamma} x_{23} \tag{3.69}$$

In the laboratory frame, this coherence represents an electric dipole oscillating at the optical frequency Ω_{13}, as can easily be checked by reversing the transformation to the interaction frame.

The effect of the laser field on the atoms is thus the conversion of the coherence between states $|2\rangle$, $|3\rangle$ into an optical coherence. The optical photon that coupled to transition $|1\rangle \leftrightarrow |2\rangle$ has been frequency shifted by the radio frequency of transition $|2\rangle \leftrightarrow |3\rangle$. In the language of nonlinear optics, this process corresponds to a resonant three-wave mixing process between the laser field, the material excitation, and the Raman field. In contrast to stimulated Raman scattering, the resulting optical polarisation is linear in the amplitude of the incident field.

Off-resonance effects

So far we have assumed that the incident probe laser beam is exactly resonant with transition $|1\rangle \leftrightarrow |2\rangle$. In practice, this is often not the case, in particular when the system being studied is inhomogeneously broadened or the laser frequency jitter cannot be neglected. We take these effects into account by adding an off-resonance term to the Hamiltonian

$$\mathcal{H}^{\mathrm{r}} = \delta\,(\sigma_2 + \sigma_3) - \omega_{12} S_x^{(12)} \tag{3.70}$$

Under these conditions, the coherence in the Raman transition evolves as

$$\dot{x}_{13} = \delta\, y_{13} - \gamma\, x_{13} \qquad \dot{y}_{13} = -\delta\, x_{13} - \frac{1}{2} \omega_{12} x_{23} - \gamma\, y_{13} \tag{3.71}$$

and reaches a steady-state value

$$x_{13\infty} = -x_{23}\,\frac{\delta\,\omega_{12}}{2(\delta^2+\gamma^2)} \qquad y_{13\infty} = -x_{23}\,\frac{\gamma\,\omega_{12}}{2(\delta^2+\gamma^2)} \tag{3.72}$$

The amplitude of the steady-state coherence thus shows the usual Lorentz dependence on the laser detuning δ. The width of the resonance line is determined by the dephasing rate γ of the coherence in the Raman transition.

To find the frequency of the Raman field, we transform the density operator back to the laboratory frame, using the transformation

$$\rho^{\mathrm{Lab}} = U\,\rho^{\mathrm{r}}\,U^{-1} \qquad \text{with} \qquad U = \exp\{-\mathrm{i}\,t\,[\Omega\,\sigma_2 - (\Omega+\delta_{23})\,\sigma_3]\} \tag{3.73}$$

The resulting frequency of the optical coherence in the Raman transition is $\Omega_{\mathrm{Raman}} = \Omega + \delta_{23}$, where Ω is the frequency of the incident laser field. If the two fields are incident on a quadratic detector, the resulting signal beats with the difference frequency $\Omega_{\mathrm{Raman}} - \Omega = \delta_{23}$. The frequency of the signal is therefore independent of the laser frequency, but its amplitude drops off with a half-width determined by the dephasing rate γ. At the same time, the phase of the signal is shifted with respect to the resonant case. This behaviour is important if the laser frequency jitter is larger than the optical linewidth. The average over the laser jitter then results primarily in a decrease of the signal amplitude, but does not affect the width of the Raman resonance.

3.5 Overdamped systems

3.5.1 Characteristics

Pressure broadening

So far we have neglected relaxation processes or at most included them as a perturbation. This corresponds to the situation where the interaction with the laser field dominates the evolution of the atomic system. The opposite limit is reached when dissipative processes dominate and the coupling to the radiation field is just a small perturbation. This situation occurs, e.g., in collisionally broadened systems, like alkali atoms in a buffer gas environment. An inert buffer gas like argon or helium is frequently added to sample cells containing atomic vapours to suppress the unwanted inhomogeneous Doppler broadening. In addition, the buffer gas reduces the mean free path of the atoms, turning the free flight atomic motion into a diffusion process and reducing the average velocity significantly. Atoms in such an environment spend significantly longer times in the region where interaction with light occurs (Dehmelt

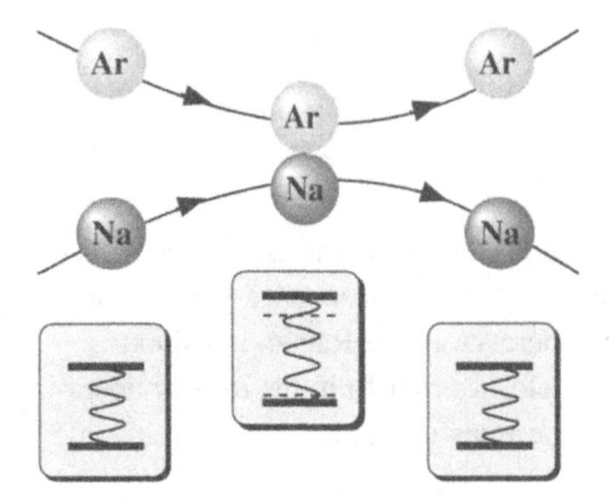

Figure 3.26. Shift of energy levels during a collision.

1957b), thus permitting more precise measurements and reducing unwanted wall collisions.

Figure 3.26 illustrates how the interaction between the collision partners shifts the energies of the atomic states during a collision. This energy shift changes the precession frequency of the induced dipole. The net effect of the collision for the individual atom is that the phase of the optical coherence acquires a shift by an unpredictable amount. For the ensemble, the collisions of its members cause a decay of the optical polarisation. Under typical experimental conditions, this dephasing process occurs on a timescale of a hundred picoseconds – significantly faster than a Rabi oscillation.

Under these conditions, the optical coherences remain small, and the evolution of the optical system is more appropriately described by rate equations than by the superposition of oscillations that is typical of the coherent limit discussed so far. While the optical coherences remain small, this is not necessarily true for sublevel coherences within the electronic ground state, in particular if the electronic state has the spherical symmetry of an s-orbital: Such spherically symmetric states are only weakly affected by the collisions. Sublevel coherences in spherically symmetric ground states have therefore long lifetimes and evolve coherently, even in pressure-broadened systems. Optical fields interacting with such systems may cause coherent dynamics within the ground state already at low laser intensities.

System

We consider the Λ-type three-level system of Figure 3.27, with two substates in the electronic ground state and a single excited state.

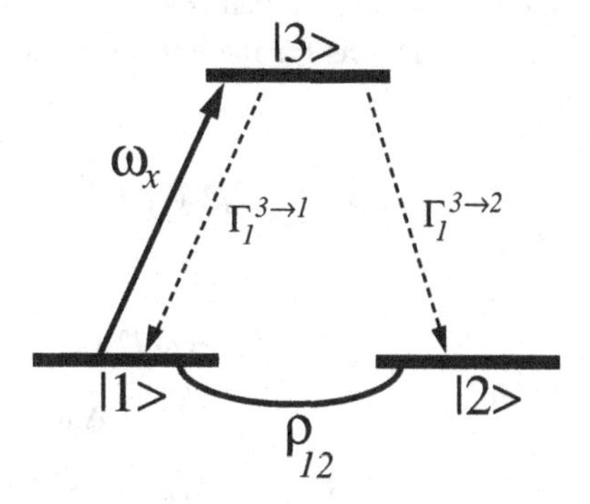

Figure 3.27. Relevant level scheme for the perturbation analysis of optical pumping. Levels 1 and 2 are substates of the electronic ground state, level 3 is an eletronically excited state. ω_x represents the optical Rabi frequency. The solid arrow indicates the transition that is driven by the laser field, and the dashed arrows indicate spontaneous decay channels.

The Hamiltonian for this system is, in the interaction representation,

$$\mathcal{H}^{\mathrm{r}} = \Delta\omega_0\ \sigma_3 - \omega_x\ S_x^{(13)} \tag{3.74}$$

where $\Delta\omega_0$ represents the offset of the laser frequency from optical resonance and ω_x the optical Rabi frequency. As usual, we expand the density operator in single-transition operators, writing the expansion coefficients as x_{ij}, y_{ij} which are defined as

$$x_{ij} = \rho_{ij} + \rho_{ji} \qquad y_{ij} = \mathrm{i}(\rho_{ij} - \rho_{ji}) \tag{3.75}$$

The x_{ij} represent the real part of the coherence while the y_{ij} correspond to the imaginary part. Physically, the coherences in the transitions $|1\rangle\leftrightarrow|3\rangle$ and $|2\rangle\leftrightarrow|3\rangle$ correspond to an induced electric dipole moment, coherence in transition $|1\rangle\leftrightarrow|2\rangle$ to the transverse components of the magnetic dipole moment.

Equation of motion

The equation of motion for this system includes, apart from the Hamiltonian contribution, relaxation effects, which we write as

$$\dot{\rho} = -\mathrm{i}\ [\mathcal{H}^{\mathrm{r}},\rho] + \hat{\Gamma}_1\rho + \hat{\Gamma}_2\rho \tag{3.76}$$

The superoperator $\hat{\Gamma}_1$ summarises the transfer of population between different levels due to relaxation processes, and $\hat{\Gamma}_2$ the dephasing of coherences. We use the notation $\Gamma_1^{3\to 1}$ for the spontaneous decay rate from level $|3\rangle$ to level $|1\rangle$ and assume a uniform decay of all optical coherences with a rate Γ_2. With these definitions, we can rewrite the equation of motion (3.76) into a set of coupled equations for the expansion coefficients (x_{ij}, y_{ij}) of the density operator:

$$\dot{\rho}_{11} = -\omega_x/2\ y_{13} + \Gamma_1^{3\to 1}\ \rho_{33} \qquad \dot{\rho}_{22} = \Gamma_1^{3\to 2}\ \rho_{33}$$
$$\dot{\rho}_{33} = \omega_x/2\ y_{13} - \Gamma_1\ \rho_{33}$$
$$\dot{x}_{12} = -\omega_x/2\ y_{23} \qquad \dot{y}_{12} = -\omega_x/2\ x_{23}$$
$$\dot{x}_{13} = \Delta\omega_0\ y_{13} - \Gamma_2\ x_{13} \tag{3.77}$$
$$\dot{y}_{13} = -\Delta\omega_0\ x_{13} + \omega_x(\rho_{11} - \rho_{33}) - \Gamma_2\ y_{13}$$
$$\dot{x}_{23} = \omega_x/2\ y_{12} + \Delta\omega_0\ y_{23} - \Gamma_2\ x_{23}$$
$$\dot{y}_{23} = \omega_x/2\ x_{12} - \Delta\omega_0\ x_{23} - \Gamma_2\ y_{23}$$

$\Gamma_1 = \Gamma_1^{3\to 1} + \Gamma_1^{3\to 2}$ is the total decay rate of the excited state and Γ_2 the dephasing rate of the optical coherences. The relaxation effects that are included here are the population decay rates of the excited state $|3\rangle$ and the optical dephasing rate Γ_2. The lifetime of the ground-state populations and the ground state coherences x_{12}, y_{12} are taken as infinite.

3.5.2 Adiabatic limit

For an evaluation of this equation of motion, it is useful to consider the limit of a dephasing rate that is large compared to the optical Rabi frequency, $\Gamma_2 >> \omega_x$. The system then settles into a quasi-stationary state on a timescale of the order of $1/\Gamma_2$. This quasi-stationary state is characterised by small optical coherences. If, in addition, $\Gamma_1\Gamma_2 >> \omega_x^2$, the population of the excited state remains small.

Figure 3.28 summarises the different timescales on a logarithmic scale. During the first part, which is of the order $1/\Gamma_2$, optical coherences such as x_{13} decay and the system settles into a quasi-equilibrium state. On a longer timescale, which is of the order Γ_2/ω_x^2, dynamics occur within the electronic ground state. This is the timescale which interests us in this context: We analyse the evolution with a time resolution that is coarse compared to $1/\Gamma_2$. This significantly reduces the complexity of the problem and permits a rather intuitive analysis of the system evolution.

To calculate the state of the system after transient effects on the short timescale have disappeared, we assume that both conditions stated above are fulfilled, i.e. the optical coherences as well as the excited state populations

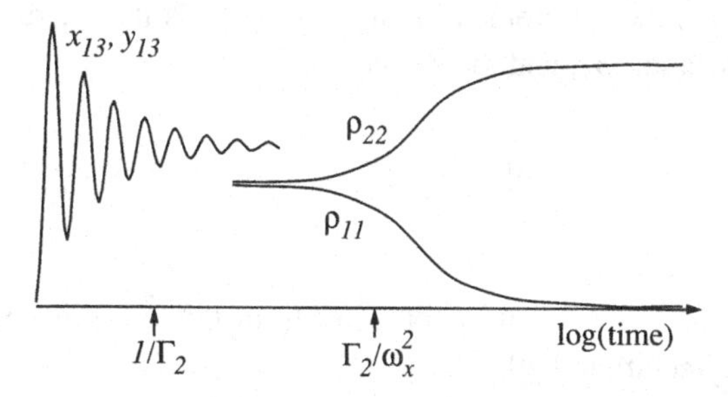

Figure 3.28. Timescales that are relevant for the dynamics of radiatively coupled systems with strong optical dephasing.

are small compared to unity. With these assumptions, it is possible to calculate the optical coherences as well as the excited state population for the quasi-stationary state. Taking the ground-state populations initially as time-independent, we find the quasi-stationary optical coherences as

$$x_{13} = \rho_{11} \frac{\delta\, \omega_x}{\Delta\omega_0^2 + \Gamma_2^2} \qquad y_{13} = \rho_{11} \frac{\Gamma_2\, \omega_x}{\Delta\omega_0^2 + \Gamma_2^2} \tag{3.78}$$

and the population of the excited state is

$$\rho_{33} = \rho_{11} \frac{\Gamma_2}{2\,\Gamma_1} \frac{\omega_x^2}{\Delta\omega_0^2 + \Gamma_2^2} \tag{3.79}$$

All three density operator components are thus proportional to ρ_{11}. This is the motivation for calling this state "quasi-stationary": The system is not exactly time-independent, but the optical coherences as well as the excited state population remain proportional to ρ_{11}. The fast motion due to Rabi flopping has decayed, but they follow the slowly changing population ρ_{11}. Neglecting the fast motion of the optical coherences, as we do here, is referred to as the adiabatic approximation, in analogy to the Born–Oppenheimer approximation in molecular physics, which separates the electronic degrees of freedom from the motion of the nuclei.

3.5.3 Optical pumping

The excited state population ρ_{33} is due to excitation from the ground-state level $|1\rangle$. Since it has a finite probability to decay to level $|2\rangle$, there is a net transfer of population from level $|1\rangle$ to level $|2\rangle$, which is known as optical pumping. The number of atoms arriving in state $|2\rangle$ is the product of the excited state population ρ_{33} and the decay rate $\Gamma_1^{3\to2}$

$$\frac{d}{dt}\rho_{22}(t) = \Gamma_1^{3\to2}\,\rho_{33} \tag{3.80}$$

We eliminate the excited state population from this equation by using the quasi-stationary solution 3.79.

$$\frac{d}{dt}\rho_{22}(t) = \Gamma_1^{3\to2}\,\rho_{33} = \Gamma_2\frac{\Gamma_1^{3\to2}}{\Gamma_1^{3\to1}+\Gamma_1^{3\to2}}\frac{\omega_x^2}{2(\Delta\omega_0^2+\Gamma_2^2)}\,\rho_{11} \tag{3.81}$$

Using the assumption that the excited state population is small, $\rho_{33} << (\rho_{11}, \rho_{22})$, we obtain an equation of motion for the ground-state populations

$$\frac{d}{dt}\rho_{11}(t) = -\frac{d}{dt}\rho_{22}(t) = -k\,\rho_{11} \tag{3.82}$$

with the rate constant

$$k = \Gamma_1^{3\to2}\frac{\rho_{33}}{\rho_{11}} = \Gamma_2\frac{\Gamma_1^{3\to2}}{\Gamma_1^{3\to1}+\Gamma_1^{3\to2}}\frac{\omega_x^2}{2(\Delta\omega_0^2+\Gamma_2^2)} \tag{3.83}$$

Instead of using the individual populations, we may express the evolution of the sublevel populations with the population difference $z_{12} = \rho_{22} - \rho_{11}$:

$$\dot{z}_{12} = \frac{d}{dt}(\rho_{22}-\rho_{11}) = 2\dot{\rho}_{22} = 2\,k\,\rho_{11}(t) = k\,(1 - z_{12}(t)) \tag{3.84}$$

Under the assumptions made here, the ground-state population is therefore pumped completely into the second level, as shown in Figure 3.29.

This is only valid as long as the relaxation of the ground state can be neglected and no other effects influence the dynamics. However, as long as additional effects, like relaxation of the ground-state sublevels or magnetic fields, are not strong enough to perturb the fast evolution of the system, the overall dynamics of the system is just the superposition of the optical pumping with the additional effects.

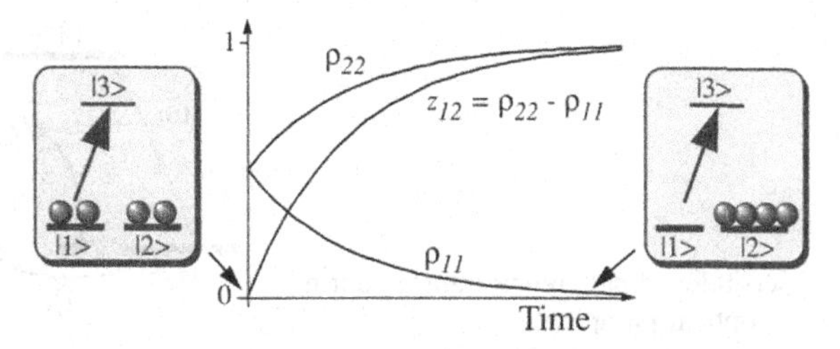

Figure 3.29. Evolution of the populations of the two ground-state sublevels and the population difference $z_{12} = \rho_{22} - \rho_{11}$, starting from thermal equilibrium.

3.5.4 Light shift and damping

Phenomenology

Besides optical pumping, which describes the effect of the optical irradiation on the ground-state populations, the light also affects the ground-state coherences x_{12} and y_{12}. Equations (3.77) and Figure 3.30 show how the optical field couples them to the optical coherences x_{23} and y_{23}. As a result, the distinction between ground-state coherences and optical coherences is no longer exact and this partial mixing affects the precession and decay of the ground-state coherences. Since the relaxation rate of the optical coherences is extremely fast compared to that of the ground-state coherences, even a small mixing between the ground-state coherences and the optical coherences leads to a damping of the ground-state coherences. In addition, the optical resonance offset, which appears as a precession of the optical coherences, also causes a precession of the ground-state coherences, which is known as light shift.

This effect was first predicted by Barrat and Cohen-Tannoudji (Barrat and Cohen-Tannoudji 1961b; Barrat and Cohen-Tannoudji 1961a; Barrat, Cohen-Tannoudji and Ribaud 1961) and observed by Arditi (Arditi and Carver 1961) and Cohen-Tannoudji (Cohen-Tannoudji 1961; Cohen-Tannoudji 1962; Dupont-Roc et al. 1967; Cohen-Tannoudji and Dupont-Roc 1972). The light shift can manifest itself either in level crossing experiments (Arditi and Carver 1961; Barrat and Cohen-Tannoudji 1961a; Barrat, Cohen-Tannoudji and Ribaud 1961; Cohen-Tannoudji 1961; Dupont-Roc et al. 1967; Mathur, Tang and Happer 1968; Cohen-Tannoudji et al. 1970a; Cohen-Tannoudji et al. 1970b; Cohen-Tannoudji and Dupont-Roc 1972; Delsart, Keller and Kaftandjian 1980; Kaftandjian, Delsart and Keller 1981) or dynamically (Rosatzin, Suter and Mlynek 1990), as discussed in Chapter 8. Its observation presented conceptual difficulties and therefore several authors worked on different formulations of the process (Pancharatnam 1966; Happer and Mathur 1967a; Happer 1970).

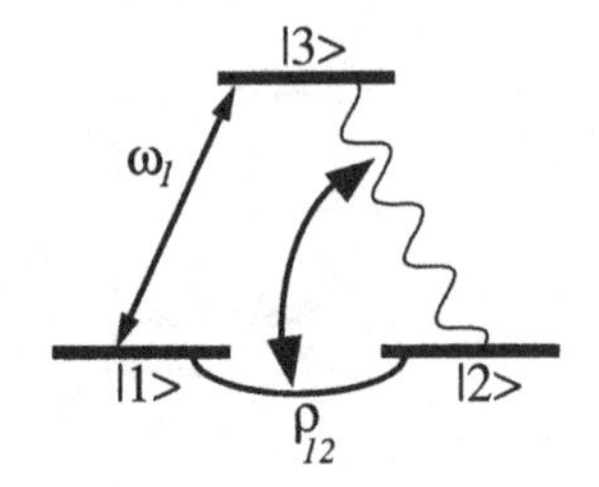

Figure 3.30. Laser-induced coherence transfer during optical pumping.

Equation of motion

We analyse the situation again by looking at the equation of motion in the adiabatic approximation. The optical coherences in the quasi-stationary state should be time-independent, provided the ground-state coherences x_{12} and y_{12} can be considered as time-independent. We look then for the solutions of

$$\begin{aligned} \dot{x}_{23} &= 0 = \omega_x/2\, y_{12} + \Delta\omega_0\, y_{23} - \Gamma_2\, x_{23} \\ \dot{y}_{23} &= 0 = \omega_x/2\, x_{12} - \Delta\omega_0\, x_{23} - \Gamma_2\, y_{23} \end{aligned} \tag{3.85}$$

Again making the assumption that the sublevel coherences are time-independent, we find the quasi-stationary optical coherences

$$x_{23\infty} = \omega_x \frac{\Delta\omega_0\, x_{12} + \Gamma_2\, y_{12}}{2(\Delta\omega_0^2 + \Gamma_2^2)} \qquad y_{23\infty} = \omega_x \frac{\Gamma_2\, x_{12} - \Delta\omega_0\, y_{12}}{2(\Delta\omega_0^2 + \Gamma_2^2)} \tag{3.86}$$

We use these quasi-stationary solutions to eliminate the optical coherences from the equation of motion for the ground-state coherences

$$\begin{aligned} \dot{x}_{12} &= -(\Gamma_2\, x_{12} - \Delta\omega_0\, y_{12}) \frac{\omega_x^2}{4(\Delta\omega_0^2 + \Gamma_2^2)} \\ \dot{y}_{12} &= -(\Delta\omega_0\, x_{12} + \Gamma_2\, y_{12}) \frac{\omega_x^2}{4(\Delta\omega_0^2 + \Gamma_2^2)} \end{aligned} \tag{3.87}$$

with no reference to the optical coherences.

Rate constants

We rewrite these equations as

$$\dot{x}_{12} = -\gamma\, x_{12} + \delta\, y_{12} \qquad \dot{y}_{12} = -\delta\, x_{12} - \gamma\, y_{12} \tag{3.88}$$

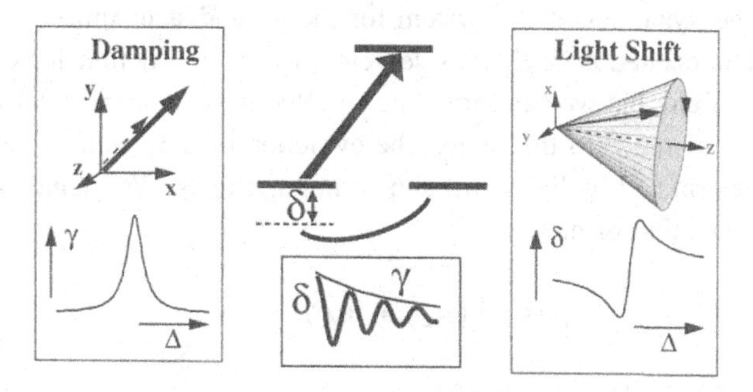

Figure 3.31. Light shift δ and damping γ of sublevel coherences, induced by a laser field coupling to an optical transition.

where

$$\delta = \Delta\omega_0 \frac{\omega_x^2}{4(\Delta\omega_0^2 + \Gamma_2^2)} \qquad \gamma = \Gamma_2 \frac{\omega_x^2}{4(\Delta\omega_0^2 + \Gamma_2^2)} \tag{3.89}$$

These equations describe a damped oscillation with frequency δ and decay rate γ. The general solution is

$$x_{12} = A\cos(\delta t - \phi)\,\mathrm{e}^{-\gamma t} \qquad y_{12} = A\sin(\delta t - \phi)\,\mathrm{e}^{-\gamma t} \tag{3.90}$$

where amplitude A and phase ϕ are determined by the initial conditions. The frequency δ represents the light shift, i.e., the angular frequency of the precession caused by the optical field, and γ the associated relaxation rate.

Figure 3.31 summarises these effects of the laser on the ground-state coherences. The field coupling to the transition from one ground-state sublevel to the excited state induces a damping γ and a precession around the z axis at a rate δ. This precession appears like a shift of the level to which the radiation field couples. The shift δ has a dispersionlike dependence on the laser frequency detuning $\Delta\omega_0$, whereas the damping effect represented by γ has an absorptionlike dependence. The two form a Hilbert transform pair, just like the absorption and dispersion functions of the optical transition. The common factor for the two terms can be interpreted as a mixing coefficient between the ground-state and optical coherences.

3.5.5 Ground-state dynamics

Since the population of the excited state as well as the optical coherences remains small, it is often interesting to disregard the existence of level $|3\rangle$ and formulate the dynamics of the system for the ground-state sublevels $|1\rangle$ and $|2\rangle$ alone. The excited state $|3\rangle$ thus serves a bootstrap function for coupling the radiation field, but we can formulate an effective evolution that no longer needs direct reference to this state. The evolution of the resulting two-level system is determined by light shift and damping effects. We include them in a modified equation of motion

$$\dot{\rho} = -\mathrm{i}\,[\mathcal{H}_{\mathrm{eff}},\rho] + \hat{P}\rho + \hat{\Gamma}\rho \tag{3.91}$$

where the effective Hamiltonian is

$$\mathcal{H}_{\mathrm{eff}} = -\delta\, S_z^{(12)} \tag{3.92}$$

with the light shift δ as calculated above. The superoperator $\hat{P}$ describes the optical pumping:

$$\hat{P}\rho = k(1 - z_{12})\, S_z^{(12)} \tag{3.93}$$

and $\hat{\Gamma}$ the relaxation effects:

$$\hat{\Gamma}\rho = -\gamma\,(x_{12}\, S_x^{(12)} + y_{12}\, S_y^{(12)}) - \gamma_0\,(x_{12}\, S_x^{(12)} + y_{12}\, S_y^{(12)} + z_{12}\, S_z^{(12)}). \tag{3.94}$$

Here, the rate γ_0 describes damping effects that do not depend on the interaction with the laser field. An important and unavoidable contribution to these effects is loss of atoms by diffusion. More details of the dynamics that result from this equation of motion are given in Chapter 8.

4
Internal degrees of freedom

The two- and three-level model systems considered in the preceding sections can describe a wide range of phenomena. On the other hand, these models are purely mathematical constructs that lack physical content. The connection to physical systems is arbitrary and must be made separately in each case. In this chapter, we take the first step in this direction: We discuss the internal degrees of freedom that an atom has available. By far the most important ones are the electric dipole moment, which is responsible for the coupling to external fields, and the angular momentum, which determines the selection rules for the optical transitions and couples to magnetic fields.

4.1 Rotational symmetry

4.1.1 Motivation

The number of energy levels that contribute to the dynamics of a quantum mechanical system is a direct measure of the number of degrees of freedom required for a full description of the system. The systems in which we are interested always include electric dipole moments – the degrees of freedom that couple to the radiation field. The second most important contribution is the magnetic dipole moment associated with the atomic angular momentum. Electric and magnetic dipole moments are those degrees of freedom that couple to external fields. Other degrees of freedom do not couple strongly to external fields but they may still modify the behaviour of the system and its optical properties.

For this step towards a more realistic picture of atomic systems, we assume that we can distinguish between internal and external degrees of freedom of the atoms. The external degrees include position and momentum of the atom. The assumption that we can discuss these variables separately implies that the position and motion of the atoms do not change significantly during absorption and emission of a photon. It allows us to describe the atom in a coordinate system that is fixed to the atom rather than to the laboratory.

Without external fields, the equations of motion that describe the behaviour of physical systems are always isotropic: there is no preferred direction in space. For a system that evolves under a linear Schrödinger equation, the spherical symmetry of the Hamiltonian implies that it will retain its initial symmetry. As in many other physical systems, consideration of the symmetry properties of a system often allows a significant reduction of the complexity of the system and sometimes leads to relevant conclusions without the need for solving the equation of motion of the system. The primary consequence of the rotational symmetry is the conservation of angular momentum. More generally, every quantity whose associated operator commutes with the Hamiltonian of the system remains invariant during the evolution.

For systems whose equations of motion are nonlinear, the evolution does not always conserve the initial symmetry. They may undergo spontaneous symmetry breaking. The best known examples are probably those of phase transitions in solid state physics, like the appearance of a magnetisation at the Curie point, which reduces the system's symmetry. Similar cases of spontaneous symmetry breaking are known in molecular systems as Jahn–Teller distortion. In atomic systems, spontaneous symmetry breaking in the absence of external fields is less common. The strong interaction between laser fields and atomic media, however, may produce systems whose evolution follows a nonlinear equation of motion and whose overall symmetry (which may be lower than spherical) can be spontaneously broken.

Group theory provides the mathematical formalism for the analysis of symmetry properties, and we will use many of its results. The goal of this chapter is not, however, a rigorous mathematical analysis of symmetry properties, but to provide a summary of the situations where the consideration of symmetry properties is helpful and to introduce some of the group theoretical tools that are useful for the analysis of specific physical systems.

4.1.2 Rotations around a single axis

Types of rotations

Rotations may act on objects or on the coordinate systems used to describe these objects. Although it appears natural to consider the coordinate system as fixed and rotate the object, it is often mathematically more convenient to apply the rotation to the coordinate system.

The two kinds of rotations are intimately connected. As Figure 4.1 shows, a clockwise rotation of an object changes its coordinates with respect to a

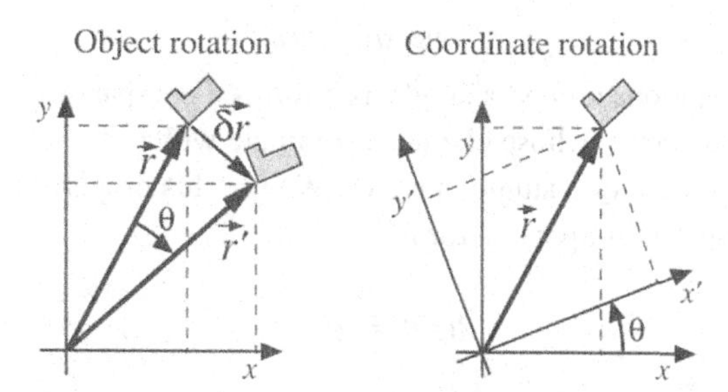

Figure 4.1. Rotation of an object (left) versus rotation of the coordinate system (right). The rotation axis z is perpendicular to the plane of the drawing.

fixed axis system in the same way as a counterclockwise rotation of the coordinate system that leaves the object fixed.

The objects whose rotation properties we discuss are wavefunctions $\Psi(\vec{r})$. We write a rotation of the wavefunction by an angle θ around the z axis as $\Psi'(\vec{r}) = P_{\mathrm{R}}(z,\theta)\ \Psi(\vec{r})$. To find its rotation properties, we start with a rotation of its argument, the coordinate vector $\vec{r}$. A rotation of the coordinate system changes it into

$$\vec{r}\,' = \begin{pmatrix} x' \\ y' \\ z' \end{pmatrix} = R(z,\theta)\vec{r} = \begin{pmatrix} \cos\theta & \sin\theta & 0 \\ -\sin\theta & \cos\theta & 0 \\ 0 & 0 & 1 \end{pmatrix} \begin{pmatrix} x \\ y \\ z \end{pmatrix} \tag{4.1}$$

The rotation of the wavefunction and the transformation of its argument depend on each other: If we rotate the object and the coordinate system simultaneously, the mathematical form of the function should remain invariant,

$$P_{\mathrm{R}}(z,\theta)\ \Psi(R(z,\theta)\vec{r}) = \Psi'(\vec{r}\,') = \Psi(\vec{r}) \tag{4.2}$$

We can use this relation to describe object rotations, whose mathematical form is not always obvious, in terms of coordinate rotations. If we multiply equation (4.2) from the left with $P_{\mathrm{R}}(z,-\theta)$, noting that $P_{\mathrm{R}}(z,-\theta)\ P_{\mathrm{R}}(z,\theta) = \mathbb{1}$, we obtain

$$\Psi(R(z,\theta)\vec{r}) = P_{\mathrm{R}}(z,-\theta)\ \Psi(\vec{r}) = \Psi(\vec{r}\,') \tag{4.3}$$

This relation is a mathematical representation of the situation represented graphically in Figure 4.1: Instead of a clockwise rotation of the object by an angle θ, we may apply a counterclockwise rotation to the coordinate system.

Group structure

The set of all rotations around a single axis forms a one-parameter Lie group, i.e., a continuous group whose elements commute with each other and can be written as the powers of a single operator. We use this property to write a rotation by an angle θ in operator form,

$$R(z,\theta) = e^{\mathrm{i}\theta J_z} \tag{4.4}$$

The operator J_z is the generator of the group. We find its matrix representation from the finite rotations written above as

$$J_z = -\mathrm{i}\,\mathrm{e}^{-\mathrm{i}\theta J_z}\frac{\partial}{\partial\theta}\mathrm{e}^{\mathrm{i}\theta J_z} = -\mathrm{i}\,R(z,-\theta)\frac{\partial}{\partial\theta}R(z,\theta) = \begin{pmatrix} 0 & -\mathrm{i} & 0 \\ \mathrm{i} & 0 & 0 \\ 0 & 0 & 0 \end{pmatrix} \tag{4.5}$$

If we write the vector $\vec{r}$ in polar coordinates (r, β, φ), we find that

$$J_z\vec{r} = \begin{pmatrix} 0 & -\mathrm{i} & 0 \\ \mathrm{i} & 0 & 0 \\ 0 & 0 & 0 \end{pmatrix}\begin{pmatrix} r\sin\beta\cos\varphi \\ r\sin\beta\sin\varphi \\ r\cos\beta \end{pmatrix} = \begin{pmatrix} -\mathrm{i}\,r\sin\beta\sin\varphi \\ \mathrm{i}\,r\sin\beta\cos\varphi \\ 0 \end{pmatrix} = \mathrm{i}\,\frac{\partial\vec{r}}{\partial\varphi} \tag{4.6}$$

The derivative on the right-hand side of this equation is known from the theory of angular momentum: The generator of the rotations around a single axis is, up to a factor $-\mathrm{i}$, the angular momentum operator in the direction of this axis.

Operators $R(z,\theta)$ with different angle θ commute and have the same set of eigenfunctions. These eigenfunctions have the general form

$$f(x, y, z) = f(r, z)\,\mathrm{e}^{\mathrm{i}m\varphi} \tag{4.7}$$

with

$$x = r\cos\varphi \qquad y = r\sin\varphi \tag{4.8}$$

A rotation of such a function by an angle θ changes it into

$$P_{\mathrm{R}}(z,\theta)\,f(\vec{r}) = f(R(z,-\theta)\vec{r}) = f(r, z)\,\mathrm{e}^{\mathrm{i}m(\varphi+\theta)} = f(\vec{r})\mathrm{e}^{\mathrm{i}m\theta} \tag{4.9}$$

For a rotation by an angle θ, these functions have eigenvalues $\mathrm{e}^{\mathrm{i}m\theta}$.

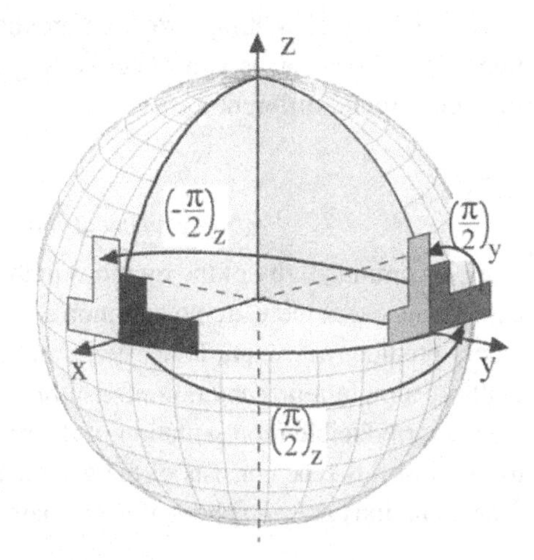

Figure 4.2. Rotation around the x axis as a sequence of three rotations around the z, y and $-z$ axes.

4.1.3 Rotations in three dimensions

Euler angles

The rotations in three-dimensional space form a group; a single rotation can transform any object into any other identical object. Three parameters are required to specify a single rotation. One possible choice uses two parameters to specify the orientation of the rotation axis and the third for the rotation angle. However, the most popular choice uses the Euler angles. They specify a sequence of three rotations around two axes. Rotating an object by

$$R_E(\alpha, \beta, \gamma)\, f(\vec{r}) = f[R(z,-\alpha)\, R(y,-\beta)\, R(z,-\gamma)\vec{\boldsymbol{r}}] \tag{4.10}$$

implies, in terms of coordinate rotations, first a rotation by $-\gamma$ around the z axis, then a rotation by $-\beta$ around the y axis, and finally a rotation by $-\alpha$ around the z axis.

This sequence of three rotations around the z and y axes permits the description of any rotation in three-dimensional space. Figure 4.2 shows as an example a $\pi/2$ rotation around the x axis: The black L-shaped object close to the x axis is rotated first by $\pi/2$ around the z axis into the dark grey object near the y axis. A $\pi/2$ rotation around the y axis brings it to the medium grey position; from here a $-\pi/2$ rotation around the z axis brings it back close to the x axis, this time in the light grey position. The same overall rotation would have resulted from a rotation by $\pi/2$ around the x axis. Mathematically, this identity can be expressed for arbitrary rotation angles θ as

$$P_R(x,\theta) = R_E(\pi/2,\ \theta,\ -\pi/2) \tag{4.11}$$

This possibility of writing a rotation around one axis by a sequence of rotations around other axes is a direct consequence of the noncommutativity of rotations around different axes.

Spherical harmonics

The noncommutativity of the rotations in three-dimensional space also has the consequence that there are no common eigenfunctions for all rotations. Group theory predicts, however, the existence of sets of functions that provide a particularly simple representation of arbitrary rotations. These sets of functions form an irreducible representation of the rotation group. The most commonly used set of this type consists of the spherical harmonics $Y_{lm}(\theta,\phi)$. The lowest order functions are, in spherical coordinates,

l	m	Y_{lm}
0	0	$\sqrt{\frac{1}{4\pi}}$
1	0	$\sqrt{\frac{3}{4\pi}}\cos\theta$
1	± 1	$\mp\sqrt{\frac{3}{8\pi}}\sin\theta\, e^{\pm i\phi}$
2	0	$\sqrt{\frac{5}{4\pi}}\frac{1}{2}(3\cos^2\theta - 1)$

and higher order functions can be found in the literature (Brink and Satchler 1962; Edmonds 1974; Weissbluth 1978). The definitions of these spherical harmonics are not unique. In particular, they can be multiplied with arbitrary phase factors without changing the structure of their rotation properties. They form the basis for the rotation matrices, however, and the rotation matrix elements depend on the phase factors chosen for the spherical harmonics. The convention used here is the Condon–Shortley convention.

Figure 4.3 shows the axially symmetric spherical harmonics, which are best known as the s, p_z and d_z^2 orbitals of the hydrogen atom.

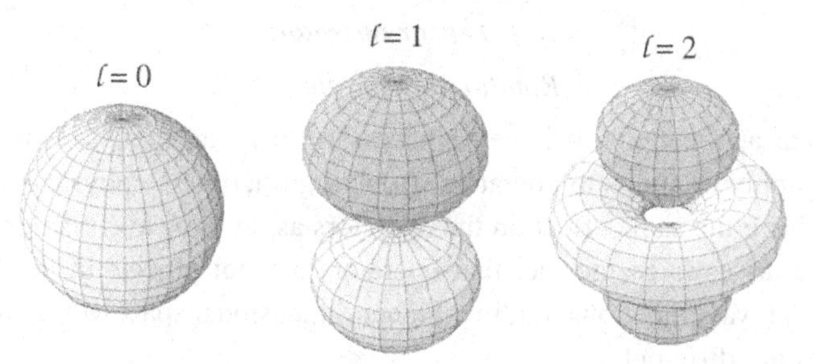

Figure 4.3. The axially symmetric spherical harmonics of rank 0, 1, 2. The drawing shows isosurfaces.

Rotation of spherical harmonics

Spherical harmonics are eigenfunctions of rotations around the z axis

$$P_R(z,\alpha)\, Y_{lm}(\theta,\phi) = e^{im\alpha}\, Y_{lm}(\theta,\phi) \tag{4.12}$$

Rotations around the y axis are described by Wigner rotation matrices, which depend on the rank l of the function

$$P_R(y,\beta)\, Y_{lm}(\theta,\phi) = \sum_{m'} d^{(l)}_{m'm}\,(\beta)\, Y_{lm'}(\theta,\phi) \tag{4.13}$$

The rotated function is thus no longer a simple multiple of the original function; instead, the rotation mixes functions with the same rank l and different order m. General rotations can be parametrised in terms of Euler angles and written as

$$\begin{aligned} R_E(\alpha,\beta,\gamma)\, Y_{lm}(\theta,\phi) &= \sum_{m'=-l}^{l} \mathscr{D}^{(l)}_{m'm}\,(\alpha,\beta,\gamma) Y_{lm'}(\theta,\phi) \\ &= \sum_{m'=-l}^{l} d^{(l)}_{m'm}\,(\beta)\, e^{i(\alpha m'+\gamma m)}\, Y_{lm'}(\theta,\phi) \end{aligned} \tag{4.14}$$

The elements of the Wigner rotation matrices $d^{(l)}_{m'm}\,(\beta)$ can be found in the literature (Brink and Satchler 1962; Edmonds 1974; Weissbluth 1978).

4.1.4 Tensor operators

Rotation of operators

In practical applications, it is never necessary to rotate functions, but often useful to apply rotations to operators. Applying rotations to physical systems does not have the same effect on the operators as on the wavefunctions. This difference arises from the fact that operators are not objects in the Hilbert space of the wavefunctions, but in a higher dimensional space whose rotation properties are different.

To find the transformation properties of the operators, we use the invariance of expectation values under rotations of the coordinate system: If the expectation value of an operator A for a state Ψ is $\langle A\rangle = \langle\Psi|A|\Psi\rangle$, we must get the same result if we perform the calculation in a different coordinate system $\langle A\rangle = \langle\Psi'|A'|\Psi'\rangle$, where the primed quantities refer to the rotated system. We may, e.g., rotate the coordinate system by $e^{-i\theta J_q}$. The wavefunction then transforms as $|\Psi'\rangle = e^{i\theta J_q}|\Psi\rangle$ and $\langle\Psi'| = \langle\Psi|e^{-i\theta J_q}$ and we must have

$$\langle A\rangle = \langle\Psi'|A'|\Psi'\rangle = \langle\Psi|\, e^{-i\theta J_q}\, A'\, e^{i\theta J_q}|\Psi\rangle = \langle\Psi|A|\Psi\rangle \quad (4.15)$$

so that the rotated operator must be

$$A' = e^{i\theta J_q}\, A\, e^{-i\theta J_q} \quad (4.16)$$

Transformations of this type were used several times in Chapters 2 and 3 to transform operators into interaction representations.

Irreducible tensor operators

As for the rotation of functions that we considered before, we may now look for operators that show particularly simple transformation properties under rotations. According to group theory, this set of basis operators must be an irreducible representation of the group. It consists of the spherical tensor operators $T_q^{(k)}$, which transform like the spherical harmonics under rotation:

$$R_E(\alpha,\beta,\gamma)\, T_q^{(k)}\, R_E^{-1}(\alpha,\beta,\gamma) = \sum_{q'=-k}^{k} \mathcal{D}_{q'q}^{(k)}(\alpha,\beta,\gamma)\, T_{q'}^{(k)} \quad (4.17)$$

Rotations around the z axis correspond again to multiplication with a phase factor

$$R_E(\alpha,0,0)\, T_q^{(k)}\, R_E(-\alpha,0,0) = e^{iq\alpha}\, T_q^{(k)} \quad (4.18)$$

The order q of the tensor operator thus appears as a multiplier of the rotation angle: The operator is unchanged for rotations by $2\pi/q$ if $q \neq 0$. Tensor operators of order $q = 0$ are invariant under rotations around the z axis: They represent axially symmetric quantities.

Important examples for spherical tensor operators are the spherical vector components, i.e., the symmetry-adapted components of vectors like the electric dipole moment. A vector is a tensor of rank 1 with the three components

$$r_0 = \sqrt{\frac{3}{4\pi}} \cos\theta, \qquad r_{\pm 1} = \mp \sqrt{\frac{3}{8\pi}} \sin\theta \, \mathrm{e}^{\pm \mathrm{i}\phi} \tag{4.19}$$

They are particularly useful in axially symmetric systems, since they are, in contrast to the cartesian components, eigenfunctions of the rotation around the z axis. In terms of cartesian components, they are

$$\mathrm{r}_0 = \sqrt{\frac{3}{4\pi}} \frac{z}{r} \qquad r_{\pm 1} = \mp \sqrt{\frac{3}{8\pi}} (x \pm \mathrm{i}y) \tag{4.20}$$

Since the generator of the rotations is the angular momentum operator J_z, they are also eigenfunctions of the angular momentum with eigenvalues $\lambda_m = \mathrm{e}^{im\alpha}$ $(m = 0, \pm 1)$.

The Wigner–Eckart theorem

The Wigner–Eckart theorem (Eckart 1930; Wigner 1931) is a very powerful tool for the evaluation of matrix elements of operators with known rotation properties. It allows us to write down the matrix elements of any tensor operator in an irreducible representation, provided we know just a single non-vanishing matrix element. The Wigner–Eckart theorem uses the invariance of expectation values under rotations: Expectation values are numbers, i.e., scalars, which cannot change when a rotation transforms the operator and the basis states simultaneously.

On a practice-oriented level, we may formulate the theorem as follows: Given an operator A that transforms like a vector under rotations, we can decompose it into three irreducible components A_0, $A_{\pm 1}$, whose matrix elements are proportional to those of the spherical vector components r_0, $r_{\pm 1}$. For arbitrary operators, we may decompose them into components $T_q^{(k)}$ whose matrix elements are

$$\langle \alpha, L, m | T_q^{(k)} | \alpha', L', m' \rangle = (-1)^{L-m} \begin{pmatrix} L & k & L' \\ -m & q & m' \end{pmatrix} \langle \alpha, L || T^{(k)} || \alpha', L' \rangle \tag{4.21}$$

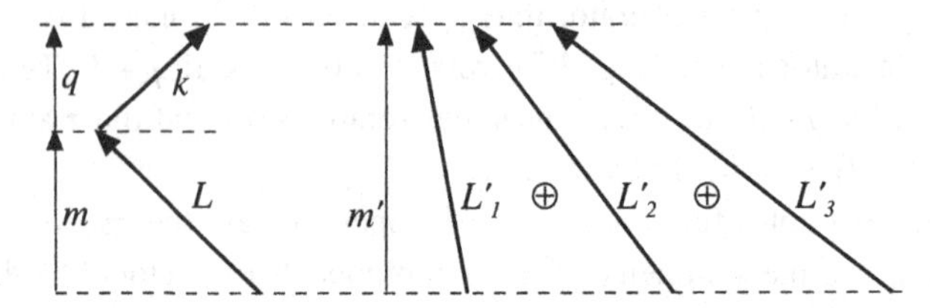

Figure 4.4. Coupling of two angular momentum vectors $|L,m\rangle$, $|k,q\rangle$ yields three possible total angular momenta $|L_i, m'\rangle$, i=1,2,3.

where the quantity in parentheses represents a $3J$ symbol. It denotes a symmetrised form of the coupling coefficients for angular momentum vectors.

Figure 4.4 shows schematically how two angular momentum vectors with length L, k, and z components m, q combine. Their sum can have a total angular momentum $|L - k| \leq L' \leq L + k$, and the z component of the vector sum must be the sum of the z components, $m' = m + q$. The $3J$ symbol determines the weight of the different possible vectors. Numerical values or the general definition of the $3J$ symbols can be found in the literature (Brink and Satchler 1962; Edmonds 1974; Weissbluth 1978). Among the most important properties of $3J$ symbols, we note here only that their value is zero unless $-m + q + m' = 0$. This property alone allows us to calculate the vast majority of matrix elements.

The $3J$ symbol contains all the orientation-dependent information of the operator. The reduced operator element $\langle\alpha,L||T^{(k)}||\alpha',L'>$ appearing in equation (4.21) contains the orientation-independent information specific for the particular operator. The reduced operator does not depend on the orientational quantum number m, but possibly on additional quantum numbers, which we summarise here with the parameter α. The Wigner–Eckart theorem cannot relieve us completely from the task of calculating matrix elements, but it directly gives us their ratios. For many purposes, this is already sufficient. In all other cases, we have to calculate only a single nonvanishing matrix element, and we have the freedom to choose the one for which the calculation is easiest. Specific examples are discussed in Section 4.4. Proofs of this theorem can be found in the literature (Brink and Satchler 1962; Edmonds 1974; Weissbluth 1978).

4.1.5 Hamiltonian and Schrödinger equations

Symmetry properties

The eigenfunctions of the free atom Hamiltonian, i.e., the atomic wavefunctions, must reflect the rotational symmetry of the Hamiltonian. Not all wavefunctions have the full rotational symmetry, but the nonsymmetric eigenfunctions appear in sets whose members form an irreducible representation of the rotation group; it is then not the individual eigenstate that reflects the symmetry of the Hamiltonian, but the symmetry related set. An important example is the collection of electron orbitals of the hydrogen atom: Only *s*-type orbitals have the full rotational symmetry. A single *p*-type orbital has lower symmetry than the Hamiltonian, but the set of three *p* orbitals restores the full symmetry. In the absence of external fields, the members of such a set are degenerate. The dimension of the irreducible representation to which they belong is a lower bound on the multiplicity of the degenerate levels. If external fields that break the rotational symmetry are present, they can lift this degeneracy.

Whenever an operator P commutes with the Hamiltonian of a quantum mechanical system, the eigenfunctions of the Hamiltonian must simultaneously be eigenfunctions of P. This theorem is so important for practical calculations that we include here a short sketch of a proof. If $[P, \mathcal{H}] = 0$, we have

$$P^{-1}\mathcal{H}\, P = \mathcal{H} \tag{4.22}$$

and

$$\langle\Psi_r|\mathcal{H}|\Psi_s\rangle = \langle\Psi_r|P^{-1}\ \mathcal{H}\ P|\Psi_s\rangle = \langle P\Psi_r|\mathcal{H}|P\Psi_s\rangle \tag{4.23}$$

for all Ψ_r, Ψ_s. This requires

$$P\Psi_s = \lambda_s(P)\ \Psi_s \tag{4.24}$$

where $\lambda_s(P)$ is the eigenvalue of Ψ_s with respect to the operator P.

Conservation laws

The commutation between P and $\mathcal{H}$ also implies that the evolution under the Hamiltonian $\mathcal{H}$ conserves the symmetry properties of the system: If it is initially in a symmetric state, it will never end up in an asymmetric state. These properties are valid for all symmetry operations, but the most important for the systems considered here are the rotational symmetries, which leave all free atom Hamiltonians invariant.

Because a single operator generates all rotations around a single axis, these symmetries are not independent of each other. A closer look shows that the relevant conservation law is associated with the generator of the rotation, i.e., the angular momentum operator. Rotational symmetry around a single axis implies conservation of the angular momentum component in the direction of the rotation axis. Full rotational symmetry implies corresponding conservation of all three components of the angular momentum. For isolated atoms, this is always the case, but external fields that break the symmetry remove the invariance of the angular momentum components perpendicular to the symmetry axis.

If only a single field is present, the system retains an axial symmetry and the Hamiltonian is invariant under rotations around the field. The eigenfunctions of the Hamiltonian must then simultaneously be eigenfunctions of the rotation around this axis, which is usually chosen as the z axis. This choice can significantly reduce the complexity of the description; it is important to realise, however, that this choice of quantisation axis is merely a mathematical convenience. In principle, the description can use any coordinate system. The most important fields that occur are optical (laser) fields and static magnetic fields. If a laser field and a magnetic field are present simultaneously and not parallel to each other, the choice of the coordinate system is quite arbitrary and depends on the details of the problem (Omont 1977).

External fields

Even in the presence of external fields, the rotational symmetry may be helpful. The system is still invariant under coordinate rotations if they rotate the system coordinates together with those of the external fields. It is therefore often useful to group the various terms of the Hamiltonian in such a way that each term is also invariant under rotations. As an example, consider the interaction between the electric field and the electric dipole moment of the atom:

$$\vec{E} \cdot \vec{\mu}_e = E_x \, \mu_x + E_y \, \mu_y + E_z \, \mu_z \tag{4.25}$$

The rotational invariance of the general formulation $\vec{E} \cdot \vec{\mu}_e$ is not readily apparent in the cartesian representation. A symmetry-adapted representation may use the basis

$$E_0 = E_z \qquad E_+ = -\frac{E_x + \mathrm{i}\, E_y}{\sqrt{2}} \qquad E_- = \frac{E_x - \mathrm{i}\, E_y}{\sqrt{2}} \tag{4.26}$$

$$\mu_0 = \mu_z \qquad \mu_+ = -\frac{\mu_x + \mathrm{i}\, \mu_y}{\sqrt{2}} \qquad \mu_- = \frac{\mu_x - \mathrm{i}\, \mu_y}{\sqrt{2}} \tag{4.27}$$

The interaction is then

$$\vec{E} \cdot \vec{\mu}_e = E_0 \, \mu_0 - E_+ \, \mu_- - E_- \, \mu_+ \tag{4.28}$$

This representation has the advantage that for rotations around the z axis the invariance is manifest not only for the sum, but for each of the three terms individually. As an example, consider

$$P_R(z,\theta)\,(E_+\,\mu_-) = (P_R(z,\theta)\,E_+)\,(P_R(z,\theta)\,\mu_-) = E_+ e^{i\theta}\,\mu_- e^{-i\theta} = E_+\,\mu_- \quad (4.29)$$

i.e., the phase factor acquired by one term during the rotation cancels the phase factor acquired by the other term. More generally, the Hamiltonian is written in the form

$$\mathcal{H} = \sum_{k/k'q} A_q^{(k)}\, B_{-q}^{(k')} \quad (4.30)$$

where A is an operator acting on the atomic system and B is a function of the spatial variables describing the field.

4.2 Angular momentum

The two most important internal degrees of freedom of atomic media are the electric dipole moment and the magnetic dipole moment associated with the atomic angular momentum. Angular momentum exists in various forms in the atomic medium, as well as in the radiation field. In the atomic system, we may distinguish between electron orbital and electron spin angular momentum as well as nuclear spin. In the radiation field, we again find spin and orbital angular momentum. We start the discussion with the radiation field.

4.2.1 Radiation field

In classical electrodynamics, electromagnetic radiation is a vector wave. The electric field at every point in space is a three-dimensional vector that varies as a function of time. This contrasts with scalar waves like sound waves, where the quantity of interest is a scalar – the pressure in the case of ordinary sound. The electric field vector cannot assume an arbitrary direction: For a plane wave propagating through an isotropic, source-free medium, the solutions of the Maxwell equations are restricted to the plane perpendicular to the direction of propagation. For a start, we will restrict ourselves to this case and discuss the basis states that span the space of degenerate solutions.

Total angular momentum

In classical mechanics, the angular momentum of a point particle is $\vec{r} \times \vec{p}$, where $\vec{r}$ is the position and $\vec{p}$ the linear momentum of the particle. Using the correspondence principle, we can apply this definition directly to the radiation field, where the linear momentum is $\vec{E} \times \vec{B}$. In classical electrodynamics, the angular momentum of a laser beam is therefore

$$\vec{M} = \frac{1}{c^2 \mu_0} \int \mathrm{d}v \, \vec{r} \times (\vec{E} \times \vec{B}) \tag{4.31}$$

where the integration runs over the volume of the laser beam. For a plane wave, the linear momentum $\vec{E} \times \vec{B}$ is parallel to the wavevector, and we expect that the angular momentum in the direction of propagation will vanish. In a beam with finite diameter, however, the wave front is necessarily curved. The linear momentum $\vec{E} \times \vec{B}$ then also has transverse components and the angular momentum can have a longitudinal component (Haus and Pan 1993).

This representation includes the internal as well as the orbital angular momentum of the light. To separate the two contributions, it is helpful to express the magnetic field $\vec{B} = \vec{\nabla} \times \vec{A}$ in terms of the vector potential $\vec{A}$. The angular momentum is then

$$\vec{M} = \frac{1}{c^2 \mu_0} \int \mathrm{d}v \left[\vec{E} \times \vec{A} + \sum_i E_i (\vec{r} \times \vec{\nabla}) A_i \right] \tag{4.32}$$

The first term of the integrand does not depend on the position $\vec{r}$ and is therefore independent of the coordinate system. It represents an internal angular momentum, i.e., a spin contribution to the total angular momentum. The second term, which depends on the coordinate system, is the analogue of the orbital angular momentum. Internal (spin) and orbital angular momentum of the light are independent quantities in the vacuum. The medium, however, may introduce a coupling between them (Dooghin et al. 1992).

Internal

Quantum mechanically, the radiation field consists of photons, which carry one unit of internal angular momentum each ($S = \hbar$). The three possible spin states of the photon, $m = 0, \pm 1$, correspond to three orthogonal polarisation states of the radiation field, which are usually labelled as π, σ_+ and σ_-, respectively. The vanishing rest mass of the photon requires that the $m = 0$ state not be occupied, which results in the classical description in the transversality condition for the optical field, $\vec{k} \cdot \vec{E} = 0$, i.e., $E_z = 0$ if the light propagates along the z axis. The possible polarisation states can be expanded in the two remaining states.

For this purpose, it is helpful to write the usual description of the optical field

$$\vec{E} = (E_x, E_y, 0) = (A_x \, \mathrm{e}^{\mathrm{i}\phi_x}, A_y \, \mathrm{e}^{\mathrm{i}\phi_y}, 0) \tag{4.33}$$

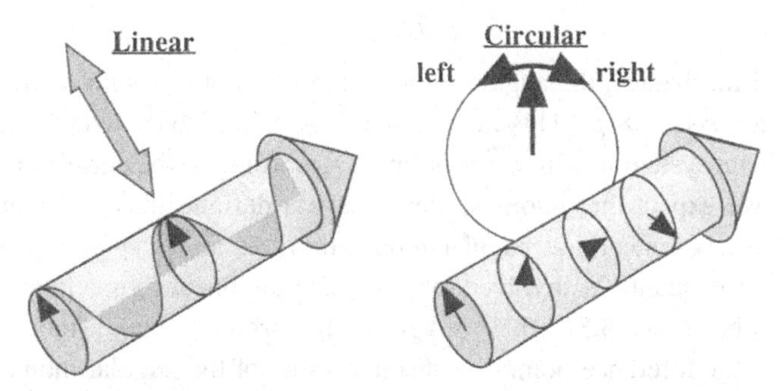

Figure 4.5. Possible polarisations of a laser beam; in the beam shown at the right-hand side, the direction of the E-field vector changes along the beam.

in a spherical basis (D'Yakonov 1965) with

$$\vec{E} = (0, E_+, E_-) = (0, A_+ \, e^{i\phi_+}, A_- \, e^{i\phi_-}) \tag{4.34}$$

In both bases, A_x and A_y (A_+ and A_-) are the amplitudes along the respective coordinate axes and ϕ_x and ϕ_y (ϕ_+ and ϕ_-) the corresponding phases. The longitudinal component E_z (E_0) vanishes for a plane wave. The amplitudes are related by

$$E_\pm = \frac{1}{\sqrt{2}} (\mp E_x + iE_y) \tag{4.35}$$

$$E_x = -\frac{1}{\sqrt{2}} (E_+ - E_-) \qquad E_y = -\frac{i}{\sqrt{2}} (E_+ + E_-) \tag{4.36}$$

The field components of the spherical basis represent circularly polarised light, where the field vector rotates along the direction of propagation as well as at a fixed location as a function of time (see Figure 4.5). They correspond to the eigenstates of the photon angular momentum.

Retardation plates, which change the polarisation of the light, change the spin angular momentum of the photons; this implies that their own angular momentum must change simultaneously. This consequence can be verified experimentally: The polarised light impinging on a retardation plate exerts a measurable torque on the material (Beth 1936; Carusotto, Fornaca and Polacco 1968). Scattering of circularly polarised light by particles also changes the spin angular momentum of the light and therefore exerts a torque on the body (Chang and Lee 1985).

Orbital

In classical mechanics, the angular momentum of an object with momentum $\vec{p}$ at position $\vec{r}$ is $\vec{r} \times \vec{p}$. This implies that it is defined only with respect to the coordinate system in which $\vec{r}$ is defined. According to the correspondence principle, we expect that photons, which have a definite linear momentum, should also have an orbital angular momentum. As an order of magnitude estimate of the quantities involved, we calculate the linear momentum of a 1 mW laser beam as $3.3 \cdot 10^{-12}$ mkg/sec. Its orbital angular momentum depends on the reference point; the absolute value of the angular momentum itself is therefore not physically meaningful but an angular momentum change, which can occur, e.g., by a displacement of a laser beam. To get an idea of the orders of magnitude, consider that the displacement of a photon by its wavelength corresponds to an angular momentum change by h. Displacing a laser beam by its diameter corresponds to a change of the orbital angular momentum in the range of 10^{-17}–10^{-13} m^{2}kg/sec.

The orbital angular momentum of a Gaussian beam, integrated over its diameter, vanishes if the symmetry axis passes through the origin of the coordinate system. Different beam shapes, however, e.g., Laguerre–Gaussian modes, carry orbital angular momentum with respect to the symmetry axis of the beam. Conversions between different beam shapes again change the orbital angular momentum and conversely the light exerts a torque on the material that initiates the conversion (Abramochkin and Volostnikov 1991; Allen et al. 1992; Kristensen and Woerdman 1994a).

4.2.2 Atomic angular momentum

In the physical situation under discussion here, the optical radiation couples to the transition from the electronic ground state to an electronically excited state. Both states may consist of a number of substates with different angular momenta, which can be attributed to different sources: orbital angular momentum of the electron, conventionally designated by the letter $\vec{L}$, spin angular momentum of the electron, designated as $\vec{S}$, and nuclear spin, designated $\vec{I}$. These angular momenta couple to a total angular momentum $\vec{F} = \vec{J} + \vec{I} = \vec{L} + \vec{S} + \vec{I}$.

Orbital

The Bohr–Sommerfeld model of the atom, where the electrons orbit the atomic nucleus in planetary orbits, clearly shows that the electrons have orbital angular momentum. The quantum mechanical formalism, where the electron orbitals are described by spatially extended wavefunctions without direct physical meaning, makes the visualisation considerably more difficult. The rotational symmetry discussed above requires, however, that the eigenfunctions of the atomic Hamiltonian simultaneously be eigenfunctions of the angular momentum operator. This forms the basis for the classification of electron orbitals as states with definite angular momentum. Depending on the principal quantum number n, states with orbital angular momentum $l = 0, \ldots, n$ exist, which are labelled as $s, p, d, \ldots$ orbitals (Weissbluth 1978). The existence of electronic orbitals with vanishing angular momentum is a direct contradiction to the Bohr–Sommerfeld model: In the classical picture, an electron with no angular momentum would fall into the nucleus.

Electron spin

To keep the formalism simple, we restrict the discussion to single electron systems, i.e., atoms with only a single electron that is not part of a filled shell. The electronic contribution to the spin angular momentum is then always $S = 1/2$, and the multiplicity of the system is $2(2L + 1)$. The total angular momentum J of the electron is the vector sum of the orbital and spin angular momentum, $\vec{J} = \vec{L} + \vec{S}$.

The two contributions to the angular momentum "feel" each other. We do not discuss the details of the interaction (Weissbluth 1978), but merely write down the form of the corresponding contribution to the Hamiltonian

$$\mathcal{H}_{\mathrm{F}} = f\vec{L} \cdot \vec{S} \tag{4.37}$$

where f is the coupling constant. This interaction lowers the symmetry of the Hamiltonian. In its absence, the Hamiltonian is invariant under rotations of $\vec{L}$ and $\vec{S}$, but in the presence of spin–orbit interaction, the Hamiltonian remains invariant only if they are rotated jointly. This lowering of the symmetry lifts the degeneracy of the levels. Their energy depends now on the relative orientation of the two component angular momenta.

Figure 4.6 illustrates the spin–orbit interaction for the case of atomic sodium. The electronic ground state is an s state and has no orbital angular momentum, whereas the excited $3p$ state has an angular momentum $L = 1$. The spin $S = 1/2$ of the electron couples with the orbital angular momentum. For parallel orientation of the two contributions, the total angular momentum is $J = 3/2$ and for antiparallel orientation $J = 1/2$. The splitting between these two states is $3f/2$, as can be found from the eigenvalues of the Hamiltonian of equation (4.37). The electronic ground state has an electronic angular momentum of $J = 1/2$, from the spin alone.

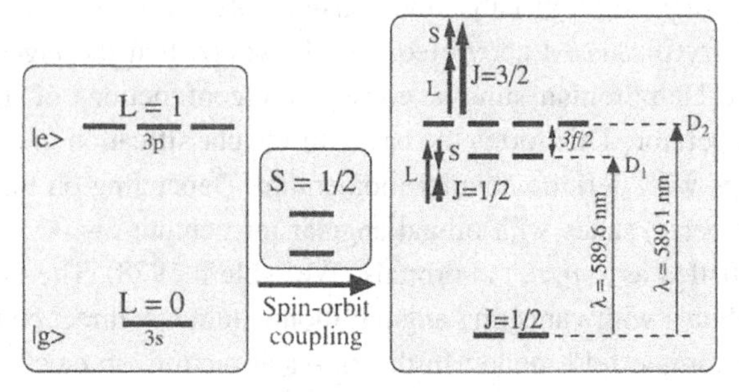

Figure 4.6. Spin–orbit interaction in atomic sodium.

The splitting between the two electronically excited angular momentum states was discovered spectroscopically a long time ago. It leads to the splitting between the D_1 and D_2 resonance lines of the alkali, which is known as fine structure. In the case of sodium, this splitting is of the order of 500 GHz and considerably larger for the heavier alkali.

Nuclear spin

Besides the electronic contribution to the total angular momentum, the spin angular momentum of the atomic nucleus can play an important role. It is usually designated as $\vec{I}$. Nuclear spin values range from 0 to 7, depending on the atomic species and isotope. The total angular momentum $\vec{F}$ of the atomic system is the sum of the electronic contribution $\vec{J}$ and the nuclear spin contribution, $\vec{F} = \vec{I} + \vec{J}$. Again, there is an interaction between the two types of angular momenta, which lifts the degeneracy of the angular momentum states, depending on the relative orientation of the two angular momenta. Again, we do not discuss the details of the interaction, but merely write the simplest form of the resulting Hamiltonian as

$$\mathcal{H}_{\text{HF}} = A\vec{J} \cdot \vec{I} \tag{4.38}$$

The hyperfine coupling constant A depends on the electronic state of the system; for the ground state of sodium, it is $A_g = 900$ MHz and for the $P_{1/2}$ excited state 95 MHz. Figure 4.7 summarises the resulting level structure.

Figure 4.7 shows how the coupling between the $J = 1/2$ electronic states with the nuclear spin results in the excited as well as in the ground state in a total of eight states that form two multiplets with total angular momentum $F = 1$ and $F = 2$. The splitting between these states is, for the chosen parameters, two times the hyperfine coupling constant A.

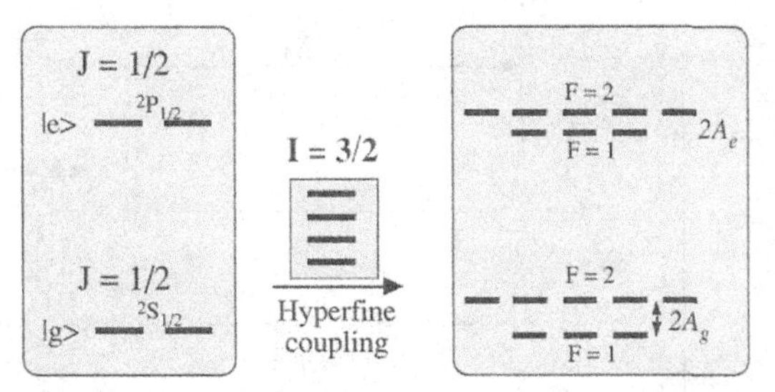

Figure 4.7. Level scheme of the Na D_1 transition showing the effect of the hyperfine interaction.

The hyperfine structure can be observed readily in high resolution laser spectra. Figure 4.8 shows the theoretical spectrum at the top. The larger splitting originates from the ground-state hyperfine splitting, the smaller from the excited state. The additional lines in the experimental spectrum at the bottom are cross-over resonances (Demtröder 1991) that appear at the mean frequency of two "real" resonances when saturation spectroscopy is used to eliminate the Doppler effect.

4.3 Multipole moments

4.3.1 Multipole expansion

Operator basis

An electronic state with a total angular momentum F has $2F + 1$ angular momentum states. A complete description of such a system, e.g., through an enumeration of its density operator elements, requires the specification of $(2F + 1)^2$ independent variables. The angular momentum, which is a vector quantity, has only three independent variables, i.e., the components in the direction of the axes x, y, and z. Most atomic systems, therefore, have additional degrees of freedom. We are free to specify the way in which we want to describe those systems by choosing a specific basis for the expansion of the density operator. One choice that is convenient whenever rotational symmetry plays a role is the set of the tensor operators $T_q^{(k)}$, which transform irreducibly under rotations. In this basis, the density operator expansion is

$$\rho = \sum_{k=0}^{2F} \sum_{q=-k}^{k} c_{kq} \, T_q^{(k)} \tag{4.39}$$

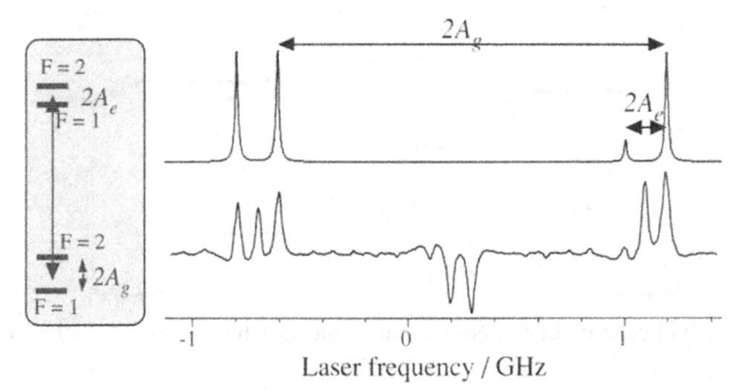

Figure 4.8. Calculated (top) and measured (bottom) spectrum of the Na D_1 transition showing the hyperfine structure of the $S_{1/2}$ ground and the $P_{1/2}$ excited state. The inset shows the relevant part of the level structure.

where the c_{kq} are the expansion coefficients. As each tensor has $2k + 1$ components, the total number of independent variables is

$$\sum_{k=0}^{2F} 2k + 1 = (2F + 1)^2 \tag{4.40}$$

These operators therefore form a complete set for the description of the system.

It may be surprising that tensors up to rank $2F$ are required for a system with angular momentum F. For a heuristic interpretation of this situation, consider the transformation properties of the density operator element $|m_F\rangle\langle -m_F|$ (see Figure 4.9) between the states with the highest and lowest quantum number.

Under a rotation $R_P(z,\phi)$ by an angle ϕ around the z axis

$$\begin{aligned} &R_P(z,\phi)|m_F\rangle\langle -m_F|R_P{}^{-1}(z,\phi) \\ &\quad = |e^{im_F\phi}\, m_F\rangle\langle e^{im_F\phi} - m_F| = e^{i2m_F\phi}|m_F\rangle\langle -m_F| \end{aligned} \tag{4.41}$$

the density operator component acquires the phase difference of the two corresponding quantum states, which reaches a maximum of $2F\phi$ for the matrix element considered. It must therefore be a component of a tensor of rank $k = 2F$. Using geometrical arguments, we note that the operator couples states with angular momentum $m_F = \pm F$. To fulfil the Schwarz inequality for the addition of angular momenta, it must have components from $F - F$ to $F + F = 0, \ldots, 2F$.

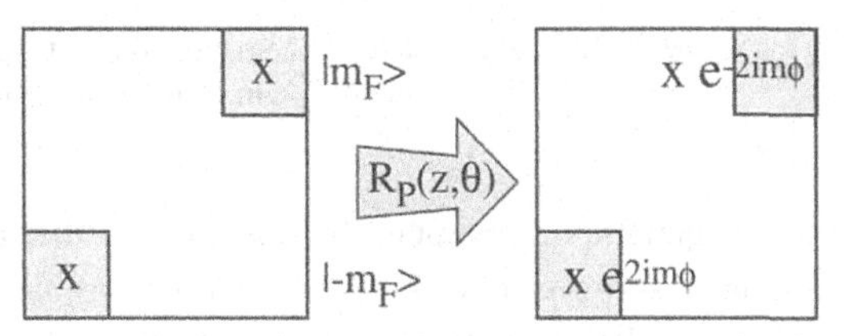

Figure 4.9. The density operator element $|m_F\rangle\langle -m_F|$ acquires a phase of $2m\phi$ during a z rotation by ϕ.

Moments

The physical interpretation of the different terms depends on the specific physical situation, not just on the rotation properties of the operators. Quite generally, however, the only variable that remains invariant under rotations is the total population of the system, which is the trace of the density operator. The total population appears in the multipole expansion of equation (4.39) as the expansion coefficient of the single tensor element with rank $k = 0$. In systems with axial symmetry, the tensor elements $T_0^{(k)}$ with order $q = 0$ are diagonal in the eigenbase of the Hamiltonian. Their expansion coefficients are linear combinations of the populations. As we have seen, the number of levels is $2F + 1$, and since k runs from 0 to $2F$, the $T_0^{(k)}$ exactly form a complete basis for the populations. Since all $T_0^{(k)}$ are invariant under rotations around the symmetry axis, this must hold also for the populations.

The three components with rank $k = 1$ and order $q = (0, \pm 1)$ are the components of the angular momentum, which is always associated with a magnetic dipole moment. The five components with $k = 2$ are known as quadrupole moment or alignment. As a tribute to their importance in magnetooptics, we include a brief discussion of their properties in the following section. The next higher moments are known as octupole and hexadecupole moments. Their orientation dependence is the same as that of the f- and g-type hydrogen orbitals.

4.3.2 Alignment

Apart from the magnetic dipole moment, the $k = 2$ tensor elements play the most important role. The $q = 0$ element, which is known as alignment, has a simple representation as a linear combination of populations.

Figure 4.10 shows such a population distribution for an $F = 1$ system: The populations of the levels with magnetic quantum numbers $m = \pm 1$ are higher than that of level $m = 0$. In systems with higher angular momentum, the contribution of sublevel populations increases proportionally to m^2. They conm = -1 0 1

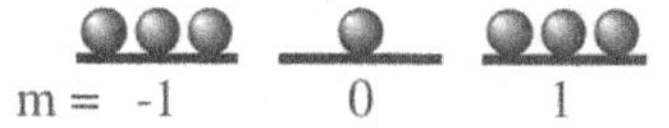

Figure 4.10. Schematic representation of the alignment $T_0^{(2)}$ in terms of populations.

tribute not only to the quadrupole moment, but also to all other axially symmetric moments. In the $F = 1$ system of Figure 4.10, it is possible that only the $m = 1$ level is populated. The populations of such a state may be expanded as

$$\begin{aligned} \begin{pmatrix} \rho_{11} \\ \rho_{22} \\ \rho_{33} \end{pmatrix} = \begin{pmatrix} 1 \\ 0 \\ 0 \end{pmatrix} &= \frac{1}{3}\begin{pmatrix} 1 \\ 1 \\ 1 \end{pmatrix} + \frac{1}{2}\begin{pmatrix} 1 \\ 0 \\ -1 \end{pmatrix} + \frac{1}{6}\begin{pmatrix} 1 \\ -2 \\ 1 \end{pmatrix} \\ &= \frac{1}{\sqrt{3}} T_0^{(0)} + \frac{1}{\sqrt{2}} T_0^{(1)} + \frac{1}{\sqrt{6}} T_0^{(2)} \end{aligned} \tag{4.42}$$

i.e., with simultaneous excitation of the dipole and quadrupole moment (alignment). The operators $T_0^{(k)}$ are normalised such that $\mathrm{tr}\{T_0^{(k)} T_0^{(k')}\} = \delta_{kk'}$.

Figure 4.11 illustrates the "shape" of the alignment, i.e., its behaviour under rotations. The alignment is invariant under all rotations around the z axis and under π rotations perpendicular to the z axis. In terms of populations, π rotations around the x or y axis exchange the populations of the $m = \pm 1$ states, which contribute equally to the quadrupole moment. Its overall orientation dependence is the same as that of a d_z^2 orbital. The other four components of the second rank tensor ($q = \pm 1, \pm 2$) involve coherences between the basis states; their orientation dependence is the same as that of the other d orbitals.

Alignment plays an important role in magnetooptic properties, since its presence makes the material birefringent. On a heuristic level, the possibility of observing alignment in an optical experiment can be understood with the following argument: The interaction of light with an optical medium involves the incident and the scattered photon. Both of them contribute one quantum of angular momentum. Together, these two photons can change the tensorial rank of a density operator component by $\Delta k = 2$. Higher order moments can be excited and detected only through an exchange of multiple photons. A more detailed discussion of this subject is presented in Section 9.3.

4.4 Interaction with external fields

4.4.1 Electric fields

Electric dipole interaction

In classical electromagnetism, Coulombs law describes the interaction of charged particles like electrons and nuclei with external fields. In the semi-

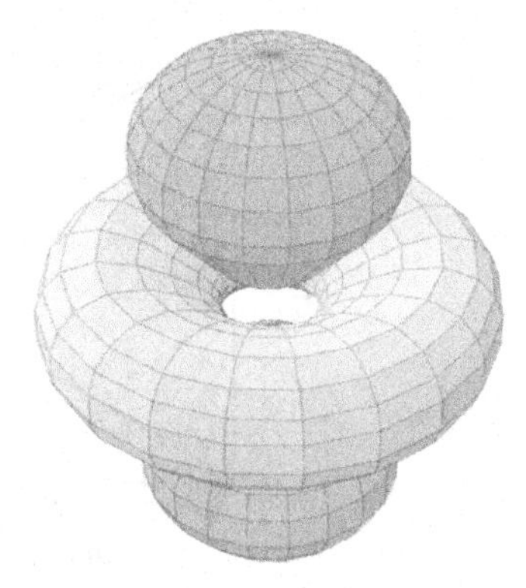

Figure 4.11. Orientation dependence of the $T_0^{(2)}$ tensor component.

classical formalism, the Hamiltonian that describes the interaction between electric fields and atoms can take either of two forms

$$\mathcal{H}_{\text{int}}^{(1)} = -e\,\vec{E} \cdot \vec{r} \qquad \mathcal{H}_{int}^{(2)} = -\frac{e}{m}\vec{p} \cdot \vec{A} \tag{4.43}$$

The first form is obtained through the correspondence principle from the classical description of the interaction between the (classical) electric field $\vec{E}$ and the atomic electric dipole $-e\vec{r}$, where $\vec{r}$ is the position operator of the (single) electron. The second form, which describes a coupling between the electronic momentum $\vec{p}$ and the electromagnetic vector potential $\vec{A}$ can be derived by quantising the equation of motion in the presence of the external vector potential. Expanding it in a multipole series, one finds that for homogeneous fields all but the lowest order term of this series vanish and the two forms of the interaction Hamiltonian become equivalent for most purposes. In very specific applications, differences between the two treatments persist that are due to different gauges of the electromagnetic fields (Lamb, Schlicher and Scully 1987).

Matrix elements

For this analysis, we consider only homogeneous fields and use the electric dipole form of the interaction Hamiltonian. In this form, the quantum mechanical part of the interaction is the position operator $\vec{r}$ of the electron. For the applications that we consider later on, it will be helpful to discuss some symmetry restrictions on the matrix elements of this operator. The first con-

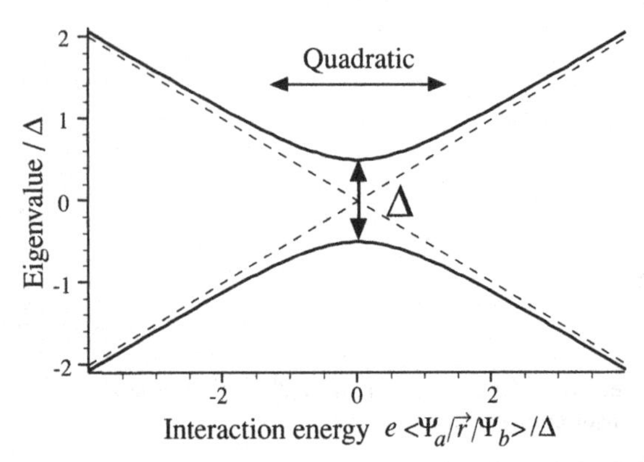

Figure 4.12. Energy of the Stark-shifted states as a function of the field strength. Δ represents the zero-field energy separation.

sideration is parity, an inversion of all spatial coordinates at the origin. Such an inversion converts $\vec{r} = \{x, y, z\}$ into $-\vec{r} = \{-x, -y, -z\}$: Polar vectors like $\vec{r}$ have negative parity; they change sign under inversion. Matrix elements $\langle\Psi_a|\vec{r}|\Psi_b\rangle$, which are scalars, must be invariant under such an operation. Since the parity of a product is the product of the parities, the parities of Ψ_a and Ψ_b must be opposite for nonvanishing matrix elements. Clearly, the matrix elements must therefore vanish for $\Psi_a = \Psi_b$. This is usually expressed by the statement that atoms have no static electric dipole moment.

For the interaction with static electric fields, the absence of diagonal elements implies that the interaction Hamiltonian contains only off-diagonal elements in the eigenbase of the unperturbed Hamiltonian. The Stark effect, i.e., the shift of energy levels by electric fields, is therefore at low fields always quadratic in the field strength. A first-order shift of the energy levels exists only if the interaction energy with the field exceeds the zero-field splitting between the two states Ψ_a and Ψ_b.

Figure 4.12 shows the transition from the quadratic to the linear Stark effect as the electric field strength increases. Here, Δ represents the energy difference between the unperturbed states Ψ_a and Ψ_b. Measurements of this type allow the determination of atomic polarisabilities (Tanner and Wieman 1988).

Electric quadrupole interaction

In nuclei with angular momentum $I > 1/2$, the distribution of the electric charge is not spherically symmetric. The interaction between the nucleus and external fields therefore includes a contribution that depends on the orientation of the atom with respect to the fields. The only relevant term is the so-called nuclear quadrupole coupling, the interaction between the nuclear quadrupole moment and the electric field gradient tensor, which are both second rank tensors. This interaction is the second term in the expansion of the interaction Hamiltonian. In the principal axis system of the electric field gradient tensor, the interaction Hamiltonian is (Cohen and Reif 1957)

$$\mathcal{H}_Q = D\left(I_z^2 - \frac{1}{3} I(I+1)\right) + \frac{\eta}{3}(I_x^2 - I_y^2) \tag{4.44}$$

The coupling constant D depends on the nuclear quadrupole moment, which is specific for every nucleus, and on the $T_0^{(2)}$ component of the electric field gradient tensor. Typical coupling strengths are of the order of a few MHz. The asymmetry parameter η indicates the deviation of the electric field distribution from axial symmetry and depends on the $T_{\pm 2}^{(2)}$ terms of the field gradient tensor.

4.4.2 Magnetic interactions

Zeeman effect

The interaction between a magnetic field $\vec{B}$ and an atom occurs through the magnetic dipole interaction

$$\mathcal{H}_m = -\vec{\mu}_m \cdot \vec{B} \tag{4.45}$$

where $\vec{\mu}_m$ is the magnetic dipole moment. In contrast to $\vec{r}$, the magnetic dipole moment is an axial vector, i.e., it has positive parity. Accordingly, the magnetic dipole moment has nonvanishing matrix elements between states of equal parity. In contrast to the electric dipole moment, this also allows nonvanishing diagonal values $\langle \Psi_a | \vec{\mu}_m | \Psi_a \rangle$. The sources of magnetic dipole moments in atomic systems are the angular momenta: electron orbital, electron spin, and nuclear spin. The interaction between the electron and the external magnetic field is

$$\mathcal{H}_m = -\frac{e\hbar}{2m} \vec{B} \cdot (\vec{L} + 2\vec{S}) = -\mu_B \vec{B} \cdot (\vec{L} + 2\vec{S}) \tag{4.46}$$

where Bohr's magneton is introduced as $\mu_B = e\hbar/2m = 9.27 \cdot 10^{-24}$ A m^2 = $9.27 \cdot 10^{-24}$ J/T = 14 GHz/T in frequency units. The factor 2 for the spin term is only an approximation for the electron g factor, but will be sufficient for our purposes.

Spin–orbit coupling

As emphasised above, the magnetic moments of the atomic system interact not only with external fields, but also with each other. The spin–orbit interaction, discussed in Section 4.2.2, does not commute with the Zeeman interaction specified above, and it is not possible to discuss the two interactions independently. In all cases relevant for us, it will be sufficient to assume that the Zeeman interaction is much weaker than the spin–orbit coupling. We may then neglect the part of the Zeeman interaction that does not commute with the spin–orbit coupling.

The remaining terms cause a splitting of the angular momentum states that depends on the relative orientation of spin and orbital angular momentum and takes the form

$$\mathcal{H}_{\mathrm{m}} = -\mu_{\mathrm{B}}\, g_J\, \vec{B} \cdot \vec{J} \tag{4.47}$$

where the Landé factor is

$$g_J = 1 + \frac{J(J+1) + S(S+1) - L(L+1)}{2\,J(J+1)} \tag{4.48}$$

This factor depends on the relative orientation of the two angular momenta $\vec{L}$ and $\vec{S}$.

As an example, consider the electronic ground state of the alkali, where $J = S = 1/2$ and $L = 0$. The Landé factor is then $g_J = 2$, as expected when the electron spin is the only source of angular momentum. In the excited $^2P_{1/2}$ and $^2P_{3/2}$ states, we find $g_J = 2/3$ and $g_J = 4/3$, respectively. This coincides qualitatively with the naive expectation that the effective interaction strength should decrease when the two magnetic moments $\vec{L}$ and $\vec{S}$ are antiparallel, compared to the case of parallel orientation. For parallel orientation, the magnitude of the coupling strength is intermediate between the values for the orbital angular momentum (i.e., $g = 1$) and for spin ($g = 2$), whereas it is smaller than unity when the two moments are antiparallel. The resulting splitting between neighbouring levels is 28 MHz/mT in the ground state, and 9.3 MHz/mT and 18.7 MHz/mT for the $^2P_{1/2}$ and the $^2P_{3/2}$ states, respectively.

Hyperfine interaction

The interaction between the electronic angular momentum and the nuclear spin does not commute with the Zeeman interaction either. The situation is not quite the same as for the spin–orbit interaction, however, since the Zee-

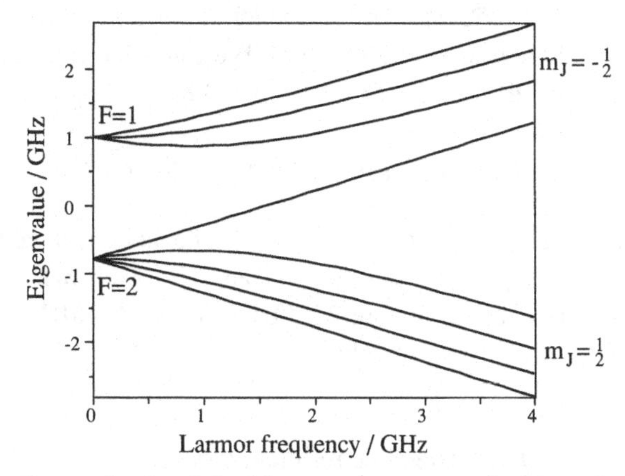

Figure 4.13. Energy levels of the Na ground state as a function of magnetic field strength.

man interaction may be smaller, comparable or even stronger than the hyperfine interaction. Figure 4.13 summarises these three regimes for the electronic ground state of atomic sodium by plotting the energy of the eight angular momentum substates of the electronic ground state.

In the figure, the magnetic field strength is measured as the Larmor frequency of an isolated electronic spin. In the case of very weak fields, the magnetic field lifts the degeneracy of the states within the two hyperfine multiplets. The energy shift is linear in the field, as the low-field Hamiltonian

$$\mathcal{H}_{\mathrm{m}} = -\mu_{\mathrm{B}}\, g_F\, \vec{B} \cdot \vec{F} \tag{4.49}$$

shows, where the Landé factor is

$$g_F = g_J \frac{F(F+1) + J(J+1) - I(I+1)}{2\, F(F+1)} \tag{4.50}$$

if we neglect the nuclear Zeeman effect. In this limit, the allowed magnetic dipole transitions between neighbouring m_F states all have the same energy separation.

This form can be simplified in the case of $J = 1/2$, which is relevant for the ground state of the alkali as well as for the $^2P_{1/2}$ excited state. We find $g_F = \pm g_J/(2I + 1)$ valid for $F = I \pm 1/2$. In this case, the coupling strength is simply reduced by the multiplicity of the nuclear spin. In simple cases, the Landé factor can be derived directly by considering the states with $|F = 2;$

$m_F = \pm 2\rangle = |m_J = \pm 1/2; m_I = \pm 3/2\rangle$, which never mix with other states. In Figure 4.13, they appear as straight lines. We can calculate the energy of those states in the product basis or in the coupled basis,

$$\mathscr{E}_{\mathrm{m}} = -\mu_{\mathrm{B}}\, g_J\, B\, m_J = -\mu_{\mathrm{B}}\, g_F\, B\, m_F \tag{4.51}$$

Using the states given above, we find $g_F = g_J/4$. The numerical values for the coupling constants $\mu_{\mathrm{B}}\, g_F$ in the Na ground state are 7 MHz/mT whereas the excited state Larmor frequencies are 2.3 MHz/mT and 4.7 MHz/mT for the $^2P_{1/2}$ and the $^2P_{3/2}$ states, respectively.

4.4.3 Magnetic resonance spectra

Quadratic Zeeman effect

For somewhat stronger magnetic fields, the energies deviate from the linear behaviour, as is clearly visible in Figure 4.13. We may improve the Hamiltonian of equation (4.49) by adding a term that is quadratic in the magnetic field

$$\mathscr{H}_{\mathrm{Z}} = -\mu_{\mathrm{B}}\, (g_F\, B\, F_z + g_F^{(2)}\, B^2\, F_z^2) \tag{4.52}$$

g_F and $g_F^{(2)}$ are the coupling constants for the linear and quadratic Zeeman interaction and the field is parallel to the z axis. The coupling constant for the quadratic term can be calculated by second-order perturbation theory as $g_F^{(2)} = (\mu_{\mathrm{B}} g_F)^2/A$. The numerical values for the Na ground state are $\mu_{\mathrm{B}}\, g_F = 7$ MHz/mT and $\mu_{\mathrm{B}}\, g_F^{(2)} = 27$ kHz mT^{-2}. The quadratic Zeeman effect makes the transitions between neighbouring Zeeman substates nonequivalent.

Figure 4.14 shows the spectrum of the transitions between neighbouring Zeeman sublevels of the Na ground state as a function of the magnetic field strength. For low fields, all transitions are degenerate. For higher fields, they become distinguishable. The linear term in the Hamiltonian of equation (4.52) determines the centre of the multiplet, whereas the term that is quadratic in the field causes the splitting between the resonance lines.

The fact that not all Zeeman coherences precess at the same Larmor frequency is a direct indication that the coupling between electron and nuclear spin is not infinitely strong. The quadratic term in the Hamiltonian is indirectly proportional to the hyperfine splitting and therefore vanishes for strong enough couplings. The differential precession of different density operator components also has a classical interpretation: The magnetic field applies a torque to the electron spin, but couples only weakly to the nuclear spin. If the coupling between them is not infinitely strong, the Zeeman interaction can excite a beat of the two spins against each other.

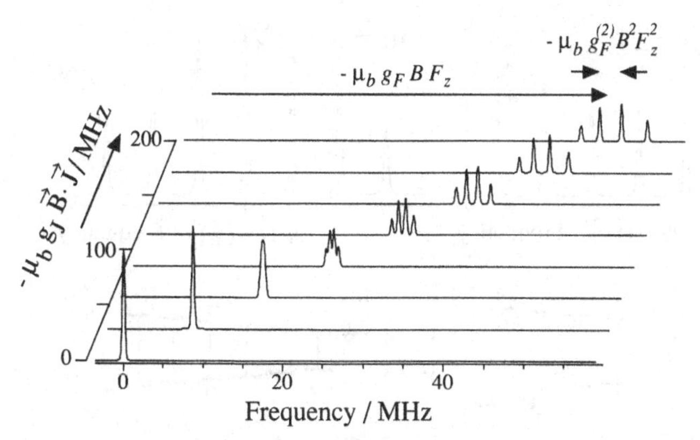

Figure 4.14. Magnetic dipole transitions between neighbouring Zeeman substates as a function of the magnetic field strength.

Hyperfine transitions

Magnetic dipole transitions can occur not only among the sublevels of one hyperfine multiplet, but also between the different hyperfine multiplets. The most important transition of this type is probably the clock transition between the two hyperfine multiplets of the Caesium ground state at 9′192′631′770 Hz. Magnetic dipole transitions can occur between states with $\Delta m_F = 0, \pm 1$, depending on the polarisation of the microwave field.

The lower half of Figure 4.15 shows the energy level scheme of the ground state of Cs in a magnetic field. The splitting between the $F = 3$ and $F = 4$ multiplets is due to the hyperfine interaction, whereas the Zeeman interaction lifts the degeneracy within the two multiplets. The Zeeman shift is greatly exaggerated in the figure. The lines connecting the different levels are allowed magnetic dipole transitions that can be excited by a microwave field polarised perpendicular (left) or parallel (right) to the static magnetic field. The magnetic field shifts the different transitions in different directions, as shown in the upper part of the figure. The only transition that is not affected by the magnetic field is the one between the two $m_F = 0$ states, which appears in the centre of the spectrum in the right-hand part of the figure. This transition is used in atomic clocks, where it is important that stray magnetic fields not shift the transition frequency.

Intermediate coupling regime

As the magnetic field strength increases, the eigenstates of the system change from the fully coupled F states into the uncoupled states where the individual angular momenta of electron and nucleus are good quantum numbers. In the

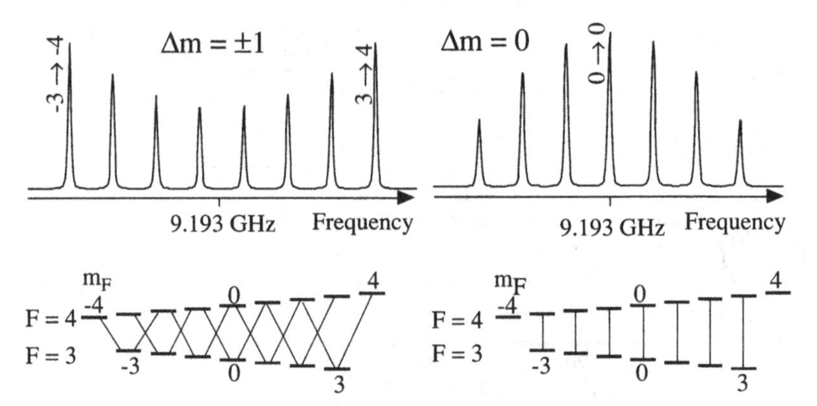

Figure 4.15. Transition frequencies between hyperfine multiplets for a microwave field polarised perpendicular (left) and parallel (right) to the static magnetic field. The lower part of the figure indicates the corresponding transitions.

intermediate field region, where the Zeeman interaction is comparable to the hyperfine coupling, the classification of the states according to the quantum numbers F, m_F breaks down, and the classification according to individual angular momenta J and I is not exact.

The Zeeman effect acts not only on the electronic ground state, but also on the excited states, where the hyperfine interaction is often smaller than in the ground state. The widths of the magnetic dipole transitions are in this case broadened by the lifetime of the excited state levels, and resolved spectra are obtained only at field strengths of several mT. At these field strengths, the Zeeman interaction is already comparable to the hyperfine interaction and the spectra become rather complicated (Tremblay et al. 1990).

Figure 4.16 shows an example of such a spectrum calculated for a field of 27.2 mT, using circularly polarised light for exciting the spectrum. Both spectra show only that range where the strongest resonance lines appear. They show clearly that at this field strength, the excited state hyperfine structure can no longer be observed and an interpretation of the spectra without explicit diagonalisation of the Hamiltonian is no longer possible.

Nuclear Zeeman interaction

Like the electronic magnetic moments, the nuclear spin couples to the magnetic field as

$$\mathcal{H}_I = -\gamma_I \vec{B} \cdot \vec{I} \tag{4.53}$$

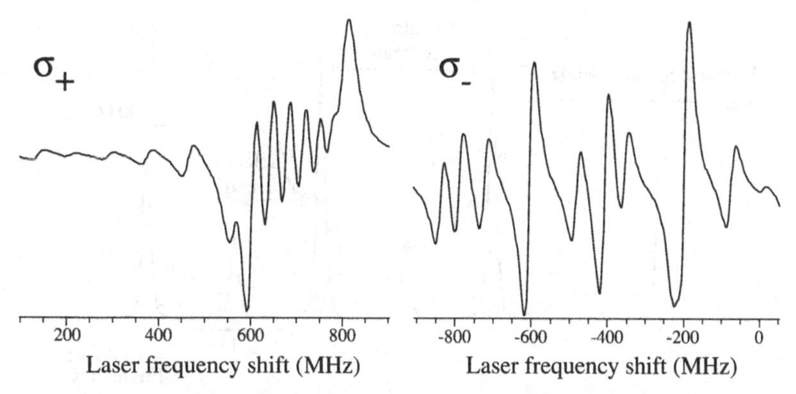

Figure 4.16. Zeeman-shifted optical transitions of the Cs D_2 resonance line between the $F = 4$ ground state and the excited states in a magnetic field of 272 G.

The gyromagnetic ratios γ_I are tabulated for all relevant isotopes. For ^{23}Na, γ_I is 11.3 MHz/T. The nuclear Zeeman interaction is typically three orders of magnitude smaller than the electron Zeeman effect. Nevertheless, it can readily be observed even at moderate field strengths of the order of a few Gauss. Since the orientation of the nuclear spin is parallel (antiparallel) to the electron spin in the $F = 2$ ($F = 1$) hyperfine multiplet, it increases (decreases) the effective gyromagnetic ratio and shifts the Zeeman transitions inside the $F = 1$ multiplet relative to those in the $F = 2$ multiplet.

Figure 4.17 shows the resulting frequencies in a field strength of 0.7 mT. The transition frequencies of the $F = 1$ multiplet are shifted from those of the $F = 2$ multiplet by the nuclear Zeeman frequency. The resonance lines are all close to 5 MHz due to the linear electron Zeeman interaction.

4.4.4 Larmor precession

The Zeeman effect not only shifts the energy levels of the different angular momentum states, it also affects the dynamics of the system. In particular, the magnetic dipole coupling causes the Larmor precession: The torque of the external magnetic field does not tilt the magnetic dipole moment towards the field direction, as one might expect, but instead forces a precession around the direction of the magnetic field. This is a consequence of the association between the magnetic dipole and the angular momentum. Torques applied to angular momenta cause a precession not a tilt, in classical mechanics as well as in quantum mechanics.

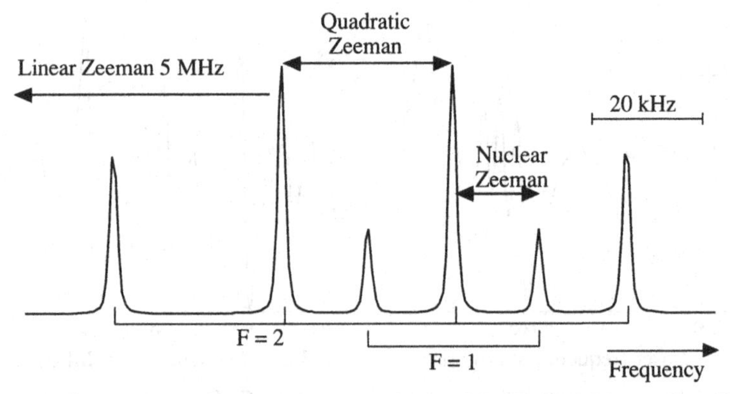

Figure 4.17. Magnetic dipole transitions in the ground state of atomic sodium. The nuclear Zeeman interaction accounts for the different transition frequencies of the $F = 2$ versus the $F = 1$ multiplet.

For a derivation of this behaviour from the Schrödinger equation, we consider a two-level system (e.g., $J = 1/2$) in a magnetic field along the z axis. The Hamiltonian is

$$\mathcal{H} = -\Omega_{\mathrm{L}}\, J_z \tag{4.54}$$

with $\Omega_{\mathrm{L}} = \mu_{\mathrm{B}}\, g\, B$, and the equation of motion

$$\dot{\rho} = -\mathrm{i}[\mathcal{H},\rho] = \mathrm{i}\, \Omega_{\mathrm{L}}\, [J_z,\rho] \tag{4.55}$$

If we expand the density operator in terms of angular momentum operators J_x, J_y, J_z,

$$\rho = \frac{1}{2}\, \mathbb{1} + m_x\, J_x + m_y\, J_y + m_z\, J_z \tag{4.56}$$

the equations of motion for the expansion coefficients are

$$\dot{m}_x = -\Omega_{\mathrm{L}}\, y \qquad \dot{m}_y = \Omega_{\mathrm{L}}\, x \qquad \dot{m}_z = 0 \tag{4.57}$$

These are essentially the Bloch equations (Bloch 1946) for the evolution of a spin 1/2 that were discussed in Chapter 2, somewhat reduced by neglecting relaxation and radio frequency fields. This simplified form has the general solution

$$\vec{m}(t) = \{m_x, m_y, m_z\}(t) = \{m_{xy} \cos(\Omega_{\mathrm{L}}\, t - \phi),\, m_{xy} \sin(\Omega_{\mathrm{L}}\, t - \phi),\, m_z(0)\} \tag{4.58}$$

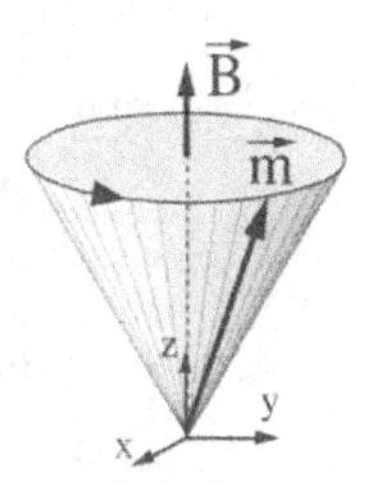

Figure 4.18. Larmor precession of the magnetic dipole $\vec{m}$ around the magnetic field $\vec{B}$.

which describes the precession of a vector in three-dimensional space. The direction of the magnetic field determines the rotation axis and the precession frequency is Ω_L. The parameters m_{xy}, ϕ, and $m_z(0)$ depend on the initial conditions. The angular momentum component parallel to the field remains time-independent, whereas the components perpendicular to the field perform a rotation.

Figure 4.18 represents this precession of the magnetisation vector $\vec{m}$ around the magnetic field graphically. The equation of motion (4.55) is the most general form for a spin 1/2 in a constant magnetic field. For higher angular momenta, $J > 1/2$, this equation of motion remains valid, as long as the linear Zeeman effect is the only relevant term in the Hamiltonian. The motion resulting from the more general cases discussed above, however, is significantly more complex and can no longer be represented by a vector in three-dimensional space.

4.5 Electric dipole transitions

For the discussion of the interaction between light and atoms, we use the electric dipole approximation. This means that we explicitly neglect the interaction of higher atomic multipoles, in particular the electric quadrupole, with external fields. Such interactions do exist and light can induce such transitions; however, the discussion of these effects lies outside the scope of this book and is not needed for the chapters that follow. Details can be found in the literature (Weissbluth 1978).

4.5.1 Angular momentum exchange

Conservation

We use the semiclassical formalism for explicit calculations, but it is often helpful to discuss certain aspects in a purely quantum mechanical context. We use this procedure here to derive the selection rules for optical transitions. As

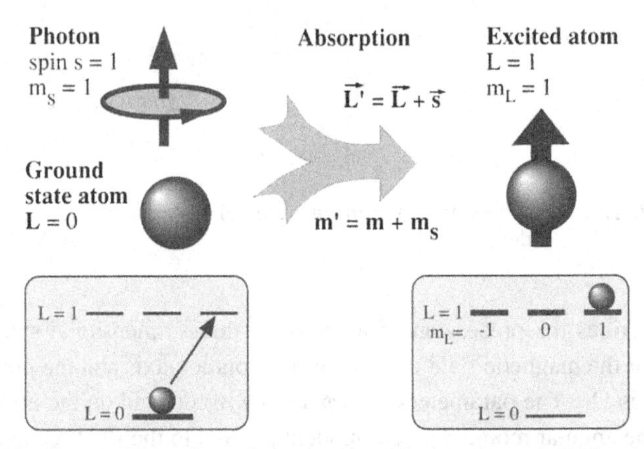

Figure 4.19. Conservation of angular momentum during absorption of a photon: The angular momentum state of the atom changes to accommodate the contribution from the photon spin.

discussed above, the photon as the carrier of the electromagnetic interaction carries an internal (spin) angular momentum of $\hbar$, which can be oriented parallel or antiparallel to the direction of propagation of the light. When photons are absorbed, their angular momentum vanishes. Since angular momentum is a conserved quantity, the atom that absorbs the photon must change its angular momentum simultaneously, much as its energy changes to accommodate the photon energy.

Figure 4.19 illustrates this for a simple example. We consider an atom that is initially in its electronic ground state, which is spherically symmetric. Its angular momentum vanishes. It absorbs a photon whose angular momentum is oriented parallel to the quantisation axis. Angular momentum conservation requires that the excited state have the same angular momentum $L = 1$ as the absorbed photon. More generally, it has to be the vector sum of the initial angular momentum of the atom and the photon. The lower part of the figure indicates the consequences for a simple atomic level scheme: The ground state is nondegenerate, whereas the excited state consists of three angular momentum substates. The interaction with the circularly polarised light populates selectively that excited state sublevel whose angular momentum corresponds to the total angular momentum, as indicated in the figure.

Selection rules

Like the conservation of energy, which restricts the final atomic state to a narrow energy interval, conservation of angular momentum restricts the possible states to those whose angular momentum is the vector sum of the initial atomic angular momentum and the absorbed photon. The interaction between the atomic system and the radiation field depends on the magnetic quantum numbers of the levels involved, as well as on the polarisation of the light. From the addition theorems for angular momenta, we know that the angular momentum L' of the final state is related to L, the angular momentum of the initial state, as $\Delta L = L' - L = 0, \pm 1$, with the additional constraint that $L' = L = 0$ is not allowed. Similarly we have the condition for the magnetic quantum number, $\Delta m_L = 0, \pm 1$. For single electron systems like the alkali, parity conservation imposes the additional constraint $\Delta L \neq 0$.

Consider as an example the transition of a single electron from the s to the p orbital. Its orbital angular momentum changes from 0 to 1. In the case of circularly polarised light, $\Delta m_L = +1$, and the excited state can only be the $|L = 1, m_L = 1\rangle$ state. For circularly polarised light interacting with an $L = 0 \leftrightarrow L' = 1$ transition, there is only a single possible transition, as in the case of Figure 4.19. The electric dipole moment between the electronic ground state and the excited state is

$$\langle \mathrm{e}; L = 1, m_L = 1|r_+|\mathrm{g}; L = 0, m_L = 0\rangle = d_L \tag{4.59}$$

while all other elements vanish.

Multiple reservoirs

When the atomic system contains several kinds of angular momentum, the polarisation that the atom receives from the laser field through the orbital angular momentum of the electron is usually shared with other reservoirs like the electron spin and the nuclear spin and couples even to the overall rotation of the sample (Woerdman, Nienhuis and Kuscer 1992; Kristensen, Eijkelenborg and Woerdman 1994). See Figure 4.20.

In most cases, the coupling between the reservoirs is stronger than the coupling to the optical field. The orbital angular momentum that couples directly to the radiation field is then no longer a good quantum number; absorption and emission of polarised light convert photon angular momentum into atomic angular momentum, which cannot be assigned to a specific reservoir. The absorption process can be considered as a multistep process, where the angular momentum is first transferred from the radiation field to the electronic orbital angular momentum reservoir, from where fine structure and hyperfine coupling distribute it into the other reservoirs. However, in the frequently encountered case where the internal interactions are stronger than the coupling

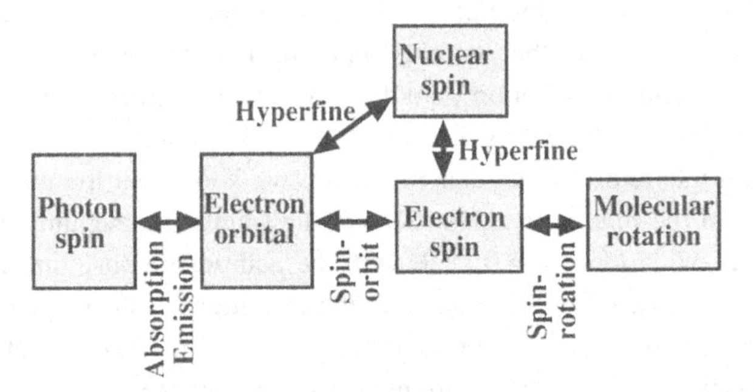

Figure 4.20. Angular momentum reservoirs and coupling between them.

to the field, this description is not very appropriate since the redistribution is faster than the transfer from the photons to the atom.

As shown in Figure 4.21, it is then meaningful to consider a unified angular momentum reservoir that includes both types of electronic angular momentum, as well as in most cases the nuclear spin. This is the angular momentum that interacts with the external field if the coupling to the field is weaker than the internal interactions and the basis states are the F, m_F states. Similarly, the selection rules are then specified in terms of total angular momentum. In those cases, where the coupling scheme of the constituent angular momenta is known, the selection rules for the coupled system can be derived directly from the known selection rules for the constituent angular momenta.

4.5.2 Spin–orbit coupling

Using the example of an $L = 0 \rightarrow L = 1$ transition of Figure 4.19, we show how the selection rules for the strongly coupled system can be derived from the known coupling scheme. If we take the electron spin S into account, the lower state splits into two degenerate substates $|J = S = 1/2,\ m_J = m_S = \pm 1/2\rangle$. In the electronically excited state, the three orbital angular momentum substates combine with the two spin states to a total of six possible states, which may be written as product states, e.g., $|L = 1,\ m_L = 1;\ S = 1/2,\ m_S = 1/2\rangle$. The coupling to the radiation field is not spin-dependent, so the transition matrix element is

$$\begin{aligned}\langle L = 1,\ m_L = 1;\ S = 1/2,\ m_S = \pm 1/2|r_+|L = 0,\ m_L = 0;\ S = 1/2,\\ m_S = \pm 1/2\rangle = \langle L = 1,\ m_L = 1|r_+|L = 0,\ m_L = 0\rangle \qquad (4.60)\\ \times \langle S = 1/2,\ m_S = \pm 1/2|S = 1/2,\ m_S = \pm 1/2\rangle = d_L\end{aligned}$$

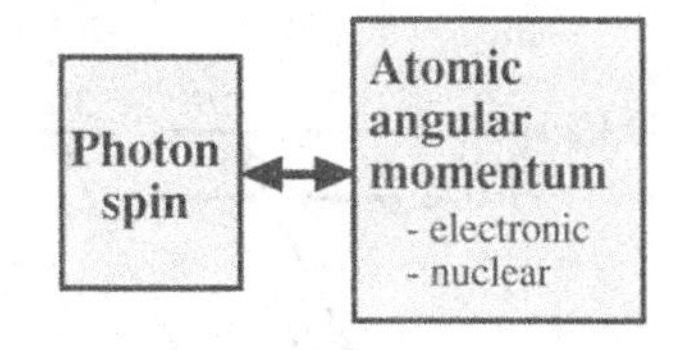

Figure 4.21. Unified reservoir of atomic angular momentum.

where the sign of m_S is the same on both sides. All other matrix elements for this component of the electric dipole moment vanish, as can be checked through application of the Wigner–Eckart theorem.

Coupled states

As discussed in Section 4.3.2, the spin–orbit coupling lifts the degeneracy of these states, as the energy depends on the relative orientation of the two angular momentum components. Energy conservation then allows only the excitation of one of the coupled states. To calculate the transition matrix element, we express the coupled states as linear combinations of the product states. This is most straightforward for the $|J = 3/2, m_J = \pm 3/2\rangle$ states, which are simultaneously product states

$$|J = 3/2, m_J = \pm 3/2\rangle = |L = 1, m_L = \pm 1; S = 1/2, m_S = \pm 1/2\rangle \quad (4.61)$$

while the $m_J = \pm 1/2$ states are formed as the symmetric superpositions of product states

$$|J = 3/2, m_J = \pm 1/2\rangle = \sqrt{\frac{1}{3}}\,|L = 1, m_L = \pm 1; \quad S = 1/2, m_S = \mp 1/2\rangle + \sqrt{\frac{2}{3}}\,|L = 1, m_L = 0; \quad S = 1/2, m_S = \pm 1/2\rangle \quad (4.62)$$

The antisymmetric superpositions become the $J = 1/2$ states

$$|J = 1/2, m_J = \pm 1/2> = \sqrt{\frac{2}{3}}\,|L = 1, m_L = \pm 1; \quad S = 1/2, m_S = \mp 1/2\rangle - \sqrt{\frac{1}{3}}\,|L = 1, m_L = 0; \quad S = 1/2, m_S = \pm 1/2\rangle \quad (4.63)$$

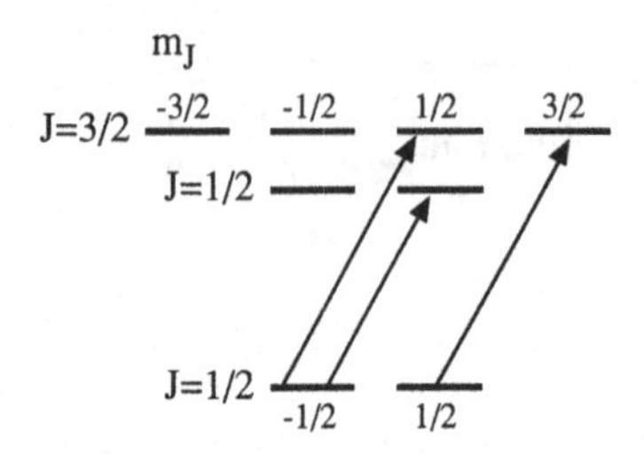

Figure 4.22. Possible transitions for σ_+ light in a single electron $s \rightarrow p$ transition.

Matrix elements

In the coupled system shown in Figure 4.22, the three σ_+ transitions that are marked by arrows conserve the angular momentum. We expect that the three corresponding matrix elements are nonzero. Using the expansions of the coupled states in the basis of product states, we can calculate the transition matrix elements through direct evaluation. To make the notation more compact, we introduce the abbreviation $|1/2,1/2\rangle = |J = 1/2,\ m_J = +1/2\rangle$ for the coupled states and $|1,1,1/2,1/2\rangle = |L = 1,\ m_L = +1;\ S = 1/2,\ m_S = +1/2\rangle$ for the product states.

We evaluate the matrix element of r_+ for the three transitions shown in Figure 4.22. For the one at the far right, we find

$$\langle \mathrm{e};\ 3/2,3/2|r_+|\mathrm{g}; 1/2,1/2\rangle = \langle \mathrm{e}; 1,1,1/2,1/2|r_+|\mathrm{g};\ 0,0,1/2,1/2\rangle = d_L \quad (4.64)$$

The leftmost transition has the matrix element

$$\begin{aligned} &\langle \mathrm{e};\ 3/2,1/2|r_+|\mathrm{g};\ 1/2,-1/2\rangle = \\ &\qquad \sqrt{\frac{1}{3}}\,\langle \mathrm{e}; 1,1,1/2,-1/2|r_+|\mathrm{g};\ 0,0,1/2,-1/2\rangle \\ &\qquad + \sqrt{\frac{2}{3}}\,\langle \mathrm{e};\ 1,0,1/2,1/2|r_+|\mathrm{g};\ 0,0,1/2,-1/2\rangle \end{aligned}$$

The second term vanishes due to the orthogonality of the spin functions and we find

$$\langle \mathrm{e};\ 3/2,1/2|r_+|\mathrm{g};\ 1/2,-1/2\rangle = \sqrt{\frac{1}{3}}\,d_L \quad (4.65)$$

and similarly

$$\langle \mathrm{e};\ 1/2,1/2|r_+|\mathrm{g};\ 1/2,-1/2\rangle = \sqrt{\frac{2}{3}}\,d_L \quad (4.66)$$

In close analogy, one finds that for the r_- (r_0) components of the electric dipole, the value of J may change by $\Delta J = 0, \pm 1$, and the magnetic quantum number m_J by -1 (0).

General form

It is not necessary to evaluate all individual matrix elements with the procedure outlined before, since group theory provides the result in closed form. The general form for the reduced matrix element for a transition from a ground state with angular momentum J_g to an excited state with angular momentum J_e is

$$|\langle e; J_e \| r \| g; J_g \rangle|^2 = (2J_g + 1)(2J_e + 1) \begin{Bmatrix} J_g & 1 & J_e \\ L_e & S & L_g \end{Bmatrix}^2 |\langle e; L_e \| r \| g; L_g \rangle|^2 \tag{4.67}$$

if the spin S is the same for both states and zero otherwise. Derivations can be found in the literature (Weissbluth 1978). Inserting the values for the D_1 transition, the $6J$ symbol evaluates to

$$\begin{Bmatrix} J_g & 1 & J_e \\ L_e & S & L_g \end{Bmatrix}(D_1) = \begin{Bmatrix} \frac{1}{2} & 1 & \frac{1}{2} \\ 1 & \frac{1}{2} & 0 \end{Bmatrix} = \frac{1}{\sqrt{6}} \tag{4.68}$$

we find the transition strength as

$$|\langle e; J_e \| r \| g; J_g \rangle|^2 = \frac{2}{3} |\langle e; L_e \| r \| g; L_g \rangle|^2 \tag{4.69}$$

and twice that for the D_2 transition.

The individual transition matrix elements can be found by applying the Wigner–Eckart theorem with

$$\langle e; J_e, m_{J_e} | r_q | g; J_g, m_{J_g} \rangle = (-1)^{J - m_J} \begin{pmatrix} J_e & 1 & J_g \\ -m_{J_e} & q & m_{J_g} \end{pmatrix} \langle e; J_e \| r_q \| g; J_g \rangle \tag{4.70}$$

where $q = 0, \pm 1$, depending on the polarisation of the light.

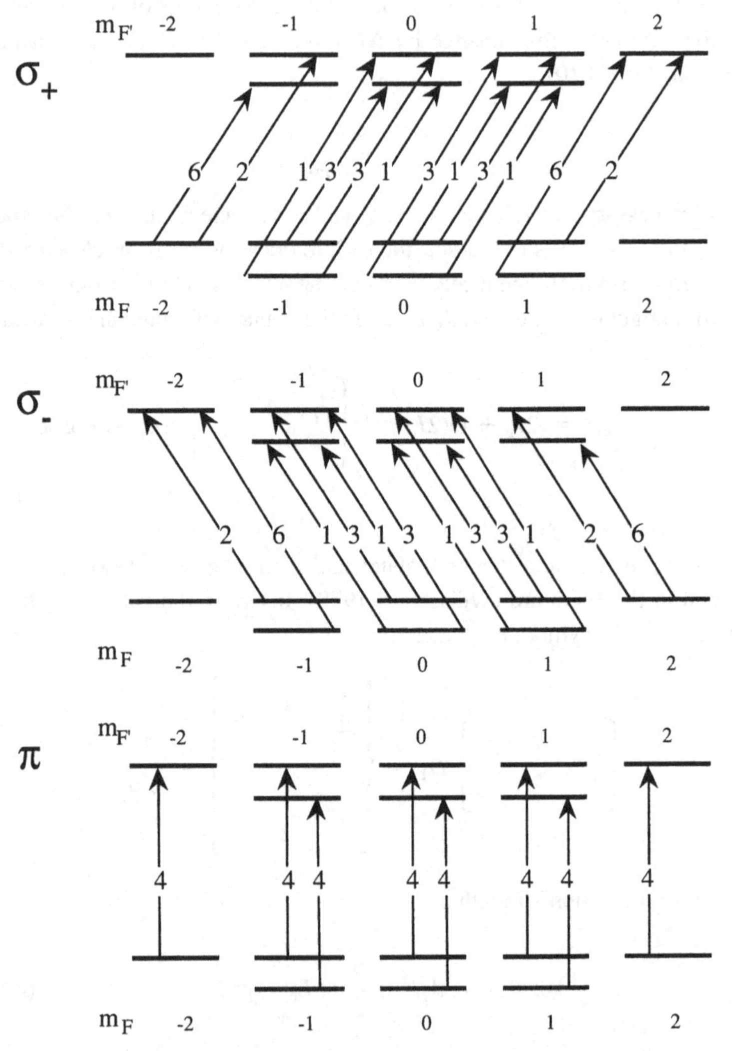

Figure 4.23. Squares of transition matrix elements for the Na D_1 transition for σ_+

4.5.3 Nuclear spin

Adding the nuclear spin to this system reiterates the procedure outlined for the electronic spin: The total angular momentum states can be obtained by diagonalising the Hamiltonian or through application of the addition theorems for angular momenta. The reduced matrix element is then obtained as

$$|\langle e; F_e||r||g; F_g\rangle|^2 = (2F_g + 1)(2F_e + 1)\begin{Bmatrix} F_g & 1 & F_e \\ J_e & I & J_g \end{Bmatrix}^2 |\langle e; J_e||r||g; J_g\rangle|^2 \tag{4.71}$$

For a numerical example, we evaluate the formula for the Na D_1 transition, where the electronic angular momentum $J = 1/2$ and the nuclear spin $I = 3/2$ add to a total angular momentum $F = 1$ or 2 in both the ground and excited states. Inserting those numbers, we find the transition strengths $|\langle e; J_e||r||g; J_g\rangle|^2$ are identical for all but the $F = 1 \leftrightarrow F' = 1$ transition, which is five times weaker. These differences between transition strengths result in different spectral intensities, as shown in Figures 4.8 and 4.23.

The individual matrix elements are again evaluated with the Wigner–Eckart theorem. The results are summarised in Figure 4.23, where the squares of the matrix elements for σ_+, σ_- and π light are indicated for the allowed transitions within the Na D_1 line.

5
Optical pumping

After the brief discussion of the internal degrees of freedom of atomic systems, we give an overview of how these variables can acquire nonvanishing average values in an ensemble of atoms. Optical pumping was one of the first experimental schemes that allowed physical systems to be pushed far from equilibrium through the interaction with light, a direct consequence of angular momentum conservation and the symmetry properties of the interaction between atoms and radiation.

5.1 Principle and overview

5.1.1 Phenomenology

Optical pumping (Happer 1972) is one of the earliest examples wherein optical radiation qualitatively modifies the properties of a material system. In its original implementation, it corresponds to a selective population of specific angular momentum states, starting from thermal equilibrium.

In the idealised process depicted in Figure 5.1, the light brings the atomic system from the initially disordered state, in which the populations of degenerate levels are equal, into an ordered state where the internal state of all atoms is the same. If we consider only the material system, it appears as if the evolution from the initially disordered state into an ordered state, where the population of one level is higher than that of another level, violated the second law of thermodynamics. This process does not proceed spontaneously, however. It is the interaction with polarised light that drives the system and increased disorder in the radiation field compensates for the increase in the population difference in the material system (Enk and Nienhuis 1992).

Conservation laws

The changes of the populations in each subsystem are accompanied by changes in energy and momentum, quantities that must be conserved in the total sys-

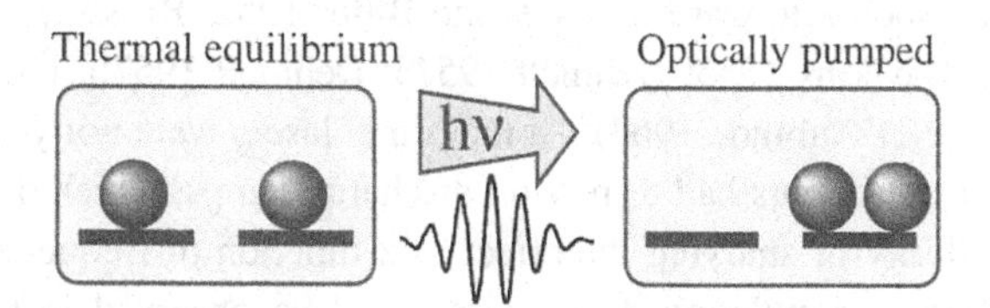

Figure 5.1. Light turns a disordered state into an ordered state.

tem. In the context discussed here, the most important conservation laws are those for energy and angular momentum. In systems with different sources of angular momentum that interact with each other, states with different total angular momentum have different energies. Transfer of population between such states changes energy and angular momentum simultaneously. The radiation field is thus a source of energy and angular momentum for the material subsystem, and a sink for entropy. In most cases, the energy aspect will turn out to be the least important, as the initial and final states may have the same energy.

The degrees of freedom that participate in optical pumping include in the material subsystem the electronic degrees of freedom and the angular momentum. In the field subsystem, the relevant variables are momentum, angular momentum, and energy. In the classical description, these quantities correspond to the direction of propagation, polarisation, and intensity of the light. Overall, the electronic degrees of freedom as well as the total photon number do not change; their role is that of a catalyst whose presence is necessary for the process and which changes temporarily, but returns to its initial state at the end.

The population differences that arise during optical pumping are determined by an equilibrium of polarising and depolarising forces. The polarising forces are primarily the interaction with the light, but under favourable conditions, anisotropic collisions can also create order. Most collisions, however, tend to decrease anisotropic polarisations. This includes collisions with the walls of a container, or with other atoms present in the system. The interaction with unpolarised light also leads to disorientation, as well as to the diffusion of polarised atoms out of the interaction region. In addition, external magnetic fields are an important contribution to the dynamics of the system. They cause Larmor precession, which may destructively interfere with the polarisation process.

5.1.2 Historical

The possibility of using the interaction with polarised light for selectively populating specific angular momentum states in atomic vapours was first suggested by Alfred Kastler (Kastler 1950). Several groups reported experimental verification soon afterward (Brossel and Bitter 1952; Brossel, Kastler and Winter 1952; Hawkins 1955; Dehmelt 1957a; Dehmelt 1957b; Karlov, Margerie and Merle-D'Aubigne 1963). At that time, lasers were not yet available and all these experiments had to rely on discharge lamps, which did not provide the possibility of studying the effect as a function of frequency. Nevertheless, significant population differences could be observed in the ground state as well as in electronically excited states. These achievements were possible only with a successful reduction of all possible disorienting interactions: Large containers had to be used to eliminate diffusion and wall collisions as disorienting effects. Multiple shielding eliminated all external magnetic fields that would have destroyed the orientation much faster than it could be established.

In 1961, Barrat and Cohen-Tannoudji put forward a comprehensive theory of the observed effects, which they called the "quantum theory of optical pumping" (Barrat and Cohen-Tannoudji 1961b; Barrat and Cohen-Tannoudji 1961a; Barrat, Cohen-Tannoudji and Ribaud 1961; Cohen-Tannoudji 1962). Besides the known population-redistribution effects, they found energy level shifts and damping effects. These effects were experimentally observed with discharge lamps as light sources. The observed effects were thus very small, of the order of one Hertz. This demonstrates not only the impressive experimental skills of these researchers, but also the potential of these methods for observing small energy differences using photons whose energies are fifteen orders of magnitude larger and whose frequency distribution was some ten orders of magnitude wider.

The development of the laser as a tuneable source of intense, monochromatic and highly directional light subsequently extended these possibilities significantly. Not only is it now possible to achieve polarisations close to unity, i.e., to bring all atoms into the same state, but this process can now be achieved in less than a microsecond, and studied as a function of frequency and intensity of the light. Various forms of optical pumping can prepare the material system selectively into specific eigenstates of different variables, not only angular momentum states. An important example is velocity-selective optical pumping (Pappas et al. 1980; Aminoff, J. Javanainen and Kaivola 1983), where the laser brings atoms with a chosen velocity into a specific internal state. Experiments of this type are important in Doppler-free spectroscopy (Demtröder 1991) and many phenomena that involve the motion of atoms, such as light-induced drift (Gel'mukhanov and Shalagin 1980; Werij et al. 1984; Mariotti et al. 1988). In solids, spectral hole burning establishes a correlation between internal degrees of freedom, such as nuclear spin states,

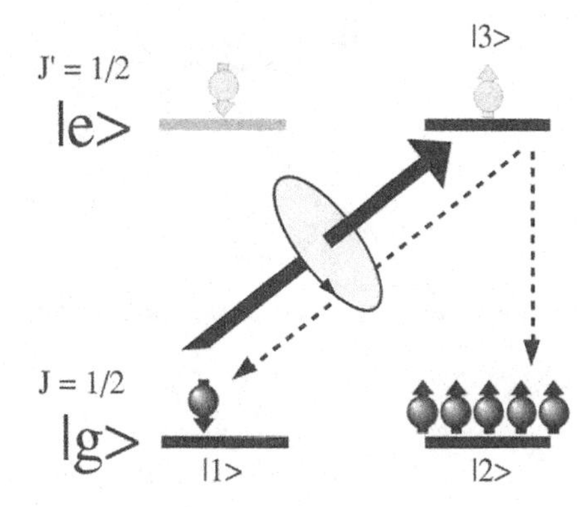

Figure 5.2. Model atomic system for the discussion of optical pumping.

with their environment, in particular crystal imperfections (Voelker 1989; Holliday and Wild 1993).

5.2 Two-level ground states

5.2.1 System

Level scheme

We start the discussion of optical pumping with a popular model atom that consists of a ground state and an electronically excited state, both with an angular momentum $J = 1/2$. This system describes an alkali atom with vanishing nuclear spin. A circularly polarised laser field interacts only with the transition from the ground-state sublevel 1 to the excited state sublevel 3, while spontaneous emission connects this excited state to both ground-state sublevels. The second excited state does not enter the dynamics of the system and we neglect it in the following. The remaining three-level system is then equivalent to the one discussed in Chapter 3. See Figure 5.2.

As we have seen during the discussion of three-level systems in the "overdamped limit," it is possible to describe the dynamics of the electronic ground state without reference to the excited state levels if the optical excitation is weak enough. The interest in the ground-state dynamics arises from its long lifetime, which allows high resolution experiments. The reduced dynamics that we use here are valid as long as the laser intensity is low enough that excited state populations as well as optical coherences remain small. As the operator basis for the Hamiltonian and density operator, we choose the three angular momentum components J_x, J_y, and J_z. Together with the unity operator, they form a basis for the expansion of all operators that act on the ground-state sublevels.

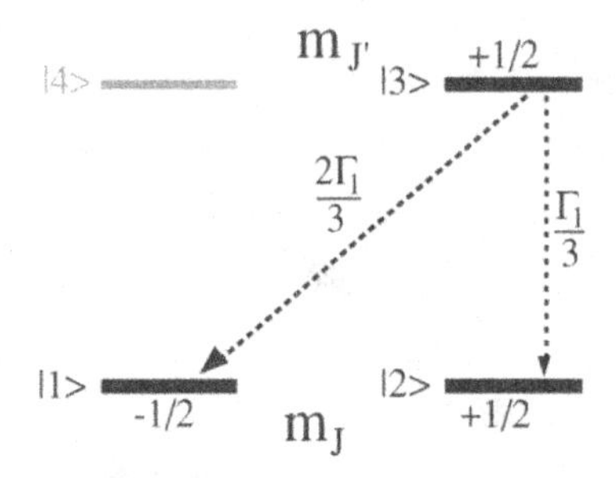

Figure 5.3. Spontaneous emission rates in $J = 1/2 - J' = 1/2$ atom.

Pumping rate

In Section 3.5, we found that optical pumping changes the populations according to

$$\dot{m}_z = \frac{\mathrm{d}}{\mathrm{d}t}(\rho_{22} - \rho_{11}) = 2\dot{\rho}_{22} = 2P_+\,\rho_{11}(t) = P_+\,(1 - m_z(t)) \qquad (5.1)$$

where the optical pumping rate is

$$P_+ = \Gamma_2 \frac{\Gamma_1^{3\to2}}{\Gamma_1^{3\to1} + \Gamma_1^{3\to2}} \frac{\omega_x^2}{2(\Delta^2 + \Gamma_2^2)} \qquad (5.2)$$

To evaluate the pumping rate for a specific system, we need to determine the spontaneous emission rates $\Gamma_1^{r\to s}$. They can be calculated from the electric dipole matrix elements by summing over all orientations and polarisations. For the present system, we find a ratio of 2:1 for the two possible transitions.

These spontaneous emission rates apply only to isolated atomic systems. If the optical pumping experiment includes buffer gas, the effect of collisions also has to be taken into account. One of the major effects is a randomisation of the angular momentum of the excited state (Franz and Franz 1966). This randomisation corresponds to a mixing of the excited state sublevels; for the spontaneous emission process, it therefore has the effect of an averaging of the transition probabilities, as shown in Figure 5.3. In the case of fast reorientation of the excited state, the two spontaneous emission rates become equal, $\Gamma_1^{3\to1} = \Gamma_1^{3\to2}$, and the pumping rate is then

$$P_+ = \Gamma_2 \frac{\omega_x^2}{4(\Delta^2 + \Gamma_2^2)} \qquad (5.3)$$

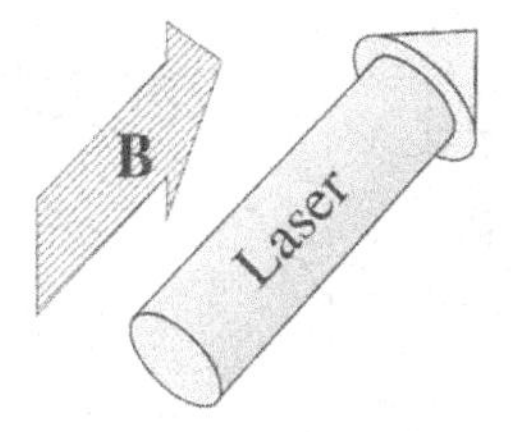

Figure 5.4. Longitudinal optical pumping.

The rate is thus proportional to the laser intensity and has the same offset dependence as the optical resonance line itself. It reaches a maximum at the optical resonance, which is the square of the optical Rabi frequency ω_x divided by four times the square of the optical dephasing rate Γ_2.

5.2.2 Longitudinal pumping

We first consider the case where a magnetic field is applied parallel to the direction of propagation of the light (see Figure 5.4). The system is then axially symmetric and the obvious choice for the quantisation axis (the z axis) is the symmetry axis. The equation of motion for the ground-state subsystem is then

$$\dot{\rho} = -\mathrm{i}[\mathcal{H}_\mathrm{L},\rho] + \hat{\Gamma}_\rho + P_+ J_z \qquad (5.4)$$

with

$$\mathcal{H}_\mathrm{L} = -\Omega_\mathrm{L} J_z \qquad (5.5)$$

the Hamiltonian describing the Zeeman interaction, as discussed in Section 4.4. $\hat{\Gamma}_\rho$ represents the relaxation superoperator, which includes all the damping mechanisms. The relaxation rate of the populations and coherences depends on the strength of the optical transitions to which the laser couples. To simplify the theoretical analysis, we assume here that population differences and coherences decay with the same rate $\gamma_{\mathrm{eff}} = \gamma_0 + P_+$, where γ_0 summarises the terms that do not depend on the optical irradiation, such as diffusion processes. This isotropic relaxation occurs in a system where collisions of the excited atoms with buffer gas atoms lead to a reorientation of the excited state magnetisation. The equation of motion can then be rewritten as

$$\dot{\rho} = -\mathrm{i}[\mathcal{H}_\mathrm{L},\rho] - \gamma_{\mathrm{eff}}\, \rho + P_+ J_z \qquad (5.6)$$

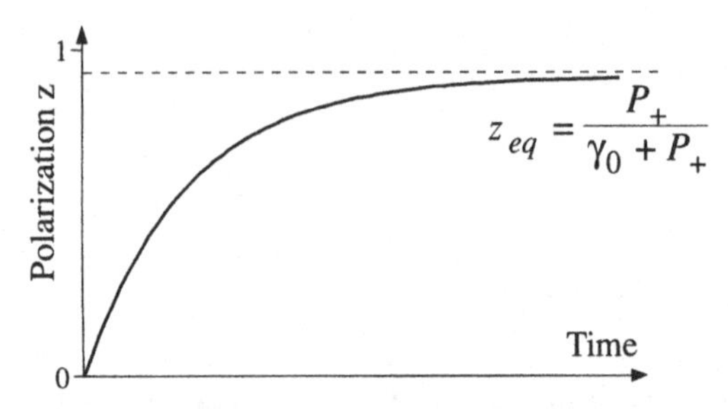

Figure 5.5. Evolution of the sublevel polarisation m_z during longitudinal optical pumping.

Starting from thermal equilibrium, the system evolves as

$$\rho(t) = J_z \, z_{\text{eq}} \, (1 - e^{-\gamma_{\text{eff}} t}) \tag{5.7}$$

where the parameter $z_{\text{eq}} = P_+/\gamma_{\text{eff}} = P_+/(\gamma_0 + P_+)$ describes the equilibrium polarisations.

Figure 5.5 shows how the optical pumping process creates polarisation along the z axis and how this component approaches the stationary value exponentially at a rate γ_{eff}. Since the density operator contains only the longitudinal polarisation J_z, it commutes with the Hamiltonian, $[\mathcal{H}_L,\rho] = 0$. The longitudinal magnetic field therefore does not affect the evolution of the system; its size is not important and may even be zero.

5.2.3 *Relaxation effects*

So far, we have discussed only the generation of order by the optical pumping process. In addition, it is necessary to take relaxation effects into account. They not only change the dynamics of the system, but also determine, together with the Hamiltonian evolution and the optical pumping, the equilibrium state of the system. Although we treat these relaxation processes only on a phenomenological level, we include here a brief summary of some of the relevant processes. They depend too strongly on the specific details of the system involved to allow a quantitative discussion; it is important to realise, however, on which parts of the system they act: some relaxation mechanisms affect only optical transitions, others only magnetic transitions or both. Some relaxation mechanisms affect only coherences (optical or magnetic), whereas others influence populations as well as coherences. For the experiments we will discuss, the most important mechanisms are:

Lifetime broadening. The electronically excited state can decay to the ground state by spontaneous emission of a photon. This process depopulates the excited state and populates the ground state. It affects the lifetime of optical population differences and coherences, as well as magnetic transitions in the excited state. Under certain conditions, spontaneous emission transfers part of the excited state polarisation to the ground state. Another important source of lifetime broadening is the removal of atoms from the interaction region by free flight (e.g., an atomic beam) or diffusion (e.g., in a gas cell in the presence of a buffer gas). This sort of lifetime broadening affects ground states as well as excited states and does not discriminate populations from coherences. It can be minimised by using a large interaction region (Dehmelt 1957b) (e.g., a glass cell with coated walls) or multiple interaction regions, as in Ramsey spectroscopy in an atomic beam (Ramsey 1990; Demtröder 1991).

Collision-induced relaxation. In a gas, collisions of the atoms with the walls of the sample cell or with other atoms and molecules present in the gas can lead to de-excitation and loss of phase of the coherence. This mechanism affects both magnetic and optical transitions, although in different ways. It also has different effects on coherences and population differences. If the electronic ground state is spherically symmetric (i.e., an s state), coherences between ground-state sublevels are largely immune to collisions. If the electron wavefunction is not spherically symmetric, however, as in the excited states of alkali atoms, relaxation of the magnetisation by collisions with buffer gas atoms can be a very efficient mechanism and often makes it impossible to observe order within the excited state sublevel structure.

Inhomogeneous effects. In the calculations we usually assume that all the parameters of the system are homogeneous throughout the sample volume. If this is not the case, the coherences of the different atoms precess out of phase with respect to each other, leading to an inhomogeneous decay of the macroscopic observables. The most important examples of inhomogeneous effects are the velocity distribution of the atoms in a gas, which leads to the Doppler broadening of the optical transitions, and strain broadening in crystals, which can affect optical as well as sublevel transitions. Inhomogeneous magnetic fields affect primarily ground-state sublevel coherences.

Fluctuating external fields. If external fields have a nonvanishing spectral density at one of the transition frequencies of the system, they can induce absorption or emission. Such an interaction drives the population difference of the transition towards zero, thus destroying order in the system. This relaxation mechanism can always be neglected for optical transitions, but is important for magnetic transitions. The spectral densities have to be evaluated in the centre of mass system of the atom, so that inhomogeneous static fields can lead to relaxation of moving atoms. Spectral densities near zero frequency can also lead to nonresonant (adiabatic) relaxation of coherences of any transition. If, as we have assumed, there is no static electric dipole moment, this mechanism mainly affects magnetic coherences.

Coupling to other systems. If the interaction between various, identical or different atoms is not negligible, our description of the total system as an ensemble of many individual subsystems is no longer valid. Since an exact description of a system of $>10^9$ atoms is not feasible, we take these effects into account in a qualitative, phenomenological way. In many cases, a qualitatively correct description of the system is obtained by adding a homogeneous relaxation mechanism to the transitions. These couplings can affect all transitions, but their effect can differ widely, depending on the nature of the interaction. These mechanisms are usually of little importance in gaseous systems, but are often significant in solids. As in the case of fluctuating external fields, these intersystem couplings can lead to relaxation effects of populations and coherences if the power spectral density of their autocorrelation function does not vanish at the transition frequency. In addition, they can cause adiabatic relaxation of coherences if the power spectral density near zero frequency is appreciable.

5.2.4 Transverse pumping

In a magnetic field parallel to the direction of the laser beam, or in zero magnetic field, optical pumping excites the atoms into states that have rotational symmetry around the direction of the laser beam. In a multipole expansion of these atomic states, only terms of order $q \neq 0$ appear. If the rotational symmetry is broken, e.g., by a magnetic field perpendicular to the laser beam (Bell and Bloom 1961b; D'Yakonov 1965; Series 1966; Pancharatnam 1968), this selection rule is lifted and it becomes possible to excite all multipole components. In a quantum mechanical description, terms with $q \neq 0$ correspond to coherent superpositions of different angular momentum eigenstates.

Equation of motion

We describe the dynamics of this system in a coordinate system whose z axis remains parallel to the laser beam and choose the x axis in the direction of the magnetic field, as shown in Figure 5.6. The equation of motion is the same as for the case of longitudinal pumping, but the Hamiltonian changes to

$$\mathcal{H}_{\mathrm{T}} = -\Omega_{\mathrm{L}} J_x \tag{5.8}$$

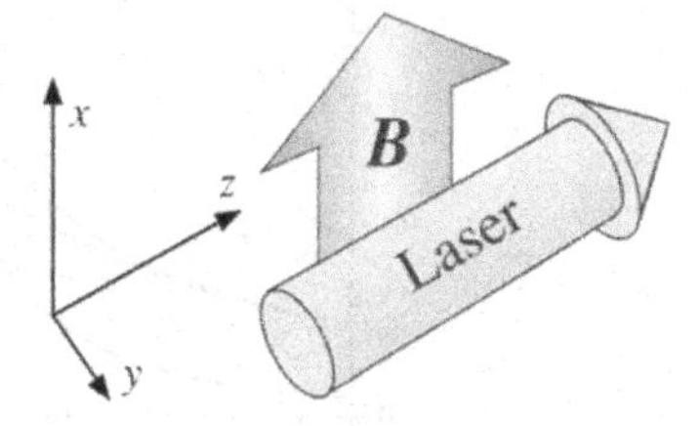

Figure 5.6. Geometry for transverse optical pumping and coordinate system used for describing the dynamics.

Since the magnetic field is no longer parallel to the laser beam, it now forces a precession of the magnetic moment around the x axis with the Larmor frequency $\Omega_L = -g_J \, \mu_B \, B$ proportional to the strength of the magnetic field B, Bohr's magneton μ_B, and the Landé factor g_J. The second and third terms of equation (5.4), which describe relaxation and optical pumping, remain the same.

It is instructive to rewrite the equation of motion in terms of the magnetisation vector $\vec{m} = \{m_x, m_y, m_z\}$, whose components are the expansion coefficients of the density operator in the basis of the spin operators:

$$\dot{\vec{m}} = \vec{\Omega} \times \vec{m} - \gamma_{\text{eff}} \, \vec{m} + \vec{P} \tag{5.9}$$

where the magnetic field (in frequency units) is $\vec{\Omega} = (-\Omega_L, 0, 0)$. This equation is quite analogous to the Bloch equation, except that the magnetisation, generated by the inhomogeneous third term in equation (4), is not aligned with the effective field. In matrix form, the equation of motion is

$$\dot{\vec{m}} = \frac{\mathrm{d}}{\mathrm{d}t} \begin{pmatrix} m_x \\ m_y \\ m_z \end{pmatrix} = \begin{pmatrix} -\gamma_{\text{eff}} & & \\ & -\gamma_{\text{eff}} & \Omega_L \\ & -\Omega_L & -\gamma_{\text{eff}} \end{pmatrix} \begin{pmatrix} m_x \\ m_y \\ m_z \end{pmatrix} + \begin{pmatrix} 0 \\ 0 \\ P_+ \end{pmatrix} \tag{5.10}$$

This form of the equation shows that the x component of the magnetisation vector $\vec{m}$ evolves independently of the y and z components.

Stationary state

After transient effects have disappeared the system ends up in a state described by the magnetisation vector

$$\vec{m}_\infty = \frac{P_+}{\Omega_L^2 + \gamma_{\text{eff}}^2} (0, -\Omega_L, \gamma_{\text{eff}}) \tag{5.11}$$

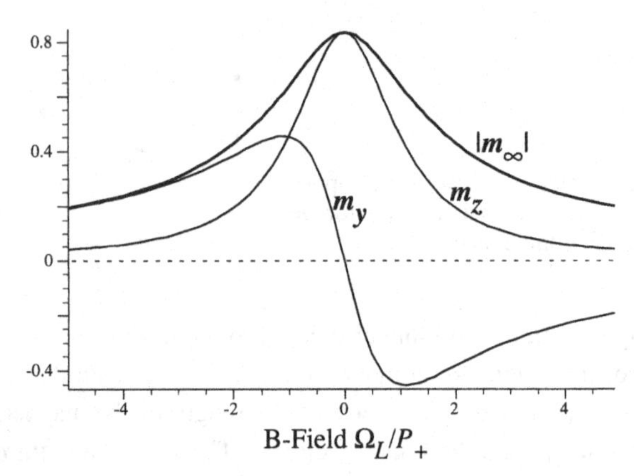

Figure 5.7. Dependence of the magnetisation on the strength of a transverse magnetic field.

Since the magnetisation is generated along a direction that does not coincide with the effective field, the equilibrium magnetisation $\vec{m}_\infty$ is not parallel to either of the two directions. Its size

$$|\vec{m}_\infty| = \frac{P_+}{\sqrt{\Omega_L^2 + \gamma_{\text{eff}}^2}} \tag{5.12}$$

is determined by the balance between the pumping rate P_+, which generates the magnetisation, and the Larmor precession Ω_L and decay rate γ_{eff}, which reorient and diminish it. The x component of the stationary magnetisation vanishes, since the equation of motion contains no term that couples it to the z direction where the magnetisation is generated. The size of the y and z components again reflects the competition between dephasing γ_{eff} and Larmor precession Ω_L: Larmor precession turns the magnetisation from the z axis, where optical pumping generates it, towards the y axis, while dephasing does not change its direction.

Figure 5.7 shows the dependence of the two nonvanishing components on the magnetic field. The polarisation of the atoms falls off rapidly when the Larmor frequency Ω_L exceeds the optical pumping rate P_+. This behaviour forms the basis of many "level crossing" experiments (Corney, Kibble and Series 1966).

Evolution

The general solution of the equation of motion (5.10) is

$$\vec{m}(t) = \sum_{i=-1}^{1} c_i \, \vec{\xi}_i \, e^{\lambda_i t} + \vec{m}_\infty \tag{5.13}$$

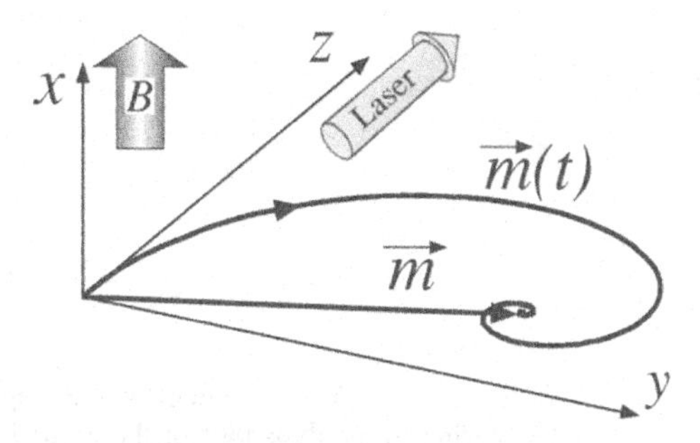

Figure 5.8. Evolution of the magnetisation during transverse optical pumping.

where the eigenvectors $\vec{\xi}_i$ and eigenvalues λ_i are

$$\begin{aligned} \vec{\xi}_0 &= (1, 0, 0) & \lambda_0 &= -\gamma_{\text{eff}} \\ \vec{\xi}_{\pm 1} &= \frac{1}{\sqrt{2}}(0, \pm \mathrm{i}, -1) & \lambda_{\pm 1} &= \pm \mathrm{i}\Omega_{\mathrm{L}} - \gamma_{\text{eff}} \end{aligned} \tag{5.14}$$

The expansion coefficients c_i are determined by the initial condition. For a sample in thermal equilibrium, the ground-state orientation vanishes, i.e., $\vec{m}(0) = 0$. The evolution is then

$$\vec{m}(t) = \vec{m}_\infty \left(1 - \cos(\Omega_{\mathrm{L}} t)\, \mathrm{e}^{-\gamma_{\text{eff}} t}\right) + \vec{m}_{\mathrm{s}} \sin(\Omega_{\mathrm{L}} t)\, \mathrm{e}^{-\gamma_{\text{eff}} t} \tag{5.15}$$

where

$$\vec{m}_{\mathrm{s}} = \{0, -z_{\mathrm{eq}}, y_{\mathrm{eq}}\} \tag{5.16}$$

is the component of the precessing magnetisation perpendicular to the initial condition.

Figure 5.8 shows the evolution for this case. The tip of the magnetisation vector moves along the spiral in the yz plane until it settles at the position indicated by the vector $\vec{m}_\infty$. It shows how the magnetisation is initially generated in the direction of the z axis and subsequently moves in the yz plane under the influence of the magnetic field.

5.2.5 Light shift

Effective Hamiltonian

As discussed in Section 3.5.4, light can not only change the populations of the various sublevels, it can also modify the dynamics by damping and light shift effects. The equation of motion (10) takes the damping into account

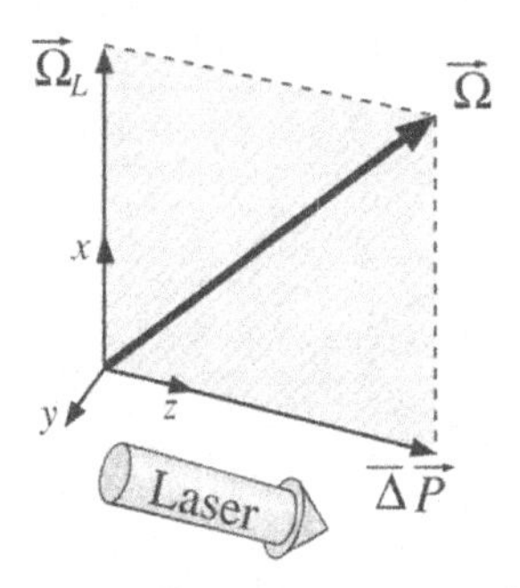

Figure 5.9. The true and effective magnetic fields appearing in the dynamics of the ground-state pumping.

through the relaxation rate γ_{eff}, which depends on the optical pumping rate. The light shift effect can be accounted for by modifying the Hamiltonian of the system. For the two-level system treated here, this requires the addition of a virtual magnetic field in the direction of the laser beam. The Hamiltonian is then

$$\mathcal{H}_{\text{LS}} = \overline{\Delta} P_+ \, J_z - \Omega_{\text{L}} \, J_x \tag{5.17}$$

where $\overline{\Delta} = \Delta\omega_0/\Gamma_2$ is the normalised optical detuning. The effective magnetic field, written in frequency units, now contains components in the x and z directions

$$\vec{\Omega} = (-\Omega_{\text{L}}, 0, \overline{\Delta} P_+). \tag{5.18}$$

As Figure 5.9 shows, the effective magnetic field $\vec{\Omega}$ now lies in the xz plane, which is spanned by the direction of the magnetic field and the laser beam. The virtual magnetic field due to the light shift couples the x and y components of the magnetisation vector in the equation of motion:

$$\dot{\vec{m}} = \frac{\text{d}}{\text{d}t}\begin{pmatrix} m_x \\ m_y \\ m_z \end{pmatrix} = \begin{pmatrix} -\gamma_{\text{eff}} & -\overline{\Delta} P_+ & \\ \overline{\Delta} P_+ & -\gamma_{\text{eff}} & \Omega_{\text{L}} \\ & -\Omega & -\gamma_{\text{eff}} \end{pmatrix}\begin{pmatrix} m_x \\ m_y \\ m_z \end{pmatrix} + \begin{pmatrix} 0 \\ 0 \\ P_+ \end{pmatrix} \tag{5.19}$$

whose stationary solution becomes

$$\vec{m}_\infty = \frac{P_+}{\gamma_{\text{eff}}(\Omega_{\text{L}}^2 + \overline{\Delta}^2 P_+{}^2 + \gamma_{\text{eff}}^2)} (-\overline{\Delta} P_+ \Omega_{\text{L}}, \gamma_{\text{eff}}\Omega_{\text{L}}, \overline{\Delta}^2 P_+^2 + \gamma_{\text{eff}}^2) \tag{5.20}$$

The stationary magnetisation has thus acquired a component in the direction of the magnetic field that is proportional to the light shift effect.

Eigenvalues and eigenvectors

The eigenvectors $\vec{\xi}_i$ and eigenvalues λ_i of the evolving magnetisation are now

$$\begin{aligned} \vec{\xi}_0 &= (-\Omega_L, 0, \overline{\Delta}P_+) & \lambda_0 &= -\gamma_{\text{eff}} \\ \vec{\xi}_{\pm 1} &= (\overline{\Delta}P_+, \mp i\Omega, \Omega_L) & \lambda_{\pm 1} &= \pm i\Omega - \gamma_{\text{eff}} \end{aligned} \tag{5.21}$$

with

$$\Omega = \sqrt{\Omega_L^2 + \overline{\Delta}^2 P_+^2} \tag{5.22}$$

Note that the eigenvectors $\vec{\xi}_0, \vec{\xi}_{\pm 1}$ given by equation (5.21) are not normalised to unit length. The precession frequency Ω depends on the Larmor frequency Ω_L and the light shift term $\overline{\Delta}P_+$. The deviation from the Larmor frequency is always positive and largest if the optical detuning is equal to the homogeneous linewidth ($\overline{\Delta} = 1$).

Evolution

Compared to the situation without light shift, the evolution changes in several ways. The main effect is that the rotation axis, which determines the eigenvector $\vec{\xi}_0$, is now in the xz plane. We may write the evolution as

$$\vec{m}(t) = \vec{m}_\infty + (\vec{m}_L + \vec{m}_C \cos(\Omega t) + \vec{m}_S \sin(\Omega t))\, e^{-\gamma_{\text{eff}} t} \tag{5.23}$$

where

$$\vec{m}_L = \frac{\overline{\Delta}P_+^2}{\gamma_{\text{eff}}\Omega^2} \begin{pmatrix} \Omega_L \\ 0 \\ -\overline{\Delta}P_t \end{pmatrix} \tag{5.24}$$

is the magnetisation component in the direction of the field $\vec{\Omega}$ and

$$\vec{m}_C = \frac{-P_+\Omega_L}{\Omega^2(\gamma_{\text{eff}}^2 + \Omega^2)} \begin{pmatrix} \overline{\Delta}P_+\gamma_{\text{eff}} \\ \Omega^2 \\ \gamma_{\text{eff}}\Omega_L \end{pmatrix} \qquad \vec{m}_S = \frac{-P_+\Omega_L}{\Omega(\gamma_{\text{eff}}^2 + \Omega^2)} \begin{pmatrix} -\Delta P_+ \\ \gamma_{\text{eff}} \\ -\Omega_L \end{pmatrix} \tag{5.25}$$

the components that precess in the plane perpendicular to the field.

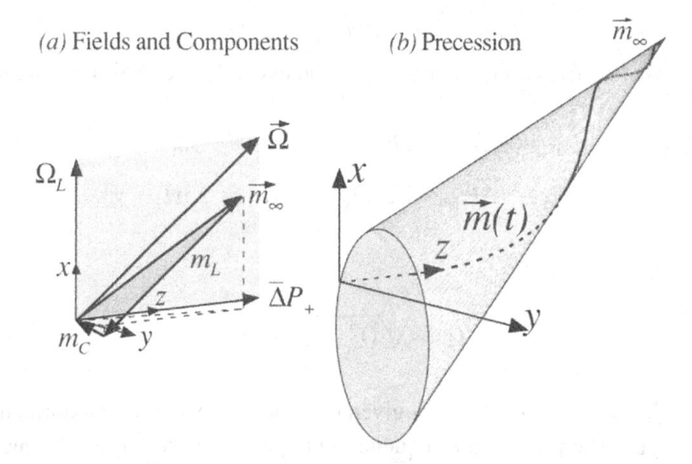

Figure 5.10. **(a)** Spatial orientation of the components of the magnetic field and the three vectors $\vec{m}_\infty$, $\vec{m}_{\mathrm{L}}$, and $\vec{m}_{\mathrm{C}}$ that add up to the initial magnetisation. **(b)** Evolution of the magnetisation vector as described by equations (5.23)–(5.25) on the surface of a cone.

Figure 5.10a shows the relation among these three vectors; the component $\vec{m}_{\mathrm{L}}$, which is parallel to the field direction, together with the orthogonal component $\vec{m}_{\mathrm{C}}$, adds up to $-\vec{m}_\infty$ for $t = 0$. In time the orthogonal component precesses around the field and the three time-dependent components decay at the rate γ_{eff}. The right-hand side of Figure 5.10 shows graphically the motion resulting from the precession and the damping of all three components. The curved line represents the tip of the magnetisation vector tracing out a curve on the surface of a cone. The tip of the cone indicates the stationary magnetisation $\vec{m}_\infty$; its location is determined by the system parameters (see equation (5.20)). The symmetry axis of the cone is parallel to the direction of the effective field $\vec{\Omega}$, and the initial condition (in our case the origin) determines the opening angle, as the corresponding point must lie on the surface of the cone. If the optical field is applied at exact resonance ($\overline{\Delta} = 0$), the virtual field due to the light shift vanishes and the cone collapses to a circle lying in the yz plane.

5.3 Modulated pumping

5.3.1 Motivation

Destructive interference

As discussed in Section 5.2.4, transverse magnetic fields can significantly reduce the efficiency of optical pumping. Possible countermeasures include increasing the pump rate through the laser intensity. In practice, however, this "solution" soon reaches limits imposed by the available laser intensity. To un-

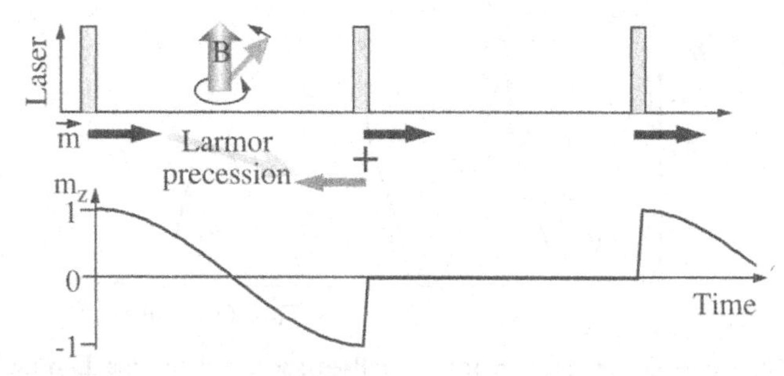

Figure 5.11. Destructive interference of magnetisation components that are created at different times and undergo Larmor precession.

derstand how the magnetic field reduces the efficiency of optical pumping and to develop a technique that avoids this problem, we first consider a discrete time evolution as shown in Figure 5.11.

The uppermost trace in Figure 5.11 shows a sequence of laser pulses that create packets of magnetisation at equidistant times. These packets of magnetisation, oriented in the z direction, are shown as black arrows in the figure. Between the pulses, the transverse magnetic field causes Larmor precession. If the distance between the pulses matches half a Larmor period, the precessing magnetisation vector reverses its direction, as the grey arrow shows. The newly created magnetisation exactly cancels the inverted magnetisation. On longer timescales, the magnetisation component m_z, shown in the trace near the bottom, then oscillates around zero.

In the case of continuous optical pumping, the magnetisation packets that are created at different times also undergo Larmor precession and interfere in a similar way. This destructive interference effect competes with the creation of magnetisation by optical pumping. The equilibrium magnetisation observed under stationary conditions thus represents a balance between these two opposing effects.

Resonant enhancement

This consideration suggests a method for restoring the efficiency of the optical pumping process: If the pulse separation is one Larmor period, the interference between the different packets becomes constructive rather than destructive. Figure 5.12 shows an experimentally observed signal that represents the z component of the magnetisation vector while a sequence of short laser pulses optically pumps the system. The delay between the pulses, indicated by arrows, is close to one Larmor period. Each pulse creates magnetisation in the same direction in which the average magnetisation vector currently points

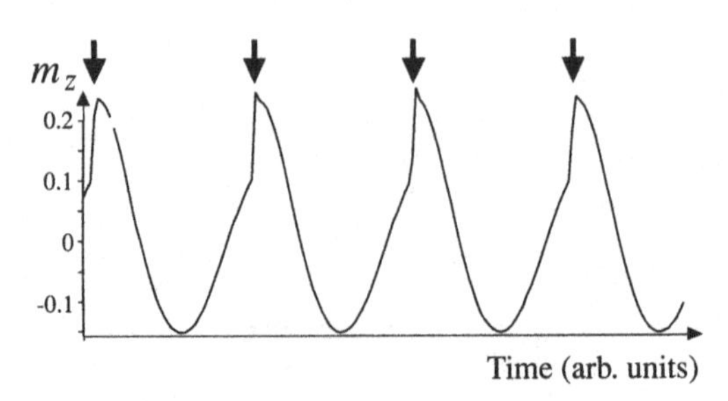

Figure 5.12. Constructive interference for pulses separated by one Larmor period. The arrows indicate the laser pulses.

when the pulse adds another packet of magnetisation. An infinite sequence of pulses drives the system towards a periodic state at approximately the unperturbed damping rate γ_0. The equilibrium polarisation depends on the average optical power of the pulse sequence and is resonantly enhanced whenever the pulse spacing is equal to the Larmor period or a multiple thereof (Fukuda et al. 1981; Mlynek et al. 1981a; Tanigawa et al. 1983; Bradley 1992).

Similar effects have been observed in nuclear magnetic resonance experiments (Morris and Freeman 1978), where sequences of pulses with small flip angles were applied to an ensemble of precessing nuclear spins. If the pulse separation is synchronised with the precession, the pulses flip the magnetisation each time by a small angle. The resulting motion of the magnetisation vector is reminiscent of Dante's passage into the underworld. In magnetic resonance, the experiment has therefore been termed DANTE.

Higher resonances

Resonant enhancement of the stationary magnetisation occurs not only when the separation between the pulses matches the Larmor period, but also whenever the magnetisation packets that are created at different times interfere constructively. For infinite pulse trains, such a positive interference occurs whenever the spacing between the pulses is an integer multiple of the Larmor period.

Figure 5.13 illustrates this behaviour. It shows the equilibrium polarisation for irradiation with an infinite sequence of equidistant pulses as a function of the pulse spacing. The enhancement of the polarisation is clearly visible whenever the pulse spacing is an integer multiple of the Larmor period. The decrease of the polarisation at longer pulse spacings is due to the decrease of the average optical power of the pulse sequence as the separation between the pulses increases.

These higher harmonics can nevertheless be useful, in particular when a pulse sequence with a spacing at the Larmor period cannot be generated for

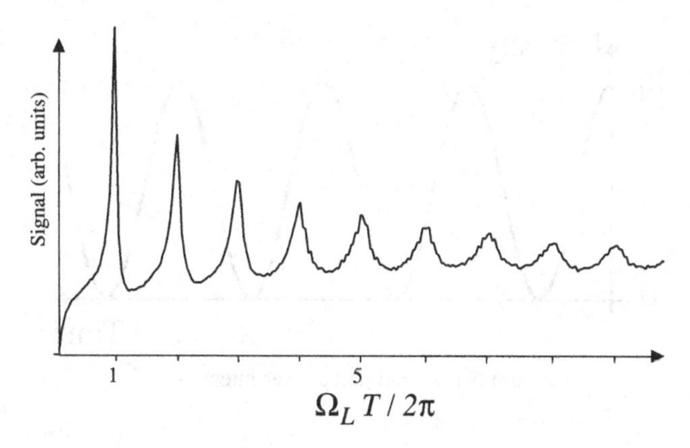

Figure 5.13. Magnetisation amplitude as a function of pulse spacing T.

technical reasons. Very high order resonances have been used in systems where the level splitting arises not from the Zeeman effect, but from the hyperfine splitting. Using the 110th harmonic, it was possible to excite coherence in the Cs ground state (Lehmitz and Harde 1986), where the level splitting is 9.2 GHz, using a mode-locked dye laser at 84 MHz. A necessary condition for using higher harmonics is that the optical pulses be short compared to the reciprocal of the splitting frequency. Measurements of GHz-frequency splittings therefore require picosecond pulses (Harde and Burggraf 1982; Bradley 1992).

Although the consideration of discrete pulses is convenient for a qualitative analysis of the situation, in practice it is often more useful to modulate continuously the optical pumping rate (Bell and Bloom 1961b; Corney and Series 1964; Suter and Mlynek 1991a; Klepel and Suter 1992), typically in a sinusoidal manner. If the peak laser power is limited, such continuous modulation schemes provide the highest pumping rates.

5.3.2 Equation of motion

For a quantitative analysis of the situation, we start from the equation of motion for the ground-state spin 1/2 system derived in Section 5.2.4. We assume that the laser intensity is modulated sinusoidally as

$$I(t) = I_0 \, (1 + \cos(\omega_m t)) \tag{5.26}$$

with a modulation frequency ω_m and mean intensity I_0. Experimentally, such a modulation can be achieved with an acoustooptic or electrooptic modulator acting on a cw-laser beam. See Figure 5.14.

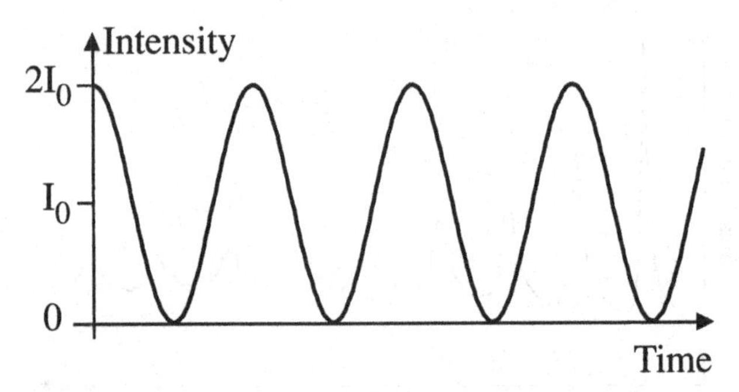

Figure 5.14. Modulated laser intensity.

The intensity modulation transfers directly to the pumping rate, which becomes

$$P(t) = P_{+0}\,(1 + \cos(\omega_m t))\,J_z \tag{5.27}$$

where P_{+0} is the pumping rate at the mean intensity I_0. Similarly, the light shift is modulated so that the Hamiltonian becomes

$$\mathcal{H}_m(t) = -\Omega_L\,J_x + 2\omega_z\,[1 + \cos(\omega_m t)]\,J_z \tag{5.28}$$

where $2\omega_z = \overline{\Delta} P_{+0}$ is the light shift at the mean intensity I_0.

Finally, the relaxation rate also becomes time-dependent, since

$$\gamma_{\rm eff}(t) = \gamma_0 + P_{+0}\,(1 + \cos(\omega_m t))$$

The equation of motion is now

$$\mathrm{d}\rho/\mathrm{d}t = -\mathrm{i}[\mathcal{H}_m(t),\rho] - \gamma_{\rm eff}(t)\,(\rho - \mathbb{1}) + P(t) \tag{5.29}$$

In this form, there is no simple analytical solution to this equation of motion. As in the case of the optical two-level system, we need to make an additional approximation.

5.3.3 Rotating frame

Transformation of the Hamiltonian

Modulated optical pumping is typically appropriate when the precession of the magnetisation is faster than the optical pumping rate. In this case, it is useful to transform the time-dependent equation of motion into a frame of refer-

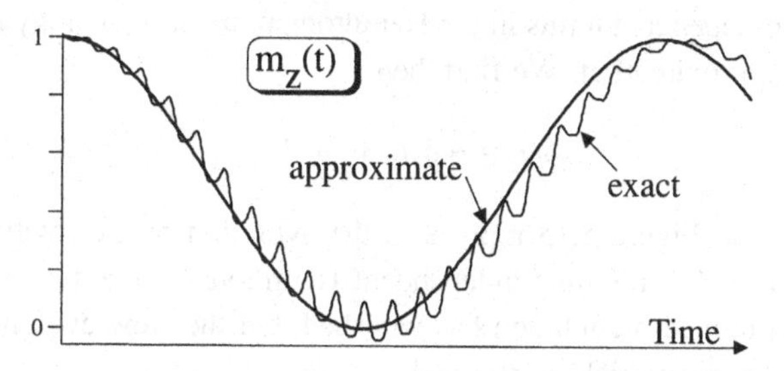

Figure 5.15. Evolution of the rotating-frame magnetisation component $m_z(t)$ from the exact equation of motion (wavy line) and the approximate solution that disregards nonresonant terms of the rotating frame Hamiltonian (thick line).

ence rotating at the modulation frequency ω_m, using the unitary operator

$$U(t) = e^{-i\omega_m t J_x} \tag{5.30}$$

The equation of motion in the rotating coordinate system is

$$d\rho^r/dt = -i[\mathcal{H}^r, \rho^r] + \hat{\Gamma}^r \rho^r + P^r \tag{5.31}$$

where the superscript indicates that the operators refer to the rotating reference frame. The density operator transforms as

$$\rho^r(t) = U^{-1}(t)\, \rho(t)\, U(t) \tag{5.32}$$

and the Hamiltonian

$$\begin{aligned} \mathcal{H}^r(t) &= U^{-1}(t)\, \mathcal{H}(t)\, U(t) - i\dot{U}^{-1}(t) U(t) \\ &= -\delta\, J_x + \omega_z[1 + 2\cos(\omega_m t) + \cos(2\omega_m t)]\, J_z \\ &\quad + \omega_z\, [2\sin(\omega_m t) + \sin(2\omega_m t)]\, J_y \end{aligned} \tag{5.33}$$

and $\delta = (\Omega_L - \omega_m)$ represents the difference between the sublevel splitting and the modulation frequency.

Figure 5.15 shows the evolution of the magnetisation as determined by this equation of motion after the initial transients have decayed. The wiggly line represents the exact solution of the equation of motion (5.29), obtained by numerical integration. If we are not interested in the fast evolution, we may neglect the time-dependent terms in the Hamiltonian, in close analogy to the rotating wave approximation. We find then

$$\mathcal{H}^{\mathrm{r}} = -\delta J_x + \omega_z J_z \tag{5.34}$$

The thick line in Figure 5.15 represents the evolution of the rotating frame magnetisation under this time-independent Hamiltonian. The fast oscillation due to the amplitude modulation has vanished, but the slow evolution under the effective magnetic field is retained.

The Hamiltonian in the rotating frame is formally the same as the unmodulated laboratory frame Hamiltonian. The main difference is that the field due to the sublevel splitting is reduced by the modulation frequency. If we set the modulation frequency to the Larmor frequency, the effective field vanishes in the rotating frame. In addition, the size of the light shift is smaller. If we compare to continuous irradiation at the maximum intensity, the light shift is reduced by a factor of four. Compared to irradiation at the same mean intensity, the reduction is a factor of two. This factor of two can be understood much in the same way as in the rotating frame transformation of the two-level atom discussed in Chapter 2: The linear oscillating field corresponds to two counterrotating components, each with half the amplitude. Only the component that rotates in the same direction as the freely precessing pseudo-spin (magnetisation in the present case) affects the dynamics of the system significantly, whereas the other component is neglected in the rotating wave approximation.

Optical pumping and relaxation

Apart from the Hamiltonian, we need to transform the optical pumping term. With the same transformation as for the Hamiltonian, we find

$$P^{\mathrm{r}}(t) = P_{+0}\frac{1}{2}\left[(1 + 2\cos(\omega_{\mathrm{m}}t) + \cos(2\omega_{\mathrm{m}}t))\,J_z + (2\sin(\omega_{\mathrm{m}}t) + \sin(2\omega_{\mathrm{m}}t))\,J_y\right] \tag{5.35}$$

and after the omission of time-dependent terms

$$P^{\mathrm{r}} = \frac{1}{2} J_z P_{+0} \tag{5.36}$$

This approximate form corresponds to the optical pumping rate that would be obtained with a continuous beam at half the mean intensity.

The third term in the equation of motion (5.29), which describes the relaxation processes, also has to be transformed into the rotating coordinate system. Since we have assumed, however, that the relaxation is isotropic, a transformation to the rotating frame does not change its form. We thus find

$$\gamma^{\mathrm{r}}_{\mathrm{eff}} = \gamma_0 + P_{+0} \tag{5.37}$$

In the general case of different relaxation rates for the three components, the time-averaged relaxation rate for the x component is the same in the rotating frame as in the laboratory coordinate system, while the y and z components become identical.

The resulting equation of motion for the system is completely analogous to that obtained with unmodulated light, except that the dynamics occur in a frame of reference rotating at the modulation frequency and the parameters are scaled. Using the mean intensity as the reference, the scaling factor is 1/2 for the light shift and optical pumping term and unity for the relaxation term.

5.3.4 Polarisation modulation

Principle

Instead of modulating the *amplitude* of the pump laser beam, it is also possible to use a constant amplitude and modulate the *polarisation* of the light between opposite circular polarisations. The modulation frequency should again be close to the Larmor frequency of the system. The polarisation that such a scheme generates in the atomic system changes sign twice during every modulation period.

Figure 5.16 shows this possibility again with a discrete model, where pulses of opposite circular polarisation, separated by half a Larmor period, optically pump the system. Again, the Larmor precession forces the magnetisation packets into precession. Since the polarisation of the pulses alternates between opposite circular polarisations, the newly created spin packets, represented by the black arrows, always point in the same direction as the magnetisation already present (represented by grey arrows).

In the following, we discuss in detail a continuous implementation of the polarisation-modulation scheme. As in the case of amplitude modulation, the polarisation modulation can resonantly increase the polarisation of the system if the modulation frequency is equal to the Larmor frequency. The maximum atomic polarisation that can be achieved with this scheme is somewhat higher than in the case of sinusoidal amplitude modulation. In addition, the alternation between opposite circular polarisations provides a higher symmetry than

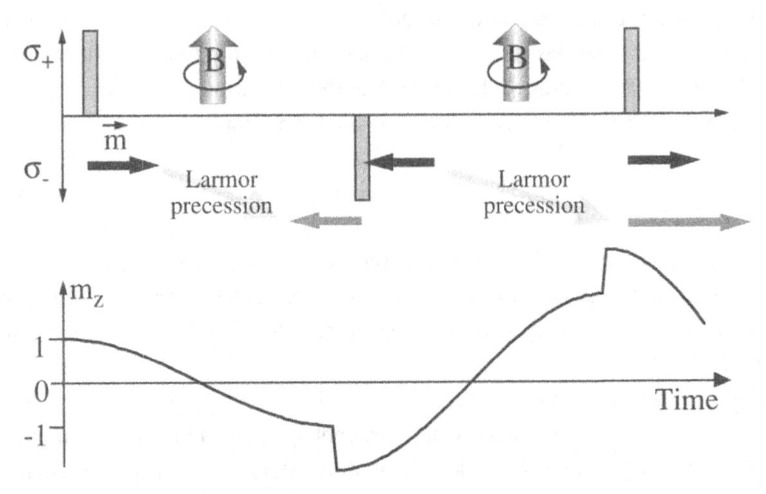

Figure 5.16. Discrete model of polarisation modulation.

the amplitude modulation of a single sense of circular polarisation. This higher symmetry can be advantageous in certain experimental situations.

Polarisation-modulated light

Changing the polarisation of the light in a time-dependent manner requires an electronically tunable retardation plate – in most cases an electrooptic modulator. We calculate the effect of the modulator on a continuous laser beam whose complex electric field amplitude $\vec{E}(t)$ at the entrance window of the modulator is

$$\vec{E}(t) = (E_{x0}, E_{y0}, 0)\ \mathrm{e}^{-\mathrm{i}(\omega_\mathrm{L}t-kz)} \tag{5.38}$$

This form leaves the polarisation of the incident light arbitrary. The z axis is the direction of propagation, whereas the x and y axes are aligned along the modulator principal axes. Behind the modulator of length L, the field is

$$\vec{E}(t,L) = (E_{x0}\ \mathrm{e}^{\mathrm{i}k_0 n_x L}, E_{y0}\ \mathrm{e}^{\mathrm{i}k_0 n_y L}, 0)\ \mathrm{e}^{\mathrm{i}\omega_\mathrm{L}t} \tag{5.39}$$

where k_0 is the vacuum wavevector and $n_{x,y}$ the indices of refraction of the principal axes.

We assume that the time-harmonic electric field with frequency ω_m modulates the index of refraction of the x direction:

$$n_x = n_{x0} + n_{x1}\cos(\omega_\mathrm{m}t) \tag{5.40}$$

If propagation effects of the modulating field can be neglected, the amplitude of the light behind the modulator is

$$\vec{E}(t,L) = (E_{x0}\, e^{ik_0 L n_{x1} \cos(\omega_m t)}, E_{y0}, 0)\, e^{i(\omega_L t + \phi)} \tag{5.41}$$

where we have assumed $n_{x0} = n_y$ (i.e., no static birefringence); ϕ represents the overall phase shift due to the propagation length. It has no effect on the behaviour of the system and we set it to zero in the following.

The quantity that determines the dynamics of the sublevel polarisation is the difference between the intensities of the circularly polarised components of the laser beam. For the general polarisation assumed here, the intensity difference can be written as

$$I_+ - I_- = \frac{1}{2}\sqrt{\frac{e_0}{\mu_0}}\,(|E_+^2| - |E_-^2|) = \frac{1}{c\,\mu_0}\,\mathrm{Im}\{E_x\, E_y{}^*\} \tag{5.42}$$

where $E_\pm = \mp (E_x \pm i\, E_y)/\sqrt{2}$ are the field amplitudes of the circularly polarised components. For the transmitted beam, we find

$$\begin{aligned} I_+ - I_- = \frac{1}{c\,\mu_0}\, E_{y0}\, E_{x0}{}^* \, [&i\, J_0(k_0 L n_{x1}) \\ &+ 2\, i \sum_{i=1}^{\infty} J_{2i}(k_0 L n_{x1}) \cos(2i\omega_m t) \\ &+ 2 \sum_{i=1}^{\infty} J_{2i-1}(k_0 L n_{x1}) \cos((2i-1)\omega_m t)] + \text{c.c.} \end{aligned} \tag{5.43}$$

where the J_i represent Bessel functions.

Modulation schemes

This form suggests two possible polarisations of the input beam: In the first scheme, the light is linearly polarised at ±45° with respect to the principal axes of the modulator. The product $E_{y0}\, E_{x0}{}^*$ of the two incident amplitudes is then real and the even order Bessel functions cancel in the sum with the complex conjugate. The intensity difference becomes an odd function of time, and the frequency spectrum contains only odd harmonics of the modulation frequency, as shown in the left-hand side of Figure 5.17.

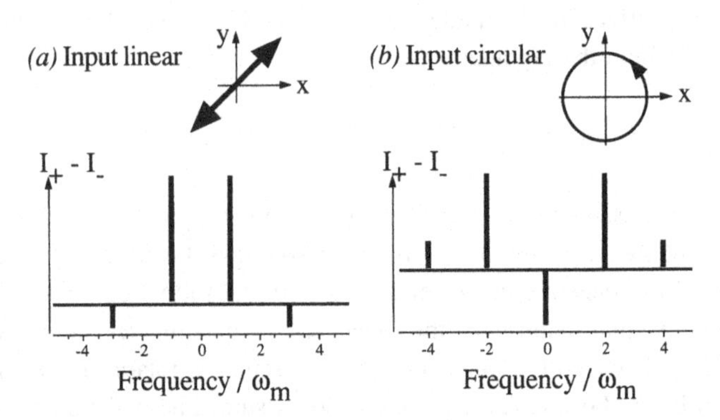

Figure 5.17. Spectral distribution of the intensity difference $I_+ - I_-$ between the two circular polarisations for linear (left) and circular (right) polarisation of the incident light.

The second scheme uses circularly polarised light at the input to the EOM. The product of the field amplitudes is then imaginary and the odd-order Bessel functions cancel. The frequency spectrum contains only the even harmonics of the modulation frequency. In both cases, the Bessel functions determine the intensities of the modulation sidebands.

Dependence on modulation amplitude

The efficiency of the optical pumping is determined by the amplitude of the sideband whose frequency matches the Larmor frequency. In the simplest case of linear input polarisation and modulation at the Larmor frequency, the efficiency is highest at the first maximum of J_1, i.e., for $n_{x1} \approx 1.84/(k_0 L)$.

Figure 5.18a illustrates the dependence of the pumping efficiency on the modulating amplitude n_{x1}. The lines represent the theoretical prediction of equation (5.43); the symbols represent data observed experimentally in sodium vapour (Klepel and Suter 1992). The left-hand side of the figure represents the case of linearly polarised light before the modulator, with the polarisation plane oriented at 45° with respect to the principal axes of the EOM. The optical pumping is then most efficient if the Larmor frequency of the system is an odd multiple of the modulation frequency. The four different curves were measured with the same setup and identical modulation frequencies by shifting the Larmor frequency through the magnetic field strength. The observed polarisations, represented by the discrete symbols, agree reasonably well with the theoretical curves. For experimental applications, the first-order harmonic will in most cases be used, since it provides the highest polarisation, but the

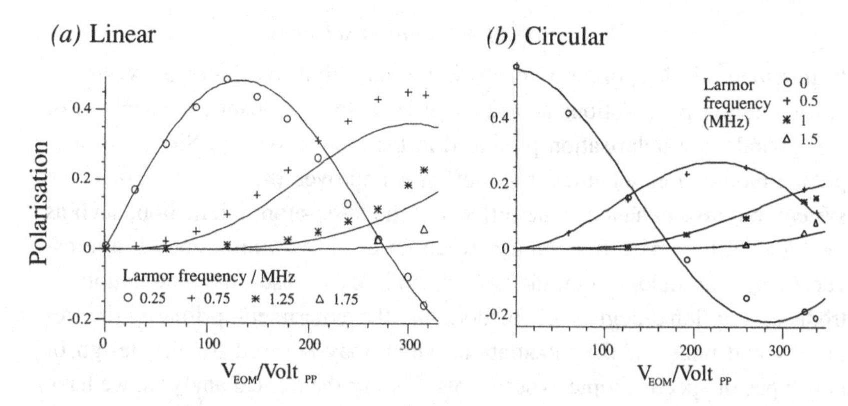

Figure 5.18. Relative polarisation of the atomic system during excitation with polarisation-modulated light for various magnetic field strengths. The solid lines represent the theoretical behaviour; the symbols correspond to the experimental points measured with four different field strengths. **(a)** The light incident on the EOM was polarised at 45° with respect to the principal axes of the modulator and the intensity was 1 mW/mm². **(b)** The light incident on the EOM was circularly polarised. The intensity of the pump laser beam was 0.5 mW/mm².

higher order harmonics may also be useful, if the desired frequency range cannot be reached directly.

If the pump beam is circularly polarised before the EOM, optical pumping is resonantly enhanced when the even harmonics of the modulation frequency match the Larmor frequency. The right-hand side of Figure 5.18 shows this behaviour, again for sodium vapour. For the second harmonic, the highest pumping efficiency occurs at $n_{x1} \approx 3.05/(k_0 L)$, i.e., at a field strength approximately 2/3 higher than for the first harmonic.

In both cases, the experimental pump laser intensity was quite low (0.5 and 1 mW/mm², respectively). This demonstrates that the modulation allows an efficient polarisation of the atoms in a transverse magnetic field even with low laser power. In addition, the low laser intensity makes certain that the observed effect is really resonant and only the nearest harmonic of the excitation is important. Under these conditions, the excitation efficiency drops rapidly if the modulation frequency deviates from resonance. The half width of the excitation bandwidth is of the order of 5 kHz. Although the modulation frequency of these demonstration experiments is quite low, higher frequencies are readily possible. Using rf electronics and tuned resonance circuits allows the excitation of magnetisation at Larmor frequencies in the MHz region and travelling wave modulators could extend this range into the GHz regime.

Discussion of modulation schemes

In this method, the polarisation of the resonant light oscillates between opposite circular polarisations at a rate that leads to a resonant enhancement of the ground-state polarisation produced in the atomic system. Similar to amplitude-modulation schemes, this not only improves the polarisation of the system, but also enhances other effects of the laser–atom interaction, such as the light shift, whose effect on the dynamics of the system is small if a transverse magnetic field acts on the system. As in these cases, the modulation introduces additional degrees of freedom into the experiment, primarily the frequency and phase of the modulation, which may be used for the design of new types of spectroscopic experiments. For our theoretical analysis, we have used a time-domain description of the field. Alternatively, it would be possible to use a frequency-domain interpretation and describe the modulation as a generation of sidebands.

For the $J=1/2 \leftrightarrow J'=1/2$ system considered here, the polarisation modulation is most efficient when it is performed between left and right circularly polarised light. This is not the only possibility, however, and other experimental situations may profit from different schemes, where the polarisation oscillates between different linear orientations or even more general circuits on the Poincaré sphere. It has been shown that such experiments can lead to geometrical effects ("Berry's phases") which depend on the trajectory on the Poincaré sphere (Pancharatnam 1956; Chiao and Wu 1986; Tomita and Chiao 1986; Berry 1987; Segert 1987; Bhandari and Samuel 1988; Chiao et al. 1988; Jiao, Wilkinson and Chiao 1989; Bose and Dutta-Roy 1991). Since the atomic system follows the polarisation of the light adiabatically, these effects should be observable in the resonant medium.

5.4 Multilevel ground states

5.4.1 Overview

The $J = 1/2$ ground state discussed in the preceding section is a convenient and very popular model system for the discussion of optical pumping effects. The success of this relatively simple model in describing many experimental results should not obscure the fact that it does not represent the real atom and therefore cannot give a full description of all experimental findings, since real atoms have significantly more complicated level structures. Some consequences of the level structures of real atoms are rather obvious – a good example is the hyperfine splitting. Other consequences, however, are more subtle and their influence on the experimental results may appear only in very specific situations.

A good example of this behaviour is the presence of a nonvanishing nuclear spin in systems with small hyperfine interactions. If the hyperfine splitting is smaller than the optical linewidth, it may appear tempting to assume that its influence on the behaviour of the system is negligible. We therefore discuss in this section the influence of the nuclear spin on the dynamics of optical pumping. As we shall see, the presence of the nuclear spin has large effects that do not vanish even if the hyperfine interaction is much smaller than the optical linewidth.

If the hyperfine splitting is large compared to the optical linewidth, there is an additional process that contributes to the evolution of the system, known as hyperfine pumping. The existence of an energy level splitting implies that even unpolarised light can create nonthermal population distributions in the ground state. This process is based on energy conservation rather than angular momentum conservation. Only ground-state atoms in one hyperfine state can absorb photons with the available energy and change into an excited state. From this state, spontaneous emission can bring it back to either of the two ground states.

The larger multiplicity implies that the description in terms of the magnetisation given above is not complete in these systems. Although angular momentum is transferred to the atoms, optical pumping also changes higher order multipole moments. Consider, e.g., a state with a total angular momentum $F = 2$. If all the population is transferred into the $m_F = 2$ substate, the system reaches its highest possible angular momentum. At the same time, it acquires alignment, i.e., quadrupole moment, as well as an octupole and hexadecupole moment. The creation of these higher multipole moments cannot be a one-step process, since no such order exists in the radiation field, but requires multiple absorption–emission processes. The vector sum of the angular momenta of different photons combines into these higher order moments.

5.4.2 Hyperfine pumping

Hyperfine pumping arises whenever atoms with nonvanishing nuclear spin interact with a light source whose spectral bandwidth is smaller than the hyperfine interaction. As shown in Figure 5.19, the hyperfine splitting makes the states $|1\rangle$ and $|2\rangle$ nondegenerate and the resonance frequencies for the transition from the two states to the excited state $|3\rangle$ differ. Depending on the hyperfine splitting, the width of absorption lines, and the laser linewidth, atoms in one ground state are more likely to be excited by the laser, whereas the probabilities for a decay from the excited state to either of the two ground states are often comparable.

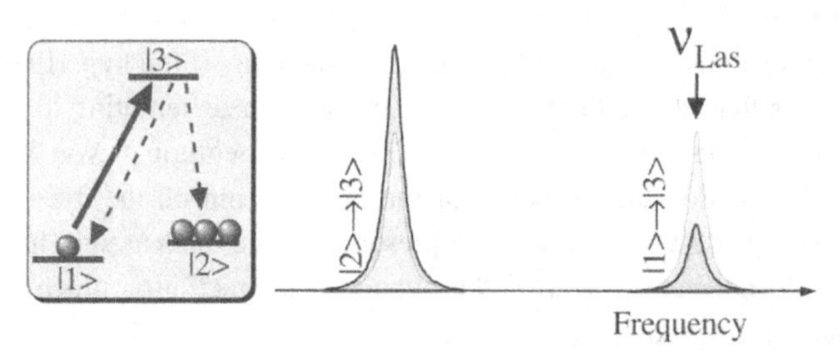

Figure 5.19. Typical situation for hyperfine pumping: The laser frequency is resonant with the transition between one hyperfine level and the excited state, but far from resonant with the other transition.

In the case of Doppler broadened systems, the hyperfine pumping depends on the atomic velocity. The absorption frequency of the atoms is determined not only by their angular momentum state, but also by their velocity component in the direction of the laser beam. In such a system, a pump laser beam excites one velocity group from one ground-state sublevel to an electronically excited state from which it may decay into either of two or more ground states. This process does not significantly affect the atomic velocity. As Figure 5.20 shows, the laser transfers atoms with a narrow velocity range from one ground state into another. This process, known as velocity-selective optical pumping (VSOP) (Pappas et al. 1980; Aminoff, J. Javanainen and Kaivola 1983), allows Doppler-free spectroscopy at laser intensities far below the intensity for optical saturation (Pinard, Aminoff and Laloe 1979; Bjorkholm, Liao and Wokaun 1982). It is also involved in many other phenomena that depend on the motion of atoms, such as light-induced drift (Gel'mukhanov and Shalagin 1980; Werij et al. 1984; Mariotti et al. 1988).

5.4.3 Degenerate multilevel systems

Multistep pumping

In systems with larger multiplicity, there is not only a single ground-state sublevel that can absorb light of a given polarisation. The atoms may then absorb several photons, each time falling back into a different state.

Figure 5.21 shows one possible history for an atom interacting with circularly polarised light: Each of the sequential absorption–emission processes increases the magnetic quantum number by one. Obviously, this is not the only possible sequence of events. In an unperturbed system, a single absorption–emission cycle with circularly polarised light can change the magnetic

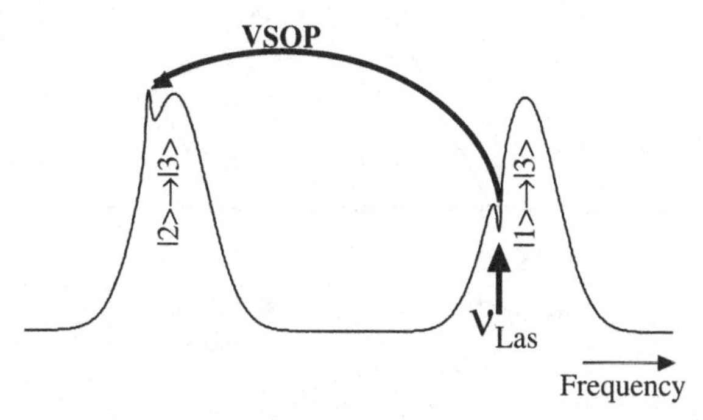

Figure 5.20. Velocity-selective optical pumping: The laser, represented by the arrow, excites one velocity group within the inhomogeneously broadened resonance line. The population of this velocity group is transferred to the transition from the other hyperfine state.

quantum number by 0, ±1 or ±2. In the presence of excited state mixing, a single absorption – mixing – emission event can transfer population between any two states. The sequence of Figure 5.21 reflects the average situation, however, and should indicate that a complete polarisation of such a system requires the absorption of many photons.

The lower part of the figure indicates how a sequence of optical pumping steps excites in the system not only angular momentum, but also higher-order multipole moments. Each step raises the order of the multipole moment by one. Although the absorption, as well as the emission, of a photon can change the angular momentum state of an atom, the emission has no preferred direction in space and its contribution to the average polarisation is usually small.

$J = 1$ *to* $J' = 0$

Optical pumping does not necessarily require circularly polarised light. Depending on the system, linearly polarised light may also induce nonthermal populations, not only in the case of hyperfine pumping and other nondegenerate systems, but also in systems with purely degenerate angular momentum states. For symmetry reasons, linearly polarised light cannot excite angular momentum in such a system, but creates alignment.

An important example is the $J = 1 \rightarrow J' = 0$ system shown in Figure 5.22. Linearly polarised light whose polarisation is parallel to the quantisation axis, interacts only with the $m_J = 0$ to $m_{J'} = 0$ transition. It therefore depopulates the $m_J = 0$ substate of the electronic ground state and increases the popula-

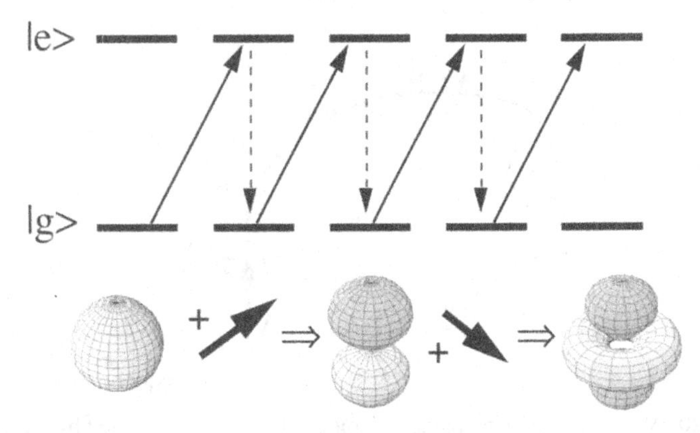

Figure 5.21. Multiple absorption processes in multilevel ground states polarise not only the dipole moment, but also, through the addition of the different dipoles, higher order moments.

tion of the ± 1 states. Obviously, the resulting state has a nonvanishing alignment. The reverse process to the one depicted in the figure, pumping with light polarised perpendicular to the quantisation axis, also generates alignment with opposite sign by selectively populating the $m_J = 0$ state at the expense of the $m_J = \pm 1$ states.

J = 0 to J′ = 1

Electronically excited states which are not populated in thermal equilibrium may be populated appreciably by sufficiently intense optical radiation. This population is in most cases selective and therefore anisotropic. The short lifetime of excited states implies that high intensity is necessary to achieve appreciable population. Since the population can be completely selective, however, selective population of excited state sublevels was observed long before the development of the laser (Hanle 1924), when the available light sources could bring only a small fraction of the total population into the excited state.

Figure 5.23 shows the process for the case of a $J = 0$ ground state and a $J' = 1$ excited state. Interaction with circularly polarised light populates the $m_{J'} = 1$ state selectively, thereby polarising the excited state dipole moment as well as its alignment. The polarisation of the excited states can be observed most efficiently through the polarisation of the fluorescence. Angular momentum conservation imposes a correlation between spatial direction and polarisation of the fluorescence. In the present example, the spin of the fluorescence photon must be oriented along the quantisation axis.

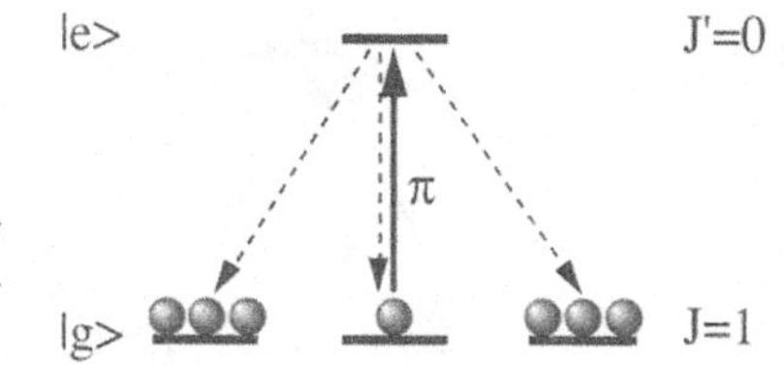

Figure 5.22. Optical pumping with linearly polarised light in a $J = 1 \rightarrow J' = 0$ system.

5.4.4 The sodium ground state

Hyperfine interaction

The $^2S_{1/2}$ electronic ground state of atomic sodium has a total of eight angular momentum substates, due to the presence of the nuclear spin $I = 3/2$. The hyperfine interaction of 1.8 GHz is, however, relatively small and is often not even observable in the optical resonance lineshape, as Figure 5.24 shows.

As the absorption spectra of Figure 5.24 show, the hyperfine splitting is readily observable in the Doppler-free spectrum. However, the Doppler-broadened background covers the excited state hyperfine splitting completely and reveals the ground-state splitting only as a shoulder. In the pressure-broadened spectrum at the top, the hyperfine splitting is observable only as a small asymmetry of the lineshape. This often serves as an argument that the nuclear spin does not have to be taken into account. As we shall see, however, it can still modify the dynamics significantly, even in the limit of a very large linewidth. To discuss these effects, we start with a quantitative analysis of the $J = 1/2$ model for the sodium ground state.

The $J = 1/2$ model

As seen in Section 3.5, optical pumping with circularly polarised light creates a population difference between the two sublevels of the $J = 1/2$ ground state. The pumping rate was given as a function of the Rabi frequency. For a quantitative analysis, we start by calculating the matrix elements of the electric dipole moment. The absolute value can be obtained, e.g., from the excited state lifetime. For the $^3P_{1/2}$ state of Na, the lifetime is $1/\Gamma_1 = 16$ ns. We calculate the dipole moment for the D_1 transition as (Demtröder 1991)

$$d^2 = \Gamma_1 \frac{3\epsilon_0 hc^3}{2\,\omega_0^3} \qquad d = 2.13 \cdot 10^{-29}\mathrm{Cm} \tag{5.44}$$

still within the $J = 1/2$ model.

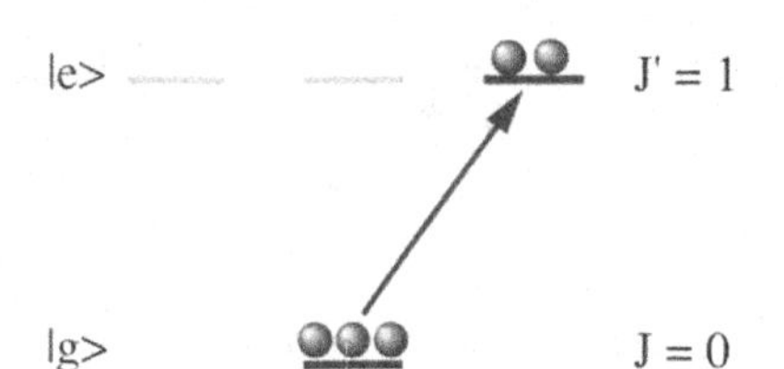

Figure 5.23. Excited state optical pumping in a $J = 0 \rightarrow J' = 1$ system.

To connect the measured laser intensity with the atomic interaction energy, we calculate the peak amplitude of a travelling electromagnetic wave of intensity I as $E_0 = \sqrt{2 I z_0}$ for linear polarisation or $E_0 = \sqrt{I z_0}$ for circular polarisation. In the rotating frame, the (constant) field strength becomes for circularly polarised light $E^r = E_0$ and the interaction energy is $E_0 d = \sqrt{I z_0}\, d = 2\omega_x$, where ω_x represents the Rabi flopping frequency. In the framework of the $J = 1/2$ model, the optical pump rate $P_{1/2}$, defined through

$$\mathrm{d}/\mathrm{d}t(\rho_{22} - \rho_{11}) = P_{1/2}(\rho_{22} - \rho_{11}) \tag{5.45}$$

is then

$$P_{1/2} = \frac{E_0{}^2 d^2}{\hbar^2 \Gamma_2} \tag{5.46}$$

For an intensity of 1 mW/mm^2, we find $P_{1/2} = 5.8 \cdot 10^5$/sec.

The full system

If the nuclear spin is taken into account, the situation is more complex, as shown in Figure 5.25. This figure also introduces a numbering scheme of the different sublevels that we shall use in the following.

One of the less obvious effects of the nuclear spin is the modification of the optical pumping rate. As discussed qualitatively above, the higher multiplicity of the ground state requires a larger number of photons for complete polarisation of the system. In addition, the individual transitions are weaker than the transitions in a hypothetical $J = 1/2$ system with vanishing nuclear spin. For the discussion of the optical effects, we assume that the laser intensity is well below the saturation intensity of the optical transition; we may then neglect the excited state populations. Furthermore, we neglect Doppler broadening. Experimentally, it is possible to satisfy both conditions by using buffer gas to achieve a homogeneous, pressure-broadened optical resonance line. Figure 5.26 shows the possible transitions between ground-state sublevels and angular momentum states of the $^2P_{1/2}$ excited state.

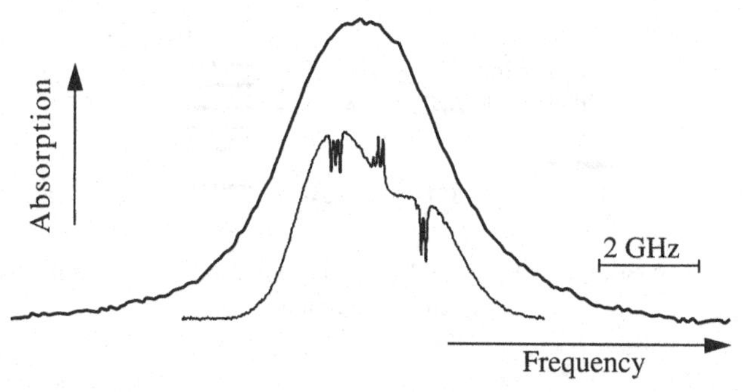

Figure 5.24. Na D_1 absorption lineshape in the presence of 200 mbar buffer gas (top) and, for comparison, a Doppler-free spectrum measured without buffer gas.

The arrows indicate those transitions, which couple to circularly polarised light propagating along the quantisation axis. The numbers close to the arrows indicate the squares of the electric dipole matrix elements; underscores indicate negative matrix elements. Each of these transitions contributes to the optical pumping and light shift effects.

Optical pumping rates

We start with the calculation of the rates for depopulation pumping, i.e., the number of atoms that are removed from a particular ground state by absorption of a photon. The loss of population from level i due to optical pumping is $\dot{\rho}_{ii}(t) = -k_i\, \rho_{ii}(t)$, where k_i is the depopulation pumping rate for the ground-state sublevel i. We express the individual rates in terms of the constant

$$k_0 = \frac{E_0{}^2\, d^2}{8\, \hbar^2\, \Gamma_2} \frac{1}{1 + (\Delta/\Gamma_2)^2} \tag{5.47}$$

where E_0 is the amplitude of the optical field, d the reduced dipole moment, Γ_2 the optical dephasing rate, and $\Delta = \omega_0 - \omega_{\mathrm{Las}}$ the detuning of the laser frequency ω_{Las} from the atomic resonance frequency ω_0. The individual rates are proportional to the squares of the dipole matrix elements. As shown in Figure 5.26, the individual rates k_i for levels $i = 1, \ldots, 8$ are

$$(k_1, k_2, \ldots, k_8) = k_0 \cdot (1, 2, 3, 0, 1, 2, 3, 4) \tag{5.48}$$

The repopulation rate, i.e., the number of atoms falling back into the individual ground states, is obtained from the conservation of the total population. In a pressure-broadened system, where the excited state reorientation is fast, every ground-state sublevel should receive 1/8 of the number of atoms

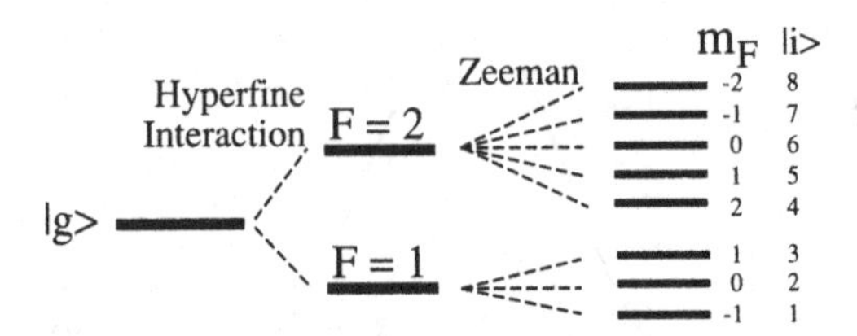

Figure 5.25. Level system of the ground state of atomic sodium taking the hyperfine structure into account.

that were excited. The equation of motion for the different populations is then

$$\dot{\rho}_{ii}(t) = -k_i\,\rho_{ii}(t) + \frac{1}{8}\sum_{i=1}^{8} k_i\,\rho_{ii}(t) \tag{5.49}$$

Starting from thermal equilibrium ($\rho_{ii} = 1/8$), irradiation with circularly polarised light changes the ground-state populations at an initial rate of

$$\dot{\rho}_{ii}(0) = \frac{k_0}{4} - \frac{k_i}{8} \tag{5.50}$$

which takes the values of $P_{1/2}/32$ (1,0,−1,2,1,0,−1,−2) for the eight sublevels. If the optical pumping is the only process driving the system, we may integrate the equation of motion (5.49) numerically, as shown in Figure 5.27, which displays the evolution of the eight sublevel populations as a function of time.

For this calculation, it was assumed that the system is initially in thermal equilibrium, corresponding to a population of 1/8 for each sublevel. As the laser beam is turned on, the three populations with the weakest coupling to the pump laser (1, 4, 5) start to increase, while two (2, 6) remain constant and the remaining three (3, 7, 8) decrease. Over a longer time, however, all populations decay towards zero, the only exception being the $m_F = 2$ level. This is the only substate not coupled to the field. In the long-time limit, all atoms end up in this single substate if relaxation effects can be neglected.

Experimental verification

In an actual experiment, it is often not possible or not interesting to observe these sublevel populations individually. The actual signal is a weighted su-

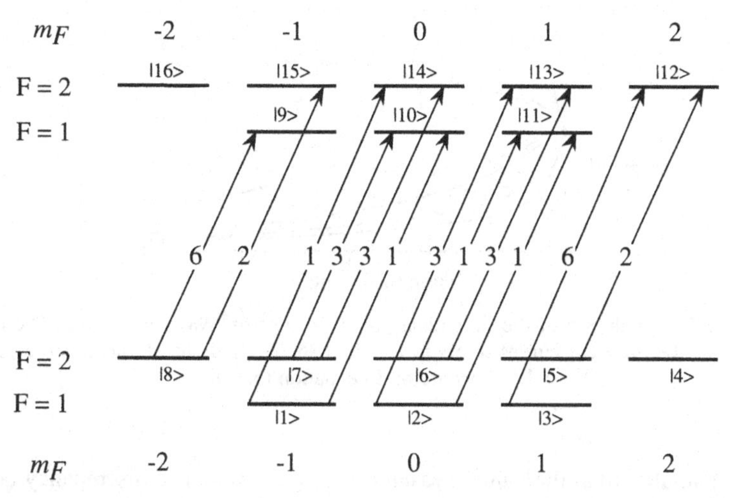

Figure 5.26. Level scheme of the Na ground state. The individual levels are labelled with the quantum number (F, m_F) and a running index starting at the lowest energy level. The arrows indicate optical transitions for circularly polarised light propagating in the direction of the quantisation axis and the numbers indicate the squares of the dipole matrix elements for these transitions.

perposition of all sublevel populations. Since these populations evolve at different rates, the overall signal becomes nonexponential. For an experimental confirmation of this behaviour, it is necessary to homogeneously illuminate the sample with a well defined intensity. This can be achieved by making the pump laser beam considerably wider than the probe beam, so that the observed signal stems only from the central part of the (Gaussian) pump beam, which is relatively homogeneous over the diameter of the probe beam, as in the example shown in the inset of Figure 5.28. The intensity at the centre of the probe beam was 23.2 mW/cm^2, corresponding to interaction energy $E_0 d/h$ of $3.5 \cdot 10^7$ rad/s. For the optical dephasing rate, a value of $\Gamma_2 = 9.4 \cdot 10^9$ rad/s was measured in the absorption spectrum.

Figure 5.28 shows the observed signal, together with the theoretical signals calculated for the $J = 1/2$ model and the full Na level system (equation (5.49)). The experimental parameters, together with theoretical analysis of the $J = 1/2$ model, predict an optical pumping rate of $P_{1/2} = 1.3 \cdot 10^5\ \text{s}^{-1}$, or a time constant $1/P_{1/2} = 7.7\ \mu$s, much shorter than the observed timescale. The curve calculated for the full system, on the other hand, reproduces the experimental signal quite well, especially the overall shape of the signal, which is clearly nonexponential. The agreement becomes almost perfect (not shown in the figure) if we account for the reflection losses from the entrance window to the sample cell by assuming that the optical beam power in the sample cell is 20

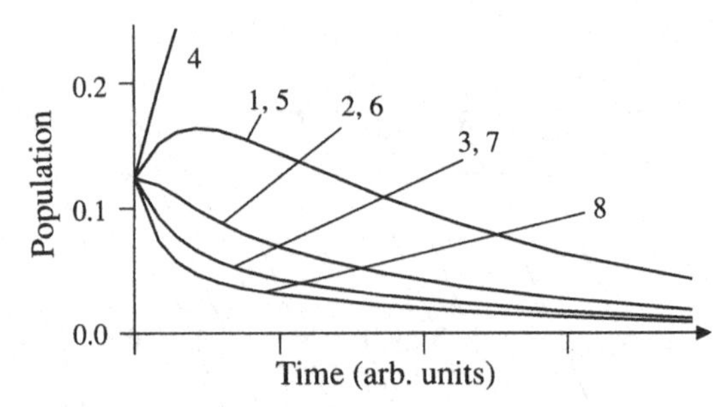

Figure 5.27. Evolution of the individual populations of the Na ground state. The labelling of the different curves refers to Figure 5.26. The individual curves were calculated according to equation (5.49).

percent smaller than the value measured before the cell. The discrepancy between the two nonexponential curves and the simple exponential predicted by the $J = 1/2$ model is surprisingly large. The time required to reach 50 percent polarisation is an order of magnitude longer for the actual level system than it would be in a true $J = 1/2$ system with no nuclear spin. In addition, the observed time dependence is not a simple exponential, indicating that the transition strengths are not equal for all transitions. The system has therefore several different optical pumping rates.

This comparison between the simple $J = 1/2$ model and the full level system reveals again that the simple model system is appropriate as long as the atomic medium is close to thermal equilibrium. As an example, the $J = 1/2$ model correctly predicts the total absorption of the Na ground state as long as the system is in internal equilibrium and the hyperfine coupling is small compared to the optical linewidth. When the system is far from equilibrium, however, as in this example of optical pumping, the effect of the nuclear spin becomes strongly apparent as the behaviour of the true system differs qualitatively from that of a hypothetical atom without nuclear spin. This behaviour does not depend on the relative size of hyperfine splitting and homogeneous linewidth, but persists even under conditions of arbitrarily high buffer gas pressure.

5.4.5 Light shift and damping

Rates

The calculation of light shift and damping effects in multilevel systems follows the same steps as for the $J = 1/2$ ground state. We add the effects of the

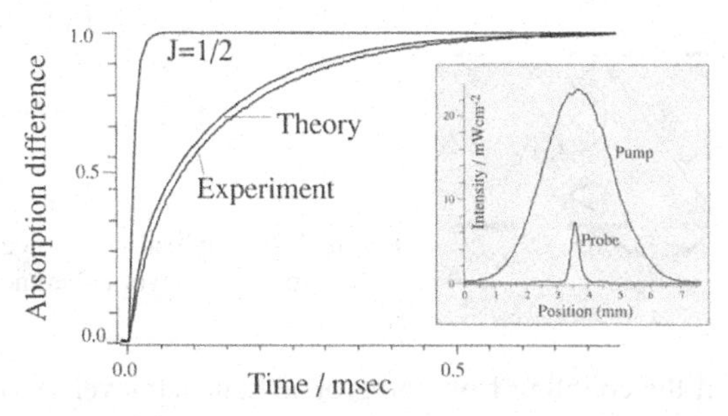

Figure 5.28. Evolution of the signal as a function of time when a circularly polarised pump beam is turned on at $t = 0$. The curve labelled "experiment" represents the experimental signal. The other two curves are theoretical functions calculated within the $J = 1/2$ model and the full Na ground state for the experimental parameters used in the actual experiment and no adjustable parameters. The inset shows the intensities of the pump and probe beam (recorded separately) at the centre of the sample cell. Both beam profiles are approximately Gaussian.

different possible transitions, which is possible as long as the intensity is low enough that the description of the dynamics in terms of light shift and damping is meaningful. We may then consider an individual optical coherence ρ_{rj} coupled to a ground-state coherence ρ_{ij} by the radiation field, as shown in Figure 5.29.

The three relevant levels represent an example of the three-level system discussed in Section 3.5. We therefore expect the ground-state coherence to evolve as

$$\rho_{ij}(t) = \mathrm{e}^{-\mathrm{i}(\delta_{ij} - i\gamma_{ij})t} \tag{5.51}$$

where light shift δ_{ij} and damping γ_{ij} can include contributions from laser field coupling to the transitions from sublevels $|i\rangle$ and $|j\rangle$ to any of the excited states. For the light shift, the net effect is the difference between the contributions by the two states, while the contributions to the damping add up:

$$\gamma_{ij} = \Gamma_2\,(k_j + k_i)/2 \qquad \delta_{ij} = \Delta\,(k_j - k_i) \tag{5.52}$$

where

$$k_i = \sum_r \frac{\omega_{ir}^2}{4(\Delta^2 + \Gamma_2^2)} \tag{5.53}$$

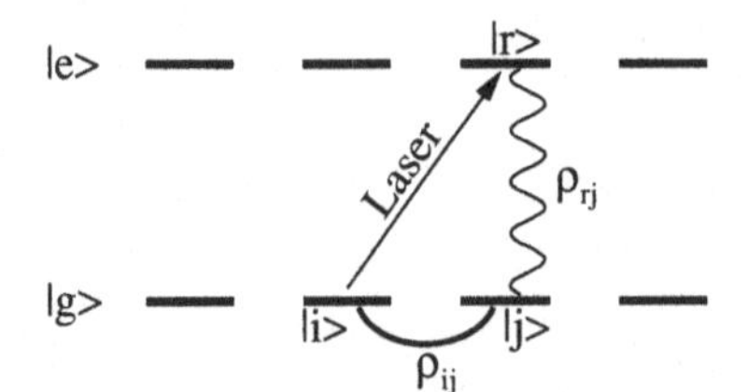

Figure 5.29. Optical coherence ρ_{rj} coupled to the sublevel coherence ρ_{ij}.

is the strength of the coupling from the ground-state sublevel $|i\rangle$ to all excited states. For the ground-state coherences of sodium, the damping rates γ_{ij} and light shifts δ_{ij} are thus proportional to the following numbers:

coh. ij	12	13	23	45	46	47	48	56	57	58	67	68	78
δ_{ij}	1	2	1	1	2	3	4	1	2	3	1	2	1
γ_{ij}	3	4	5	1	2	3	4	3	4	5	5	6	7

The light shift effect is proportional to the magnetic quantum number m_F and thus again equivalent to a magnetic field in laser beam direction. In general, light shift effects also contain contributions corresponding to electric quadrupole fields (Barrat and Cohen-Tannoudji 1961b; Barrat and Cohen-Tannoudji 1961a). These contributions have different signs for interactions with different hyperfine multiplets of the excited state. If, as in the present calculation, the excited state hyperfine interaction is small compared to the width of the resonance line, the different contributions cancel. Physically, the line broadening that leads to the cancellation arises from collisions of the atoms with the buffer gas that lead to a pressure broadening much bigger than the excited state hyperfine splitting. These collisions also cause a rapid reorientation of the excited state atoms.

5.4.6 Diamagnetic ground states

A frequently held misconception about optical pumping is that it should be applicable only to paramagnetic systems. The reason for this argument is that the optical radiation does not couple to the nuclear spin, but only to the electron. This is certainly correct, but the conclusion is too fast. A well-known example is the polarisation of noble gases (Happer et al. 1984). These experiments use optical pumping of alkali atoms, usually rubidium, in the presence of Xe gas. The two kinds of atoms collide and can associate for several microseconds in weakly bound van der Waals complexes. During this time, the

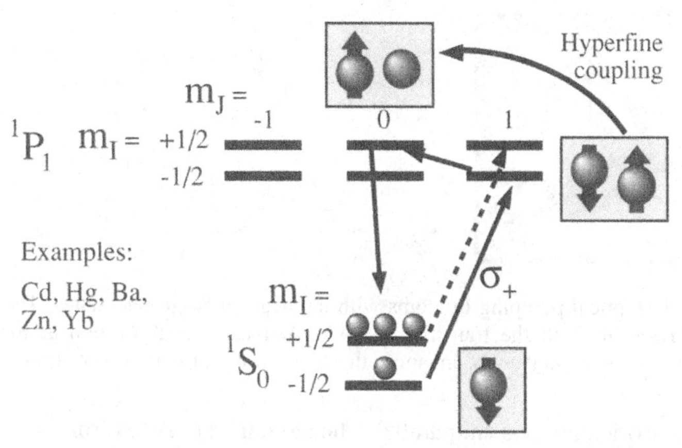

Figure 5.30. Schematic representation of the relevant processes that allow optical pumping of diamagnetic ground states.

two spins couple, mainly by magnetic dipole–dipole interaction. This coupling allows simultaneous spin flips of the two species, thereby transferring polarisation from the Rb atoms to the Xe nuclear spin. The process is slow but can yield polarisations close to unity. The resulting polarised atoms can be condensed into a liquid (Candela, Hayden and Nacher 1994) or frozen, and experiments can then be performed on an almost completely polarised solid (Cates et al. 1990).

Direct polarisation

It is not always necessary to use this indirect procedure, for it is also possible to pump atomic species with a diamagnetic ground state directly (Lehmann 1964). We discuss this possibility in a system with vanishing electron spin, as shown in Figure 5.30. The ground state is a $^1S_0$ state, and the excited state a $^1P_1$ state. The important point therefore is not the presence of an electronic spin, but of an orbital angular momentum in one of the two states. Since this is always the case for an allowed optical transition, this is not really a restriction. We assume for simplicity a nuclear spin $I = 1/2$.

The excitation of the atoms is independent of the nuclear spin; both ground-state sublevels therefore have the same probability to absorb a photon. If the light is circularly polarised, the absorption process brings it into the $m_{J'} = 1$ state, conserving the nuclear spin state. The excited state population is therefore asymmetric, its orbital angular momentum is polarised. The hyperfine coupling between the electronic and nuclear dipole moments allows an exchange of polarisation between the orbital angular momentum and the nuclear

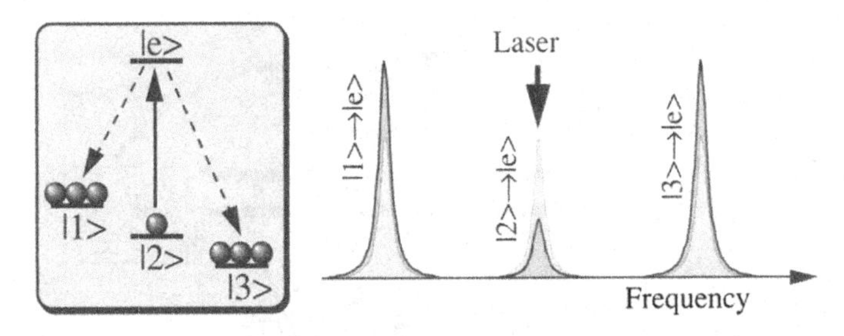

Figure 5.31. Optical pumping in atoms with nondegenerate ground states. The laser, which is resonant with the transition $|2\rangle \rightarrow |e\rangle$, reduces the absorption at the laser frequency and simultaneously enhances the absorption at the two other transitions.

spin if their orientation is antiparallel. The atom then changes from the $|m_{J'} = 1, m_I = -1/2\rangle$ state to the $|m_{J'} = 0, m_I = +1/2\rangle$ state. From there, it may return to the electronic ground state by spontaneous emission. The spontaneous emission process does not involve the nuclear spin and the atom ends up in the $m_I = +1/2$ state. An absorption-emission cycle therefore has a nonvanishing probability for transferring an atom from the nuclear spin state $m_I = -1/2$ into the $m_I = +1/2$ state. If the atom is initially in the $m_I = +1/2$ state, it can still absorb light, as indicated by the dashed arrow, but the excited atom cannot change its orientation and spontaneous emission returns it to the initial state. Overall, irradiation with polarised light increases the nuclear spin polarisation of the system. This possibility can be used, e.g., to prepare spin-polarised targets for nuclear physics experiments (Burns et al. 1977).

5.4.7 Spectral holeburning

Principle

Spectral hole burning (Voelker 1989; Holliday and Wild 1993) is a closely related process that occurs in inhomogeneously broadened optical systems whenever an atomic (or molecular) species has more than one nondegenerate ground state.

Figure 5.31 illustrates the principle: An atom has three nondegenerate ground states and a single excited state. If the laser linewidth is smaller than the ground-state sublevel splitting, it can be tuned to a single optical transition – in this example to the transition $|2\rangle \rightarrow |e\rangle$. As long as there is a finite probability that the system will decay from the excited state $|e\rangle$ to the other ground-state sublevels, the laser field depletes state $|2\rangle$ and increases the populations of the other levels. The absorption at the laser frequency therefore decreases, while the absorption at the other frequencies increases, as the right-hand part of the figure shows.

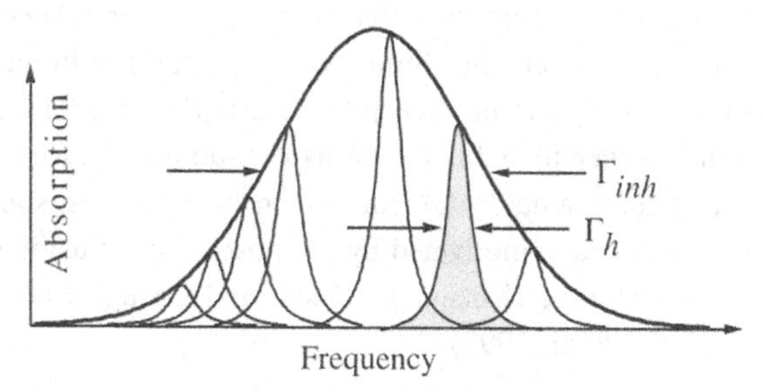

Figure 5.32. Inhomogeneous broadening of optical resonance lines, by superposition of homogeneous resonance lines associated with atoms in different environments. The homogeneous linewidth is Γ_{h} and the inhomogeneous width Γ_{inh}

Inhomogeneous broadening

Hole burning occurs when this optical pumping proceeds in solid materials, where the optical resonance lines experience strong inhomogeneous broadening. Figure 5.32 shows how this effect arises: Atoms at different positions in the solid environment experience different interactions with their environment. This interaction shifts their absorption frequencies, and the absorption lineshape of the sample, which is a weighted superposition of the contributions from all atoms, becomes much wider than the linewidth of the individual atom. Even in the case of rare earth ions, where this broadening effect is minimal, the widths of the optical resonance lines are several GHz – much larger than either the laser linewidth, the homogeneous optical linewidth, or the separation between the ground-state sublevels.

In such a system, the laser is always resonant with all possible transitions for some group of atoms. If the laser linewidth is narrower than the separation between their ground states, it optically pumps the resonant atoms into different ground states, as discussed above. The resulting reduced absorption at the laser frequency is referred to as a "hole" that the laser burns into the inhomogeneously broadened resonance line. Typical measurements of this effect use a pump-probe configuration, where a strong laser field burns a hole into the absorption line, while the frequency of a weaker second laser is scanned over the inhomogeneously broadened line close to the pump laser frequency. Since the number of atoms in the ground state remains constant, it finds not only regions of decreased absorption, but also frequencies where the absorption increases, corresponding to enhanced populations. The two regions of modified absorption are referred to as holes and antiholes, respectively.

Since the frequency separations between holes and antiholes correspond to sublevel splitting, the technique of spectral hole burning allows us to measure these splittings with a resolution not affected by the inhomogeneous broadening. In most cases, the laser jitter limits the frequency resolution. In commercially available ring dye lasers, the jitter is less than 1 MHz, and many orders of magnitude lower in stabilised lasers (Szabo and Kaarli 1991; Mitsunaga, Uesugi and Sugiyama 1993). Interest in hole-burning spectroscopy has increased in recent years, motivated by the possibility of using these systems for optical data storage (Mitsunaga, Yano and Uesugi 1991; Holliday and Wild 1993; Kohler et al. 1993).

6
Optically anisotropic vapours

From a macroscopic viewpoint, atomic vapours appear homogeneous and isotropic. As we have seen during the discussion of the microscopic dynamics that light induces in atomic systems, however, the interaction with light as well as external fields can break the rotational and translational symmetries and make the system both inhomogeneous and anisotropic. This microscopic anisotropy appears directly in the optical properties of the system. This chapter discusses the propagation of light in such a medium and describes experiments that analyse the transmitted light to obtain information about the microscopic state of the medium.

6.1 Isotropic atoms

6.1.1 The Lorentz–Lorenz model

Outline

The interaction of electromagnetic radiation with matter, in particular light, has inspired philosophers of nature for many centuries and has led to heated debates like the one between Newton and Goethe. The original interest was with the difference in absorption of various materials as a function of the optical wavelength and with the refraction of light, i.e., its dispersion.

The first theoretical analysis that connected these macroscopic effects to microscopic properties of the material was the Lorentz–Lorenz theory of dispersion (Lorentz 1880; Lorenz 1881), which was put forward shortly after the Maxwell equations and published in the year after Maxwell's death. (See Figure 6.1.) This theory models the material as a collection of dipoles, driven by the electromagnetic wave. We give here a brief summary of the theory, since its physical content is still the foundation of today's description although the mathematical formalism has changed significantly. It shows that absorption and dispersion are two manifestations of the same physical process.

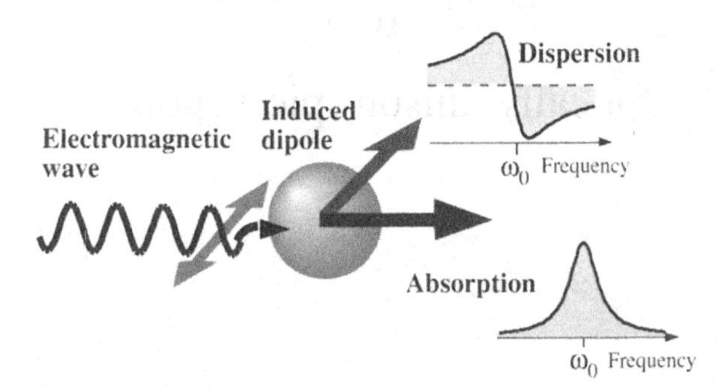

Figure 6.1. The Lorentz–Lorenz theory of dispersion.

Driven dipoles

An electric field interacting with a polarisable medium induces in the individual atom or molecule an electric dipole moment $\vec{\mu}_e = \alpha\vec{E}$, where α is the polarisability of the medium. As we want to discuss low-density atomic vapours, we may omit local field corrections and assume that the field acting on the atom is equal to the externally applied field. Using a one-dimensional coordinate system fixed at the atomic nucleus, we write the equation of motion for the electron as

$$F = m\,\ddot{x} = -U\,x - 2\,\gamma\,m\,\dot{x} - e\,E_0\,e^{i\omega t} \tag{6.1}$$

where U is a constant that characterises the atomic potential, m and $-e$ the mass and charge of the electron, γ a phenomenological damping constant, and E_0 the amplitude of a monochromatic light wave with angular frequency ω.

We seek a stationary solution of the form

$$x(t) = x_0\,e^{i\omega t} \tag{6.2}$$

Inserting this ansatz into the equation of motion, we find the amplitude

$$x_0 = \frac{e\,E_0}{m}\,\frac{1}{\omega^2 - U/m - 2\,i\,\gamma\,\omega} \tag{6.3}$$

We restrict the analysis to the resonant case where the frequency ω of the light is close to the eigenfrequency $\omega_0 = \sqrt{U/m}$ of the system, $\omega \sim \omega_0$. In terms of the laser detuning $\Delta = \omega_0 - \omega$, the oscillation amplitude is then

$$x_0 = \frac{e\,E_0}{2\,m\,\omega}\frac{-1}{\Delta + \mathrm{i}\,\gamma} = \frac{-e\,E_0}{2\,m\,\omega}\frac{\Delta - \mathrm{i}\,\gamma}{\Delta^2 + \gamma^2} \tag{6.4}$$

The real part is the component that oscillates in phase with the applied field, whereas the imaginary part is the out-of-phase component.

In a homogeneous medium, the induced dipoles form a macroscopic polarisation

$$P_\omega = (\epsilon - 1)\,\epsilon_0\,E_0 = \chi\,\epsilon_0\,E_0 = N\,\alpha\,E_0 = -N\,e\,x_0 \tag{6.5}$$

The subscript ω indicates that the induced polarisation oscillates at the frequency of the incident wave. Since the polarisation is proportional to the field strength, the more useful quantity is the optical susceptibility

$$\chi = \frac{N\,e^2}{2\,m\,\epsilon_0\,\omega}\frac{\Delta - \mathrm{i}\,\gamma}{\Delta^2 + \gamma^2} \tag{6.6}$$

i.e., the proportionality constant between the optical polarisation and the amplitude of the optical field. Both are assumed to be monochromatic and the susceptibility depends on their frequency through the laser detuning Δ.

Propagation

The optical polarisation acts as the source of an optical wave that propagates together with the incident wave. Interference between these two waves causes dispersion (i.e., changes in the velocity of propagation) and absorption (i.e., attenuation). To describe these effects, we start from Maxwell's equations for a homogeneous, isotropic medium. We look for harmonic plane-wave solutions with the wavevector parallel to the z direction

$$\vec{E} = \mathrm{Re}[\{E_x, E_y, 0\}\,\mathrm{e}^{\mathrm{i}(\omega t - k_z z)}] \qquad \vec{D} = \epsilon_0(1+\chi)\,\vec{E}$$

$$\vec{H} = \mathrm{Re}[\{H_x, H_y, 0\}\,\mathrm{e}^{\mathrm{i}(\omega t - k_z z)}] \qquad \vec{B} = \mu_0\,\vec{H} \tag{6.7}$$

where we have assumed that the material is nonmagnetic ($\mu = 1$). Solving Maxwell's equation

$$\vec{\nabla} \times \vec{E} = -\frac{\partial}{\partial t}\vec{B} \tag{6.8}$$

for the magnetic field we find

$$\begin{pmatrix} H_x \\ H_y \\ 0 \end{pmatrix} = \frac{k_z}{\mu_0 \omega} \begin{pmatrix} -E_y \\ E_x \\ 0 \end{pmatrix} \tag{6.9}$$

and inserting this relation into

$$\vec{\nabla} \times \vec{H} = \frac{\partial}{\partial t} \vec{D} \tag{6.10}$$

we obtain a one-dimensional wave equation that contains only the electric field amplitudes:

$$\begin{pmatrix} E_x \\ E_y \\ 0 \end{pmatrix} \frac{k_z^2}{\mu_0 \ \omega} = \begin{pmatrix} E_x \\ E_y \\ 0 \end{pmatrix} \omega \ \epsilon_0 \ (1 + \chi) \tag{6.11}$$

This equation requires that the wavevector is

$$k_z = \frac{\omega}{c} \sqrt{1 + \chi} = k_0 \sqrt{1 + \chi} \approx k_0 \left(1 + \frac{\chi}{2}\right) \tag{6.12}$$

where $k_0 = \omega/c$ is the vacuum wave vector. The approximate form holds for low-density atomic vapours ($\chi << 1$).

Inserting this solution into the ansatz, we find

$$\vec{E} = \mathrm{Re}\left[\begin{pmatrix} E_x \\ E_y \\ 0 \end{pmatrix} \mathrm{e}^{\mathrm{i}\omega t} \exp\left(-\mathrm{i} \ k_0 \ z\left(1 + \frac{\chi}{2}\right)\right)\right] = \mathrm{Re}\left[\begin{pmatrix} E_x \\ E_y \\ 0 \end{pmatrix} \mathrm{e}^{\mathrm{i}\omega t} \ \mathrm{e}^{-\mathrm{i}nk_0 z} \ \mathrm{e}^{-\alpha z/2}\right] \tag{6.13}$$

where $n = 1 + \mathrm{Re}\{\chi\}/2$ is the refractive index and $\alpha = -k_0 \ \mathrm{Im}\{\chi\}$ the absorption coefficient. This relation between the optical susceptibility χ, which can be derived from the microscopic properties of the material, and the coefficients α and n, which describe the propagation of light in this medium, was the main result of the Lorentz–Lorenz theory of dispersion.

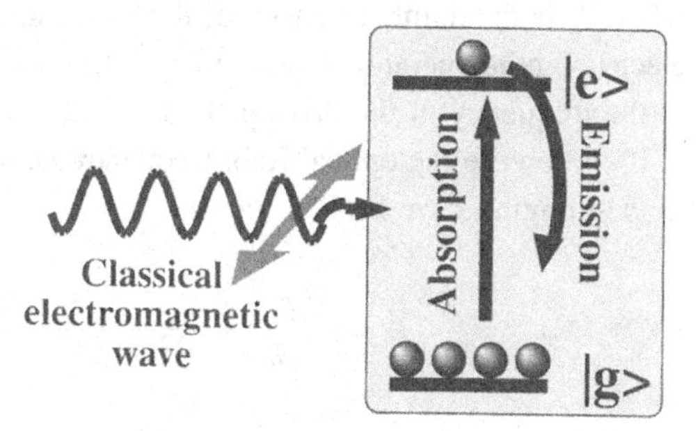

Figure 6.2. The semiclassical model of matter–radiation interaction treats the field classically, the matter quantum mechanically.

6.1.2 Semiclassical theory

Obviously, the original form of this theory could not survive the development of quantum mechanics. Today the fundamental theory for the description of light and its interaction with material systems is quantum electrodynamics, or QED. In many cases, however, it is sufficient to treat the electromagnetic field as a classical quantity and to use quantum mechanics only for the description of the material system. This semiclassical approximation is very closely related to the original Lorentz–Lorenz theory. The main difference lies in the derivation of the susceptibility χ from the microscopic model. In the simplest case, the semiclassical theory replaces the classical oscillating dipoles of the Lorentz–Lorenz theory by quantum mechanical two-level systems. See Figure 6.2.

Microscopic model of susceptibility

The semiclassical model uses the two-level system discussed in Chapter 2. We draw directly from those results and use the stationary solution of the driven two-level atom, equation (2.53):

$$(s_x,s_y,s_z)_\infty = \frac{1}{\Gamma_2^2 + \Delta^2 + \omega_x^2\Gamma_2/\Gamma_1} (\Delta\ \omega_x,\ \omega_x\ \Gamma_2,\ \Gamma_2^2 + \Delta^2) \qquad (6.14)$$

The transverse components s_x, s_y describe the two quadrature components of an oscillating electric dipole moment, which is in complex notation $s_x + \mathrm{i}\ s_y$. The macroscopic polarisation of the medium is then

$$\vec{P} = N\ \mu_e \frac{\omega_x}{\Gamma_2^2 + \Delta^2 + \omega_x^2\Gamma_2/\Gamma_1} (\Delta + \mathrm{i}\ \Gamma_2) \qquad (6.15)$$

where N is the number density of the atoms and μ_e the matrix element of the electric dipole operator. The oscillation frequency of this polarisation is equal to the frequency of the driving field.

If we rewrite the optical Rabi frequency $\omega_x = E\,\mu_e$ in terms of the electric field strength E, we find

$$\vec{P} = \frac{N\,E\,\mu_E^2}{\hbar}\,\frac{\Delta + \mathrm{i}\,\Gamma_2}{\Gamma_2^2 + \Delta^2 + E^2\,\mu_e^2\Gamma_2/\Gamma_1}$$

An important difference between this result and the classical theory is the behaviour for strong fields: When the third term in the denominator matches the other two terms, the polarisation reaches a maximum. For even stronger fields, the denominator becomes more important and the polarisation starts to decrease. This saturation effect does not occur in the classical dipole, which can become arbitrarily large.

To find the optical susceptibility, we consider the low field limit $E^2\mu_e^2 << \Gamma_1\Gamma_2$, where the polarisation is proportional to the field

$$\vec{P} = \chi\,\epsilon_0\,E \approx \frac{N\,E\,\mu_e^2}{\hbar}\,\frac{\Delta + \mathrm{i}\,\Gamma_2}{\Gamma_2^2 + \Delta^2}$$

We solve this equation for the susceptibility

$$\chi = \frac{N\,\mu_e^2}{\hbar\,\epsilon_0}\,\frac{\Delta + \mathrm{i}\,\Gamma_2}{\Gamma_2^2 + \Delta^2} \tag{6.16}$$

The expression (6.16) for the atomic susceptibility can be used to describe the propagation in exactly the same way as outlined in the context of the Lorentz–Lorenz model.

As seen in Section 6.1, the real part of the susceptibility determines the refractive index of the medium. Figure 6.3 represents the real part as a solid line, showing the well-known dispersive behaviour. The imaginary part determines the absorption coefficient; Figure 6.3 represents it as a dashed line, which shows the usual Lorentzian dependence on the laser detuning Δ. The main differences between the classical and the semiclassical model arise from the limitation on the size of the atomic dipole in the quantum mechanical case, which leads to saturation effects, and from the possibility of inverting the populations of a quantum mechanical system. In such an inverted medium, the polarisability can be negative (Steinberg and Chiao 1994b; Steinberg and Chiao 1994a).

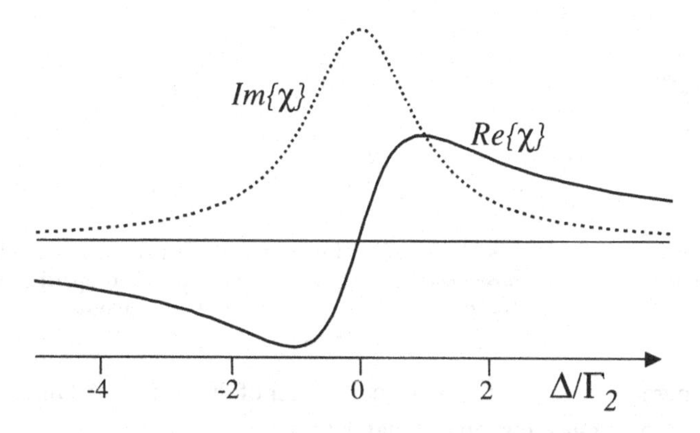

Figure 6.3. Complex susceptibility as a function of laser detuning Δ.

6.2 Anisotropic media

6.2.1 Introduction

Preliminaries

As emphasised before, the main deficiency of the two-level model for the description of atom–radiation interaction is lack of an orientation dependence of the interaction. Physically, the interaction between the electric dipole moment and the field involves three components. A quantum mechanical system with only two energy levels, however, cannot describe a three-component polar vector like the electric dipole moment, whose diagonal elements must vanish. This is not a major problem when the atomic medium is close to thermal equilibrium. A weakly perturbed atomic medium is isotropic and in most cases allows a description as a two-level system. Interactions with light or external fields can, however, break this symmetry. In the general anisotropic medium, the response to incident fields of different orientation must be analysed separately.

Nonequilibrium populations, as well as coherences between sublevels, can change the optical properties of the medium and in particular make it anisotropic. The optical properties of the medium depend on the details of the level structure and on the state of the system, i.e., on the individual populations and sublevel coherences. A completely general treatment is therefore not very informative. Instead, we use here as a specific example the $J = 1/2$ to $J' = 1/2$ level system discussed before. We assume that only the ground state is populated, but treat the population difference and the coherence analytically. In particular we do not make the assumption of cylindrical symmetry, which is often used in this context. As usual, we use the basis of angular mo-

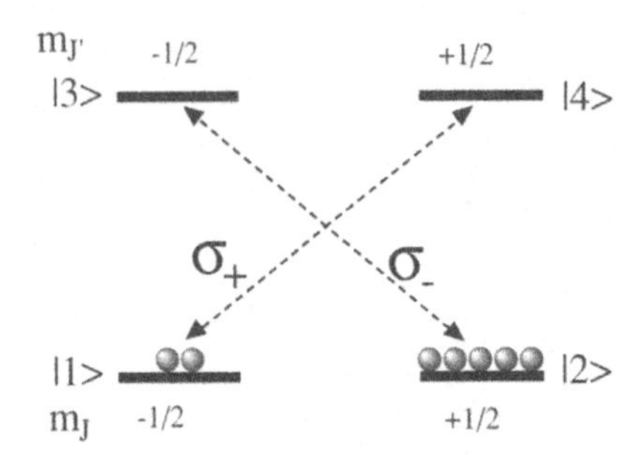

Figure 6.4. Effect of sublevel polarisation on the optical properties of an atomicvapour.

mentum operators to expand the density operator of the system. All three components of the angular momentum may be nonzero.

Circular dichroism and birefringence

As a simple and intuitive example of how the existence of a sublevel polarisation can affect the optical properties of a medium, we include a qualitative analysis for the standard level system shown in Figure 6.4.

In the example of Figure 6.4, where all atoms are in the electronic ground state, left circularly polarised light interacts only with the atoms in the $-1/2$ substate, whereas the opposite polarisation interacts only with the atoms in the $+1/2$ state. The absorptivity as well as the refractivity of an atomic medium is proportional to its density. Only the atoms in the sublevels to which the light couples contribute to the effective atomic density. A sublevel polarisation like the one shown makes the medium appear denser to σ_- light than to σ_+ light. If all atoms are in level $|2\rangle$, only σ_- light interacts with the medium, which appears completely transparent to σ_+ light. This difference in absorptivity for the two circular polarisations is known as circular dichroism. The size of the effect is directly proportional to the population difference between the two ground-state sublevels.

Not only does the absorption of the medium depend on the population difference, but the dispersion is also proportional to the population of the sublevel to which the light couples. The polarised medium exhibits circular birefringence. Linearly polarised light that propagates through such a medium becomes elliptically polarised by the differential absorption while the differential dispersion rotates its plane of polarisation.

This consideration directly shows the effect of a longitudinal (i.e., parallel to the laser beam) sublevel polarisation. If the cylindrical symmetry is broken, either by a magnetic field perpendicular to the laser beam, or by the presence of multiple laser beams, the ground-state polarisation is not necessarily

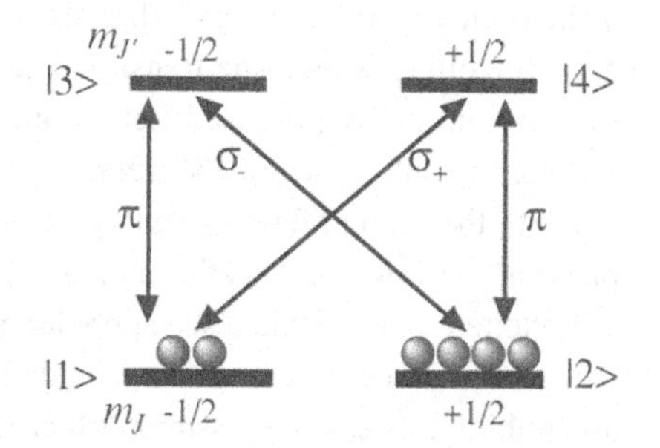

Figure 6.5. Schematic representation of the atomic level system interacting with theoptical field. Ground and excited states are both assumed to have an angular momentum $J = 1/2$. The quantisation axis is parallel to the k vector of the travelling wave.

parallel to the probe laser beam, and the influence of transverse components must also be taken into account. We include the effect of arbitrary ground-state polarisations by calculating a susceptibility tensor for the medium. With the help of this susceptibility tensor, it is then possible to calculate the propagation of light with arbitrary polarisation.

Apart from the phenomenological interest in the optical properties of polarised atomic vapours, we emphasise here the possibility of learning from the optical properties more about microscopic processes in the atomic medium. An alternative possibility for this is the detection of the fluorescence scattered from the atoms. Observation of the polarisation of scattered light (Kastler 1967) primarily provides information on the excited state. The polarisation depends only on the population of the excited state and the decay rates for the individual transitions to the ground state. Since the characteristics of the radiation that optically pumps the system are known, however, indirect information about the ground state is also available.

6.2.2 System response

Outline

For the analysis of the optical properties of polarised atomic media, we use a perturbation expansion, this time around the (nonthermal) state of the system, which interacts with a probe beam of small intensity. We first calculate the linear response of the atoms to an electromagnetic wave and subsequently the propagation of the wave through the medium. Only the optical coherences are affected in this approximation, while the populations and the Zeeman coherences remain invariant. We approximate the beam profile by a plane wave and use a one-dimensional model for describing the beam propagation.

For the calculation, we use the atomic model system represented in Figure 6.5. We assume that the atom interacts with the radiation field only by two sets of eigenstates of the free-atom Hamiltonian, which are separated roughly by the optical photon energy. The quantisation axis is parallel to the k vector of the travelling wave. The transitions labelled $\sigma_\pm$ correspond to the interaction with circularly polarised light while the transitions labelled π couple to the longitudinal modes of the field.

Due to their long lifetimes, multipole moments in the electronic ground state appear at the lowest intensities and are easier to study than those in electronically excited states. Excited state populations contribute to the signal in a closely analogous manner, but we disregard their presence for the calculation of the susceptibility, effectively assuming that the excited state population vanishes. We describe the light as a weak electromagnetic wave whose frequency is close to the resonance frequency of the optical transition. The experiments we have in mind are pump and probe experiments where a strong pump beam polarises the medium and a second, usually much weaker, laser beam probes the resulting dynamics of the coherences between the different ground-state sublevels (Mlynek and Lange 1979). The intensity of the probe beam must be low enough that it does not significantly affect the sublevel dynamics.

For this calculation, we do not take into account the presence of magnetic fields. Although it is well known that longitudinal as well as transverse fields affect the polarisation and intensity of the transmitted light, these linear magnetooptic effects can be separated from the effects discussed here and will be analysed separately in Section 6.2.3. They are related to the level structure of the system and to external fields but not to the microscopic state of the system.

Formalism

As usual, we parametrize the ground-state density operator by the magnetisation components

$$\vec{m} = (m_x, m_y, m_z) = (\rho_{12} + \rho_{21}, -\mathrm{i}(\rho_{12} - \rho_{21}), \rho_{22} - \rho_{11}) \qquad (6.17)$$

where ρ_{ij} represent the density operator elements and the substates are labelled as in Figure 6.5. The density operator components are then

$$\rho_{11} = (1 - m_z)/2 \qquad \rho_{22} = (1 + m_z)/2 \qquad \rho_{12} = \rho_{21}{\times} = (m_x + i\, m_y)/2 \qquad (6.18)$$

With the quantisation axis parallel to the k vector of the field, m_z represents a ground-state polarisation parallel to the direction of propagation, whereas m_x and m_y are the orthogonal components.

The Hamiltonian includes, besides the free atom Hamiltonian, the interaction between the atom and the applied laser field. With the conventional choice of phases for the individual states (Weissbluth 1978), the matrix representation of the electric dipole moment operator is

$$\mu_x = \mu_e \begin{pmatrix} 0 & 0 & 0 & 1 \\ 0 & 0 & 1 & 0 \\ 0 & 1 & 0 & 0 \\ 1 & 0 & 0 & 0 \end{pmatrix} \quad \mu_y = \mu_e \begin{pmatrix} 0 & 0 & 0 & i \\ 0 & 0 & -i & 0 \\ 0 & i & 0 & 0 \\ -i & 0 & 0 & 0 \end{pmatrix}$$

$$\mu_z = \mu_e \begin{pmatrix} 0 & 0 & -1 & 0 \\ 0 & 0 & 0 & 1 \\ -1 & 0 & 0 & 0 \\ 0 & 1 & 0 & 0 \end{pmatrix} \tag{6.19}$$

where μ_e represents the reduced matrix element.

Equation of motion

The atom interacts with a time-harmonic field of arbitrary direction, $\vec{E} = \{E_x, E_y, E_z\}$. The elements of the Hamiltonian in the rotating wave approximation are

$$\mathcal{H}^r = \begin{pmatrix} 0 & 0 & -\beta_0 & \beta_+ \\ 0 & 0 & \beta_- & \beta_0 \\ -\beta^*{}_0 & \beta^*{}_- & \Delta & 0 \\ \beta^*{}_+ & \beta^*{}_0 & 0 & \Delta \end{pmatrix} \tag{6.20}$$

with $\beta_\pm = \mu_e\,(E_x \pm iE_y)$, $\beta_0 = \mu_e\,E_z$ and Δ represents the detuning of the laser frequency ω below the atomic resonance frequency ω_0. From the Schrödinger equation

$$\dot{\rho} = -i[\mathcal{H},\rho] \tag{6.21}$$

we derive equations of motion for the optical coherences

$$\begin{aligned} \dot{\rho}_{13} &= i/2\,[\beta_- m_+ - \beta_0(1 - m_z)] & \dot{\rho}_{14} &= i/2\,[\beta_+(1 - m_z) + \beta_0 m_+] \\ \dot{\rho}_{23} &= i/2\,[\beta_-(1 + m_z) - \beta_0 m_-] & \dot{\rho}_{24} &= i/2\,[\beta_+ m_- + \beta_0(1 + m_z)] \end{aligned} \tag{6.22}$$

where $m_{\pm} = m_x \pm im_y$. For this perturbation expansion, we consider only the effect of the field on the optical coherences and assume that the intensity is low enough that the ground-state populations and coherences are not significantly affected.

To calculate the steady-state optical polarisation, we have also to include the effects of relaxation and the detuning of the laser frequency from optical resonance. Within the approximations made, laser detuning and relaxation affect all four optical coherences in the same way:

$$\dot{\rho}_{ij} = (\pm \, \mathrm{i} \, \Delta - \Gamma_2) \, \rho_{ij} \tag{6.23}$$

where Δ represents the resonance detuning and Γ_2 the optical dephasing rate. The plus sign applies above the diagonal, the minus sign below. To first order in the optical field, the coherences evolve towards the steady-state values

$$\begin{aligned} \rho_{13\infty} &= 1/2 \, [\beta_- m_+ - \beta_0(1 - m_z)] \, g(\Delta) \\ \rho_{14\infty} &= 1/2 \, [\beta_+(1 - m_z) + \beta_0 m_+] \, g(\Delta) \\ \rho_{23\infty} &= 1/2 \, [\beta_-(1 + m_z) - \beta_0 m_-] \, g(\Delta) \\ \rho_{24\infty} &= 1/2 \, [\beta_+ m_- + \beta_0(1 + m_z)] \, g(\Delta) \end{aligned} \tag{6.24}$$

where $g(\Delta)$ represents the lineshape function

$$g(\Delta) = \frac{-\Delta + \mathrm{i}\Gamma_2}{\Delta^2 + \Gamma_2^2} \tag{6.25}$$

which describes a complex Lorentzian.

Susceptibility

In Cartesian coordinates, the optical polarisation is

$$\vec{P} = N \, \mathrm{tr}\{\vec{\mu}_\mathrm{e} \, \rho_\infty\} \tag{6.26}$$

Using the stationary solution to the equation of motion, we find

$$\vec{P} = \begin{pmatrix} P_x \\ P_y \\ P_z \end{pmatrix} = \epsilon_0 \, \chi_0 \, (\Delta) \left[\begin{pmatrix} E_x \\ E_y \\ E_z \end{pmatrix} + \mathrm{i} \begin{pmatrix} m_y E_z - m_z E_y \\ m_z E_x - m_x E_z \\ m_x E_y - m_y E_x \end{pmatrix} \right] \tag{6.27}$$

where $\chi_0(\Delta) = N \mu_e^2 g(\Delta)/(h\epsilon_0)$ represents the susceptibility of the unpolarised medium ($\vec{m} = 0$). If we rewrite this expression as

$$\vec{P} = \epsilon_0 \chi_0(\Delta) (\vec{E} + \mathrm{i} \vec{m} \times \vec{E}) \tag{6.28}$$

it becomes apparent that the effect of the sublevel polarisation $\vec{m}$, represented by the last term in equation (6.28), is a rotation of the optical polarisation with respect to the electric field. In tensor notation, the optical susceptibility is $\vec{P} = \epsilon_0 \chi(\Delta) \vec{E}$ with

$$\chi(\Delta) = \chi_0(\Delta) \begin{pmatrix} 1 & -\mathrm{i}\, m_z & \mathrm{i}\, m_y \\ \mathrm{i}\, m_z & 1 & -\mathrm{i}\, m_x \\ -\, m_y & \mathrm{i}\, m_x & 1 \end{pmatrix} \tag{6.29}$$

The main effect of the population difference m_z is an apparent change of the number of atoms that the circular polarisation components "see." This leads to circular dichroism and circular birefringence for a wave propagating parallel to the quantisation axis. In the susceptibility tensor, this is reflected by the element that connects the x and y components. The primary effect of the sublevel coherence m_x, m_y is a coupling between the transverse and longitudinal field modes. In the presence of sublevel coherences, the transverse field components E_x, E_y induce polarisation parallel to the direction of propagation. The m_x, m_y components therefore connect the transverse components to the longitudinal direction.

6.2.3 Magnetooptic effects

Besides nonthermal polarisation of the material, there are other possible causes of optical anisotropy in atomic vapours. The most important are the so-called magnetooptic effects, the changes in the optical properties due to static magnetic fields. These effects were first observed during the nineteenth century and provided some of the most important clues to the structure of atoms as the building blocks of matter, as well as to the nature of the interaction between electromagnetic waves and matter. We discuss these effects independently of the sublevel polarisation considered so far, assuming for this purpose that the system is close to thermal equilibrium.

Faraday

Phenomenologically, the Faraday effect is a rotation of the plane of polarisation of the light, which propagates in the direction of a magnetic field, as

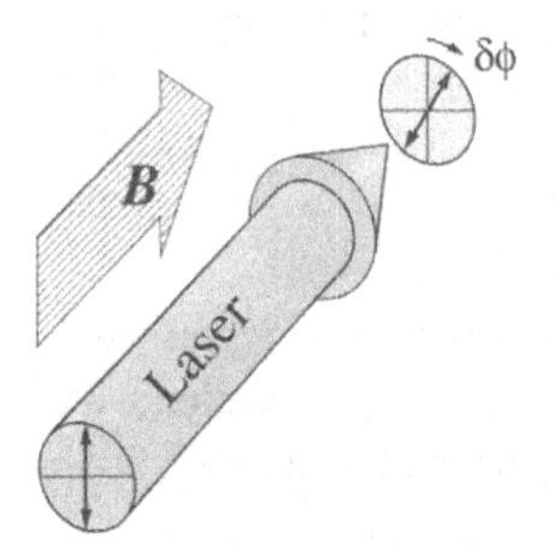

Figure 6.6. Faraday effect: A magnetic field parallel to the direction of propagationrotates the polarisation of the light.

shown in Figure 6.6. The rotation angle is proportional to the magnetic field strength and to the length of the interaction region and it depends on the material. The effect is well known in dielectric solids, e.g., in glass containing suitable metallic components, but we restrict ourselves to resonant atomic vapours, where it was first investigated by Macaluso and Corbino (Macaluso and Corbino 1898; Macaluso and Corbino 1899; Rosenfeld 1929). Very similar effects are found in molecules (Serber 1932).

For a microscopic analysis of the effect, we choose the model system of Figure 6.7, consisting of a $J = 0$ ground state and a $J' = 1$ excited state. The magnetic field shifts the energy levels of the excited state by $-B\,\mu_B\,g_J\,m_J$. The resonances for $\sigma_\pm$ light, which couple the electronic ground state to different excited states, as shown in Figure 6.7, now have slightly different resonance frequencies. The separation between the two curves is for the chosen system $2\,\hbar\,B\,\mu_B\,g_J$. The difference between the refractive indexes for left and right circularly polarised light makes the sample circularly birefringent. As discussed in Section 6.3.2, circular birefringence rotates the plane of polarisation of transmitted light by an angle

$$\delta\phi = \frac{\omega\,L}{2\,c}\,(n_+ - n_-) \tag{6.30}$$

where ω is the optical frequency, L the sample length, and $n_\pm$ the refractive indexes for circularly polarised light. The difference between the two curves shown at the bottom of Figure 6.7 is proportional to the Faraday rotation angle.

Other contributions to the Faraday rotation include population differences between magnetic substates whose energy is shifted by the magnetic field. In contrast to the effect discussed above, this contribution to the Faraday rota-

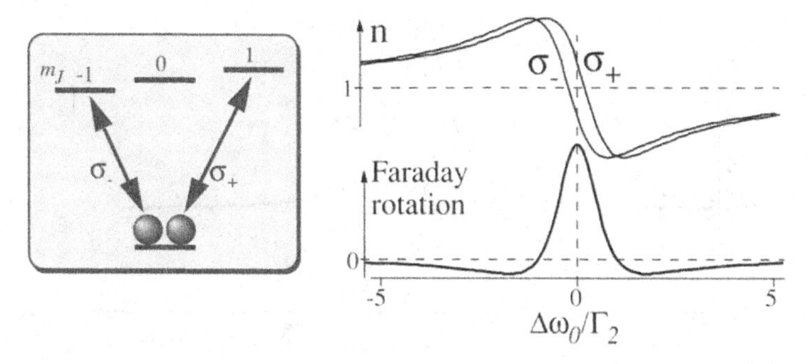

Figure 6.7. $J = 0 \rightarrow J' = 1$ model atom for the analysis of the resonant Faraday effect and refractive index n for circularly polarised light in a longitudinal magnetic field. The difference between the two curves yields the Faraday rotation angle shown at the bottom.

tion is temperature-dependent (Rosenfeld 1929). It becomes important when the magnetic-field induced energy level splitting is no longer negligible compared to the thermal energy kT. Another effect that contributes to the Faraday effect is the mixing between hyperfine states when the Zeeman interaction is no longer negligible compared to the hyperfine splitting of the ground or excited state (Chen, Telegdi and Weis 1987).

Assuming that the system remains close to thermal equilibrium, as we did here, the Faraday effect is independent of the laser intensity. At higher intensities, one observes with the same experimental setup much stronger effects, which depend on the laser intensity (Ducloy, Gorza and Decomps 1973; Gawlik et al. 1974a; Gawlik et al. 1974b). This nonlinear Faraday effect (Drake, Lange and Mlynek 1988; Barkov, Melik-Pasheayev and Zolotorev 1989; Chen, Telegdi and Weis 1990b; Zetie et al. 1992; Kanorsky et al. 1993; Weis, Wurster and Kanorsky 1993) differs qualitatively from the linear effect: The laser light drives the system far from equilibrium through an optical pumping process. The atomic medium acquires population differences and sublevel coherences in the electronic ground-state and, under still higher optical intensity, also in the excited state. In our formalism, the analysis of the optical anisotropies due to the order in the ground-state sublevels covers this situation. In such an experiment, the single laser beam combines the pump and probe functions: It establishes a polarisation in the medium, as described in Chapter 5, and it probes the optical anisotropy, as this section discusses.

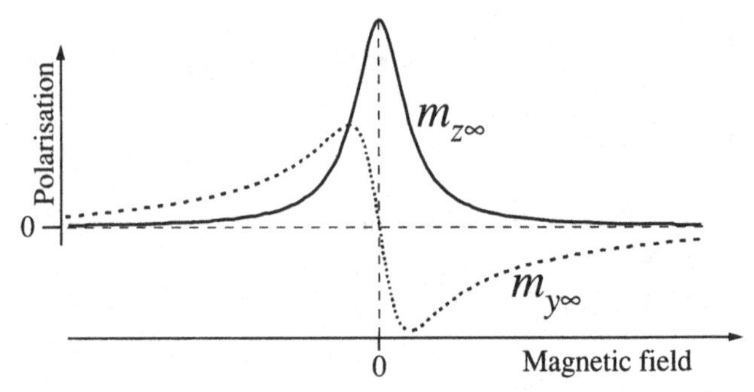

Figure 6.8. Magnetic field dependence of the two components of the excited state polarisation.

Hanle effect

As emphasised before, the conservation of angular momentum requires that the fluorescence from an anisotropic excited state be polarised. This is the case for free atoms. In 1924, however, Hanle observed (Hanle 1924) that a magnetic field can destroy the polarisation of the fluorescence.

For an analysis of this situation, we use the $J = 1/2 \rightarrow J' = 1/2$ system and consider only the excited state. If the magnetic field is applied in the x direction and the optical irradiation populates level $|3\rangle$ at a constant rate P, the equations of motion for the excited-state part of the density operator are in our usual notation

$$\dot{m}_y = -\Gamma_1\, m_y - \Omega_L\, m_z \qquad \dot{m}_z = -\Gamma_1\, m_z + \Omega_L\, m_y + P \tag{6.31}$$

where Ω_L is the Larmor frequency and Γ_1 the spontaneous emission rate. The stationary solution of this equation is

$$m_{y\infty} = \frac{-P\,\Omega_L}{\Gamma_1^2 + \Omega_L^2} \qquad m_{z\infty} = \frac{\Gamma_1\, P}{\Gamma_1^2 + \Omega_L^2} \tag{6.32}$$

As Figure 6.8 shows, the two components depend on the magnetic field strength like a complex Lorentzian. The longitudinal component reaches a maximum when the magnetic field vanishes. In this situation, the levels are degenerate, and the measurement of the polarisation as a function of the magnetic field strength is often referred to as a level crossing experiment. Depending on the details of the experimental setup, the measured signal is proportional to $m_{z\infty}$ or to $m_{y\infty}$.

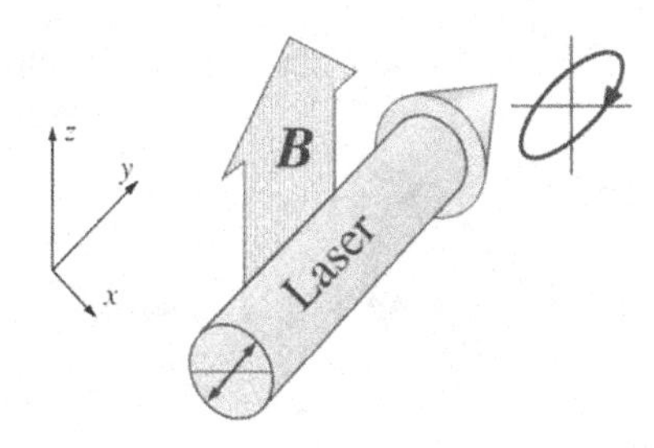

Figure 6.9. Voigt effect: a magnetic field perpendicular to the direction of propagation makes the medium birefringent.

Voigt

Closely related to the Hanle effect is the Voigt effect (Voigt 1908; Colegrove et al. 1959): Again, a magnetic field is applied perpendicular to the incident laser beam. If the laser beam is polarised parallel to the magnetic field, it induces in the medium an optical polarisation along the field direction, which is not affected by the presence of the field. A laser beam polarisation perpendicular to the field, however, induces an electric dipole moment perpendicular to the field. The magnetic field forces a precession of this electric dipole moment (see Figure 6.9).

To see how the magnetic field causes the precession of the electric dipole moment, consider the matrix representation of the two operators:

$$\mu_x = \frac{\mu_e}{\sqrt{2}} \begin{pmatrix} \cdot & 1 & \cdot & 1 \\ 1 & \cdot & \cdot & \cdot \\ \cdot & \cdot & \cdot & \cdot \\ 1 & \cdot & \cdot & \cdot \end{pmatrix} \qquad J_z = \begin{pmatrix} \cdot & \cdot & \cdot & \cdot \\ \cdot & 1 & \cdot & \cdot \\ \cdot & \cdot & 0 & \cdot \\ \cdot & \cdot & \cdot & -1 \end{pmatrix} \tag{6.33}$$

where we have assumed that the optical field is polarised along the x axis. Clearly, the two operators do not commute. The Zeeman interaction therefore forces a precession of the induced electric dipole moment. In close analogy to the Hanle effect, the stationary optical polarisation is reduced by a factor

$$v_x = \frac{\Gamma_2^2}{\Gamma_2^2 + \Omega_L^2} \tag{6.34}$$

Since the susceptibilities in x and z directions are different, the medium is optically anisotropic as soon as the magnetic field is not zero. This optical anisotropy causes a rotation of the plane of polarisation as well as a differential absorption for the two orthogonally polarised components of the light. In addition, it leads to a displacement of a laser beam propagating through the

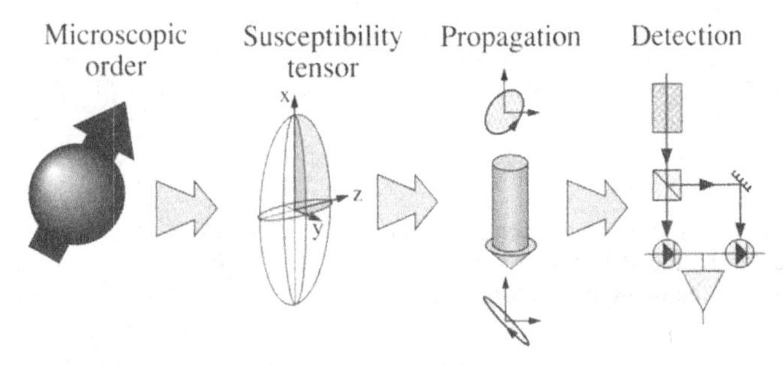

Figure 6.10. Propagation of light in the anisotropic medium is one link in the chain between the microscopic order and the measurement.

medium (Schlesser and Weis 1992), as we will discuss in detail in the following section. As in the case of the Faraday effect, there are nonlinear analogues to the Voigt effect (Drake, Lange and Mlynek 1988; Chen, Telegdi and Weis 1990a; Schuller et al. 1991), which we do not include in this analysis of linear magnetooptic effects, but discuss instead in terms of the order generated in the angular momentum multiplets of the atomic medium.

6.3 Propagation

The anisotropic optical susceptibility is the primary signature of the microscopic order that is present in the atomic medium. To measure this susceptibility, one sends a laser beam with known characteristics like intensity, direction, and polarisation through the medium and measures the changes in these characteristics when the laser beam emerges behind the sample. In this section, we discuss the effect that the susceptibility has on the propagation of the light.

This is one step between the microscopic state of the system, represented by the left-hand part of Figure 6.10, and the actual measurement, which is depicted to the right. This step does not involve quantum mechanics. We use Maxwell's equations to describe the propagation of the light. The optical polarisation calculated in the preceding sections appears here as a source term in Maxwell's equation. Whenever possible, we treat the propagation generally, but we again use our standard $J=1/2 \leftrightarrow J'=1/2$ model whenever an explicit example helps.

6.3.1 Eigenpolarisations of plane waves

Ansatz

Using the susceptibility tensors calculated above, we solve Maxwell's equations for the propagation of electromagnetic waves in the anisotropic medium.

We start again from Maxwell's equations (6.8), (6.10), and assume that a plane wave

$$\vec{E} = \begin{pmatrix} E_x \\ E_y \\ E_z \end{pmatrix} e^{i(\omega t - k_z z)} \tag{6.35}$$

propagates along the z axis. In contrast to the previous section, we do not assume that the field is transverse, but allow for a nonzero component along the z axis. In addition, the susceptibility of the medium may now be a tensor,

$$\vec{D} = \vec{\vec{\chi}} \cdot \vec{E} \tag{6.36}$$

As in the isotropic case, we insert this ansatz into Maxwell's equation to eliminate the magnetic field. This leads us to a modified equation for the electric field amplitudes

$$\begin{pmatrix} E_x \\ E_y \\ 0 \end{pmatrix} \left(\frac{c\,k_z}{\omega}\right)^2 = \vec{E}(1 + \chi_0) - i\,\chi_0\,\vec{E} \times \vec{m} \tag{6.37}$$

which have two independent solutions

$$\begin{pmatrix} E_x \\ E_y \\ E_z \end{pmatrix}_{\pm} = E_0 \begin{pmatrix} 1 + \chi_0\left(1 + i\,\dfrac{m_x m_y}{m_z}\right) \\ i\left(\pm\, r_1 - \chi_0 \dfrac{m_x^2 - m_y^2}{2m_z}\right) \\ \dfrac{\chi_0}{1 + \chi_0}\left[\pm\, m_x r_1 + i m_y (1 + \chi_0) - \chi_0\, m_x \dfrac{m_x^2 + m_y^2}{2m_z}\right] \end{pmatrix} \tag{6.38}$$

Here, E_0 represents the (arbitrary) amplitude and we use the abbreviation

$$r_1 = \sqrt{(1 + \chi_0)^2 + \chi_0^2\left(\frac{m_x^2 + m_y^2}{2m_z}\right)^2} \tag{6.39}$$

The wavevectors corresponding to the eigenpolarisations are

$$k_{\pm z}^2 = \left(\frac{\omega}{c}\right)^2 \left[1 + \chi_0 + \frac{\chi_0}{1 + \chi_0}\left(\pm\, m_z r_1 - \chi_0 \frac{m_x^2 + m_y^2}{2}\right)\right] \tag{6.40}$$

In the general case, the wavevector therefore depends on all three components of the ground-state polarisation.

Low density limit

If the medium is a low-density atomic gas, where $\chi_0 << 1$, these formulas can be simplified considerably by neglecting terms of order > 1 in χ_0. The wavevector becomes

$$k_z \cong \frac{\omega}{c}\left(1 + \chi_0 \frac{1 \pm m_z}{2}\right) \tag{6.41}$$

and the eigenpolarisations

$$\begin{pmatrix} E_x \\ E_y \\ E_z \end{pmatrix}_{\pm} \cong \begin{pmatrix} 1 \\ \mathrm{i}(\pm 1 - \chi_0 m_{\pm}^2/2m_z) \\ \pm \chi_0 m_{\pm} \end{pmatrix} \tag{6.42}$$

Propagation

These solutions represent the eigenpolarisations of the medium: A field that is initially a multiple of one of these solutions remains unchanged during propagation, apart from multiplication with a complex number. We may use two real parameters to describe their propagation completely: the index of absorption α and the index of refraction n. As the form of the wavevector shows, the indexes differ for the two polarisations whenever m_z is nonzero. Since the eigenpolarisations are very close to the circular polarisations, we will refer to them as circular polarisations. We write α_+ (α_-) for the absorption coefficient for right (left) circularly polarised light and n_+ (n_-) for the index of refraction. From the expression for the wavevector, we obtain

$$\alpha_{\pm} = \alpha_0(1 \pm m_z) \qquad n_{\pm} - 1 = (n_0 - 1)(1 \pm m_z) \tag{6.43}$$

where α_0 and n_0 represent the coefficients of the unpolarised medium.

We now have to calculate the complex amplitude of the probe laser beam as it passes through the test region. We write $E_+(0)$ ($E_-(0)$) for the amplitude of the left (right) circularly polarised light as it enters the sample region. After passing a distance L through the sample, the amplitude of the + component becomes

$$E_+ = E_+(0)\, \mathrm{e}^{-\alpha_+ L/2}\, \mathrm{e}^{-\mathrm{i} n_+ k_0 L/2} \tag{6.44}$$

and for the − component accordingly. Phase and amplitude of the light behind the sample therefore contain information about the polarisation of the medium. With an appropriate experimental setup, it is possible to extract this information from either the absorption or the dispersion, i.e., from either the amplitude or the phase of the transmitted light.

6.3.2 Arbitrary polarisation

Field amplitude

In most experiments, the incident light is not an eigenpolarisation of the medium, but a superposition of the two eigenpolarisations. To analyse their propagation, we approximate the eigenpolarisations by the circular polarisations and expand the incident light as

$$\vec{E}(0) = \begin{pmatrix} E_+(0) \\ E_-(0) \end{pmatrix} \tag{6.45}$$

Since Maxwell's equations are linear, the light emerges behind the sample of length L as

$$\vec{E}(L) = \begin{pmatrix} E_+(0)\ \mathrm{e}^{-\alpha_+ L/2}\ \mathrm{e}^{-\mathrm{i}n_+ k_0 L/2} \\ E_-(0)\ \mathrm{e}^{-\alpha_- L/2}\ \mathrm{e}^{-\mathrm{i}n_- k_0 L/2} \end{pmatrix} \tag{6.46}$$

To find the effects that the medium has on the light, we express the absorption coefficients and indexes of refraction according to equation (6.46) as a function of m_z

$$\vec{E}(L) = \begin{pmatrix} E_+(0)\ \mathrm{e}^{-\alpha_0(1+m_z)L/2}\ \mathrm{e}^{-\mathrm{i}(1+(n_0-1)(1+m_z))k_0 L/2} \\ E_-(0)\ \mathrm{e}^{-\alpha_0(1-m_z)L/2}\ \mathrm{e}^{-\mathrm{i}(1+(n_0-1)(1-m_z))k_0 L/2} \end{pmatrix}$$

$$= \begin{pmatrix} E_+(0)\ \mathrm{e}^{-\alpha_0 m_z L/2}\ \mathrm{e}^{-\mathrm{i}(n_0-1)m_z k_0 L/2} \\ E_-(0)\ \mathrm{e}^{+\alpha_0 m_z L/2}\ \mathrm{e}^{+\mathrm{i}(n_0-1)m_z k_0 L/2} \end{pmatrix} \mathrm{e}^{-\alpha_0 L/2}\,\mathrm{e}^{-\mathrm{i}n_0 k_0 L/2} \tag{6.47}$$

where we have written the effect of the unpolarised medium as a common factor for both polarisations: the attenuation by $\mathrm{e}^{-\alpha_0 L/2}$ and the phase shift by $\mathrm{e}^{-\mathrm{i}n_0 k_0 L/2}$ are just the effects that we found in the isotropic medium. The differential effects are written inside the parentheses. The first exponential term describes the absorption change (circular dichroism), the second the modification of the dispersion (circular birefringence). A nonvanishing population difference m_z affects the two polarisation components in opposite directions. For the E_+ component, the absorption and dispersion increase when the population difference m_z is positive, and for the opposite polarisation they de-

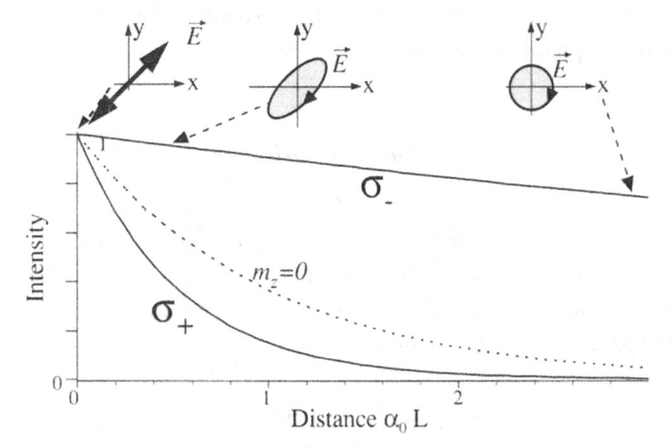

Figure 6.11. Change of the polarisation of light as it propagates through a medium with circular dichroism. The ground-state polarisation is $m_z = 0.9$. The dashed line shows the intensity of the transmitted light as a function of sample length in an unpolarised sample. The insets at the top represent the polarisation of initially linearly polarised light at three discrete points.

crease. This is exactly what the qualitative expectation from the modified populations would have predicted.

Circular dichroism

Figure 6.11 summarises the effect of the circular dichroism on polarisation and amplitude of light propagating through the medium. The two full lines show the intensity change of circularly polarised light in a sample whose ground state has a polarisation $m_z = 0.9$. The dashed line shows for comparison the intensity of light propagating through an unpolarised medium. While σ_+ light is attenuated almost twice as fast as in an unpolarised medium, σ_- light experiences only very little absorption. Light that is initially linearly polarised, as represented by the leftmost inset at the top of the figure, loses the σ_+ component as it propagates through the sample and becomes almost purely circularly polarised after a distance $L \approx 2/\alpha_0$, as indicated by the inset on the right-hand side.

Birefringence

A difference in the refractive indexes for the two circularly polarised components is best measured through a rotation of the plane of polarisation when linearly polarised light propagates through the medium. Linearly polarised light in a circular basis has equal absolute values for both amplitudes,

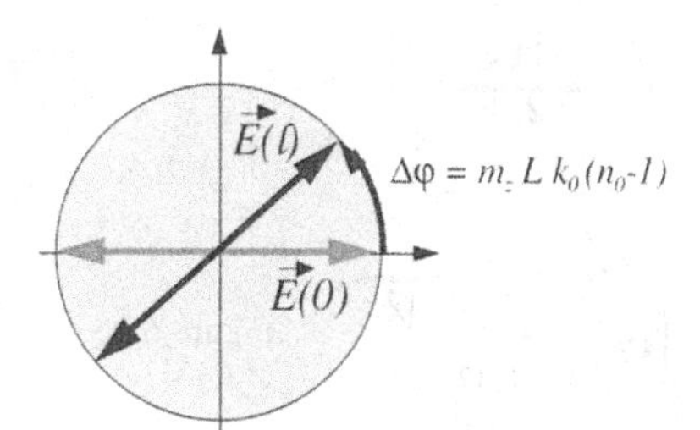

Figure 6.12. Rotation of linearly polarised light during passage through the sample.

$$E_{\mathrm{lin}} = a \begin{pmatrix} e^{i\varphi/2} \\ e^{-i\varphi/2} \end{pmatrix} \tag{6.48}$$

The phase difference φ between the two amplitudes indicates the orientation of the plane of polarisation of the light. According to equation (6.48), the two circular polarisation components acquire different phases during propagation through the polarised atomic medium. Behind the sample, the phase difference is

$$\varphi(L) = \varphi(0) + m_z\, k_0\, L\, (n_0 - 1) \tag{6.49}$$

As Figure 6.12 shows, this phase difference appears as the rotation angle of the plane of polarisation of linearly polarised light that passes through the sample. If the other parameters are known, the orientation of linearly polarised light behind the sample directly measures the polarisation of the medium.

As we have seen during the discussion of the magnetooptic effects, optically anisotropic media can also exhibit linear, as opposed to circular, birefringence. Since linear birefringence is the situation most commonly found in crystal optics, it seems not necessary to analyse this case in detail. It is formally closely related to circular birefringence. If the incident light is in a superposition of the eigenpolarisations, it evolves into a different polarisation state during its traversal of the medium. The most common case in this respect is where the incident light is polarised at 45° to the directions of the eigenpolarisations. For a phase difference of $(n_x - n_y)\, k_0\, L = \pi/2$, corresponding to a quarter-wave plate, the polarisation of the transmitted light is circular. In the case of linear birefringence, the effect of the medium on the light depends on the orientation of the polarisation plane when the incident light is linearly polarised.

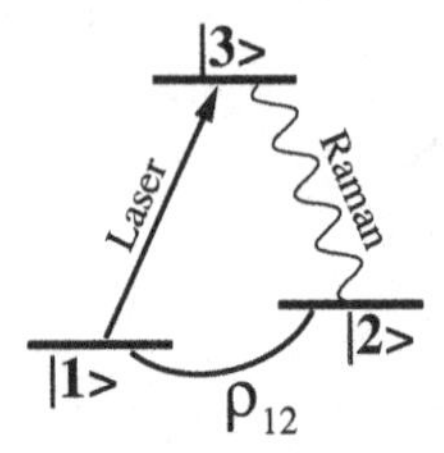

Figure 6.13. Coherent Raman scattering by a sublevel coherence ρ_{12}.

6.3.3 Coherent Raman scattering

Qualitative outline

So far, we have analysed the effect of the optically anisotropic medium on the propagation of light, using a terminology derived from crystal optics. This is of course not the only possibility, and historically other concepts have also proved helpful for understanding the observed effects. One important example is coherent Raman scattering. This method of describing the observed effects is especially attractive when the sublevels involved with the optical response are not degenerate: In those cases, the observed signal becomes time-dependent. This implies that the optical field contains several frequency components. Furthermore, the observation of order in atomic or molecular media is not restricted to the angular momentum substates, but is applicable to any type of multilevel system that allows Raman transitions to occur (Walmsley, Mitsunaga and Tang 1988).

As discussed in Section 3.4.2, in the context of three-level systems, a coherent superposition of two sublevels allows coherent Raman scattering. In Figure 6.13, light interacting with transition $|1\rangle \leftrightarrow |3\rangle$ excites an optical polarisation not only in this transition, but also in transition $|2\rangle \leftrightarrow |3\rangle$. This polarisation is the source of a Raman wave, which propagates together with the incident laser field. Since the optical susceptibility of the medium also accounts for the Raman polarisation, we can use it to describe the coherent Raman process.

Wave mixing

The conventional description of Raman scattering uses the terminology of wave-mixing processes. In the case of coherent Raman scattering, two optical waves couple to an excitation of the medium. We disregard issues of phase matching and assume that the Raman field propagates together with the original wave. A wave that is polarised along the x coordinate axis causes, together with the anisotropic part of the susceptibility (equation (6.29)) an optical polarisation

$$P_y = \mathrm{i}\ \epsilon_0\ \chi_0(\Delta)\ m_z\ E_x \tag{6.50}$$

along the y axis. The polarisation of the Raman wave is thus orthogonal to that of the incident wave. The field component polarised along the y axis also generates an optical polarisation

$$P_x = -\mathrm{i}\ \epsilon_0\ \chi_0(\Delta)\ m_z\ E_y \tag{6.51}$$

in the direction of the x axis. Together, these two equations describe the coupling between two orthogonally polarised waves. The polarisation in z direction causes an exchange of energy between the waves that are polarised in x and y direction.

We calculate the effect of this coupling on the propagation of the two waves starting from a one-dimensional wave equation

$$\partial E_r/\partial z = -2\pi\ \mathrm{i}\ k\ P_r \tag{6.52}$$

where E_r indicates the slowly varying field amplitude of polarisation direction r, P_r the corresponding optical polarisation of the medium, k the wavevector, and z the coordinate in the propagation direction. We limit the analysis to the dispersive case, where the absorption of the light is negligible. The susceptibility is then real and the coupled wave equations become

$$\frac{\partial E_x}{\partial z} = -c_\mathrm{m}\ E_y \qquad = \frac{\partial E_y}{\partial z} = c_\mathrm{m}\ E_x \tag{6.53}$$

where the coupling constant is, for a $J = 1/2$ ground state, $c_\mathrm{m} = -2\pi\ k\ \epsilon_0\ \chi_0(\Delta)\ m_z$

For a wave that is initially polarised along the x axis, equation (6.53) has the obvious solution

$$E_x = E_\mathrm{in}\cos(c_\mathrm{m}\ z) \qquad \text{and} \qquad E_y = E_\mathrm{in}\sin(c_\mathrm{m}\ z) \tag{6.54}$$

where E_in represents the amplitude of the incident field. The energy oscillates between the two waves with a wavevector that is proportional to the polarisation of the medium. For the polarisation of the light, this corresponds to a rotation in the xy plane. This result is identical to what we obtained with the analysis in Section 6.3.2, using the expansion in the eigenpolarisations. The difference between the two descriptions lies only in the choice of the basis states, which are circular or linear polarisations.

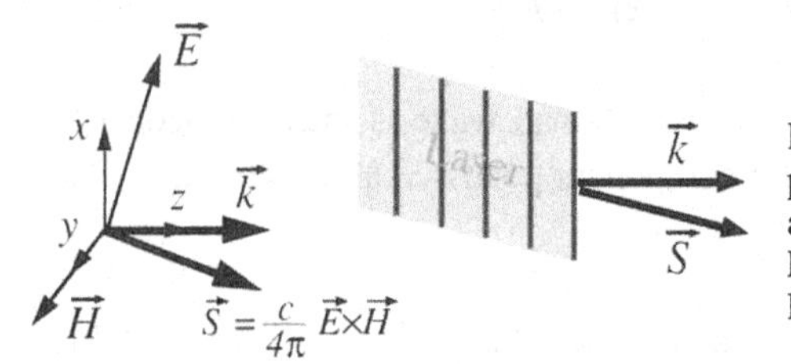

Figure 6.14. The longitudinal component of the electric field implies a transverse component of the Poynting vector $\vec{S}$, which is no longer parallel to the wavevector $\vec{k}$.

6.3.4 Transverse effects

Poynting vector

If the medium is not cylindrically symmetric ($m_\pm \neq 0$), the eigenpolarisations become slightly elliptical and the longitudinal component E_z becomes nonzero; both effects scale with χ_0 and the transverse magnetisation component $m_\pm$. The existence of a longitudinal field component causes the Poynting vector to deviate from the direction of the k vector and a laser beam propagating through such a medium is laterally displaced (Figure 6.14).

To estimate the size of this effect, we calculate the Poynting vector in the linear regime as

$$\frac{4\pi}{c}\vec{S} = \vec{E} \times \vec{H} = \frac{k_z}{\mu_0 \omega} \begin{pmatrix} E_x \\ E_y \\ E_z \end{pmatrix} \times \begin{pmatrix} -E_y \\ E_x \\ 0 \end{pmatrix}$$

$$= \frac{k_z}{\mu_0 \omega} \begin{pmatrix} -E_x E_z \\ -E_y E_z \\ E_x^2 + E_y^2 \end{pmatrix} \approx \frac{k_z (E_x^2 + E_y^2)}{2\mu_0 \omega} \begin{pmatrix} \mp \chi_0 m_\pm \\ \mp \mathrm{i}\chi_0 m_\pm \\ 2 \end{pmatrix} \tag{6.55}$$

The angle between the Poynting vector and the wavevector is of the order of $\chi_0\, m_\pm$. This angle is proportional to the atomic density, the transition strength, and the transverse ground-state orientation. A laser beam propagating through a completely polarised medium of length L is displaced by a distance of the order of $\chi_0 L$. The displacement is therefore proportional to the atomic density. The effect cannot be made arbitrarily large, however, by increasing the particle density, since the beam is attenuated in the medium by $\sim\exp(-\mathrm{Im}[\chi_0]L/\lambda)$. Retaining significant intensity requires $\mathrm{Im}[\chi_0] \lesssim \lambda/L$ so that the observable beam displacement δs is limited to $\delta s \leq \lambda$ $(\mathrm{Re}[\chi_0]/\mathrm{Im}[\chi_0])$.

Interpretation

The transverse displacement can, at least qualitatively, be understood as arising from the conservation of angular momentum. As seen in Section 3.5.4, the light has two effects on the transverse components of the atomic angular momentum: a damping effect, which is proportional to the absorption from the optical field, and the light-shift contribution, which is proportional to the difference between the intensities of right and left circularly polarised components and to the index of refraction of the medium. The damping effect leads to a reduction of the atomic angular momentum while the light shift rotates it around the direction of the laser beam. Since the presence of sublevel coherence does not change the polarisation of the optical field (k_z does not depend on $m_{\pm}$) and therefore does not change the *internal* angular momentum of the photons, the total angular momentum must be conserved by a change in the *external* angular momentum of the laser beam, i.e., by a lateral displacement. Conservation of angular momentum during the interaction of light with atoms differs somewhat for the transverse and longitudinal components:

Source magnetisation	Destination	Observable as
longitudinal	photon spin	polarisation change
transverse	photon orbital angular momentum	beam displacement

Figure 6.15 illustrates how the same argument also provides the direction of the displacement: Consider an atomic vapour that is completely polarised in the y direction. Resonant light incident along the z axis causes a reduction of this magnetisation, irrespective of the polarisation, thereby reducing the y component of the atomic angular momentum. The laser beam must therefore pick up angular momentum parallel to the y axis. As its linear momentum is along the z axis, the additional angular momentum along the y axis corresponds to a displacement in the positive x direction, independent of the polarisation of the light, and proportional to the imaginary part of the susceptibility.

In the dispersive regime, the interaction of the light with the atomic vapour primarily causes a light shift, i.e., a rotation of the angular momentum around the z axis. If the atomic system has a nonvanishing component along the y axis, the light shift generates in first order an angular momentum component

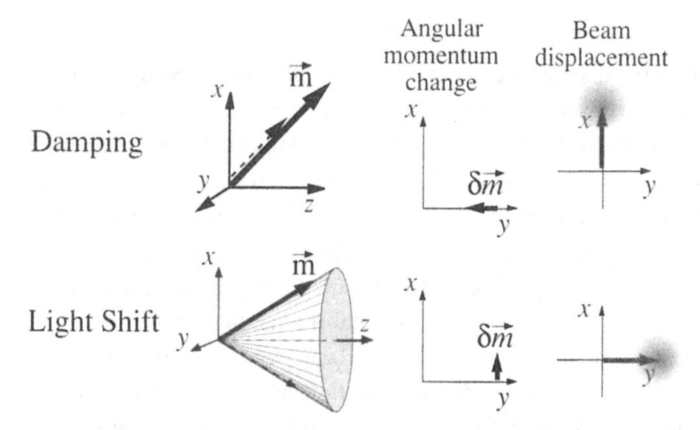

Figure 6.15. Effect of damping and light shift on the atomic medium. The second column from the right shows the change $\delta\vec{m}$ of a magnetisation that is originally oriented along the y axis. The rightmost column indicates the direction of the beam displacement that conserves the total angular momentum.

along the x axis, which must be compensated for by a beam displacement in the y direction. The size of the displacement must be proportional to the real part of the susceptibility. Since the light shift changes sign with the laser beam polarisation, the associated beam displacement must have opposite signs for the two circular polarisations, as predicted by equation (6.55).

Under typical experimental conditions, the size of the displacement is only a fraction of the optical wavelength. It is therefore not a major perturbation of an experiment, but can be observed with a suitable experimental setup (Blasberg and Suter 1992). Such displacements originate not only from transverse components of the magnetisation, but also from other effects that cause longitudinal components of the E field to appear. One example that we have met is the Voigt geometry, where a transverse magnetic field couples the two components of the optical polarisation perpendicular to the field. If the optical wave has a component perpendicular to the field, it also experiences a lateral displacement, which is in this case proportional to the strength of the static magnetic field (Schlesser and Weis 1992).

6.4 Polarisation-selective detection

6.4.1 Fundamentals

Motivation

Since the microscopic order in the atomic medium modifies the light propagating through the medium, a measurement of these amplitudes can provide information about the system. We can use the results on the propagation of the light to calculate its state behind the sample as a function of the internal state of the medium and look for a detection scheme that provides us with easily interpretable and undistorted connection to the magnetisation, thus reversing the various steps sketched in Figure 6.10.

The order we wish to observe is associated with population differences and coherences between angular momentum substates, whose energy differences are in the radio-frequency range of the spectrum. For the observation of this order, we use radiation whose frequency is in the optical range, between 10^{14} and 10^{15} Hz. The time-dependence of the optical field and of the optical coherences is therefore many orders of magnitude faster than the processes that we wish to examine. There are two possible paths to bridge this gap. We may calculate the optical properties of a system whose time-dependence we neglect temporarily, and when we have found a signal as a function of the microscopic order, we "unfreeze" the system to obtain the time-dependence of the signal. This approach may be considered as a "time-domain picture," as it yields a signal that depends on time in the same way as the system itself.

There is, however, another possible approach, which is often used in the literature. It emphasises the coupling between different waves, which may have different frequencies and it may therefore be considered as a frequency-domain approach. We include an analysis along these lines to show the connection between the approach that we use here and that part of the published literature that uses the other approach. In addition, the frequency-domain approach is somehow complementary and can improve our understanding of the relevant processes. In this formalism, the interaction of the time-dependent order in the system with the probe laser beam couples the incident laser beam to a Raman field whose frequency is shifted by the system frequency. We must then detect the amplitude of the frequency-shifted field and the frequency difference between the two fields. The technique that achieves both these tasks is called heterodyne detection.

Heterodyne detection

The heterodyne detection of a weak optical field by superimposing a local oscillator on it was discussed briefly in Chapter 2. In close analogy, we write the complex amplitude of the laser field at the detector position as

$$E_{\mathrm{L}} = E_0 \, \mathrm{e}^{\mathrm{i}\omega_{\mathrm{L}} t} \tag{6.56}$$

and the scattered Raman or signal field as

$$E_R = E_0 \, c_{Sys} \, e^{i(\omega_L + \Omega_L)t} \tag{6.57}$$

where we have assumed that the Raman field is proportional to the laser field and the proportionality constant c_{Sys} contains the information about the microscopic state of the medium in which we are interested. The frequency shift Ω_L is a difference between two sublevel energies.

Photodetectors are quadratic detectors, i.e., their signal is proportional to the square of the electric field. If the laser and Raman fields overlap completely, the detector signal is

$$s = (E_L + E_R)\,(E_L + E_R)^* \tag{6.58}$$

Using the expansions (6.57) for the field amplitudes, we find

$$\begin{aligned} s &= (E_0 + E_0 \, c_{Sys} \, e^{i\Omega_L t}) e^{i\omega_L t} \, (E_0 + E_0 \, c_{Sys} \, e^{i\Omega_L t})^* e^{-i\omega_L t} \\ &= |E_0|^2 \, (1 + c_{Sys} \, e^{i\Omega_L t}) \, (1 + c_{Sys}{}^* \, e^{-i\Omega_L t}) \\ &= |E_0|^2 \, (1 + |c_{Sys}|^2 + \mathrm{Re}\{c_{Sys}\} \cos\Omega_L t + \mathrm{Im}\{c_{Sys}\} \sin\Omega_L t) \end{aligned} \tag{6.59}$$

The interference term between laser and Raman field oscillates at the system frequency Ω_L. The two quadrature components are directly proportional to the real and imaginary parts of the system parameter c_{Sys}. The interference between the two fields provides us with a signal that contains only the difference frequency; in other words, the laser field serves as a local oscillator for shifting the frequency $\omega_L + \Omega_L$ of the Raman field from the optical range to the radio frequency Ω_L.

This detection scheme has several remarkable properties. First, the interference term is linear in the system parameter c_{Sys}. For weak coupling between laser and Raman waves, $c_{Sys} << 1$, as is typically the case in dilute systems, the heterodyne signal is much stronger than the directly detected Raman signal. This so-called heterodyne advantage can provide a sensitivity considerably higher than if the signal only were detected (Levenson and Eesley 1979). In addition, the subtraction of the laser frequency from the signal formed with the same laser beam eliminates laser frequency jitter as a possible line-broadening mechanism. Although the detection uses optical radiation, the resolution is therefore not limited by the laser linewidth but can be many orders of magnitude higher.

Although such measurements are possible and actually have been used (Dehmelt 1957a), it is possible to improve this scheme significantly. The main problem arises from the first term in equation (6.59), which is proportional to the laser intensity but does not depend on the system. For a perfectly stable laser, it contributes an offset to the signal that is easily eliminated. If the laser amplitude fluctuates, however,

$$E_L = E_0[1 + \delta(t)] \tag{6.60}$$

these fluctuations appear directly in the observed signal

$$s_\delta = |E_0|^2 \left(1 + 2\left[\mathrm{Re}\{\delta(t)\} + \mathrm{Re}\{c_{Sys}\}\cos\Omega_L t + \mathrm{Im}\{c_{Sys}\}\sin\Omega_L t\right]\right) \tag{6.61}$$

where we have assumed that the fluctuations and the signal are small,

$$|\delta| << 1 \qquad |c_{Sys}| << 1 \tag{6.62}$$

which is the only relevant case. In this scheme, the signal to noise ratio is the ratio $|c_{Sys}/\delta|$ between the coupling constant of the system and the noise fraction of the laser beam. This ratio does not increase with the laser intensity and the heterodyne advantage is lost. A possible improvement that reduces the effect of laser amplitude fluctuations without requiring a stabilised laser is a balanced detection scheme.

Balanced detection

Balanced detection schemes subtract two partial signals in such a way that the unwanted background is eliminated while the signal contributions add. The principle of the method is illustrated in the right-hand part of Figure 6.16. A beam splitter combines laser and Raman fields and two photodiodes detect the two output ports of the beam splitter. For a 50/50 beam splitter, the fields that fall on the detectors are

$$E_1 = \frac{1}{\sqrt{2}}(E_L - \mathrm{i}\,E_R) \qquad E_2 = \frac{1}{\sqrt{2}}(E_L + \mathrm{i}\,E_R) \tag{6.63}$$

The phase shift between the two fields occurs during reflection on the optically denser medium. The two partial signals are

$$s_{1,2} = I_0\left(\frac{1}{2} + \mathrm{Re}\{\delta(t)\} \pm \mathrm{Im}\,\{c_{Sys}\}\cos\Omega_L t \pm \mathrm{Re}\{c_{Sys}\}\sin\Omega_L t\right) \tag{6.64}$$

which is valid for ($|\delta| << 1$, $|c_{Sys}| << 1$). I_0 represents the intensity of the local oscillator beam. The difference between the two partial signals becomes

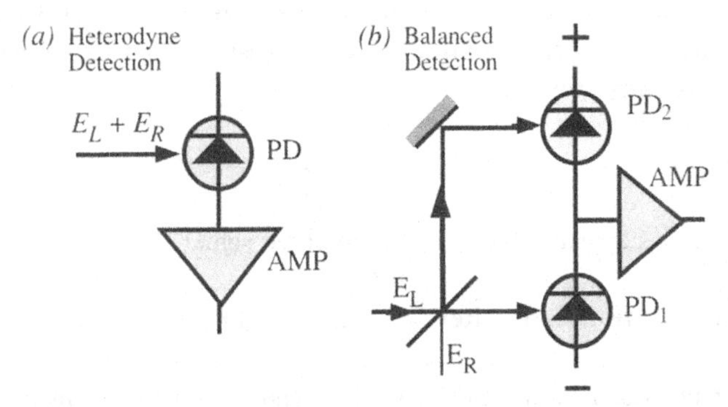

Figure 6.16. Comparison between single channel detection (left) and a balanced detection scheme (right). PD = photodiode, AMP = amplifier.

$$s_1 - s_2 = 2\, I_0\, (\mathrm{Im}\{c_{\mathrm{Sys}}\} \cos\Omega_{\mathrm{L}} t + \mathrm{Re}\{c_{\mathrm{Sys}}\} \sin\Omega_{\mathrm{L}} t) \tag{6.65}$$

The difference signal contains only the two terms that carry the signal in which we are interested. The elimination of the local oscillator background makes this scheme very sensitive. The local oscillator can be chosen arbitrarily large to enhance the detector sensitivity without introducing an offset to the signal. Fluctuations of the local oscillator amplitude appear in this case only as fluctuations in the detector sensitivity. This should be compared to the case of single channel detection, where fluctuations in the local oscillator contribute to the background, which may be much larger than the signal.

Polarisation-selective detection

To apply this scheme to the detection of microscopic order in atomic systems, we replace the beam splitter of the general setup by a polarising beam splitter. It separates the incident beam into two orthogonally polarised beams. The balanced detector then measures the intensity difference between two orthogonally polarised components of the incident beam. This beam splitter separates the incident light into its linearly polarised components, but it is also possible to split it into circularly polarised components. For this purpose, a quarter-wave plate, placed before the beam splitter and oriented at 45° to its major axes, transforms the circularly polarised components into linearly polarised components, which are then again separated by the beam splitter.

These two decompositions are complementary in the sense that they measure different optical parameters of the laser beam. The information contained

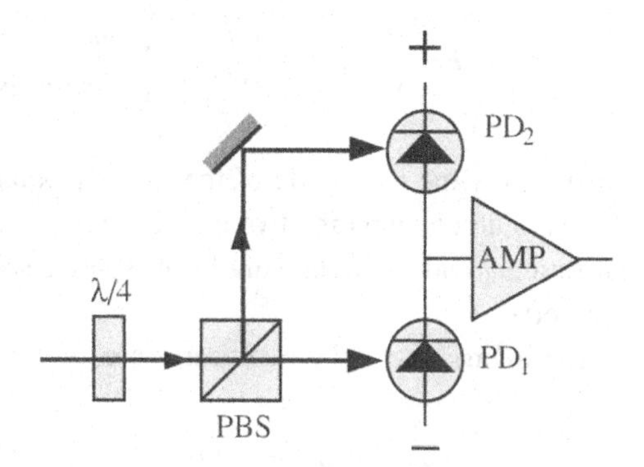

Figure 6.17. Optical beam path for polarisation selective absorption measurement. The $\lambda/4$ retardation plate is rotated 45° with respect to the beam splitter.

in the transmitted light of a single beam contains four independent real-valued parameters, which may be attributed to the amplitude and phase of the two orthogonal polarisation components. In practice, a measurement of the optical phase is often impractical, as it would require the knowledge of the optical path length to a fraction of a wavelength. Practical implementations therefore deal with three independent parameters, which are usually classified as the Stokes parameters (Born and Wolf 1986). In the case of the $J = 1/2$ ground-state system discussed here, we analyse two optical parameters, the difference between absorption and dispersion of the eigenpolarisations. Both parameters provide the same information about the microscopic system, but the sensitivity of the two schemes differs, depending on the detuning of the laser frequency from optical resonance.

6.4.2 Detection schemes

Absorptive detection

If the transmitted laser beam passes through the setup shown in Figure 6.17, the retardation plate, together with the beam splitter, separates the light into two components whose intensity is equal to the intensity of the circularly polarised components. The intensity of each beam is proportional to the input intensity times the attenuation for the corresponding circular polarisation by the sample. On the photodiodes, this intensity is converted into a photocurrent. The two currents are subtracted and amplified to provide a signal proportional to the difference of the two intensities.

For a mathematical analysis, we assume that the probe laser beam that is incident on the sample is linearly polarised. Behind the sample, in the basis of the circularly polarised states, the field is, in a circular basis,

$$\vec{E}(L) = E_0 \frac{1}{\sqrt{2}} \begin{pmatrix} e^{-\alpha_0 m_z L/2} & e^{-i(n_0-1)m_z k_0 L/2} \\ e^{+\alpha_0 m_z L/2} & e^{+i(n_0-1)m_z k_0 L/2} \end{pmatrix} e^{-\alpha_0 L/2}\, e^{-i n_0 k_0 L/2} \tag{6.66}$$

where E_0 is the amplitude of the linearly polarised incident light. The retardation plate converts the two circular polarisations into orthogonal linear polarisations, and the polarising beam splitter separates them into the two output ports.

The intensities of the two beams are

$$I_1 = \frac{1}{2} I_0\, e^{-\alpha_0(1+m_z)L} \qquad I_2 = \frac{1}{2} I_0\, e^{-\alpha_0(1-m_z)L} \tag{6.67}$$

with I_0 the intensity of the incident beam. The difference of the two partial signals is

$$\Delta I = I_0\, e^{-\alpha_0 L} \sinh(m_z\, \alpha_0\, L) \tag{6.68}$$

For small signals, $m_z \alpha_0 L << 1$, it is useful to expand this expression in a power series with respect to m_z. Since all even order terms vanish, the linear term

$$\Delta I_1 = m_z\, (-I_0\, e^{-\alpha_0 L}\, \alpha_0 L) \tag{6.69}$$

is often a good approximation for the exact signal. This detection scheme apparently allows a direct measurement of the polarisation component m_z through the change in the differential absorption. It is background-free and the resulting signal is directly proportional to the magnetisation component m_z, weighted with the absorption coefficient α_0 times the interaction length L of the sample and attenuated by the absorption of the isotropic sample $e^{-\alpha_0 L}$.

Dispersive detection

Instead of measuring the difference in absorption, it is also possible to measure the difference in dispersion between right and left circularly polarised light. The experimental setup remains almost the same as the one shown in Figure 6.17, except that the retardation plate labelled $\lambda/4$ is replaced by a half-wave plate. This rotates the polarisation of the beam so that for an unpolarised medium, the two partial beams have equal intensity. The beam splitter separates the beam into its linearly polarised components.

The fields at the output ports of the beam splitter are

$$E_{1,2} = E_+ \pm i\, E_-$$

$$= \frac{E_0}{\sqrt{2}}\, \mathrm{e}^{-(\alpha_0 + \mathrm{i}\, k_0\, n_0)L/2}\, \mathrm{e}^{-(\alpha_0 + \mathrm{i}\, k_0 (n_0 - 1)) m_z L/2} \qquad (6.70)$$

$$\pm\, \mathrm{i}\, \mathrm{e}^{(\alpha_0 + \mathrm{i}\, k_0 (n_0 - 1)) m_z L/2}$$

and the intensities

$$I_{1,2} = I_0 \frac{1}{2} \mathrm{e}^{-\alpha_0 L} \left[\cosh(\alpha_0\, L\, m_z) \mp \sin(k_0\, L\, m_z (n_0 - 1))\right] \qquad (6.71)$$

Taking the difference, we obtain the signal

$$\Delta I = I_1 - I_2 = -I_0\, \mathrm{e}^{-\alpha_0 L} \sin(\delta_0 m_z) \qquad (6.72)$$

where $\delta_0 = (n_0 - 1)\, k_0\, L$ is the phase shift due to the unpolarised medium. We can again make a linear expansion

$$\Delta I_1 = m_z\, \delta_0\, I_0\, \mathrm{e}^{-\alpha_0 L} \qquad (6.73)$$

This detection scheme also provides a signal proportional to the sublevel polarisation m_z along the beam direction, but proportional to the dispersion δ_0 of the unpolarised medium, again attenuated by the absorption of the unpolarised medium $\mathrm{e}^{-\alpha_0 L}$.

Both detection schemes have similar properties, except for their different dependence on the laser detuning. The dispersive scheme is advantageous for measurements far from resonance, since its sensitivity drops off more slowly as a function of optical detuning, whereas the absorptive scheme allows measurements near the centre of the optical resonance line. Both detection schemes discussed here refer to spin-1/2 systems where the observable of interest is always a component of the magnetisation.

High-order effects

So far we have considered the linear regime, where the observed signal is proportional to the polarisation m_z. If the propagation length is long enough, the linear expansion may not be sufficient. Since a long propagation distance in the absorptive case leaves little signal to be detected, we treat here only the

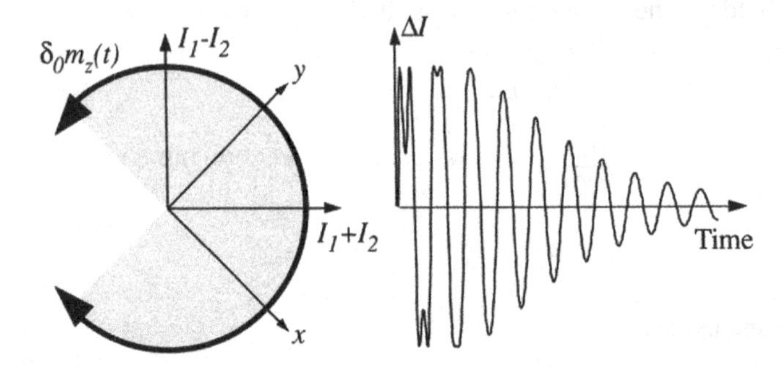

Figure 6.18. Illustration of the high-order effects for finite propagation distances. The left-hand part shows the motion of the polarisation plane of the light in the xy plane. The directions x and y indicate the principal axes of the polarising beam splitter. The right-hand side shows the resulting signal for a time-dependent polarisation m_z.

dispersive case, neglecting absorption. Higher order effects manifest themselves most clearly when the polarisation is time-dependent, like

$$\Delta I = -I_0 \sin(\delta_0 m_0 (cos \Omega_L t) e^{-\gamma t}) \tag{6.74}$$

Here, m_0 represents the amplitude of the magnetisation at the beginning of the experiment. It subsequently precesses at the Larmor frequency Ω_L and decays with the damping rate γ. The signal can never exceed the intensity I_0 of the incident beam, so we expect distorted signals when the argument of the sine function approaches $\pm\ \pi/2$.

This is most easily appreciated by considering that the effect of the sublevel polarisation on the light is a rotation of the polarisation plane of the light. Since the detection scheme measures not the rotation angle, but rather the projection onto the direction labelled $I_1 - I_2$ in the left-hand part of Figure 6.18, the signal distorts as soon as the polarisation exceeds this direction. The right-hand part of the figure shows the resulting signal as a function of time. Initially, the polarisation plane rotates by more than $\pi/2$ and the signal shows significant distortion. As the polarisation decays, the signal approaches the linear regime and provides a faithful representation of the polarisation.

If we expand the signal in a Taylor series

$$\Delta I = I_0 \left(\delta_0 m_z + \frac{1}{6} (\delta_0 m_z)^3 + \cdots \right) \tag{6.75}$$

and insert again the time-dependent polarisation

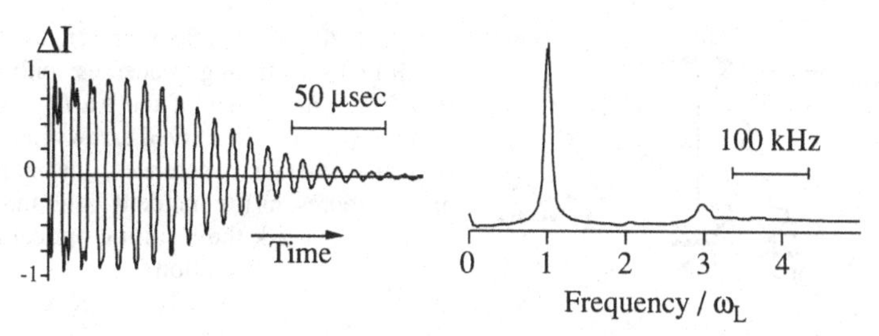

Figure 6.19. Experimental signal recorded with the dispersive detection scheme described in the text. The Raman spectrum on the right was obtained by Fourier transformation of the time-domain data.

$$m_z(t) = m_0(\cos\Omega_L t)e^{-\gamma t} \tag{6.76}$$

we see that the resulting signal should have frequency components not only at the fundamental Ω_L but also at its odd harmonics. If we describe this process as an example of coherent Raman scattering, as discussed in Section 3.4.2, we have to refer to the higher harmonics as multiple scattering processes (Suter and Marty 1993a).

Figure 6.19 shows an experimental example. The data were recorded at a particle density where the optical path length δ_0 is larger than unity, resulting in a rotation of the polarisation plane that exceeds $\pi/2$. The left-hand side contains the observed time-domain data. The flat envelope for early times, together with the indentations at the signal maximum, shows where the polarisation plane rotates by more than $\pi/2$. The right-hand side shows the Fourier transform of the same data. In addition to the usual line at the Larmor frequency, we also find the third harmonic, which may be attributed to multiple Raman scattering. With modified detection geometries, it is also possible to select even order harmonics. However, since the zeroth-order harmonic contains the unmodulated laser beam, such detection geometries have inherently lower signal to noise ratios.

6.4.3 Observables in multilevel ground states

The $J = 1/2$ model used here is only a simplified model that allows an easy visualisation of the basic phenomena. The model is surprisingly effective in explaining the observed phenomena, as long as the details of the level structure are not important. Some of the experiments, however, require a more detailed analysis. This section gives a brief survey of the quantities that are important in such experiments.

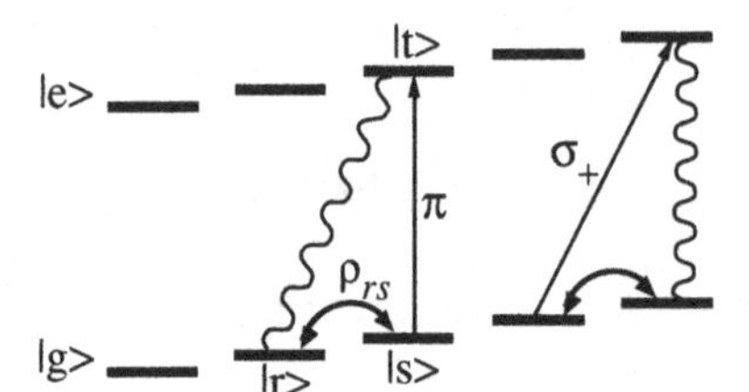

Figure 6.20. Schematic representation of the Raman scattering processes; only the two $F = 2$ multiplets of the Na D_1 transition are shown. The curved arrows indicate two ground-state coherences, the straight arrows represent the incident light and the wavy lines mark the resulting optical polarisation.

Order and observables

Different types of order within the sublevel structure can contribute to the observable signal. We may distinguish population differences and coherences between sublevels, being aware that this distinction depends on the (arbitrarily chosen) coordinate system. The contribution of the populations is easily taken into account by weighting the corresponding transition probabilities with the populations. The contribution of the coherences is somewhat more complicated, but can be included by evaluating the formalism developed in Section 3.4.2 for the description of three-level systems: In the low intensity limit, where we neglect modifications of the populations and sublevel coherences due to the interaction with the probe laser beam, we may evaluate contributions from different three-level systems separately, as shown schematically in Figure 6.20.

Figure 6.20 summarises the processes that contribute to the susceptibility. The sublevels of the electronic ground state and the electronically excited state are assumed to be eigenfunctions of the angular momentum operator F_z. The curved arrows indicate sublevel coherences ρ_{rs} between neighbouring angular momentum states, i.e., between states with $\Delta m_F = 1$. The straight lines indicate the interaction with the incident light that may be σ or π polarised. If a coherence exists between levels $|r\rangle$ and $|s\rangle$, an optical field that couples state $|s\rangle$ to the excited state $|t\rangle$ excites not only an optical coherence in transition $|s\rangle \leftrightarrow |t\rangle$, but, as seen in Chapter 3, also a coherence in transition $|r\rangle \leftrightarrow |t\rangle$. Whereas the incident field is π-polarised, the coherence in transition $|r\rangle \leftrightarrow |t\rangle$ corresponds to σ-polarisation. The sublevel coherence couples the incident light to an optical polarisation in a different direction, indicating that the medium is optically anisotropic. The reverse process is also possible, as shown in the right-hand part of Figure 6.20, where the σ_+ light interacting with the sublevel coherence generates π-polarised optical polarisation.

Similarly, coherence between two levels separated by $\Delta m = \pm 2$ can turn σ_+ light into σ_- and vice versa, as shown in Figure 6.21. Such behaviour is the signature of linear birefringence. Higher order coherences with $\Delta m > 2$,

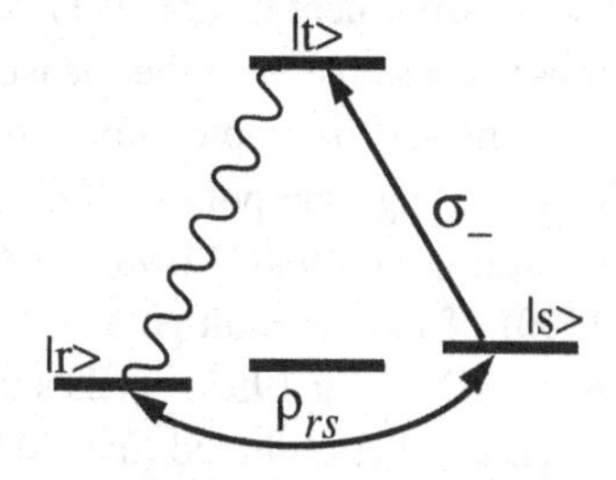

Figure 6.21. Linear birefringence due to $\Delta m = \pm 2$ coherence.

however, cannot contribute to the optical properties: Light interacting with them generates optical coherences with $\Delta m > 1$, which do not correspond to dipole radiation.

Symmetry and selection rules

As discussed in Section 5.2.4, transverse optical pumping can excite all possible multipole components in a given energy level scheme. Several of these sublevel coherences contribute to the anisotropic susceptibility of the medium and can therefore be observed optically (Dehmelt 1957a; Partridge and Series 1966; Happer and Mathur 1967b; Pancharatnam 1968; Mlynek et al. 1983; Wong et al. 1983). Direct optical observation of atomic multipole moments is possible and has been reported for dipole moments (Dehmelt 1957a; Happer and Mathur 1967b; Pancharatnam 1968; Lange and Mlynek 1978) and quadrupole moments (Partridge and Series 1966; Pancharatnam 1968; Mishina, Fukuda and Hashi 1988; Appelt, Scheufler and Mehring 1989). Higher moments are not accessible in direct optical experiments (Cohen-Tannoudji 1962; Pancharatnam 1968), as just shown. This is strictly true only for low intensity (linear) observation schemes. Using nonlinear schemes, observors reported indirect evidence for the presence of moments up to hexadecupole moments (Ducloy, Gorza and Decomps 1973; Gawlik et al. 1974a; Gawlik et al. 1974b).

When different multipole moments contribute to the signal, the precession frequency may help to disentangle their contributions (Dehmelt 1957a; Happer and Mathur 1967b; Pancharatnam 1968; Lange and Mlynek 1978). Even so, a complete characterisation of such an anisotropic atomic state may be rather demanding, especially in systems with nonvanishing nuclear spin. A basis of irreducible tensor operators for an atomic state with total angular momentum F contains operators of rank $0 \leq k \leq 2F$ with a total of $(2F + 1)^2$ components. A complete characterisation of the atomic state therefore requires the measurement of $(2F + 1)^2$ independent variables, significantly more than those accessible with the measurements described so far.

Using spectroscopic information from experiments of this type, it is possible to study, e.g., the polarisability (Kulina and Rinkleff 1982) and the relaxation (Ghosh et al. 1985; Lowe and Norberg 1987; McLean, Hannaford and Lowe 1990) of atomic multipoles. The long lifetime of ground-state multipole moments (Dehmelt 1957b) makes the natural width of the Raman resonances very small and the observed linewidths are independent of the width of the optical transition and the laser linewidth. Under optimised experimental conditions, the observed linewidths can fall into the sub-Hertz range (Cohen-Tannoudji 1961).

6.4.4 The sodium ground state

System

For a specific example, we turn to the ground state of sodium, which consists of the eight sublevels shown in Figure 6.22. If the laser frequency is close to the D_1 transition, the excited state also consists of eight sublevels.

For the description of this system, we need a density matrix ρ whose dimension is 16×16. Using the labelling of Figure 6.22, the ground-state populations are $\rho_{ii}(1 \leq i \leq 8)$ and the ground-state coherences with $\Delta m_F \leq 2$ that contribute to the optical properties are the elements ρ_{12}, ρ_{13}, ρ_{23} $(F = 1)$ and ρ_{45}, ρ_{46}, ρ_{47}, ρ_{48}, ρ_{56}, ρ_{57}, ρ_{58}, ρ_{67}, ρ_{68}, and ρ_{78} $(F = 2)$. To calculate the optical susceptibility of this system, we consider the equation of motion

$$\dot{\rho}(t) = -\mathrm{i}\, [\mathcal{H}, \rho(t)] + \hat{\Gamma}\, \rho(t) \tag{6.77}$$

where the superoperator $\hat{\Gamma}$ describes the relaxation processes and the Hamiltonian $\mathcal{H}$ contains the interaction with the field. In this context, the only relevant contribution is the dephasing of the optical coherences ρ_{ij}, which we assume to be a Markovian process described by the single rate constant Γ_2. The Hamiltonian $\mathcal{H}$, written in the rotating reference frame, includes the optical detuning and the interaction of the atomic dipole moment with the laser field.

Since we are interested in the optical properties of the system as a function of the internal coordinates, we consider the contributions from sublevel populations and coherences separately.

Sublevel coherences and optical polarisation

The contributions of the coherences to the optical susceptibility can be evaluated individually in the weak field limit by searching the stationary solution of the equation of motion

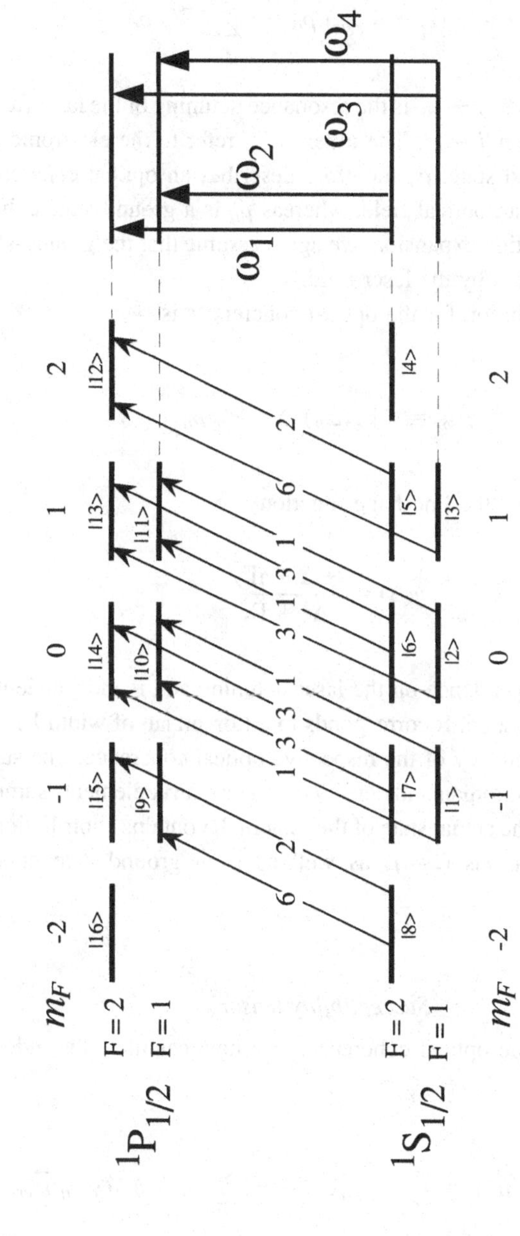

Figure 6.22. Numbering scheme for the energy levels as used in the text, together with elements of the dipole moment operator for σ_+ transitions. The numbers are the squares of the dipole matrix elements and negative matrix elements are underlined. The right-hand part introduces a numbering scheme for the four possible hyperfine transition frequencies.

$$\dot{\rho}_{ir} = -(\Gamma_2 - \mathrm{i}\,\Delta_{ir})\,\rho_{ir} - \mathrm{i}\sum_j \mathcal{H}_{jr}\,\rho_{ij} \tag{6.78}$$

where $\Delta_{ir} = (\mathcal{H}_{ii} - \mathcal{H}_{rr}) - \omega_L$ is the resonance detuning of the laser frequency ω_L from the transition $|i\rangle \leftrightarrow |r\rangle$. The indexes i, j refer to the electronic ground state, r to the excited state. ρ_{ir} therefore describes an optical coherence and $\mathcal{H}_{jr}$ the coupling to an optical field, whereas ρ_{ij} is a ground-state coherence. Within this perturbation expansion, we again assume that the ground-state coherence is not affected by the laser field.

The stationary solution for the optical coherences is

$$\rho_{\infty ir} = -\mathrm{i}\; g(\Delta_{ir}) \sum_j \mathcal{H}_{jr}\,\rho_{ij} \tag{6.79}$$

where $g(\Delta)$ represents the lineshape function

$$g(\Delta) = \frac{-\Delta + \mathrm{i}\Gamma_2}{\Delta^2 + \Gamma_2^2} \tag{6.80}$$

that contains the dependence on the laser detuning and is independent of the actual state of the system. It corresponds to a Lorentzian of width Γ_2, centred at the transition frequency of the respective optical coherence. The sum over the products of Hamiltonian and density operator matrix elements summarises the dependence on the actual state of the system. It contains contributions from ground-state populations ($i = j$) as well as from ground-state coherences ($i \neq j$).

Susceptibility tensor

From the steady-state optical coherences, we now calculate the induced polarisation as

$$\vec{P} = \mathrm{tr}\{\rho_\infty\,\vec{\mu}_e\} = \sum_{ir} \rho_{ir\infty}\,\vec{\mu}_{eri} = -\mathrm{i}\sum_{ijr} g(\Delta_{ir})\,\mathcal{H}_{jr}\,\rho_{ij}\,\vec{\mu}_{eri} \tag{6.81}$$

We summarise the result by ordering it into contributions from different density operator components. Using cartesian coordinates, the tensor components are

$$
\begin{aligned}
\chi^{xx}_{yy} = {} & \pm \frac{g_3 - g_4}{8}(\rho_{13} + \rho_{31}) + \frac{g_3 - g_4}{24}(\rho_{11} - 2\rho_{22} + \rho_{33}) \\
& + (5g_3 + g_4)\frac{\rho_{11} + \rho_{22} + \rho_{33}}{6} \pm \frac{g_1 - g_2}{8} \\
& \times \left[\left(\sqrt{6}\,\rho_{46} + 3\rho_{57} + \sqrt{6}\rho_{68}\right) + \left(\sqrt{6}\,\rho_{64} + 3\rho_{75} + \sqrt{6}\rho_{86}\right)\right] \\
& - \frac{g_1 - g_2}{8}(2\rho_{44} - \rho_{55} - 2\rho_{66} - \rho_{77} + 2\rho_{88}) \\
& + \frac{g_1 + g_2}{2}(\rho_{44} + \rho_{55} + \rho_{66} + \rho_{77} + \rho_{88}) \\
\mathrm{i}\chi^{xy}_{yx} = {} & \frac{-g_3 + g_4}{8}(\rho_{13} - \rho_{31}) \pm \frac{5g_3 - g_4}{8}(\rho_{11} - \rho_{33}) + \frac{g_1 - g_2}{8} \\
& \times \left[\left(\sqrt{6}\rho_{46} + 3\rho_{57} + \sqrt{6}\rho_{68}\right) - \left(\sqrt{6}\rho_{64} + 3\rho_{75} + \sqrt{6}\rho_{86}\right)\right] \\
& \pm \frac{g_1 + 3g_2}{8}\left[(2\rho_{44} + \rho_{55} - \rho_{77} - 2\rho_{88})\right]
\end{aligned}
\tag{6.82}
$$

where the functions g_i represent Lorentzians, as defined in equation (6.80), centred at the positions of the four hyperfine transitions of Na. The numbers are defined in Figure 6.22. The significantly larger number of independent elements in this susceptibility tensor indicates that the nuclear spin – the only difference between the $J = 1/2$ model and the present system – does have a significant influence on the optical properties of the system. On the other hand, we would expect that we recover the simpler model in the limit of vanishing hyperfine interaction.

Comparison

An analysis of the susceptibility tensor, equation (6.82), shows that the contribution from the higher order components of the density operator, not present in the $J = 1/2$ model, have opposite sign when the laser couples to the $F = 1$ excited state, compared to the case where it couples to the $F = 2$ state. The frequency-dependent prefactor of these terms in the susceptibility tensor, which is $(g_1 - g_2)$ *or* $(g_3 - g_4)$, is evidence for this sign change. In the limit of small excited state hyperfine splitting, the two resonances coincide and cancel.

This also implies that the integral of the corresponding lineshape must vanish, as the bottom curve in Figure 6.23 shows. The frequency shift between

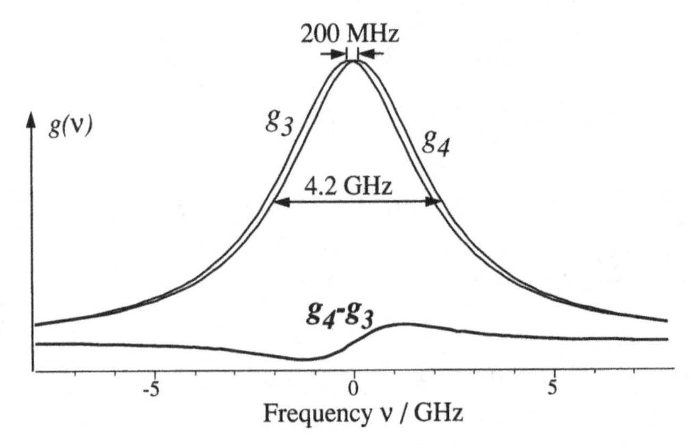

Figure 6.23. Frequency dependence of the higher order contributions to the optical susceptibility. The two upper curves show two pressure-broadened optical resonance lines separated by the excited state hyperfine splitting; the curve at the bottom represents the difference between the two.

the two resonance lines, the excited state hyperfine splitting, is close to 200 MHz, small compared to the width of the pressure-broadened resonance lines. The full width of 4.2 GHz assumed in this spectrum corresponds to a buffer gas pressure of 200 hPa, a value typical for high-resolution experiments on ground-state sublevels. In the limit of small hyperfine interaction, where the influence of the nuclear spin on the optical properties should vanish, the two resonance lines cancel exactly, and the contributions from higher order density operator components to the observable signal vanish.

Wide line limit

For the important situation of pressure-broadened optical transitions, it is useful to consider the limit of a linewidth that is large compared to the hyperfine splitting, in particular compared to the excited state hyperfine splitting. The collisions which cause the pressure broadening also lead to excited state reorientation, which should further reduce signal components that depend on differences between excited state energies. Overall, we expect these contributions to be negligible when the buffer gas pressure exceeds a few kPa.

If we neglect all terms that are proportional to differences between two absorption lines, i.e., $(g_1 - g_2)$ *and* $(g_3 - g_4)$, the susceptibility tensor becomes

$$\chi_{yy}^{xx} = \frac{(5g_3 + g_4)}{6} (\rho_{11} + \rho_{22} + \rho_{33})$$

$$+\frac{g_1+g_2}{2}(\rho_{44}+\rho_{55}+\rho_{66}+\rho_{77}+\rho_{88})$$

$$\mathrm{i}\chi^{xy}_{yx} = \pm\frac{5g_3-g_4}{8}(\rho_{11}-\rho_{33}) \pm\frac{g_1+3g_2}{8}[(2\rho_{44}+\rho_{55}-\rho_{77}-2\rho_{88})] \tag{6.83}$$

The diagonal elements of the susceptibility tensor, which are just the total populations for the two hyperfine multiplets, are identical for both axes, indicating that the medium has no linear birefringence in this limit. Correspondingly, the diagonal elements do not affect the polarisation of the light, but they cause an overall attenuation and a phase shift. For a polarisation-selective detection, they are essentially invisible. The off-diagonal elements, however, cause the circular birefringence. The weighted sums of the populations in the off-diagonal elements represent the expectation value of the operator F_z. The two terms collect the contributions from the two hyperfine multiplets. Physically, the situation is almost identical to the $J = 1/2$ model: The z components of the magnetisation cause circular birefringence, while the total populations lead to an overall attenuation. The only difference is that the contributions from the two hyperfine multiplets are separated by the ground-state hyperfine interaction.

The anisotropic contributions to the diagonal terms xx and yy, which are present in the full form and indicate linear birefringence, originate from second rank tensor elements. Tensor elements of rank $k > 1$ depend on the presence of the nuclear spin and vanish in the limit of small hyperfine interaction, due to the cancellation effects discussed before. This is compatible with the purely circular birefringence that we found in the $J = 1/2$ model.

7
Coherent Raman processes

After the sequence of introductory chapters that collected the necessary tools, we are now ready to discuss complete experiments. The first type of experiment we consider is the creation and detection of order in multilevel atoms through coherent Raman processes. The atoms that we use to study these processes are rare earth ions in a crystal matrix. Enclosing them in a crystal allows long interaction times, but also makes it necessary to consider their interaction with neighbouring atoms.

7.1 Overview

7.1.1 Raman processes

Introduction

In the preceding chapters, we mentioned several types of Raman processes. Their common feature is a resonant change of the energy of the photons that interact with the material system. The energy of the scattered photons may be lower (Stokes process) or higher (anti-Stokes) than that of the incident photons. The energy difference is transferred to the material system, where it must match an energy level separation. The photon energy itself, however, does not have to match exactly a transition frequency of the medium. This is commonly expressed by the statement that the Raman scattering proceeds through a virtual state, represented by the dashed line in Figure 7.1. The presence of a real state of the atom, indicated by the full line, nevertheless increases the coupling efficiency, as discussed in Chapter 3.

The earlier sections on three-level effects and optical anisotropy dealt with the mathematical formalism of Raman processes, using generic level systems to describe them. Here we discuss some examples of physical systems that exhibit the predicted effects. The systems we use to describe the effects are ionic solids containing rare earth ions. The electronic structure, with the va-

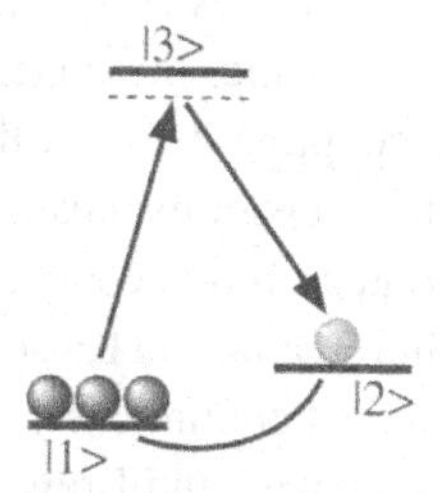

Figure 7.1. Raman excitation of a transition between two ground-state sublevels.

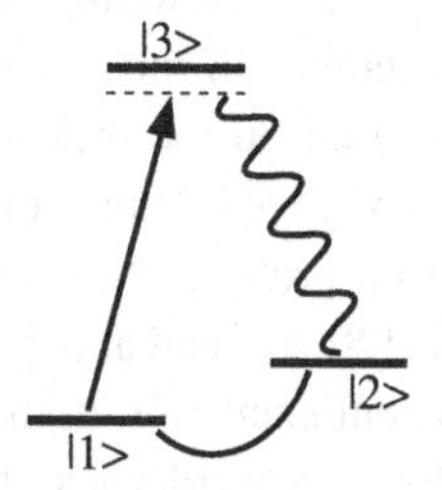

Figure 7.2. Coherent Raman scattering: The curved line represents the sublevel coherence, the straight arrow the incident laser field, and the wave the resulting Raman field.

lence electrons relatively well shielded by the s electrons of the higher shells, makes the interaction of these ions with their environment relatively small. This good isolation from other degrees of freedom makes rare earth ions an attractive tool for detailed studies with high resolution. Nevertheless, we will also have to discuss the interaction of the ions with their environment. For many applications, this interaction is the primary motivation for the study.

We first include a brief summary of those physical processes most important for the experiments we describe later.

Excitation and detection

The Raman effect can be used to excite transitions between atomic states not directly coupled to the radiation field. In such a situation, the laser field (arrows in Figure 7.1) excites a coherent superposition of the two states through the two-photon transition enhanced by the presence of the excited state level $|3\rangle$. The resulting coherence between states $|1\rangle$ and $|2\rangle$ is represented by the curved line in Figure 7.2. In the examples discussed, the states $|1\rangle$ and $|2\rangle$ are angular momentum substates of either the electronic ground state or an electronically excited state, and the energy difference between them is typically in the radio frequency range.

The second type of process that we discuss is coherent Raman scattering, represented schematically in Figure 7.2. A coherent excitation exists in the medium, in our example in the sublevel transition between states $|1\rangle$ and $|2\rangle$, represented by the curved line. A laser field (arrow) that couples to an adjacent optical transition ($|1\rangle \leftrightarrow |3\rangle$) transfers part of the sublevel coherence into the third transition, where it becomes an optical polarisation (wavy line) and therefore the source of an optical wave.

In both cases, excitation and detection, two optical waves couple to a material excitation. In the case of coherent Raman scattering, the coherence is already present in the medium and enhances the coupling between the laser and Raman waves. Under these conditions, the laser beam excites a Raman field in first order in the incident field, and the Raman mode contains equal amounts of Stokes and anti-Stokes components (Giordmaine and Kaiser 1966). The coherent excitation in the medium can be prepared continuously, e.g., by a radio frequency (Mlynek et al. 1983; Mitsunaga, Kintzer and Brewer 1984) or microwave irradiation (Bloembergen, Pershan and Wilcox 1960; Holliday et al. 1990), or in a transient mode by laser pulses (Bassini et al. 1977), frequency switching of the laser (Brewer and Genack 1976), and Stark switching of the molecular resonance frequency (Brewer and Shoemaker 1971; Brewer and Shoemaker 1972; Shoemaker and Brewer 1972; Brewer and Hahn 1973; Brewer 1977a; van Stryland and Shoemaker 1979). In the transient mode, the coherent excitation occurs before the start of the coherent Raman scattering process.

Some applications

In many cases of practical interest, the Raman-shifted field propagates parallel to the laser field. The two fields interfere on the detector and the laser beam acts as a local oscillator for heterodyne detection of the Raman field. The resulting Raman beats (Brewer and Hahn 1973; Wong et al. 1983) contain information about the frequency shift and the amplitude of the Raman field. In the case of transient excitation, the order in the medium decays after the end of the pulse under the influence of radiative decay and collision processes. The loss of coherence from the system reduces the efficiency of the Raman process and the observed signal shows the same decay as the coherence. Raman beats therefore provide information not only on the resonance frequencies, but also about various types of collision processes. On the other hand, laser frequency fluctuations and inhomogeneous broadening effects, such as Doppler broadening, do not affect the beat signal. The method is therefore ideal for spectroscopic applications like the determination of ground-state relaxation rates (McLean, Hannaford and Lowe 1990). Under favourable conditions, suitable detection schemes can provide this information separately for individual multipole moments (Vedenin et al. 1989).

Coherent Raman scattering was originally proposed to study vibrational transitions of molecules (Giordmaine and Kaiser 1966; Laubereau, Wochner and Kaiser 1976) and has been applied to study the rotation of molecules like CS_2 (Heritage, Gustafson and Lin 1975). In atomic vapours, coherent Raman scattering has been applied primarily to the study of sublevel dynamics in alkali (Lange and Mlynek 1978; Grison et al. 1991). In electronic ground states, transitions between Zeeman sublevels are usually of primary interest (Suter and Marty 1993a), but in electronically excited states, superpositions between states with considerably larger splittings, extending to several hundred GHz (Burggraf, Kuckartz and Harde 1986), have been investigated. The method therefore allows the investigation of very fast processes, but in electronic ground states, it can also achieve very high resolution. This possibility makes it particularly well suited for studying slow processes, like velocity-changing collisions (Mlynek et al. 1984; Tamm, Buhr and Mlynek 1986).

Applications to solids have concentrated on ionic crystals, in particular rare earth ions at low concentrations (Mlynek et al. 1983; Shelby et al. 1983; Wong et al. 1983; Shelby and Macfarlane 1984; Erickson 1985b; Erickson 1987; Walmsley, Mitsunaga and Tang 1988; Blasberg and Suter 1993; Blasberg and Suter 1994), but also in ruby at liquid He temperature (Endo et al. 1982). In the latter case, the Raman beats were associated with the superhyperfine structure due to the Cr–Al interaction. In these systems, the ions that are the subjects of the study are present at concentrations of typically 0.1%. They are thus best described as independent ions that interact individually with the electromagnetic field.

7.1.2 Electronic structure of rare earth ions

An ion that has formed the subject of many of these studies is $^{141}Pr^{3+}$. Half of its 56 electrons fill the shells $n = 1, \ldots, 3$, whereas 26 of the remaining electrons fill the 4*s*, 4*p*, 4*d*, 5*s*, and 5*p* orbitals. The final two electrons occupy 4*f* orbitals. Obeying Hund's rules, the lowest energy state of this system is a triplet state, and the angular momentum of the two electrons has the highest possible value $L = 5$.

Figure 7.3 shows how the spin–orbit interaction lifts the degeneracy of the possible angular momentum states. It makes the $^3H_4$ state the one with the lowest energy; the states with higher angular momentum J have energies that are some 60,000 GHz higher (Wyborne 1965). In the free ion, spherical symmetry makes the nine substates with $m_J = -4, -3, \ldots, 4$ degenerate.

In the crystal, the environment has lower than spherical symmetry and the energies of the states depend on the interaction with neighbouring charges. Depending on the symmetry of the atomic site, all angular momentum substates may become nondegenerate, as in the case represented in Figure 7.3. The separation between the crystal field states depends on the host material, but is typically of the order of several hundred to a thousand GHz. The elec-

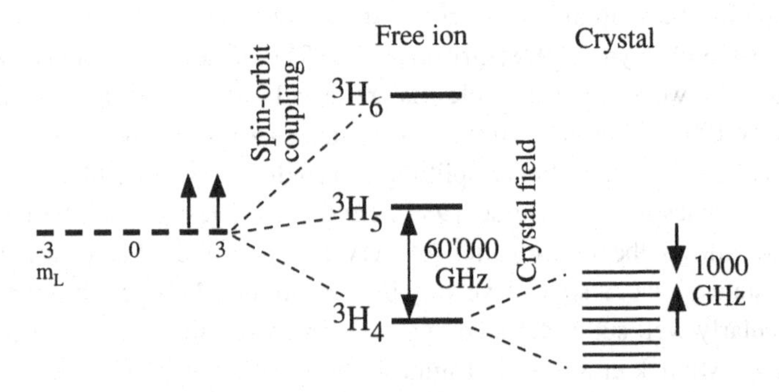

Figure 7.3. Effect of spin–orbit coupling and interaction with the crystal field on the energy level structure of Pr^{3+}.

trostatic interaction that lifts the degeneracy couples to the position operator of the electrons, which does not commute with the angular momentum operator. If the symmetry is low enough that all states become nondegenerate, the expectation value of the angular momentum operator components $\langle\Psi|J_\alpha|\Psi\rangle$ vanishes for the eigenstates Ψ of the crystal field. This effect is known as "quenching" of the angular momentum.

7.1.3 Nuclear spin states

Quadrupole splitting

The nuclear spin of the isotope ^{141}Pr is $I = 5/2$. Nuclei with spin $I > 1/2$ have a nonspherical distribution of the nuclear charge, which makes the electrostatic interaction with inhomogeneous external fields depend on the orientation of the nucleus. We describe the interaction by the Hamiltonian

$$\mathcal{H}_Q = \frac{\omega_Q}{2} \left\{ I_z^2 + \frac{\eta}{3} (I_x^2 - I_y^2) \right\} \tag{7.1}$$

where the coupling constant

$$\omega_Q = 6\, e\, Q\, V_{zz}/(4I\,(2I-1)) \tag{7.2}$$

is proportional to the nuclear quadrupole moment Q and the field gradient (EFG) tensor element V_{zz} (Cohen and Reif 1957). The asymmetry parameter

$$\eta = \frac{V_{xx} - V_{yy}}{V_{zz}} \tag{7.3}$$

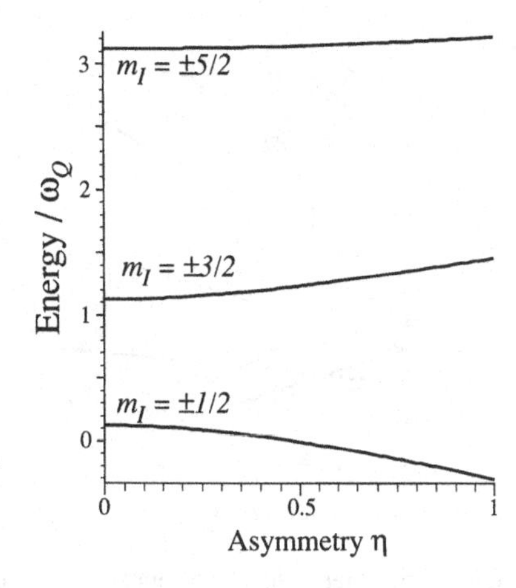

Figure 7.4. Energies of the eigenstates of the nuclear quadrupole Hamiltonian $\mathcal{H}_Q$ as a function of the asymmetry parameter η.

describes the deviation of the EFG tensor from axial symmetry. The form (7.1) of the Hamiltonian puts the quantisation axis parallel to the z axis of the EFG tensor.

Figure 7.4 shows the energies of the eigenstates of the quadrupole Hamiltonian. For a symmetric tensor ($\eta = 0$), the eigenstates of the Hamiltonian are simultaneously eigenstates of the nuclear spin operator I_z. The $\pm\, m_I$ states are pairwise degenerate and separated in energy by $\omega_Q/2\ (m_I^2 - m_{I'}^2)$. For a spin $I = 3/2$, the separation between the levels is ω_Q and $2\omega_Q$. For an asymmetric tensor, where $\eta \neq 0$, the eigenstates of the Hamiltonian are no longer eigenstates of I_z and the labels of the states in Figure 7.4 are no longer accurate. The asymmetric part of the quadrupole interaction mixes the states within the two sets $m_I = (5/2, 1/2, -3/2)$ and $m_I = (3/2, -1/2, -5/2)$, but they remain doubly degenerate.

In the rare earth ions, there is a second term in the Hamiltonian that has the same form as the nuclear quadrupole operator. It originates from the hyperfine interaction and is known as pseudo-quadrupole interaction. Since the crystal field quenches the electronic angular momentum $\vec{J}$, the hyperfine interaction $\vec{J} \cdot \vec{I}$ vanishes in first order. To second order, however, the hyperfine interaction appears as a Hamiltonian term quadratic in the nuclear spin coordinates I_α. It has the same form as the nuclear quadrupole interaction and cannot be separated from the true quadrupole interaction. We consider them together as an effective quadrupole interaction. In zero magnetic field, the effective quadrupole interaction separates the six nuclear spin states into three groups of doubly degenerate levels.

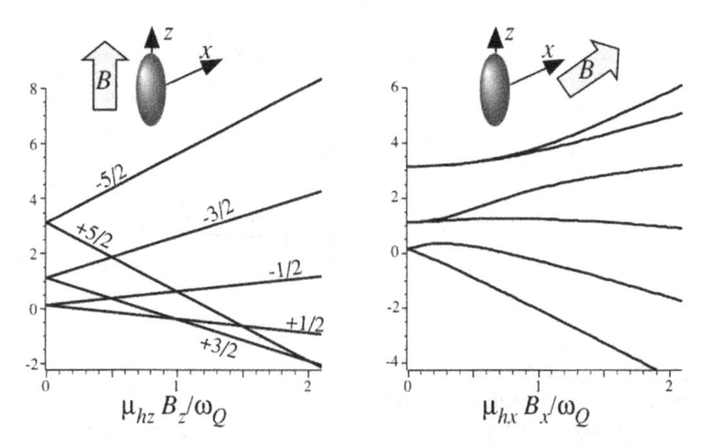

Figure 7.5. Energy levels of nuclear spin states for an axially symmetric quadrupole tensor. A magnetic field is applied parallel (left) and perpendicular (right) to the symmetry axis of the quadrupole tensor.

Enhanced Zeeman effect

The crystal field quenches not only the hyperfine interaction, but also the electron Zeeman interaction. Whereas the second-order contribution to the Zeeman interaction is too small to be observable, the cross-term between the Zeeman and hyperfine interactions is linear in the magnetic field strength and has the same form as the nuclear Zeeman interaction. We combine the two in an enhanced nuclear Zeeman interaction Hamiltonian

$$\mathcal{H}_Z = \sum_{\alpha = xyz} B_\alpha \, \mu_{h\alpha} \, I_\alpha \tag{7.4}$$

written in the principal axis system of the Zeeman tensor, which may differ from that of the effective quadrupole tensor. The enhanced nuclear gyromagnetic ratios $\mu_{h\alpha}$ depend on the level structure of the system.

The effect of the Zeeman interaction on the nuclear spin states depends on the relative orientation between the magnetic field and the principal axes of the quadrupole tensor.

Figure 7.5 shows the effect of a magnetic field along (left) and perpendicular (right) to the symmetry axis of an axially symmetric quadrupole tensor, for a Zeeman tensor that is aligned with the quadrupole tensor. In both cases, the magnetic field lifts the degeneracy of all six nuclear spin states. In the longitudinal case, the Zeeman interaction commutes with the quadrupole interaction and the shift of the energy levels is proportional to the field strength.

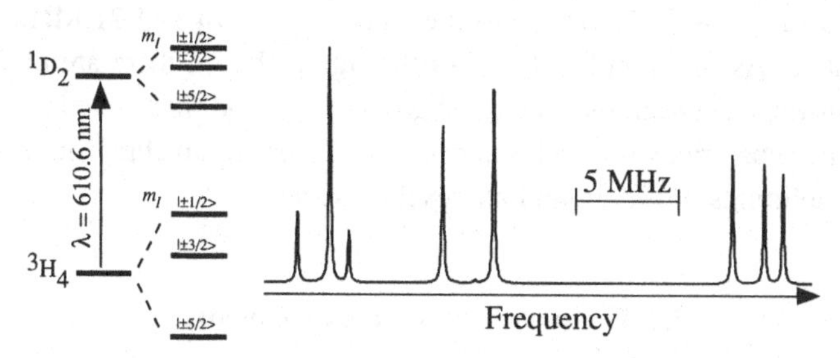

Figure 7.6. Possible transitions between the nuclear spin states of the $^3H_4$ ground state and the $^1D_2$ excited states.

If the field is perpendicular to the quadrupole tensor, the two interactions do not commute. The $m_I = \pm 3/2$ and $\pm 5/2$ states, between which the magnetic dipole moment has no matrix elements, split only in second order in the field.

Transitions

The experiments under discussion were all performed on the transition between the $^3H_4$ ground state and the electronically excited $^1D_2$ state of Pr^{3+}. In the host material $YAlO_3$, this transition occurs at the wavelength $\lambda = 610.6$ nm. According to the selection rules discussed in Section 4.5, such a transition is doubly forbidden. The orbital angular momentum changes by $\Delta L = 3$ and the spin state of the electrons is not the same. In solids, however, these selection rules are not strictly valid, since the eigenstates of the Hamiltonian, including the crystal field interaction, are only approximate eigenstates of the angular momentum operator. The transition matrix element between the electronic ground state and the excited state is therefore small but not zero.

The zero field splitting of the nuclear spin states is considerably larger than typical laser linewidths. Since the coupling to the laser field does not involve the nuclear spin, we would expect that the optical transitions occur only between identical nuclear spin states in the two electronic states. Since the nuclear quadrupole tensors for the ground and excited states are not aligned with each other, however, the nuclear spin part of the excited state eigenfunctions has nonvanishing overlap with all nuclear spin sublevels of the electronic ground state. Optical transitions are therefore possible from all six electronic ground states to all electronically excited states.

The left-hand side of Figure 7.6 shows the relevant part of the energy level structure of $Pr^{3+}:YAlO_3$. The splitting is due to the enhanced nuclear quadrupole interaction and the separation between the energy levels is 0.92 MHz,

1.56 MHz, and 2.48 MHz in the excited state and 7, 14 and 21 MHz in the ground state. As the right-hand side of the figure shows, there are nine possible transitions between the electronic ground state and the excited state sublevels. The separations between the resonance lines are algebraic sums of the sublevel splittings in the ground and excited states.

7.2 Frequency-domain experiments

7.2.1 Spectral holeburning

Inhomogeneous broadening

In solid materials, the energies of atomic states depend on their interactions with the environment, in particular the crystal field. In a perfect crystal, these interactions are identical and all atoms have the same resonance frequency. Real crystals always have defects, however, which change the crystal field in their vicinity, thus shifting the atomic resonance frequencies. The observed lineshape therefore always represents an average over the distribution of atomic environments. In the crystals that we consider, the observed ions substitute other ions. The statistics of the doping therefore also contribute to the inhomogeneous broadening: Different Pr ions have different numbers of other Pr ions in their neighbourhood. The resulting distribution of resonance frequencies depends on the crystal but is at least a few GHz wide. In such a resonance line, irradiation with a narrowband laser excites not all the atoms present in the sample, but only those in a particular environment, whose resonance frequency is close to the laser frequency.

If the homogeneous linewidth Γ_h and the laser jitter are both smaller than the separation between the energies of the nuclear spin states, the laser frequency for any one atom is resonant with only one of the possible transitions that connect the nuclear spin states of the electronic ground state to those of the electronically excited states. As briefly discussed in Section 5.4, this situation can lead to optical pumping. The laser depletes those ground states to which it couples, since the atoms can decay into different ground states.

Pump-probe experiments

These population changes are best observed in pump and probe experiments. The pump laser beam changes the populations of all atoms in resonance with it and a second laser beam probes the modified populations. Its intensity should be low enough that it does not significantly affect the populations. It finds modified absorption whenever the difference between the pump and probe

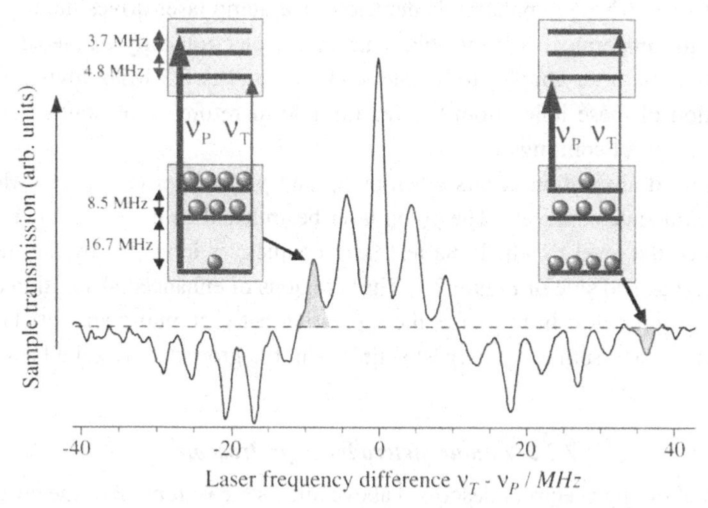

Figure 7.7. Spectral holeburning in Pr^{3+}:LaF_3.

laser frequencies are equal to the energy difference between two nuclear spin states.

Figure 7.7 shows an example of such a holeburning spectrum of Pr^{3+} in LaF_3. In this system, the splittings between the nuclear spin substates are 8.5 and 16.7 MHz in the electronic ground state and 3.7 and 4.8 MHz in the excited state. For this experiment, the frequency of the pump laser was constant, while the frequency of a test laser was scanned over the spectrum. The solid line represents the transmitted laser power as a function of the frequency difference between the two laser beams. When the curve rises above the dashed reference line, it indicates increased transmission; a signal below the dashed line corresponds to increased absorption.

When the two frequencies coincide, the test laser is resonant with the same atoms as the pump laser, which has pumped those atoms into other states. The probe laser therefore meets strongly depleted populations, as evidenced by the large increase of the sample transmission when the probe beam frequency comes close to the pump frequency near the centre of the spectrum. When the difference between the two laser frequencies is close to 4 or 8 MHz, the absorption is also strongly reduced. This situation corresponds to the case where pump and probe laser couple again to the same atoms, but through different optical transitions. Both laser fields couple to the same ground-state sublevel, but to different excited state sublevels, as indicated by the left inset in Figure 7.7. In this case, both lasers couple to the lowest energy state of the electronic ground state, whose population is depleted. The pump laser drives the transition to the uppermost nuclear spin state of the electronically excited state, while the test laser couples to a state whose energy is 8.5 MHz lower. The separation of these holes from the central hole therefore corresponds to excited state level splittings.

Enhanced absorption occurs when pump and probe laser couple to different ground-state sublevels. The pump laser beam has then increased the population of the level to which the test laser couples, as indicated by the inset in the right-hand side of Figure 7.7. These regions of enhanced absorption are known as antiholes. In this case, the separation between pump and test laser frequency is the sum of the largest splittings in the ground and excited states.

7.2.2 Raman heterodyne spectroscopy

The holeburning spectrum described above allows measurement of the energy level splitting of nuclear spin states within the electronic states. The resolution of this experiment is dominated by the frequency jitter of pump and test laser. In the example shown in Figure 7.7, the width of each resonance line is of the order of 1 MHz. In the following, we discuss experimental schemes that provide significantly higher spectral resolution. Like spectral holeburning, these methods must rely on multiphoton processes to obtain spectral information on the sublevel transitions. The main difference that allows this increased resolution is that they use coherent processes, in particular coherent Raman scattering, to obtain the spectroscopic information. They therefore involve the interaction with coherences between nuclear spin states, in contrast to spectral hole burning, where pump and probe laser interact only with the populations of the spin states.

Basic experiment

One implementation of coherent Raman scattering is the Raman heterodyne experiment, where the coherent excitation of the medium is prepared by a radio frequency field, which is applied simultaneously with a laser beam.

Figure 7.8 illustrates the principle of the experiment: The laser creates a population difference by spectral hole burning between two ground-state sublevels. For a very rough formal description of this experiment, we write the density operator of this three-level model system due to the effect of the laser alone as

$$\rho_{\text{las}} = \sigma_2 \tag{7.5}$$

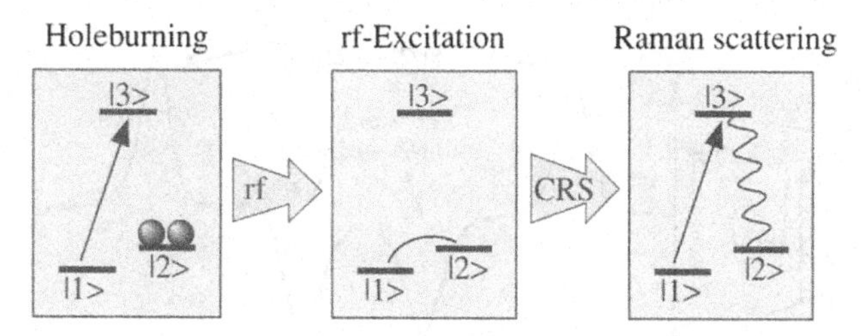

Figure 7.8. Processes involved in a Raman heterodyne experiment.

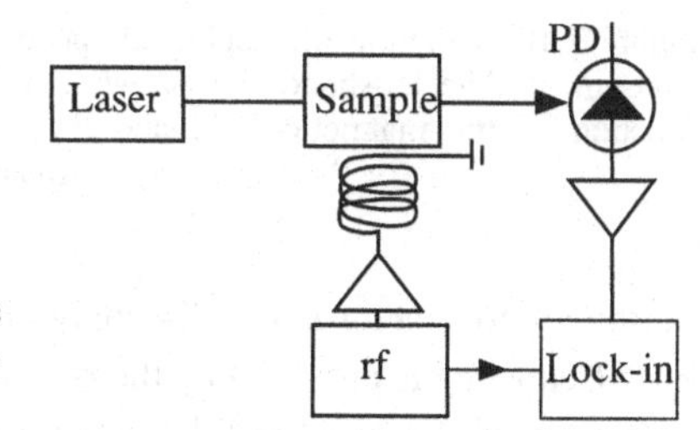

Figure 7.9 Schematic representation of the experimental setup for the Raman heterodyne experiment.

implying that only level $|2\rangle$ is populated. The application of the rf field to transition $|1\rangle \leftrightarrow |2\rangle$ converts the population difference between the two states into coherence ρ_{12}. For a weak rf field that does not significantly change the populations, the coherence obeys the equation of motion

$$\dot{\rho}_{12} = \omega_1 - (\gamma - \mathrm{i}\,\delta)\,\rho_{12} \tag{7.6}$$

written in a reference frame that rotates at the radio frequency; ω_1 is the Rabi frequency, γ the relaxation rate, and $\delta = \omega_0 - \omega_{rf}$ the frequency difference between the transition frequency ω_0 and the radio frequency ω_{rf}. The stationary value becomes

$$\rho_{12} = \omega_1 \frac{\gamma + \mathrm{i}\,\delta}{\gamma^2 + \delta^2} \tag{7.7}$$

if saturation is negligible. As seen in Section 6.3.3, coherent Raman scattering can directly measure this sublevel coherence.

Figure 7.9 shows the setup for such an experiment. It includes, besides the laser beam, a radio frequency source. The amplified radio frequency signal creates an alternating magnetic field in a solenoid coil that is ideally wound around the sample. The beat signal between Raman and laser beam observed

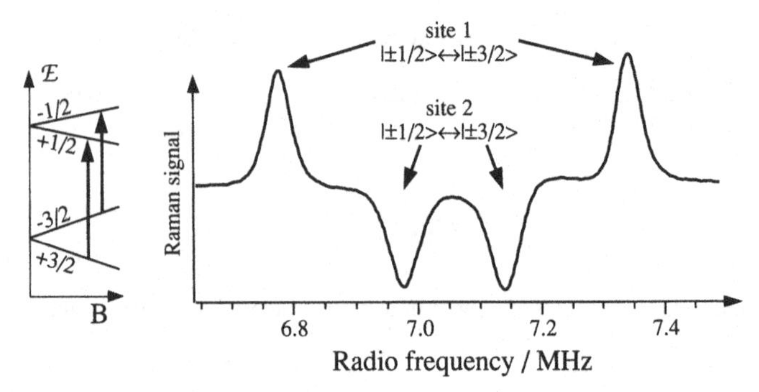

Figure 7.10. Example of a sublevel spectrum recorded with the Raman heterodyne experiment. The left-hand side shows the energies of four nuclear spin states as a function of the magnetic field and the two magnetic dipole transitions between them.

on the photodetector is usually analysed with a lock-in detector whose reference frequency is provided by the rf source. A precise control of the laser frequency is not required in this experiment, since it is always resonant with a group of atoms. In this experiment, the laser beam has three distinct effects: it creates a population difference between the sublevels, it probes the sublevel coherence through coherent Raman scattering, and it serves as the local oscillator for the heterodyne detection.

The ground state of Pr^{3+}

Figure 7.10 shows an example of a spectrum that can be obtained with this method. The system used for this experiment was the rare earth ion Pr^{3+}, doped at a concentration of 0.1% in the host material $YAlO_3$. For the experiment, the Raman heterodyne signal was measured as a function of the radio frequency. Resonance lines occur whenever the radio frequency matches a magnetic dipole transition between two nuclear spin states. The four resonance lines in this spectrum correspond to transitions from the states with $m_I = \pm 1/2$ to $m_I = \pm 3/2$. For each transition, two lines appear in the spectrum, since the Pr ions occupy two inequivalent sites in the crystal. This experiment thus allows a precise determination of the energy level separation between the different nuclear spin states.

After the first demonstration (Mlynek et al. 1983; Wong et al. 1983), this experiment was applied to the study of various materials like Pr^{3+}:$YAlO_3$ (Erickson 1979; Macfarlane, Shelby and Shoemaker 1979; Glaser, Wäckerle and Dinse 1985; Erickson 1993; Mitsunaga, Uesugi and Sugiyama 1993),

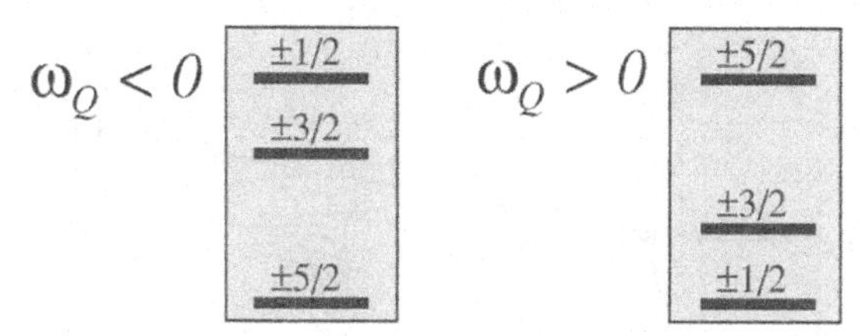

Figure 7.11. Energy level scheme for negative (left) and positive (right) quadrupole coupling.

Pr^{3+}:LaF_3 (Erickson 1977; Macfarlane, Shelby and Shoemaker 1979; Whittaker and Hartmann 1982; Reddy and Erickson 1983; Wald, Hahn and Lukac 1992), Pr^{3+}:YAG (Shelby et al. 1983), Pr^{3+}:$LiYF_4$, (Erickson 1985a), Pr^{3+}:Y_2SiO_5 (Holliday and Wild 1993), Eu^{3+}:$LiYF_4$ (Erickson 1985b), Eu^{3+}:Y_2SiO_5 (Mitsunaga, Yano and Uesugi 1991; Yano, Mitsunaga and Uesugi 1991), and Eu^{3+}:$YAlO_3$ (Erickson 1987), but also to ruby (Szabo, Muramoto and Kaarli 1990) and diamond (Manson, He and Fisk 1990). The technique is a precise and versatile method for the measurement of sublevel splittings. The spectrum does, however, provide only the absolute value, not the sign of the quadrupole interaction. Like conventional magnetic resonance, it cannot distinguish between the two energy level schemes shown in Figure 7.11.

The two energy level schemes correspond to the same absolute value of the coupling constant but opposite sign. Both systems yield the same Raman heterodyne signal, although the sequence of the energy levels is opposite.

7.2.3 Triple resonance

The additional information on the sequence of the energy levels becomes available if two independent beams share the different functions of the laser light. One beam establishes the population difference by spectral hole burning, while the second beam drives the Raman process and serves as the local oscillator. The frequencies of the two laser beams must be controlled independently.

As Figure 7.12 shows, the two beams may still originate from the same laser if acoustooptic modulators shift their frequencies. As the pump laser beam prepares only the populations of the system, no phase matching is required between the two beams and they may overlap in the sample at a small angle.

Figure 7.13a summarises at what frequencies of pump and probe laser beam it is possible to observe signals with the setup shown in Figure 7.12. In this

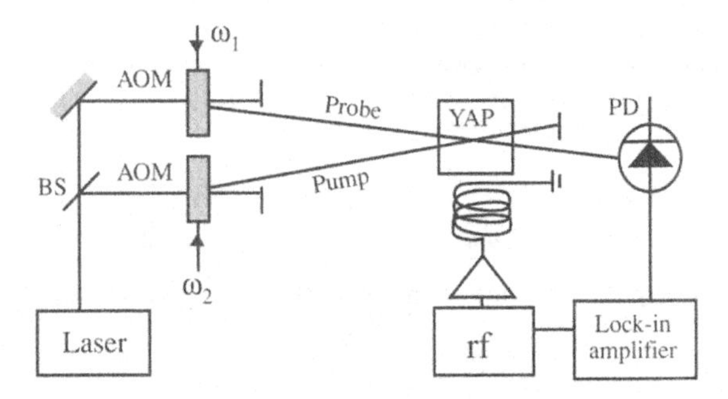

Figure 7.12. Setup for two-beam Raman heterodyne experiment. AOM = acoustooptic modulator, BS = beam splitter, PD = photodiode.

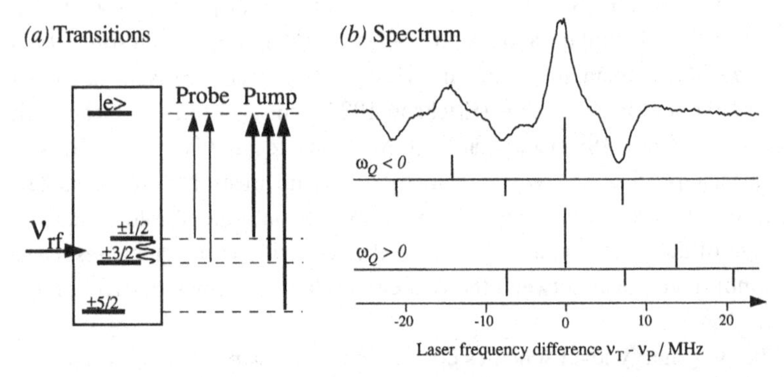

Figure 7.13. **(a)** Possible transitions in two-beam Raman heterodyne spectroscopy. **(b)** Signal of a two-beam Raman heterodyne experiment together with theoretical stick spectra for both signs of the quadrupole coupling.

example, the radio frequency is tuned to the transition between the $\pm 1/2$ and the $\pm 3/2$ states. The Raman heterodyne experiment finds a signal whenever the probe laser beam couples one of these two states to an excited state. For simplicity, we neglect the splitting within the excited state. At the same time, the pump laser beam must deplete one of the three ground-state sublevels of the same atom by spectral hole burning. The difference between pump and probe laser beam must therefore be equal to an energy separation between ground-state sublevels.

We expect triple resonance signals whenever the two laser beams and the radio frequency field are resonant with the same group of atoms. In particular, the difference between pump and probe laser frequency must match the energy separation between two nuclear spin states. Compared to hole burning spectra, where the laser frequencies must satisfy the same condition, two-beam

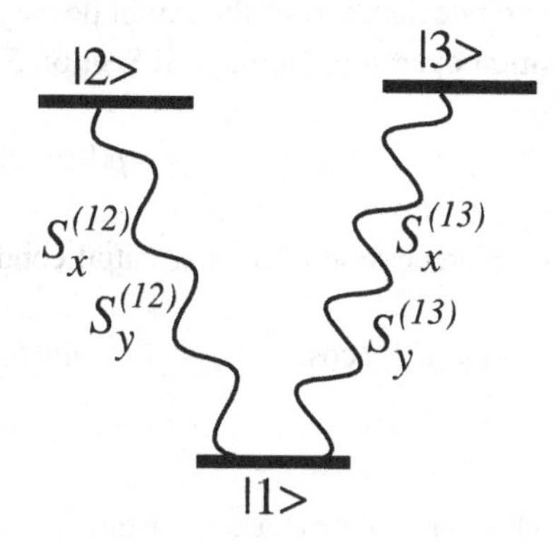

Figure 7.14. Level scheme for the analysis of photon echo modulation experiments.

Raman heterodyne spectroscopy has the additional selection rule that the probe laser does not detect the transitions from levels $m_I = \pm 5/2$ to the excited state.

Figure 7.13b shows the two-beam Raman heterodyne signal measured as a function of the difference between the two laser frequencies. The top trace shows the observed spectrum, while the two stick spectra indicate where we expect to find resonance lines in the case of negative (centre) and positive (bottom) quadrupole coupling constant. Clearly, the spectrum for negative coupling constant is compatible with the experimental spectrum, whereas the assumption of a positive coupling constant cannot explain the observed signal.

7.3 Time-resolved experiments

7.3.1 Photon echo modulation

As shown in Section 2.3.1, a laser pulse that exchanges the populations of two states (a "π pulse") also inverts the phase of a coherence between these states. If this coherence evolves under an inhomogeneous Hamiltonian, its subsequent evolution reverses the destructive interference of signal contributions from atoms in different environments. This refocusing process is observable as a "photon echo." In multilevel systems, it is also possible to excite photon echoes. If the refocusing pulse couples to more than one transition, however, it cannot initiate a complete refocusing.

Evolution

For an analysis of the relevant processes, we assume that a laser pulse prepares coherence, represented by the wavy lines in Figure 7.14, in the optical transitions between the ground state $|1\rangle$ and two excited states $|2\rangle$ and $|3\rangle$.

The relevant part of the initial density operator is, in terms of the single transition operators, defined in Section 3.3.

$$\rho(0) = S_x^{(12)} + S_x^{(13)} \tag{7.8}$$

Free precession turns this initial condition into

$$\rho(T) = S_x^{(12)} \cos(\delta_{12}T) - S_y^{(12)} \sin(\delta_{12}T) + S_x^{(13)} \cos(\delta_{13}T) - S_y^{(13)} \sin(\delta_{13}T) \tag{7.9}$$

where δ_{12} and δ_{13} are the Bohr frequencies of the two transitions.

If the refocusing pulse acted only on the transition $|1\rangle \leftrightarrow |2\rangle$, it would invert the $S_y^{(12)}$ term, initiating a refocusing process. The discussion of three-level systems showed, however, that pulses that couple to more than one transition also initiate an exchange of coherence between the different transitions. The closest approximation to an ideal refocusing pulse is described by the unitary operator

$$U = \exp\left\{\frac{\pi}{\sqrt{2}} (S_y^{(12)} - S_y^{(13)})\right\} \tag{7.10}$$

It converts the density operator into

$$\begin{aligned}\rho(T+) &= U^{-1}\, \rho(T)\, U \\ &= \frac{1}{2}\{(S_x^{(12)} + S_x^{(13)})(\cos(\delta_{12}T) + \cos(\delta_{13}T)) \\ &\quad + (S_y^{(12)} + S_y^{(13)})(\sin(\delta_{12}T) + \sin(\delta_{13}T))\}\end{aligned} \tag{7.11}$$

and additional terms, which are not of interest in this context.

Figure 7.15 illustrates the effect of the pulse. During the free precession period before the refocusing pulse, the coherences in the two transitions acquire phases $\delta_{12}T$ and $\delta_{13}T$ that are proportional to the two optical frequencies δ_{12} and δ_{13}. The pulse inverts the phases and exchanges coherence between the two transitions. The coherence in transition $|1\rangle \leftrightarrow |2\rangle$, e.g., depends now on both precession frequencies. A measurement of the coherence would therefore display interference between the two contributions. However, if the experiment is performed in an inhomogeneously broadened system, the inhomogeneous decay also washes out the interference effect.

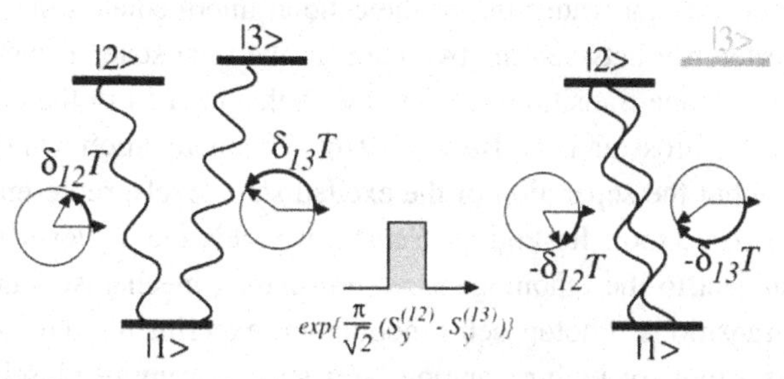

Figure 7.15. Conversion of optical coherences by the laser pulse. The final state does not show the coherence in transition $|1\rangle \leftrightarrow |3\rangle$, which is also present.

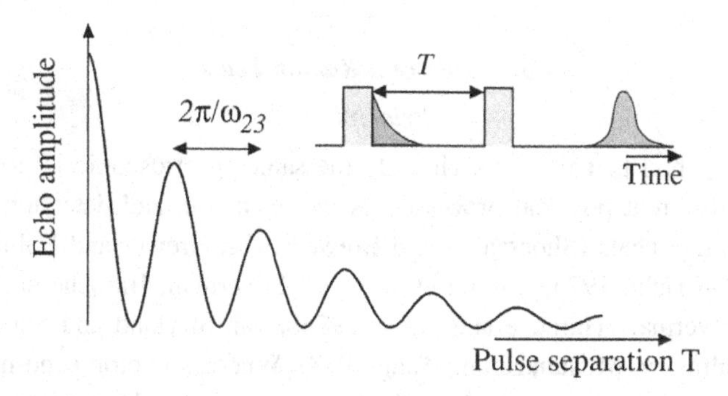

Figure 7.16. Echo intensity as a function of the separation T between the excitation and refocusing pulse.

Echo formation

As in the two-level case, the free precession after the refocusing pulse reverses the inhomogeneous decay. After an additional precession period of duration T, the optical coherences become

$$\rho(T+T) = \frac{1}{2}\{(S_x^{(12)} + S_x^{(13)})\,(1 + \cos(\delta_{23}T)) + (S_y^{(12)} - S_y^{(13)})\,\sin(\delta_{23}T)\} \tag{7.12}$$

where $\delta_{23} = \delta_{13} - \delta_{12}$ is the energy separation of the two excited states. A complete refocusing occurs only if the two excited states are degenerate or if the free precession period T is a multiple of the inverse energy separation. As Figure 7.16 shows, the echo intensity shows a modulation as a function of the pulse separation. The superimposed decay is due to the homogeneous decay of the optical coherences, e.g., by spontaneous emission.

We may consider the vanishing of the echo at intermediate times as a destructive interference between the two paths to the final state. The coherence that originated in one transition interferes with that created in the other transition (Schenzle, Grossman and Brewer 1976). The modulation, which carries information about the separation of the excited state levels, represents a possible spectroscopic tool. Its Fourier transform yields the sublevel spectrum and is not subject to the inhomogeneous broadening mechanisms that affect the optical transitions. Photon echo modulation experiments have therefore been used as a tool for high-resolution laser spectroscopy of closely spaced levels (Chen, Chiang and Hartmann 1980; Whittaker and Hartmann 1982; Glaser, Wäckerle and Dinse 1985; Szabo 1986).

7.3.2 Coherent Raman beats

Principle

Another method that provides virtually the same spectroscopic information, but uses different physical processes, is the excitation and detection of coherent Raman beats (Shoemaker and Brewer 1972; Brewer and Hahn 1973; Brewer and Hahn 1975; Bassini et al. 1977; Silverman, Haroche and Gross 1978a; Silverman, Haroche and Gross 1978b; van Stryland and Shoemaker 1979; Walmsley, Mitsunaga and Tang 1988). Whereas photon echo modulation involves only coherences in optical transitions, the coherent Raman beat method makes the coherence in an rf transition visible. Coherent Raman beats involve the same physical processes as coherent Raman scattering: creation of a coherent excitation in the material and scattering of a laser field from this excitation. The difference between these two experiments is the relative timing of the two processes: In coherent Raman scattering, they occur simultaneously, in steady state, but in the Raman beat experiment, they occur sequentially, during distinct time periods. The coherent excitation of the medium is prepared in a first step, as shown in Figure 7.17.

Coherent Raman beats can be observed in Λ- or V-type systems. In Figure 7.17, we use a Λ-type three-level system to illustrate the relevant processes. A pump laser pulse, which is resonant with both optical transitions of the three-level system, excites a coherent superposition of the two ground states, indicated by the curved line in the second inset at the bottom of the figure. When the probe laser beam is turned on and tuned close to the transition $|1\rangle \leftrightarrow |3\rangle$, it scatters by partially converting the sublevel coherence ρ_{12} into optical coherence in the second optical transition. The resulting Raman field propagates together with the incident laser beam. When they fall on a detector, the interference between the two fields generates a beat signal at the dif-

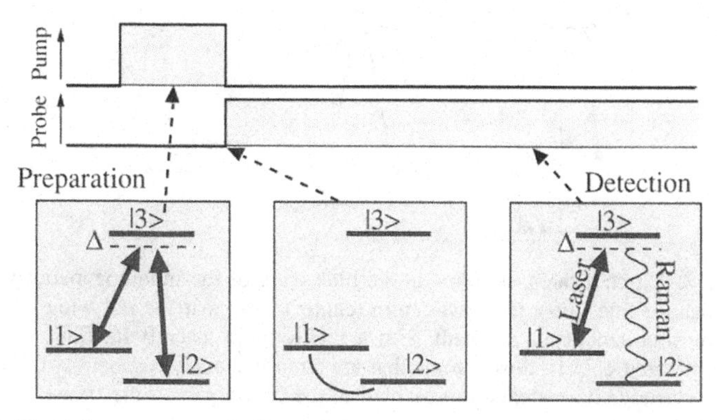

Figure 7.17. Relevant stages during coherent Raman scattering with transient excitation by a laser pulse.

ference frequency ω_{12}. The resulting signal in the form of a free induction decay can be Fourier transformed to obtain a spectrum of the sublevel transitions.

This separation of the excitation and detection processes in the time domain suggests that the relevant processes be discussed separately. Chapter 3 contains their mathematical analysis. Here we discuss some experimental aspects as well as the issues that arise when turning from the idealised picture of a three-level system to a real physical system. Examples and experimental results will be discussed in the following section.

Inhomogeneous broadening

An important difference between the mathematical analysis of Chapter 3 and real physical systems is the occurrence of inhomogeneous broadening. In the systems under discussion here, the inhomogeneously broadened optical resonance line is considerably wider than either the laser linewidth or the separation between the sublevels that we wish to observe. As discussed above, the evolution of the sublevel coherence that the laser excites in the atoms does not depend on this inhomogeneous broadening. The laser frequency, however, determines which atoms are resonant with the laser. In a pump-probe experiment such as a coherent Raman beat experiment, the pump laser beam excites selectively those atoms that are resonant with it. The second laser beam, which probes the atoms, is always resonant with some atoms, but produces a signal only if both laser beams are resonant with the same group of atoms. The difference between the two laser frequencies must therefore match a com-

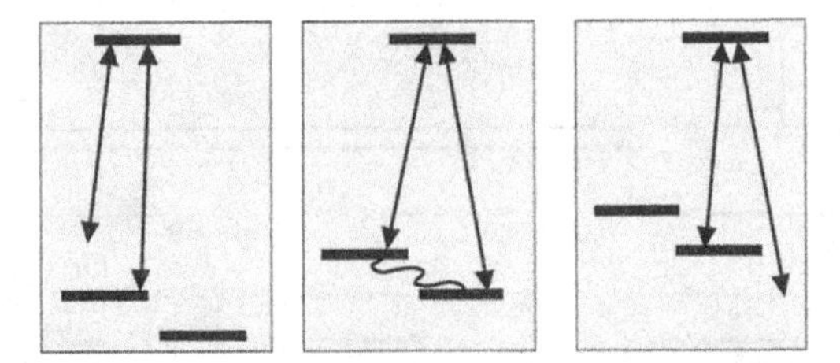

Figure 7.18. Three groups of atoms in the blue wing of the inhomogeneously broadened resonance line (left), near the centre (centre), and near the red wing of the line (right) are simultaneously resonant with a bichromatic laser field. Sublevel coherence is generated only in those atoms that are simultaneously resonant with both frequency components, represented in the centre part of the figure.

bination of energy differences in the atomic system, exactly as in the case of spectral hole burning.

As discussed in Section 3.3.6, efficient excitation of Raman beats requires a bichromatic laser field. In an inhomogeneous resonance line, this requires matching of both frequencies, as shown in Figure 7.18. Since the splitting between the nuclear spin sublevels is much smaller than the inhomogeneous broadening of the optical resonance line, a monochromatic laser field is resonant with all possible transitions – not for a single atom, but for different groups of atoms. The difference between two optical transitions, however, is almost identical for all atoms. As Figure 7.18 shows, a bichromatic laser field is resonant with three groups of three-level atoms within the inhomogeneously broadened resonance line. Excitation of sublevel coherence, however, occurs only in the central group, where both components of the bichromatic field are simultaneously resonant with the same set of atoms.

Detection of Raman beats

Efficient detection of a coherence between two levels through a resonant Raman process requires a probe laser beam to be resonant with one of the two optical transitions to which the excitation field couples; the frequency of the probe laser beam must match that of one frequency component of the excitation field, as shown in Figure 7.19.

In the situation drawn in Figure 7.19, the probe laser is resonant with the transition $|1\rangle\leftrightarrow|2\rangle$, exciting the Raman field in transition $|1\rangle\leftrightarrow|3\rangle$. In these systems, where the laser linewidth is smaller than the homogeneous optical linewidth, the usual symmetry between Stokes and anti-Stokes transitions is

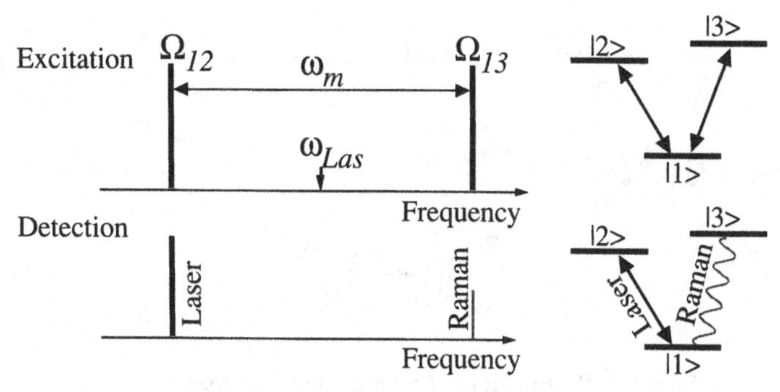

Figure 7.19. Frequency spectrum of excitation and detection fields.

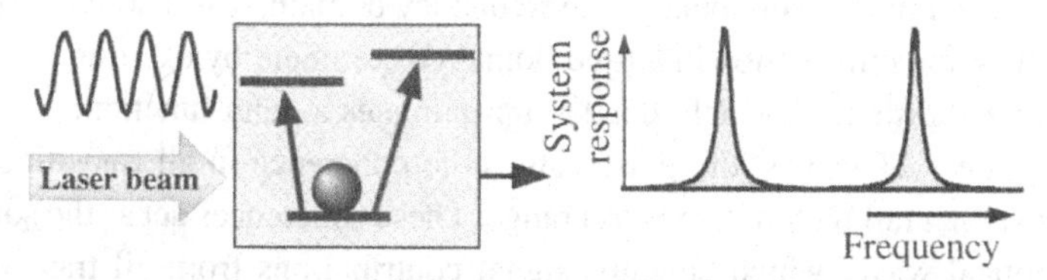

Figure 7.20. Frequency-domain spectroscopy: The spectrum is obtained as the response of the system to a monochromatic wave with slowly changing frequency.

broken and we observe only one of the two cases. In the situation depicted in the figure, the Raman field appears at the higher frequency, but if the probe laser were tuned to this frequency, the Raman field would appear at the lower frequency and we could observe a Stokes shift.

7.3.3 Time-domain spectroscopy

Principle

Coherent Raman beats, as well as photon echo modulation, generate signals that depend on the energy separation between different substates and on the lifetime of the coherences, either in an optical transition or in a sublevel transition. This is generally the information that spectroscopic experiments try to extract. It is therefore not difficult to understand that the main applications of both techniques are spectroscopic experiments. In traditional spectroscopic experiments, one measures the response of a material system – in our case ensembles of atoms – to a continuous train of monochromatic radiation, like laser light.

In the example of Figure 7.20, the absorption of laser light shows two distinct resonances. From the frequency of those resonances, we infer the energy of two transitions between an electronic ground state and two excited states.

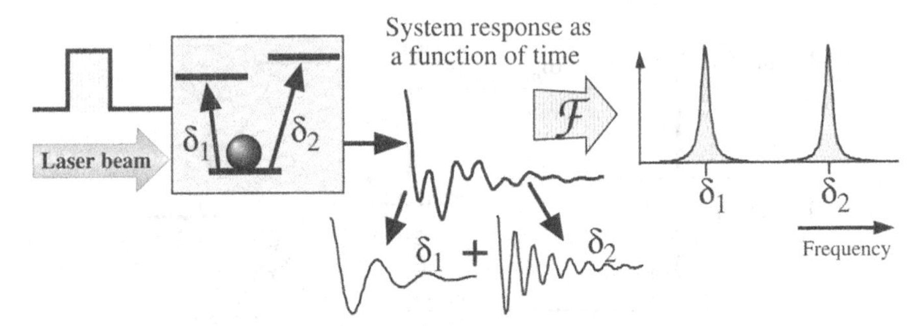

Figure 7.21. Principle of time-domain spectroscopy.

In contrast to this measurement in the frequency domain, it is also possible to obtain the same information in a time-domain experiment by exciting the system with a short pulse, which ideally approximates a delta function.

As Figure 7.21 shows, the pulse can excite coherence in all transitions in the system that fall within its spectral range. These coherences act as the source of an optical wave, which contains signal contributions from all transitions excited by the pulse. In our example with two optical transitions, the signal contains two frequency components, as shown in the figure.

$$s(t) = (e^{i\delta_1 t} + e^{i\delta_2 t})e^{-\gamma t} \tag{7.13}$$

For an easier interpretation of the signal, it is usually advantageous to Fourier transform the data. In the present example, we obtain a spectrum

$$f(\omega) = \mathcal{F}\{s(t)\} = g(\delta_1) + g(\delta_2) \tag{7.14}$$

where $g(\delta)$ represents the usual Lorentzian lineshape function. The resulting spectrum, represented in the right-hand part of Figure 7.21, is equivalent to the spectrum obtained in the frequency domain with cw excitation.

This principle is well known in magnetic resonance spectroscopy (Abragam 1961), where it is today the standard method (Ernst, Bodenhausen and Wokaun 1987). In the optical domain, the measurement of time-domain signals (Brewer and Shoemaker 1972) is considerably more demanding experimentally and still represents the exception rather than the rule. The same method can also be applied, however, to sublevel transitions, where the evolution is slow and the experimental requirements are easier to fulfil. The level splittings measured in such an experiment may be due to the Zeeman effect, but in other systems, they may be zero-field splittings. This potential is similar to the

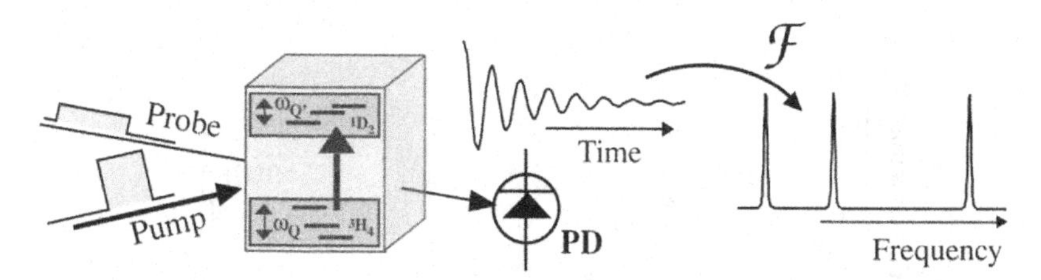

Figure 7.22. Schematic representation of time-domain spectroscopy using a coherent Raman beat experiment.

method of quantum beats, but it is applicable to electronic ground states as well as electronically excited states. One example is the measurement of the hyperfine splittings in the ionic solid Pr^{3+}: YAG at low temperature (Shelby et al. 1983). Here the FID signal contained several frequency components that could be extracted from the time-resolved data by subsequent Fourier transformation of the signals.

Experimental considerations

In the case of coherent Raman beats, the pulsed excitation of the system uses a laser pulse to prepare a coherent superposition of two sublevels, either in the ground or excited state.

As Figure 7.22 illustrates, a second laser beam probes the system response through a coherent Raman scattering process and the photodetector produces the time-dependent signal, as discussed in Section 7.3.2. By Fourier transforming the recorded time-domain signal, one can recover the frequency spectrum, which contains resonance lines at the transition frequencies between the sublevels that form the coherent superposition.

Figure 7.23 shows the spectrum that can be obtained by Fourier transformation of a coherent Raman beat following excitation with a laser pulse. The resonance lines near the origin of the frequency axis indicate transitions between the excited state sublevels; the resonances at 7, 14 and 21 MHz correspond to ground-state splittings. As discussed in Section 3.3.6, efficient excitation of the sublevel coherence requires that the laser field excite both optical transitions. The decreasing intensity of the lines at higher frequency indicates that excitation of sublevel coherences with a monochromatic laser pulse becomes inefficient at higher sublevel splittings. This off-resonance effect affects not only the excitation process, but also the detection process: Most atoms in which the pump laser pulse prepares a sublevel coherence are far from res-

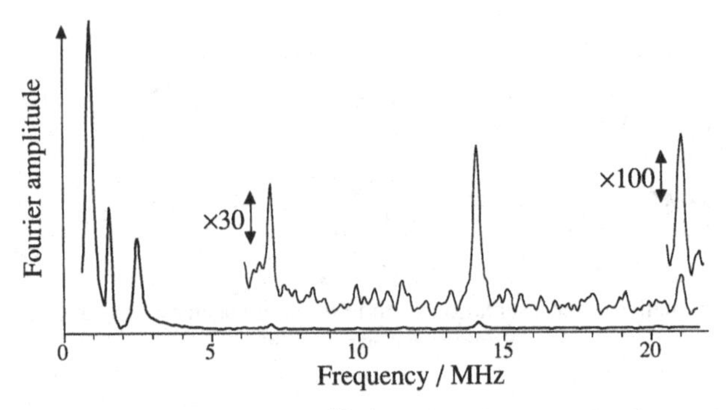

Figure 7.23. Sublevel spectrum of Pr^{3+}:$YAlO_3$ obtained by Fourier transformation of coherent Raman beats that were excited with a monochromatic laser pulse of duration 0.9 μsec.

onance with the probe laser, whose frequency was the same as the pump laser frequency. As discussed in Section 3.4.2, a probe laser that is detuned from the optical resonance yields a signal at the same beat frequency, but with a smaller amplitude.

One possibility of compensating for this loss of efficiency is to increase the optical power, using short, intense pulses. On the other hand, such experiments are often performed in ionic solids with inhomogeneously broadened optical transitions where the pulses should selectively excite only a narrow portion of the inhomogeneously broadened optical transition. This additional constraint can be met only with continuous lasers. With the intensities available from commercial cw ring lasers, it is not possible to excite systems with energy level splittings larger than a few MHz.

Bichromatic field

As pointed out in Section 3.3.6, this problem can be overcome by using a bichromatic laser field for the excitation pulse. The two frequency components for bichromatic irradiation can be generated by different methods. Special lasers have been constructed that generate two frequencies, and it is possible to use two independent lasers whose frequency difference may be controlled by observing the beat frequency that the bichromatic field generates on a quadratic photodetector.

If the frequency differences between the two optical transitions are small, it is also possible, and often advantageous, to generate the two frequencies by intensity modulation of a single laser beam, using acoustooptic or electrooptic modulators. The modulated field has the form

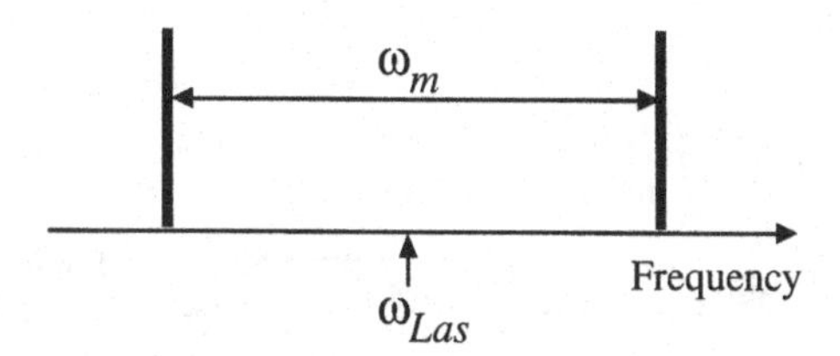

Figure 7.24. Frequency spectrum of a sinusoidally modulated laser field.

$$E = \frac{E_0}{2}\{\exp[i(\omega_{\text{Las}} - \omega_{\text{m}}/2)t] + \exp[i(\omega_{\text{Las}} + \omega_{\text{m}}/2)t]\} \qquad (7.15)$$

written in complex notation. Here, ω_{Las} is the laser frequency and ω_{m} the modulation frequency. The frequency spectrum, which is represented in Figure 7.24, contains two components, as desired.

Control of the two relevant frequencies, ω_{Las} and ω_{m}, changes directly the two parameters "laser frequency mismatch" and "difference frequency mismatch," as discussed in Section 3.3.6, describing the Raman excitation process.

Implementation

Apart from the intensity modulation, the pump laser beam must also be chopped to generate the required pulses. While the modulation produces a laser intensity

$$I_{\text{bc}}(t) = \frac{I_0}{2}[1 + \cos(\omega_{\text{m}}t)] \qquad (7.16)$$

the chopping modulates this intensity with a rectangular window, as shown in Figure 7.25.

The upper trace shows the pump laser intensity, which is $I_{\text{bc}}(t)$ during the excitation period and zero otherwise. In the figure, the test laser is switched on only after the end of the excitation period. This is not a requirement, but it reduces spectral holeburning by the test laser, which can be significant at low temperatures. In principle, the same laser beam can serve as the pump and as the probe. Since the test laser frequency is different from the carrier frequency of the excitation field, and the two intensities may differ substantially, it is in practice easier to use two separate beams. The probe laser frequency can then be kept fixed, shifted from the carrier frequency of the pump laser by half the modulation frequency. The easiest method for shifting the frequency uses an acoustooptic modulator.

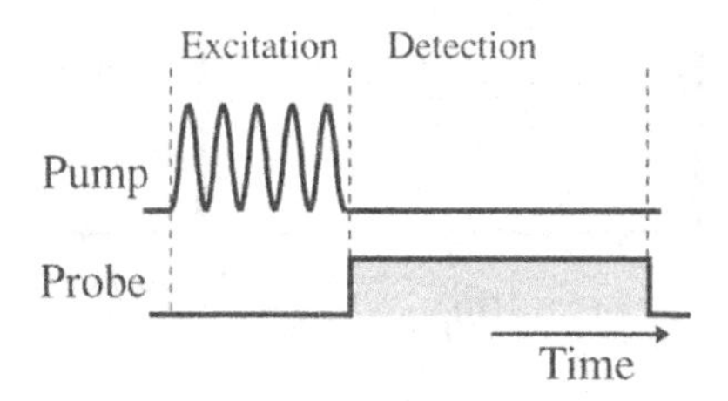

Figure 7.25. Pump and probe laser intensity during a coherent Raman beat experiment.

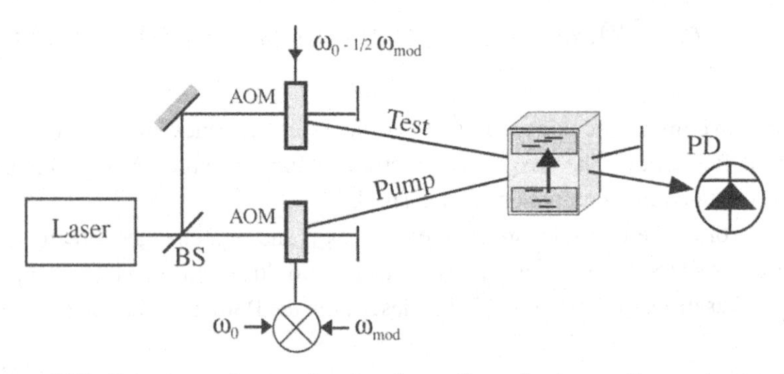

Figure 7.26. Experimental setup for the observation of coherent Raman beats with bichromatic excitation.

Figure 7.26 shows a possible experimental setup for the observation of coherent Raman beats with bichromatic excitation. The intensity of the pump laser beam (typically 200–300 mW/mm^2) is modulated sinusoidally with an acoustooptic (or electrooptic) modulator and gated with an acoustooptic modulator that produces pulses with a duration between 0.1 and 100 μsec. The weaker probe beam (<1 mW/mm^2) is switched on immediately after the pump pulse by a second acoustooptic modulator. The rf carrier frequency of this modulator is adjusted to shift the test laser frequency to one of the two modulation sidebands of the pump laser field. A fast photodiode detects the beat signal. The time-domain signal is Fourier transformed to obtain the sublevel spectrum.

7.3.4 Examples

To illustrate the method, we include a few examples measured in the system Pr^{3+}:$YAlO_3$.

Spectra

In the $^1D_2$ excited state of this system, the splittings between the sublevels are 0.9, 1.6, and 2.5 MHz, as shown in the energy level scheme (see Figure 7.27). These splittings are sufficiently small to allow excitation of the sublevel coher-

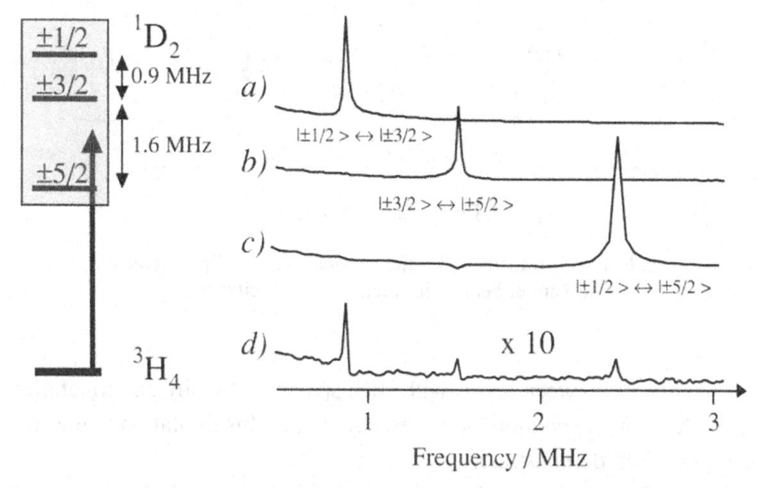

Figure 7.27. Spectra of the excited state splitting of Pr^{3+}:$YAlO_3$.

ences with short laser pulses ($\tau \lesssim 1$ μsec). Trace (d) at the bottom of Figure 7.27 shows an example of the Fourier transformed Raman beat signal obtained with monochromatic excitation. All three expected signals are visible in the spectrum. The decrease of signal amplitude at the higher frequencies, however, illustrates the limitations of the monochromatic excitation. Traces (a) to (c) demonstrate how bichromatic excitation, combined with frequency-shifted detection, can overcome this problem and provide a significantly better signal to noise ratio. The difference is particularly striking for the line near 2.5 MHz.

All four traces were measured with the same laser intensity. The pulse duration was optimised independently for all four experiments. In the case of single-frequency excitation, the optimal pulse length was 0.9 μs, whereas it was between 20 and 28 μs for the three experiments with bichromatic excitation. The three resonance lines were measured in three separate experiments, with the modulation frequency set close to the beat frequency, since the width of the spectrum that can be excited is limited to the inverse of the pulse length as in the case of monochromatic excitation. For wider spectra, it is necessary to record the spectrum in pieces, as shown here. The observation of all three resonance lines in the spectrum shows that the coherent Raman beat method is not subject to the same selection rules as the direct observation in magnetic resonance experiments. Direct detection would find resonances only at the magnetic dipole transitions $|\pm 1/2\rangle \leftrightarrow |\pm 3/2\rangle$ and $|\pm 3/2\rangle \leftrightarrow |\pm 5/2\rangle$, which appear at 0.9 and 1.6 MHz.

In the electronic ground state of Pr^{3+}:$YAlO_3$, the quadrupole splittings are significantly larger: 7, 14, and 21 MHz. Continuous lasers cannot efficiently

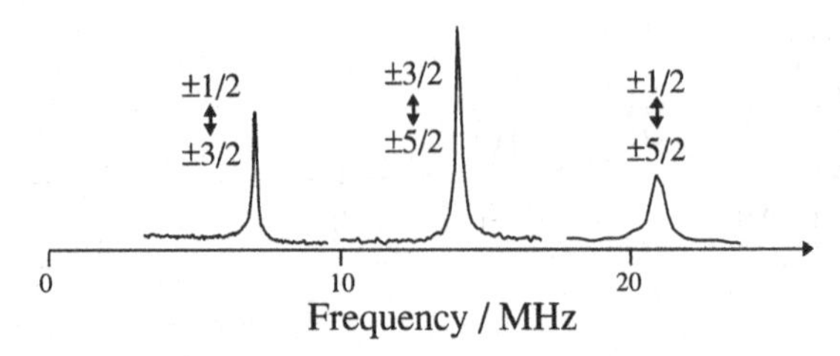

Figure 7.28. Nuclear spin transitions in the ground state of Pr^{3+} detected by coherent Raman beats with bichromatic excitation.

excite coherences between levels with such spacings (Shelby and Macfarlane 1984). Bichromatic excitation, however, overcomes this limitation as the spectra of Figure 7.28 demonstrate.

Again, the three resonance lines were measured separately, with the modulation frequency set close to each of the sublevel transition frequencies. As in the excited state, all three resonances can be observed with comparable intensity. See Figure 7.28.

Frequency dependence

In the theoretical analysis of the Raman excitation process with a bichromatic laser field, we found that the excitation efficiency is maximised when the two frequency components match the optical transition frequencies. We analysed two cases of frequency mismatch, corresponding to "difference frequency mismatch" and laser frequency mismatch. In an inhomogeneously broadened system, laser frequency mismatch is always present for most atoms; this parameter is therefore not under direct experimental control. The difference frequency mismatch, however, can be controlled easily through the modulation frequency that generates the two sidebands. When it equals the sublevel transition frequency, both components of the bichromatic laser field are in resonance with an optical transition of the three-level atom which yields the highest excitation efficiency.

Figure 7.29 shows the efficiency of the Raman excitation process as a function of the modulation frequency for different pulse durations. The dots represent experimental data measured on the $\pm 3/2 \leftrightarrow \pm 5/2$ nuclear spin transition of the excited state, which has a splitting of 1.56 MHz. For a given pulse length the modulation frequency was varied for successive experiments over the range between 1.2 and 2.0 MHz, using a pump intensity of 0.3 W/mm^2. The beat signal was Fourier transformed and the Fourier amplitude at the sub-

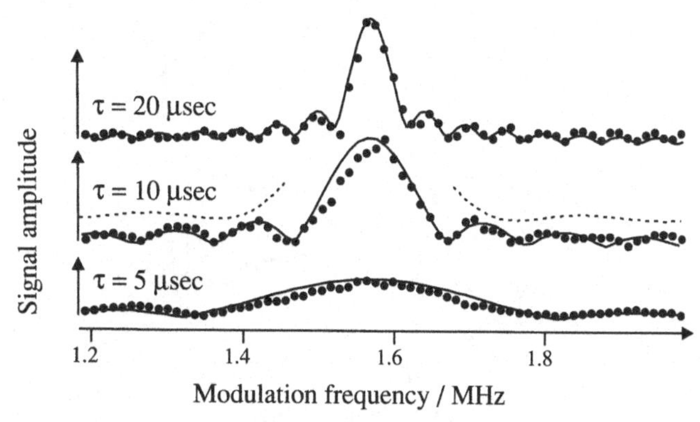

Figure 7.29. Dependence of the excitation efficiency on the modulation frequency for different pulse durations τ.

level transition frequency was measured. When the modulation frequency matches the energy separation between the two excited state sublevels, the amplitude of the beat signal is larger, as predicted by the theory. In addition to the central resonance line, characteristic side peaks occur in the spectrum, whose position depends on the pulse duration.

The lines compare these experimental with theoretically expected data. The theoretical behaviour of equations (3.48)–(3.49), evaluated for an optical Rabi frequency of 15 kHz and a pulse duration of $\tau = 10$ μsec, gives the behaviour indicated by the dashed line. The dependence of the excitation efficiency on the modulation frequency is clearly different for this dashed line than for the experimental data. The dependence on the pulse length (not shown in the figure) also differs from the experimentally observed behaviour.

The reason for this discrepancy is that we have not taken the inhomogeneous broadening of the optical transitions into account. The broadening, together with the frequency jitter of the two laser beams, averages over a range of atoms not exactly resonant with the laser, i.e., over atoms with different laser detunings δ_L. Taking this effect into account by averaging over the inhomogeneous resonance line changes the predicted behaviour from the dashed line to the full line in Figure 7.29, which is in good agreement with the experimentally observed behaviour.

A comparison of the data for different pulse durations shows that the sublevel coherence reaches its maximum at $\tau = 20$ μs, which corresponds to a quarter period of the Rabi frequency. With increasing pulse length, the width of the central resonance line decreases in inverse proportion to the pulse length until it approaches a value of 20 kHz, which is of the order of the optical Rabi frequency.

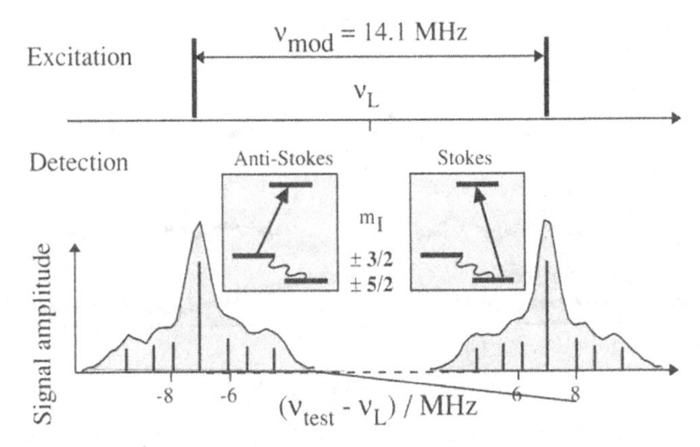

Figure 7.30. Signal amplitude as a function of the probe laser frequency.

Laser-frequency dependence

With sublevel splittings of several MHz, as in this case, precise control of the laser frequency is important not only for the excitation process, but also for the detection process. Although the frequency of the observed beat signal does not depend on the laser frequency, the amplitude of the signal drops off quickly with increasing laser detuning. Figure 7.30 demonstrates this effect by showing the detected signal amplitude as a function of the probe laser frequency.

For this example, the pump laser intensity was modulated at 14 MHz, thus exciting sublevel coherences between the $\pm 3/2$ and $\pm 5/2$ nuclear spin states of the electronic ground state. The two curves show the signal obtained with the probe laser frequency close to the two sidebands at $\omega_{pump} \pm 7$ MHz. The range, over which the Raman signal has appreciable intensity, corresponds to the laser frequency jitter, which is of the order of 1 MHz in this experiment. In addition, the unresolved quadrupole splitting of the excited state causes a line broadening, as indicated by the stick spectra.

The two insets show that the detecting laser beam couples to two different transitions at the two frequencies. For the resonance line at the left, the probe laser frequency was at the lower frequency sideband, exciting a Raman field at the higher frequency sideband. This signal therefore corresponds to anti-Stokes scattering. In the right-hand side of the figure, the frequencies of probe laser and Raman field are reversed. The use of a laser field whose frequency distribution is narrower than the sublevel splitting therefore results in an asymmetry for the Stokes and anti-Stokes processes, which can be controlled through the laser frequency. The two lines have very similar intensities, indicating that the symmetry between Stokes and anti-Stokes processes remains if we consider all probe laser frequencies.

For these examples, intensity modulation techniques generated the side-bands with splittings from 0.9 to 21 MHz. Excitation of systems with even wider sublevel splittings may be possible by superimposing laser beams that have been frequency shifted. With acoustooptic or electrooptic modulators, frequency shifts up to several GHz are possible. In principle, it should even be possible to use laser beams derived from independent sources, provided the frequency difference remains sufficiently stable.

8
Sublevel dynamics

After the discussion of ionic solids we shift our attention to atomic vapours. Here, we discuss the dynamics of microscopic order within the angular momentum substates under the influence of light and magnetic fields. For this purpose, we combine optical pumping as outlined in Chapter 5 to prepare the ordered states, and polarisation-selective detection introduced in Chapter 6 for its observation. In contrast to the introductory sections, we no longer discuss these dynamics as mathematical models, but for specific physical systems, using experimental data to illustrate the theoretical description. The systems we use for this illustration are vapours of alkali metal atoms, in particular sodium. These one-electron systems allow the most direct application of the theoretical models developed above. In addition, we concentrate on electronic ground states, whose long lifetimes permit long observation times and correspondingly precise observations.

8.1 Experimental arrangement

8.1.1 General considerations

Laser-induced dynamics

This chapter surveys the coherent evolution of coherences between angular momentum sublevels. Optical pumping excites this microscopic order, and it evolves under the influence of external magnetic fields and the laser radiation. The primary goal of this section is to show how the mathematical models developed in the preceding sections apply to real physical systems. We discuss how the observed signals arise and by which parameters the experimenter can control the dynamics of these systems. Using readily available experimental tools, it is possible to prepare the atomic system into virtually any microscopic state (Bhaskar 1993), and to observe its evolution in great detail (Suter and Klepel 1992a).

As we wish to observe the evolution of the system under the influence of external fields, we discuss primarily time-resolved experiments, which investigate transient phenomena. This approach contrasts with the time-independent observations of traditional magnetooptics (Voigt 1908; Corney 1977). The properties observed in those experiments depend only on stationary states of the microscopic system. In the time-resolved experiments, they form the long-time limit to which the system evolves when the transient effects disappear. In specific cases, there is an additional relation between time-resolved and stationary experiments: Signals from stationary systems that are a function of frequency or magnetic fields are the Fourier transforms of suitable time-resolved experiments. We will exploit this possibility for doing spectroscopy in the time domain.

The subject of our interest is microscopic order in the form of population differences or coherences. In different bases, any density operator component can appear as a diagonal or off-diagonal element. In the first case, we usually refer to it as populations, in the second case as coherences. The only difference between the two is the choice of the basis states, which is somewhat arbitrary. To avoid complications and to simplify the language, we will often refer to all types of microscopic order as coherences, implying that this term can also refer to diagonal elements of the density operator. Coherences can exist between either different Zeeman substates or hyperfine substates, but we will concentrate on Zeeman transitions, which are experimentally easily accessible and whose splitting is under experimental control through the magnetic field strength.

Systems

The specific systems that we use to illustrate these principles are atomic vapours. Atoms of alkali like sodium or cesium or of rare earths with a single valence electron like Sm or Yb provide systems that are simple enough to be studied in detail experimentally as well as theoretically. For most of the discussion that follows, we concentrate on sodium vapour, a system that has been studied extensively in the past. The relatively simple level structure and the availability of efficient laser sources in the spectral range of the strong D lines near 589 nm make it a very convenient system for this type of study. Compared to other frequently used alkali atoms like Rb or Cs, the ground-state hyperfine splitting is small enough that a narrowband laser can be simultaneously resonant with both hyperfine components in a slightly pressure-broadened system, thus minimising hyperfine pumping.

We concentrate the discussion on electronic ground states whose long lifetimes permit extended and correspondingly precise observations. In thermal vapours, where the atoms fly freely, the time that the atoms need to pass through the laser beam limits the observation time. This time, which is of the order of a microsecond for typical conditions, can be extended by adding an inert buffer gas to the sample, typically a noble gas like helium or argon. The collisions between the alkali atoms and the buffer gas atoms change the atomic motion from free flight to diffusion. In a few hundred mbar of buffer gas, the time they spend within the experimentally accessible region increases by several orders of magnitude. The electronic ground state of the alkali, which has a spherical electron distribution, is relatively insensitive to collisions and coherences are thus only weakly affected (Dehmelt 1957b; Legowski 1964; Franz and Volk 1976). In recent years, laser cooling of atoms (Chu 1991; Aspect 1992; Cohen-Tannoudji 1992b) has opened a possible alternative to the use of buffer gas, although it requires a significantly higher experimental effort.

A homogeneous environment facilitates detailed comparison between theoretical predictions and experimental results. In these experiments, the largest inhomogeneous interaction is the Doppler broadening of the optical transitions, which is much larger than the homogeneous linewidth of the free atoms. The addition of buffer gas to the cell eliminates this problem, if the mean free path of the metal atoms between collisions with the buffer gas becomes smaller than an optical wavelength. The atomic motion becomes diffusive and the Doppler shift gets correspondingly smaller. Simultaneously, the collisions increase the optical dephasing rate, thus increasing the homogeneous optical linewidth.

Another possible source of inhomogeneity is absorption. The laser beam becomes weaker as it propagates through the sample. This effect can be minimised by using low particle densities. In the experiments described in this section, the particle density was low enough that the absorption in the linear regime was less than 50%. At these relatively low densities, the absorption of the pump laser field, which operates in the nonlinear regime, may become very small. Its intensity is then almost constant over the sample length, and the observed quantities are linear functions of the microscopic variables. These conditions allow a considerable simplification of the data analysis. In addition, the relatively low density eliminates complications due to interactions between the different atoms or multiple interaction of individual photons with the atoms, such as radiation trapping (Holstein 1951; Schiffer et al. 1994).

8.1.2 Setup

All experiments discussed here use two separate laser beams in a pump-probe configuration. A strong pump beam drives the system far from equilibrium, inducing the dynamics to be studied. The second beam, the probe beam, should

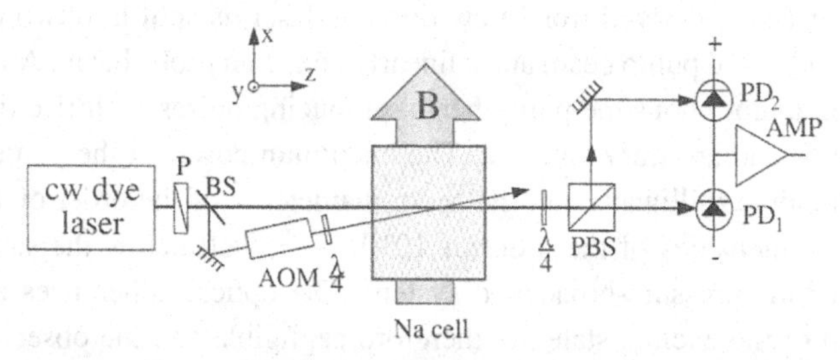

Figure 8.1. Typical experimental setup for the observation of optically excited spin transients in an atomic vapour. P = polariser, BS = beam splitter, AOM = acoustooptic modulator, $\lambda/4$ = retardation plate, PBS = polarising beam splitter, PD = photodiode, AMP = amplifier, B = magnetic field.

measure the optical properties in the sample without influencing its dynamics or equilibrium state. For this purpose, its intensity must be significantly lower than that of the pump beam. Compared to a single-beam arrangement, this setup provides better control of the experimental parameters. Ideally, the two laser beams are derived from different lasers and are therefore completely independent. If only a single laser is available, the use of two independent beams still permits independent control of the amplitudes and polarisations of the two beams and, in a limited range, their frequencies.

Figure 8.1 shows a typical experimental setup for the study of alkali vapours. It consists of the laser system and optical path, the sample cell, magnetic field coils, and the polarisation-selective detection system. Electronic components needed for timing the experiment, signal acquisition, and data analysis are not included in the figure.

The metal to be studied is placed in a sealed glass cell or in a ceramic tube that can be evacuated and filled with a buffer gas. The tube or cell is heated until the vapour pressure of the metal reaches the desired level. The buffer gas, which suppresses Doppler broadening, is typically 200 mbar of Ar. The collisional broadening of the optical resonance lines by the buffer gas results in a total optical linewidth of $2\Gamma_2 = 4.2$ GHz, larger than the hyperfine interaction. Three pairs of coils in Helmholtz geometry generate the external magnetic field. The three pairs are oriented along the coordinate axes, allowing the generation of a field in arbitrary direction. For most experiments, the direction of the magnetic field (applied plus earth) is perpendicular to the laser beam direction. The experiment thus represents an example of transverse optical pumping. The strength of the magnetic field is typically in the range of one μT to one mT, corresponding to Larmor frequencies between some 7 kHz and 7 MHz.

The laser beam, derived from a cw ring dye laser, is split into two parts, a circularly polarised pump beam and a linearly polarised probe beam. An acoustooptic modulator chops the pump beam, producing pulses with rise times of about 100 ns and arbitrary duration. The maximum power in the pump beam is a few hundred milliwatts and the beam diameter is of the order of 1 mm^2, resulting in intensities of the order of 10^5 Wm^{-2}, well below the saturation intensity of the pressure-broadened system. The optical coherences and the population of the excited state are therefore negligible and the observed signals reflect dynamics of the electronic ground state. The total power of the probe beam is some ten microwatts, too low to affect the dynamics of the system significantly. The two beams intersect in the sample at a small angle of $\sim 0.5°$. This allows one to separate them behind the sample and block the pump beam, while the probe beam passes into the polarisation-selective detector, which splits it into the two circularly polarised components. The difference between these intensities forms the signal. As described in Section 6.4, the difference of the two photocurrents is then directly proportional to the z component of the magnetisation.

8.1.3 Historical overview

After the seminal suggestion of Alfred Kastler (Kastler 1950), that it should be possible to polarise atoms by irradiation with polarised light, several groups started to apply this technique to atomic beams and vapours (Brossel, Kastler and Winter 1952; Hawkins 1955; Kastler 1967; Happer 1972). A decade later, light shift effects were discovered as a lifting of degeneracies in zero magnetic field (Arditi and Carver 1961; Barrat and Cohen-Tannoudji 1961a; Barrat, Cohen-Tannoudji and Ribaud 1961; Cohen-Tannoudji 1962; Pancharatnam 1966; Dupont-Roc et al. 1967; Mathur, Tang and Happer 1968; Happer 1970; Cohen-Tannoudji and Dupont-Roc 1972). The earliest experiments detected the microscopic order through the polarisation of the fluorescence, but it was soon realised that the polarisation of the sample modified the absorption (Dehmelt 1957a; Partridge and Series 1966; Series 1966) and dispersion (Happer and Mathur 1967b; Pancharatnam 1968; Bicchi, Moi and Zambon 1979), which permitted the use of transmitted light as a probe of the microscopic system.

The possibility of using transmitted light as a probe of *transient* phenomena in atomic ground states was first demonstrated for sodium by Lange and Mlynek (Lange and Mlynek 1978). Other groups used the technique to study thallium (Dohnalik et al. 1984), samarium (Lowe et al. 1987), cesium (Mishina, Fukuda and Hashi 1988), rubidium (Appelt, Scheufler and Mehring

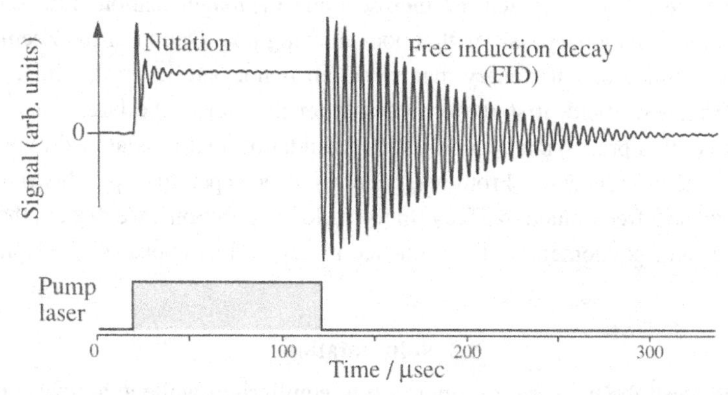

Figure 8.2. Response of sodium atoms to a pulse of circularly polarised light. Experimental parameters: pump laser power = 15 mW, laser detuning $\Delta/2\pi = -0.5$ GHz.

1989), and other systems. Besides the observation of transients following a single pulse, two-pulse experiments were performed which excite echo phenomena (Tanigawa et al. 1983; Mishina et al. 1987; Rosatzin, Suter and Mlynek 1990a; Suter, Rosatzin and Mlynek 1991c; Hashi, Fukuda and Tanigawa 1992). Besides these purely optical experiments, radio-frequency–optical double-resonance experiments are also possible, either in a stationary mode (Mlynek et al. 1984; Tamm, Buhr and Mlynek 1986; Mlynek et al. 1988) or as time-resolved experiments (Erickson 1990; Erickson 1991).

8.1.4 Phenomenology

Most experiments use a laser pulse or a sequence of several pulses to excite the system, and a continuous probe laser beam to monitor its evolution. Figure 8.2 shows a typical response of the system to a pulse of circularly polarised laser light propagating perpendicular to the magnetic field.

The system is initially in thermal equilibrium so that the signal, which is proportional to the magnetisation component in the direction of the laser beam, vanishes. When the laser is turned on, it creates magnetisation in the sample, which starts to precess around the effective field. This precession of the magnetisation around an effective field was first discovered in nuclear magnetic resonance (Torrey 1949), where it became known as spin nutation. After the introduction of lasers and coherent optical spectroscopy, nutation signals were also observed in optical transitions (Hocker and Tang 1968; Hoff, Haus and Bridges 1970; Brewer and Shoemaker 1971). In the present case, we deal with a spin nutation that is excited by the laser pulse. The precession appears as an oscillation of the signal, which is damped by the optical pumping.

On a timescale of a few tens of microseconds, the magnetisation of the sample reaches a stationary state with a nonvanishing polarisation of the Zeeman substates. Since this stationary magnetisation is not parallel to the direction of the magnetic field, it starts to precess after the end of the laser pulse. In the signal, this precession appears as a sinusoidal oscillation, damped by various dephasing processes. From magnetic resonance spectroscopy, this signal is known as a free induction decay. In the following sections, we discuss these two transient phenomena in the sequence in which they appear in the signal.

8.2 Spin nutation

We first describe the system's approach to equilibrium while a circularly polarised laser beam drives the system. For this discussion, we consider only the electronic ground state and use the $J = 1/2$ model.

8.2.1 Signal

Time dependence

During the laser pulse, the pump laser generates magnetisation perpendicular to the magnetic field. This is the situation of transverse optical pumping described in Section 5.2.4. With the experimental arrangement described in Section 6.4, the observable signal is proportional to the magnetisation component m_z in the direction of the laser beam. Using the solution of the equation of motion derived in Section 5.2.5, we write the time dependence of this component as

$$m_z(t) = [m_{\mathrm{T}} \cos(\Omega t + \phi) + m_{\mathrm{L}}]\, \mathrm{e}^{-\gamma_{\mathrm{eff}} t} + m_{z\infty} \qquad (8.1)$$

where the amplitudes of the oscillatory and the background components are

$$m_T = \frac{-P_+ \Omega_{\mathrm{L}}^2}{\Omega^2 \sqrt{\Omega^2 + \gamma_{\mathrm{eff}}{}^2}} \qquad m_{\mathrm{L}} = -\frac{\overline{\Delta}^2 P_+^3}{\gamma_{\mathrm{eff}} \Omega^2} \qquad (8.2)$$

The phase of the oscillatory signal is

$$\tan \phi = \frac{\Omega}{\gamma_{\mathrm{eff}}} \qquad (8.3)$$

and the stationary magnetisation

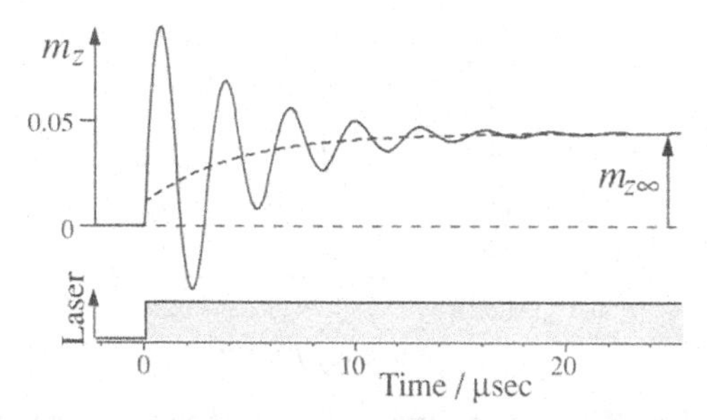

Figure 8.3. Nutation signal as a function of time. The laser pulse starts at time $t = 0$. The solid line shows the magnetisation component m_z, the dashed line the longitudinal part of it.

$$m_{z\infty} = \frac{P_+}{\gamma_{\text{eff}}}\left(1 - \frac{\Omega_L^2}{\Omega_L^2 + \overline{\Delta}^2 P_+^2 + \gamma_{\text{eff}}^2}\right) \tag{8.4}$$

As discussed in Section 5.2.5, the magnetisation vector moves on the surface of a cone. We can decompose this vector into three components that appear separately in the signal, distinguishable by their characteristic time dependence, The first term, m_T, originating from the magnetisation orthogonal to the effective field, oscillates at the frequency Ω and decays at a rate γ_{eff}. The second term, m_L, arises from the magnetisation component parallel to the effective field and the symmetry axis of the cone. It does not oscillate, but decays at the same rate as the oscillating part. The third, time-independent term, $m_{z\infty}$, represents the stationary ground-state magnetisation.

Approach to equilibrium

The stationary value of the ground-state magnetisation, $m_{z\infty}$, depends on two competing effects that are represented by two different terms in equation (8.4): The first term, P_+/γ_{eff}, is the ground-state orientation as it would result from longitudinal optical pumping. The second term describes the reduction of the polarisation by the precession around the tilted effective field. The presence of the magnetic field thus always leads to a decrease of the ground-state orientation; in the limit of strong magnetic fields, $\Omega_L{}^2 >> (P_+{}^2, \overline{\Delta}^2 P_+^2, \gamma_{\text{eff}}^2)$, the stationary value of the z magnetisation vanishes. This can be understood as a destructive interference effect, as discussed in Section 5.3, in the context of modulated optical pumping.

Figure 8.3 shows the magnetisation component m_z as a function of time. The relevant parameters used for the calculation are $\overline{\Delta} = 2$, $P_+ = 2 \cdot 10^5$

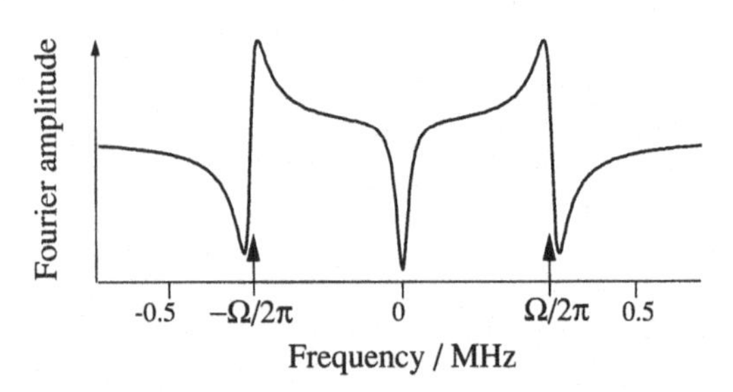

Figure 8.4. Fourier transform of the data shown in Figure 8.3.

$\sec^{-1}$, $\Omega_L/2\pi = 318$ kHz, $\gamma_0 = 3.3 \cdot 10^4 \sec^{-1}$. The solid line represents the magnetisation component $m_z(t)$ as a function of time. The dashed curve shows separately the contribution from the exponentially decaying longitudinal magnetisation, which appears as the second term in equation (8.1). The precession of the magnetisation components perpendicular to the magnetic field appears as an oscillation superimposed onto the exponential background. After some 20 μsec of optical pumping, the system approaches the stationary value $m_{z\infty}$.

The separation into different components can also be achieved in the measured signal, using Fourier transformation of the data. Figure 8.4 shows the Fourier transform of the baseline corrected time-domain signal displayed in Figure 8.3. In this representation, the longitudinal and transverse (with respect to the effective field) components of the magnetisation appear as separate resonances at $\omega = 0$ and $\omega = \pm\Omega$, respectively. The longitudinal component has the shape of an absorption signal, whereas the lineshape of the transverse component is almost antisymmetric (dispersionlike). The slight deviation is measured by the mixing angle ϕ, defined in equation (8.3).

8.2.2 Experimental control

According to equations (8.1)–(8.4), the signal should depend on several experimentally controllable parameters. We summarise its dependence on the optical parameters, the intensity and frequency of the laser.

Laser intensity dependence

Equations (8.1)–(8.4) predict that amplitude, phase, and damping rate of the nutation signal should depend on the laser intensity. Figure 8.5 shows a set of experimental signals for various intensities. They were recorded with a

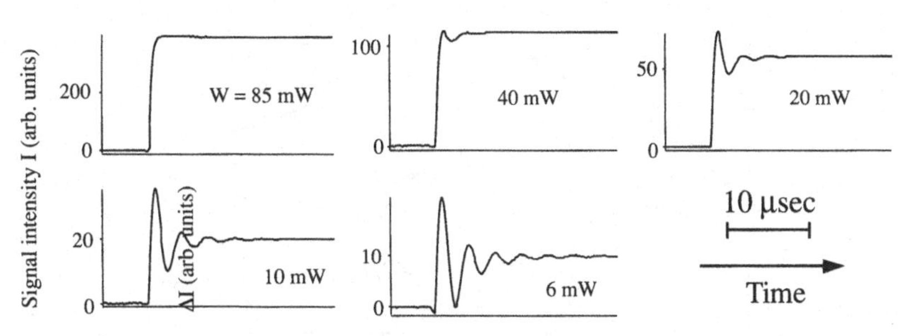

Figure 8.5. Nutation signal as a function of laser power at a resonance detuning $\Delta/2\pi = 1.5$ GH.

pump laser diameter of 1.1 mm. The laser powers given in the figure in mW are therefore almost equivalent to average intensities measured in kW/m^2. The corresponding optical Rabi frequencies are $<10^8$ Hz, well below the optical saturation intensity. For this experiment, the laser frequency was tuned to 1.5 GHz below resonance, at $\overline{\Delta} = 0.7$. When the laser field is switched on, the sublevel polarisation starts to build up and at the same time precesses around the effective field. At low laser intensities, the precession dominates, as the large (compared to the stationary magnetisation) oscillations show. The dominance of the precession leads to a small stationary magnetisation. Higher laser intensities lead not only to a higher equilibrium polarisation, but also to a faster damping of the transient nutation.

The increase in laser intensity should also increase the precession frequency $\Omega = \sqrt{(\Omega_L^2 + (\overline{\Delta}\, P_+)^2}$. Since this effect is relatively small with these experimental parameters, and obscured by the associated damping, the increase in the precession frequency is not readily seen in these figures. The measurements shown in Figure 8.5 used Gaussian laser beams, so that the laser intensity was not homogeneous over the sample. As a result, the damping of the oscillations by the optical pumping leads to a nonexponential decay of the signal, with the signal components from the centre of the pump beam decaying faster than the signal contributions from the regions with lower intensities.

Laser frequency dependence

A detuning of the laser frequency from the atomic transition frequency reduces the optical pumping rate, like a decrease of the intensity. In addition, the laser detuning also changes the light shift effect. At resonance, the light shift vanishes; at the normalised detuning $|\overline{\Delta}| = 1$, it reaches a maximum, and

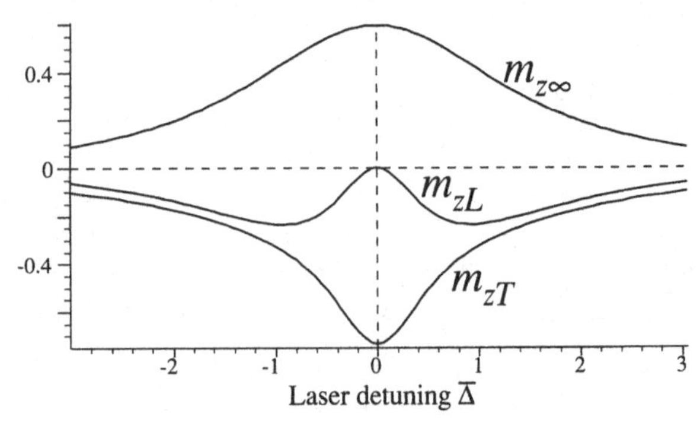

Figure 8.6. Dependence of the three signal contributions on the normalised laser detuning.

for larger detunings it decreases as $1/\overline{\Delta}$. Looking at the signal as a function of laser detuning, we expect an overall decrease of the signal level, similar to the decrease with lower laser intensity. The difference between the two cases arises from the different behaviour of the light shift with detuning, which decreases linearly, whereas the pumping rate decreases quadratically with Δ.

This difference becomes directly visible if we consider the three signal components separately: the stationary part $m_{z\infty}$, the component in the direction of the effective field, m_{zL}, and the component perpendicular to it, m_{zT}. Figure 8.6 summarises their dependence on the laser detuning. The general decrease for larger detuning is easily understood as the optical pumping efficiency decreases. The vanishing of the longitudinal component at resonance follows directly from the vanishing light shift on resonance. Without the light shift effect, the effective field is perpendicular to the laser beam direction. The projection of a magnetisation component along the field onto the z direction therefore vanishes on resonance. The difference between the components thus directly reflects the orientation of the effective field.

Figure 8.7 shows the time-dependent signals for a number of laser detunings that decrease from the top to the bottom. The left-hand column shows the experimental data, which were recorded with 20 mW pump power; the right-hand column was calculated with equations (8.1)–(8.4). The theoretical model predicts that the signals should not depend on the sign of the optical detuning; experimentally, one observes a small asymmetry of the signal amplitude, which is probably due to the unresolved hyperfine structure.

The theoretical as well as the experimental data indicate that the pump rate decreases with increasing detuning. With the pump rate, the signal intensity and the damping rate decrease. Apart from the amplitude, the phase of the nu-

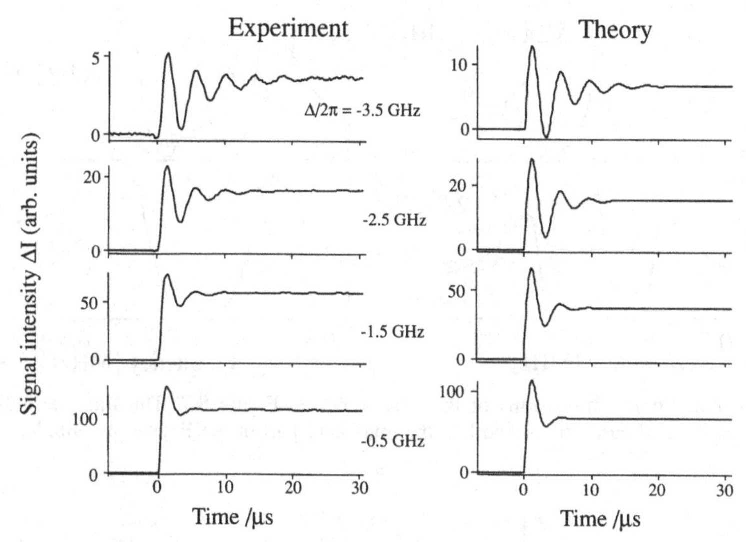

Figure 8.7. Nutation signal as a function of optical detuning. The left-hand column shows the experimental data, which were recorded with a pump beam power of 20 mW. The right-hand column represents theoretical calculations.

tation signal also depends on laser intensity and laser detuning. From the time-domain signals of Figure 8.7, a verification of these predictions is not straightforward. It becomes significantly easier in the Fourier transform of the time-domain signals, shown in Figure 8.8.

This representation again separates the longitudinal and transverse components m_{zL} and m_{zT}. The transverse component, which precesses around the effective field, appears at the frequency Ω. The longitudinal component, which is always aligned with the effective field, appears at zero frequency. The presence of a longitudinal component is direct evidence of the light shift effect. It vanishes near the centre of the optical resonance line. The larger width of the different frequency components at small detunings indicates the faster damping by the more efficient optical pumping.

8.3 Free induction decay

8.3.1 Theory

Qualitative outline

The optical pumping drives the system towards a stationary state, the equilibrium state of the system in the presence of the laser field. After the end of the laser pulse, this state is no longer the equilibrium position of the system

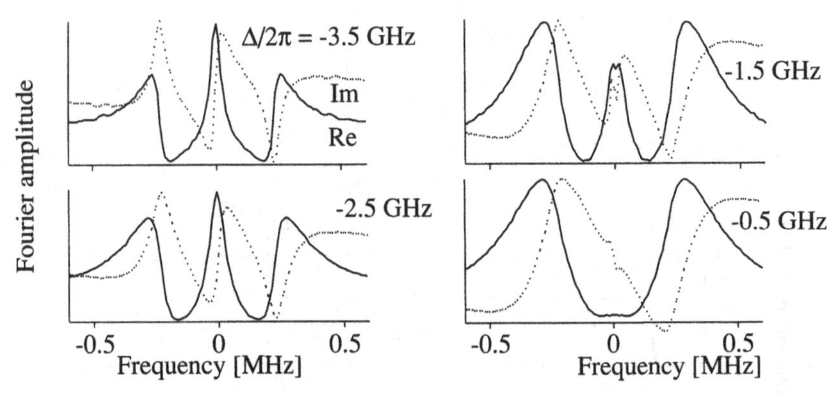

Figure 8.8. Fourier transforms of the data shown in Figure 8.7. The full line indicates the real part, the dashed line the imaginary part of the Fourier amplitude.

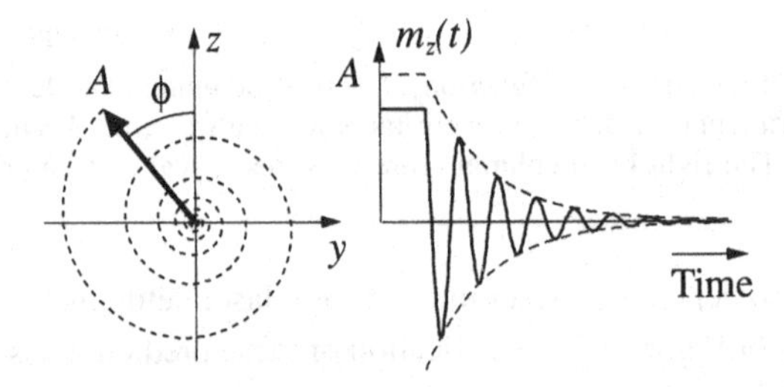

Figure 8.9. Projection of the magnetisation into the yz plane (left) and corresponding detector signal (right). The length A of the magnetisation vector determines the envelope of the FID signal (dashed line) whereas the polar angle ϕ determines the phase.

and the density operator does not commute with the Hamiltonian. The Zeeman coupling with the magnetic field forces the atomic magnetic dipole to precess around the magnetic field, which points along the x axis in our usual coordinate system. In the experiment described above, the observable signal is proportional to the z component of the magnetisation. The time-dependent signal after the end of the pump pulse therefore depends on the magnetisation component perpendicular to the field.

Figure 8.9 illustrates this connection: The arrow in the left-hand part of the figure represents the magnetisation vector at the end of the laser pulse, projected into the yz plane. Amplitude A and phase ϕ appear as its polar coordinates. The dashed spiral indicates the subsequent precession and decay. The right-hand part of the figure represents the time-dependent signal. The dashed line shows the amplitude, i.e., the length A of the vector, whereas the full line is the projection of the vector onto the z axis.

Quantitative

For a mathematical analysis, we write the projection of the precessing magnetisation into the yz plane as

$$m_z(t) = A \cos(\Omega_L t + \phi)\, e^{-\gamma_0 t} \tag{8.5}$$

where the amplitude is

$$A = \sqrt{m_y(0)^2 + m_z(0)^2} = \frac{P_+}{\gamma_{eff}} \frac{\sqrt{\gamma_{eff}^2 \Omega_L^2 + (\overline{\Delta}^2 P_+^2 + \gamma_{eff}^2)^2}}{\Omega_L^2 + \overline{\Delta}^2 P_+^2 + \gamma_{eff}^2} \tag{8.6}$$

and the phase

$$\tan\phi = \frac{m_y(0)}{m_z(0)} = \frac{\gamma_{eff}\Omega_L}{\overline{\Delta}^2 P_+{}^2 + \gamma_{eff}^2} \tag{8.7}$$

As shown in the left-hand part of Figure 8.9, the parameters A and ϕ represent the polar coordinates of the initial magnetisation vector in the yz plane. In the FID signal (right-hand side of Figure 8.9), they appear as amplitude and phase.

The FID signal depends on the laser intensity and on the laser detuning. To make the dependence more visible, we consider the limiting case where the optical pumping rate is large compared to the relaxation, $P_+ >> \gamma_0$. This condition is usually fulfilled as long as the irradiation frequency is not too far from resonance. We then have

$$A = \frac{p\sqrt{1 + p^2}}{1 + p^2 + \overline{\Delta}^2} \qquad \tan\phi = \frac{1}{p} \tag{8.8}$$

where the dimensionless parameter

$$p = \frac{\omega_1^2}{4\,\Gamma_2\,\Omega_L} \tag{8.9}$$

is the ratio of the on-resonance optical pumping rate to the Larmor frequency. In the limit considered here, the phase of the FID signal depends only on p,

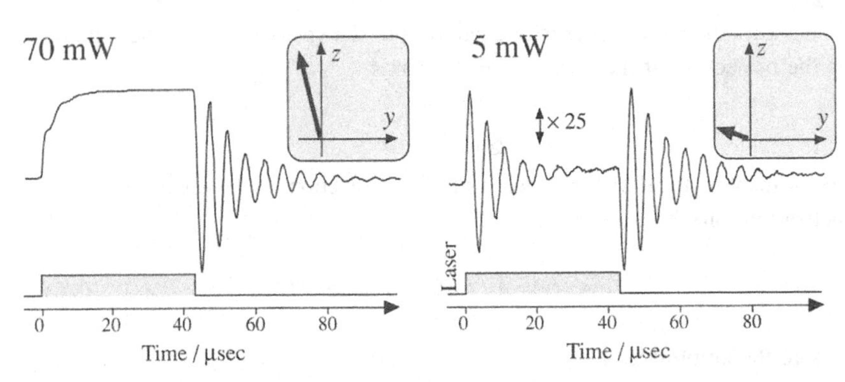

Figure 8.10. Signal observed during and after a pump laser pulse at two different laser intensities. The insets indicate qualitatively the orientation and size of the equilibrium magnetisation during the pulse.

whereas the amplitude falls off quadratically with the normalised optical detuning $\overline{\Delta}$.

8.3.2 Experimental control

Laser intensity dependence

As with the nutation signal, we expect the signal amplitude to increase with the laser intensity. This qualitative expectation agrees with the predictions of equations (8.5)–(8.7) and is corroborated by the experimental evidence shown in Figure 8.10.

Figure 8.10 shows two examples of experimental signals at different laser intensities, together with the time-dependent amplitude of the optical pump beam (Rosatzin et al. 1990a). The insets show qualitatively the orientation of the magnetisation in the *yz* plane after the optical pulse. The variation of the signal amplitude with the laser intensity is evident from the different scale of the two signals. Also evident is the strong variation of the phase with the laser intensity. At high laser intensity, represented in the left-hand part, the stationary magnetisation is large and almost parallel to the laser beam. The signal starts like a cosine function after the end of the pulse. In the case of lower intensity, the initial magnetisation is almost parallel to the *y* axis and its amplitude is much smaller. The signal is correspondingly smaller and starts as $\sin(\Omega_L t)$ with oscillations much larger than the stationary signal during the pulse.

Figure 8.11 compares the experimentally observed variation of amplitude and phase as a function of the laser intensity to the theoretical prediction of

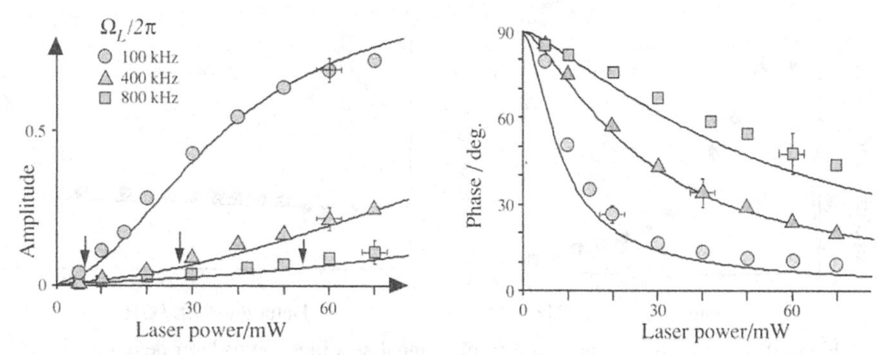

Figure 8.11. Amplitude and phase of FID signal as a function of laser power for three magnetic field strengths.

equations (8.5)–(8.7). The three sets of data were taken at different magnetic field strengths. For high enough laser power, when the optical pump rate P_+ exceeds the Larmor frequency Ω_L, the equilibrium magnetisation is almost parallel to the z axis ($\phi \sim 0$) and the amplitude can reach values near unity. If the laser intensity decreases, the amplitude of the ground-state polarisation decreases and tilts towards the y axis. For the observed signal, this corresponds to an increase of the phase towards 90°. The resonance detuning of the laser was $\Delta/2\pi = 9.5$ GHz in all measurements.

Laser frequency dependence

Amplitude and phase of the FID signal depend not only on the laser intensity, but also on the resonance offset of the laser and the strength of the magnetic field. According to the simplified equation (8.8), the dependence of the FID phase on the optical detuning should be very small, whereas the amplitude has a Lorentzian dependence. It reaches a maximum of $p/\sqrt{1+p^2}$ on resonance ($\overline{\Delta} = 0$) and falls off to half this value at $\overline{\Delta} = \pm\sqrt{1+p^2}$. This is the usual dependence of resonant processes on a frequency detuning. In this case, the resonance frequency is zero. Effects of this type have been studied extensively in traditional magnetooptics, where the polarisation of the system reaches a maximum when the transverse magnetic field vanishes. Experiments that measure this effect through some stationary property of the medium are known as level crossing experiments (Corney 1977).

So far, we have discussed the dependence of the microscopic order on various parameters. If the laser frequency changes, this affects not only the microscopic system, but also the detection efficiency. The sensitivity of the de-

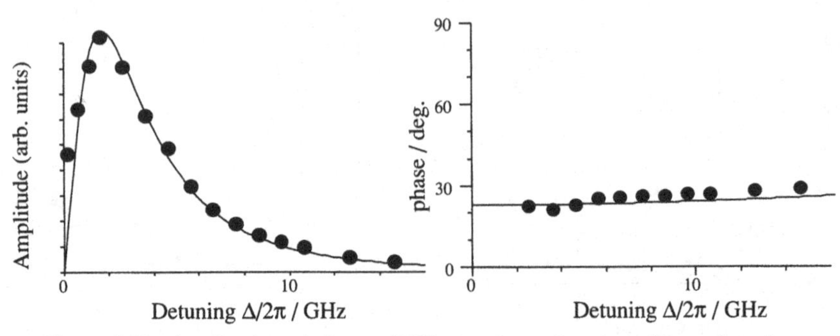

Figure 8.12. Amplitude and phase of FID signal as a function of laser detuning.

tection schemes considered here falls off with the detuning of the probe laser frequency. In the case of absorptive detection, the sensitivity reaches its maximum at resonance, but in the case of dispersive detection, the sensitivity on resonance vanishes. The observed signal is the product of the microscopic quantity m_z and the detection sensitivity, which in this case is proportional to $\overline{\Delta}/(1 + \overline{\Delta}^2)$. We illustrate this with the data of Figure 8.12, which were measured with the dispersive detection scheme described in Section 6.4.2.

The left-hand side of the figure shows the dependence of the signal amplitude on the laser detuning. According to equation (8.6), the magnetisation m_z reaches its maximum on resonance, but the detection sensitivity vanishes there. The line in Figure 8.12 which represents the theoretically expected laser-frequency dependence of the signal amplitude takes the detection sensitivity into account. The signal phase, which appears in the right-hand part of the figure, should be almost independent of the laser frequency. In this case, the detection sensitivity does not affect the measured value. A special case arises in zero magnetic field. The Hamiltonian then commutes with the magnetisation created by the optical pumping, so that no precession occurs and the magnetisation remains along the laser beam. The phase of the FID is then always zero, independent of the laser intensity.

8.4 Spin echoes

8.4.1 Introduction

An important experimental tool for investigating the interaction of matter and radiation is the ability to rearrange the order present in the system by converting populations into coherences, coherences from one transition into another and coherences into populations. As discussed in the theoretical section, this can be achieved in purely optical experiments either by applying strong pulses of

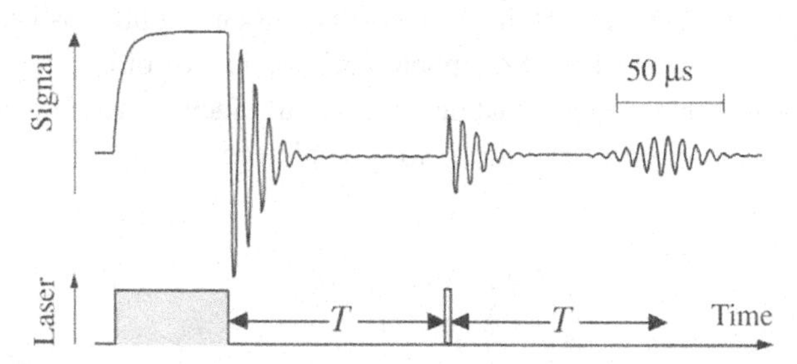

Figure 8.13. Two-pulse experiment in an inhomogeneous magnetic field. The lower trace represents the amplitude of the pump beam whereas the upper trace shows the observed signal.

polarised light to the optical transitions or, in the low-power regime, by using the light shift of off-resonance optical radiation to apply virtual magnetic fields to the spin system. This section discusses one possible application of such light-induced virtual fields, the refocusing of spin-coherence in an inhomogeneous magnetic field (Fukuda, Yamada and Hashi 1983; Rosatzin, Suter and Mlynek 1990; Suter, Rosatzin and Mlynek 1991; Tanigawa et al. 1991).

Echo phenomena are some of the most prominent features in coherent spectroscopy. After Hahn's first report on spin echoes in nuclear magnetic resonance (Hahn 1950), this phenomenon has found numerous applications (Ernst, Bodenhausen and Wokaun 1987). Echo phenomena were also observed in various other fields such as electron spin resonance (Ponti and Schweiger 1994). Photon echoes and three-level echoes are examples from optical spectroscopy discussed in previous sections of this book. The echoes discussed here are spin echoes, but the refocusing is initiated without radio frequency irradiation. Instead, two pulses of polarised light couple to an optical transition to prepare the coherence and to initiate the refocusing process.

Mechanism

As discussed in Section 3.5.4, the virtual field induced by the laser pulse points in the direction of the laser beam and acts like a pulsed magnetic field. In the arrangement discussed here, the direction of the laser beam is orthogonal to the direction of the magnetic field; the vector sum of the static magnetic field and the pseudo-field induced by the light shift add up to an effective field whose direction can be adjusted within the upper half of the xz plane by variations of the laser intensity and laser detuning.

Figure 8.13 summarises the experiment under discussion. An initial pump laser pulse polarises the atomic medium. After the end of the pulse, the spins dephase in an inhomogeneous magnetic field. A second, off-resonant laser pulse creates an effective field that initiates the refocusing process leading to the spin echo. As in the case of the photon echo, this type of echo is induced by laser pulses. The coherence that defocuses and refocuses here, however, is a magnetic dipole moment and the echo is a spin echo.

8.4.2 Mechanism

Defocusing

To calculate the evolution during a two-pulse experiment, we assume that the spin system is initially in a state of homogeneous polarisation,

$$\rho(0) = J_z \cos\phi - J_y \sin\phi \tag{8.10}$$

Only the component perpendicular to the magnetic field is of interest in this context and the FID signal is described by equation (8.5). After a time T, the spins have evolved into a state

$$\rho(T) = \mathrm{e}^{-\gamma_0 T} \left(J_z \cos(\Omega_L T + \phi) - J_y \sin(\Omega_L T + \phi)\right) \tag{8.11}$$

Depending on their position in the inhomogeneous magnetic field, different spins precess at different Larmor frequencies Ω_L. The total signal is the weighted integral of the expectation value $\langle J_z \rangle$ over the sample. The weighting function is the distribution of field strengths, whose width we call $\delta\Omega_L$. The observable signal vanishes after a time of the order of the inverse width $1/\delta\Omega_L$. This situation is closely analogous to the excitation of photon echoes discussed in Section 2.3.3.

Figure 8.14 illustrates how the differential precession of the individual spins causes destructive interference and a decay of the overall signal: Immediately after the pulse, all spins are in phase, as the solid arrow in the leftmost figure indicates. After the precession in the inhomogeneous field, the spins in a stronger than average field have precessed farther (dashed arrow), while those in a weak magnetic field are lagging behind (dotted arrow), compared to the average spin (solid arrow).

Refocusing

The second pulse of length τ rotates the magnetisation around the effective field $\Omega = \{\Omega_L, 0, \overline{\Delta}P_+\}$. At the same time, optical pumping and damping effects drive the system towards the equilibrium position. We may summarise these two effects by writing the density operator after the pulse as

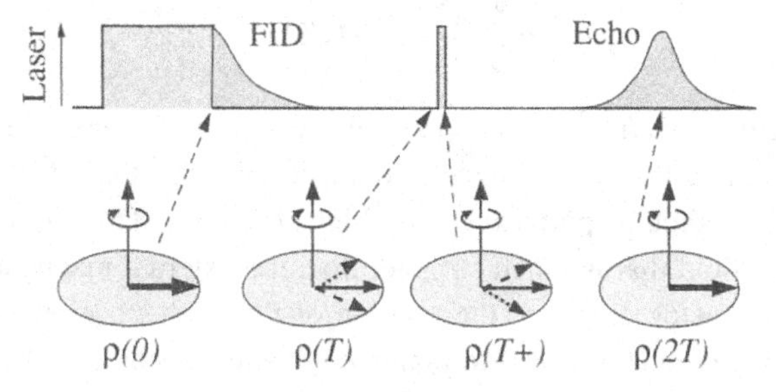

Figure 8.14. Mechanism of light-shift–induced spin echo formation.

$$\rho(T+) = \rho_{eq} + e^{-\gamma_{eff}\tau}\, U_P\, (\rho(T) - \rho_{eq})\, U_P^{-1} \tag{8.12}$$

where

$$U_P = \exp(-i\tau(\Omega_L J_x + \overline{\Delta P}_+ J_z)) \tag{8.13}$$

describes the rotation induced by the effective field and τ represents the pulse duration.

We can simplify the calculation of the echo amplitude considerably if we neglect the effect of the inhomogeneous part of the magnetic field during the pulse. This is justified in most experimental situations, where the inhomogeneity of the magnetic field is small compared to the inverse of the pulse duration. Ideally, the rotation occurs by 180° around the z axis. As Figure 8.14 shows, such a rotation exchanges the positions of the "fast" and "slow" spins. The dashed spin, which is the one in the strongest field, now lags behind, while the dotted spin, which experiences the weakest field, is ahead of the others. It thus appears plausible that they will eventually end up back in phase, as they continue to precess in the same magnetic field.

A full refocusing would be obtained if it were possible to phase-invert the density operator, i.e., to generate a density operator ρ_{inv} that results from the substitution $\Omega_L \to -\Omega_L$ in $\rho(T)$:

$$\rho_{inv} = e^{-\gamma_0 T}\, (J_z \cos(\Omega_L T - \phi) + J_y \sin(\Omega_L T - \phi)) \tag{8.14}$$

In reality, it is only possible to approximate this process. We can measure the efficiency q of the echo pulse by calculating the projection of the resulting density operator onto this phase-inverted part ρ_{inv}; it depends on the optical pumping rate and the resonance offset as

$$q = \left(\frac{\overline{\Delta}P_+}{\Omega}\right)^2 \sin^2(\Omega\tau/2)\ \mathrm{e}^{-\gamma_{\mathrm{eff}}\tau} \tag{8.15}$$

The first factor $(\overline{\Delta}P_+/\Omega)^2$ is the square of the sine of the angle between the effective field and the static magnetic field; it measures the tilt of the effective field, which should approach $\pi/2$ for high efficiency. This requires a light-shift–induced field that is much stronger than the external magnetic field.

The second factor describes the rotation of the magnetisation vector. The rotation angle $\Omega\tau$ must be 180° to obtain complete refocusing. The exponential term takes the damping effect of the pulse into account. Unit efficiency, i.e., a perfect echo pulse, results for large light shift $\overline{\Delta}P_+ >> (\Omega_L, \gamma_{\mathrm{eff}})$ and flip angle $\Omega\tau = (2n + 1)\pi$ with n integer.

8.4.3 Control parameters

Pulse duration

The laser pulse causes the magnetisation to precess around the effective field, which lies in the xz plane. A simple way of observing this precession is the measurement of the echo amplitude as a function of the length of the second pulse. We would expect to find an oscillatory dependence of the echo amplitude as a function of the pulse length. A rotation angle that is an odd multiple of π should provide the highest refocusing efficiency, whereas an even multiple should bring the system back to its initial state, in which case the echo amplitude vanishes. In addition, the damping effect of the pulse should cause an overall decrease of the echo amplitude.

The left-hand side of Figure 8.15 summarises this dependence graphically. For a π pulse, approximately one quarter of the original signal should refocus as an echo with the parameters used for the calculation. The longer pulses, where the rotation angle is an odd multiple of π, also cause a partial refocusing, but the damping effect of the pulse becomes more significant. The right-hand side of the figure shows the experimentally observed refocusing efficiency for the ground state of sodium. Compared to the theoretical curve in the left-hand part, the echo amplitude never vanishes for pulses with finite duration. We can trace this difference to the inhomogeneous laser field. At the centre of the laser beam, the intensity is higher than in the outer parts. Working with a Gaussian beam, it is never possible to find a pulse duration that corresponds to a 2π rotation for all the atoms that contribute to the signal. An inhomogeneous beam always refocuses some atoms and the signal remains nonzero for all pulse durations. The theoretical curve in the right-hand part of the figure takes this inhomogeneity into account; it appears to

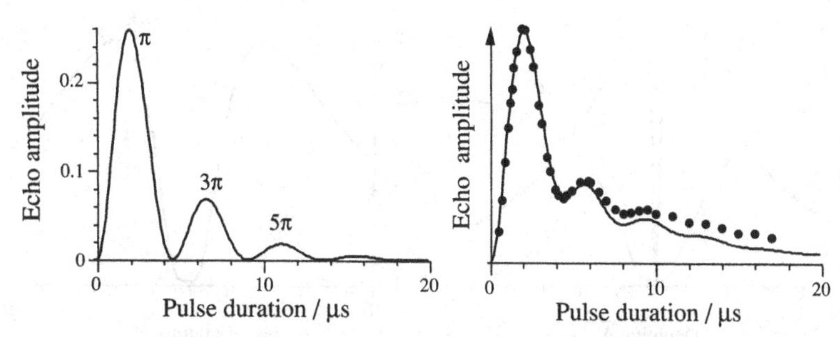

Figure 8.15. Echo intensity as a function of the length of the refocusing pulse. The left-hand side shows the theoretical calculation for a homogeneous laser field. Parameters: $\Omega_L = 1$, $\gamma_0 = 0.1$, $\omega_1^2/\Gamma_2 = 5$, $\overline{\Delta} = 5$. The right-hand side compares the experimental data with a theoretical fit, taking the inhomogeneous laser field into account. Experimental parameters: $\Omega_L/2\pi = 91$ kHz, $\Delta/2\pi = 15$ GHz, pump beam power = 130 mW.

give a satisfactory description of the experimental results, represented by the circles.

Laser frequency dependence

Apart from the pulse duration, the intensity and frequency of the light are the most important control parameters. The best refocusing is possible with the highest laser intensity. For the laser detuning, however, an optimal value exists, which depends on the available laser intensity. The laser detuning affects the refocusing efficiency, as calculated in equation (8.15), in two different ways: through the tilt angle of the effective field and through the ratio of the effective field strength Ω to the damping rate γ_{eff}. Both should be as high as possible to achieve good refocusing efficiency. These conditions cannot be satisfied simultaneously, since a large optical pump rate required for $\overline{\Delta} P_+ >> \Omega_L$ also implies a significant damping rate γ_{eff}. The different detuning dependence of the light shift term and the damping rate (Δ^{-1} vs. Δ^{-2}), however, makes it possible to achieve high efficiency asymptotically by using strong off-resonant radiation.

Figure 8.16 shows how the laser detuning affects these two ratios. The left-hand part of the figure shows them individually. The ratio $\overline{\Delta} P_+/\Omega$ of light shift to total field strength Ω vanishes at the optical resonance, reaches its maximum at $\overline{\Delta} = 1$, and falls off for larger detunings, where the light shift becomes too weak. The second curve in the left-hand part of the figure represents the damping during a π pulse, which must be longer far from reso-

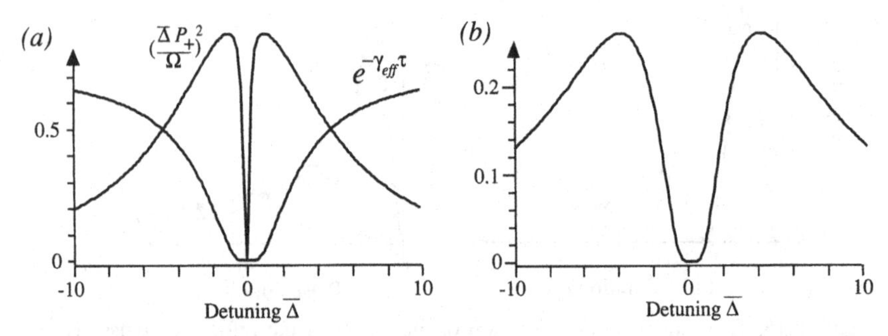

Figure 8.16. Efficiency of the refocusing pulse as a function of optical detuning. The parameters used were $P_+ = 5$, $\Omega_L = 1$, $\gamma_0 = 0.1$. **(a)** Maximum without attenuation and damping term plotted separately. **(b)** Total efficiency including damping.

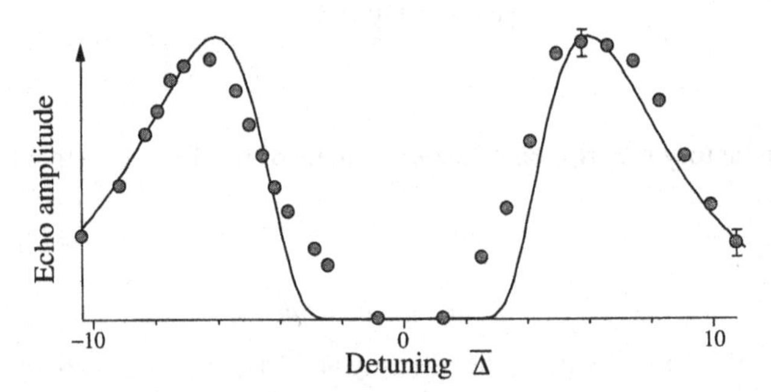

Figure 8.17. Measured vs. calculated echo amplitude as a function of the optical detuning $\bar{\Delta}$.

nance than close to the centre. If the laser frequency is close to the centre of the resonance line, this damping effect destroys the sublevel coherence almost completely and the echo intensity becomes very small. The right-hand part of the figure shows the product of the two, which is the expected refocusing efficiency. Clearly, the optimal detuning is larger than the width of the resonance line.

Figure 8.17 compares this calculated echo amplitude with experimental data obtained from the sodium ground state. These data represent the variation of the refocusing efficiency and the degree of polarisation achieved by the first laser pulse, as well as the detection efficiency. Since the most interesting region in this experiment is far from resonance, it is advantageous to use a dispersive detection scheme. In all cases, the echo amplitude vanishes on resonance, reaches a maximum and decreases at larger detunings, where the laser is too far from resonance.

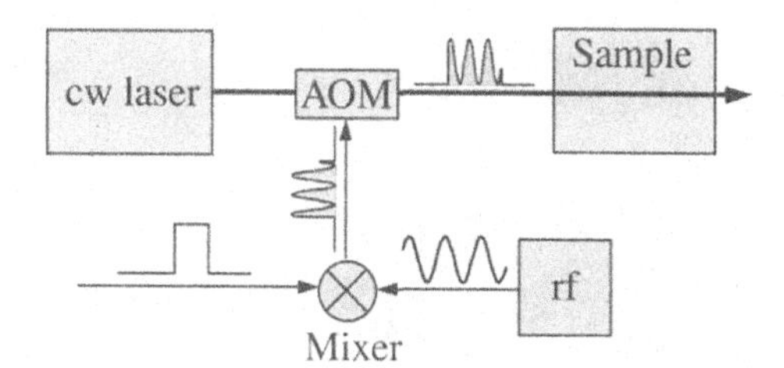

Figure 8.18. Experimental scheme for modulated optical pumping.

8.5 Modulated excitation

In transverse optical pumping experiments, the ratio between optical pumping rate P_+ and light shift $\overline{\Delta P}_+$ to the Larmor frequency Ω_L strongly affects the efficiency of the excitation and refocusing process. Optical pumping with continuous lasers, which typically produce less than one watt of output power, therefore reaches only a small equilibrium polarisation if the transverse field is stronger than approximately one Gauss. As discussed in Chapter 5, on optical pumping, this limitation can be overcome not only by more powerful lasers, but also by modulating the intensity or polarisation of the laser light at the Larmor frequency of the spins. In the following, we discuss possible implementations of such modulation schemes. As in the Raman experiments discussed in Section 5.3, we use not only continuously modulated light, but also pulses of modulated light.

Figure 8.18 shows how these kinds of pulses can be generated experimentally: An rf mixer multiplies the rectangular pulse shape with a continuous rf source. The resulting signal drives an acoustooptic modulator, which generates the desired optical pulses.

8.5.1 Laboratory-frame detection

As discussed in Section 5.3, on modulated optical pumping, the main effect of the modulation is a reduction of the effective Larmor frequency by the modulation frequency. This does not imply that the signal observed in an experiment with modulated optical pumping is the same as that in a low magnetic field. The reason is that the equations of motion were derived in a time-dependent interaction representation, which corresponds to a coordinate system rotating around the magnetic field. In this rotating frame, the system evolves as if the magnetic field were small. The probe laser beam, however, returns a signal proportional to the magnetisation component in the laboratory frame of reference. In this frame, the magnetisation generated by the optical pumping rotates at the modulation frequency.

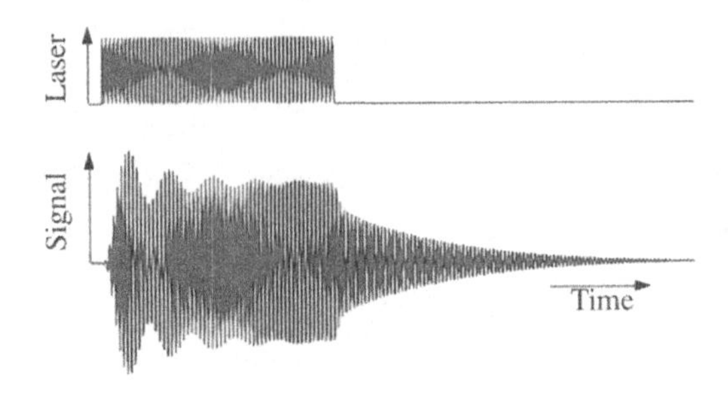

Figure 8.19. Optical pumping with a modulated laser beam.

Signal

To calculate the signal, we transform the rotating-frame density operator into the laboratory frame

$$\begin{aligned}\rho^{\mathrm{Lab}} &= \mathrm{e}^{-\mathrm{i}\omega_{\mathrm{m}}tJ_x}\,\rho^{\mathrm{r}}\,\mathrm{e}^{\mathrm{i}\omega_{\mathrm{m}}tJ_x}\\ &= m_xJ_x + J_y\,(m_y\sin(\omega_{\mathrm{m}}t) - m_z\sin(\omega_{\mathrm{m}}t))\\ &\quad + J_z\,(m_z\sin(\omega_{\mathrm{m}}t) + m_y\cos(\omega_{\mathrm{m}}t))\end{aligned} \tag{8.16}$$

The magnetisation components m_x, m_y, m_z refer to the rotating coordinate system. In terms of these coefficients, the observed signal, which is the expectation value of J_z, becomes

$$S(t) = 2\,\mathrm{tr}\{\rho^{\mathrm{Lab}}(t)\,J_z\} = m_z\cos(\omega_{\mathrm{m}}t) + m_y\sin(\omega_{\mathrm{m}}t) \tag{8.17}$$

The observed signal therefore contains contributions from both rotating-frame components. If the rotating-frame magnetisation is stationary, the signal is modulated at the frequency ω_{m} of the pump laser modulation.

Figure 8.19 shows an experimental signal recorded with modulated optical pumping. The top trace represents the pump intensity. The laser beam is modulated and pulsed. The lower trace shows the response of the system. During the modulated pulse, the magnetisation in the system builds up and starts to precess. As in the case of dc excitation, the system approaches a stationary state in which a precession at the modulation frequency occurs. The beat signal visible during the initial phase is due to interference between the eigenfrequency of the spins and the modulation frequency.

The fast oscillation of the signal, which reflects the modulation of the pump laser beam, makes digitisation of the signal quite demanding. The modulation frequency is often in the MHz range, requiring correspondingly fast digitisa-

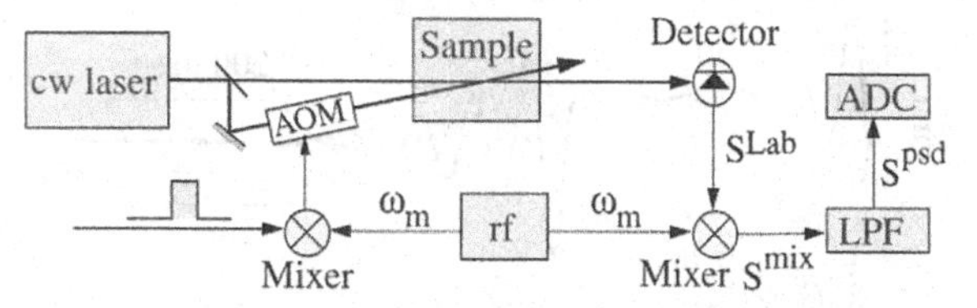

Figure 8.20. Experimental scheme for modulated optical excitation with phase-sensitive detection. AOM = acoustooptic modulator; LPF = low pass filter; ADC = analog-to-digital converter.

tion. For transients that may last several milliseconds, tens of thousands of data points must be stored. Furthermore, the fast oscillation tends to obscure other details of the evolution of the system. As it does not contain any information about the system, it is useful to remove this fast variation by phase-sensitive detection.

8.5.2 Phase-sensitive detection

Signal

Such a demodulation can be achieved by mixing the observed signal with the radio frequency signal that modulates the pump laser beam. Figure 8.20 shows a simple experimental scheme for this purpose.

The mixing of the signal with the local oscillator frequency transforms the signal into

$$S^{\mathrm{mix}}(t) = \cos(\omega_{\mathrm{m}}t + \phi)\,[m_z \cos(\omega_{\mathrm{m}}t) + m_y \sin(\omega_{\mathrm{m}}t)] \qquad (8.18)$$

which contains signal components at frequencies 0 and $2\omega_{\mathrm{m}}$. The low pass filter removes the frequency components at $2\omega_{\mathrm{m}}$ and leaves the low frequency part

$$S^{\mathrm{psd}} = \frac{1}{2}\,[m_z \cos\phi + m_y \sin\phi] \qquad (8.19)$$

Depending on the phase ϕ of the local oscillator, the demodulated signal is equal to the rotating frame magnetisation components m_z or m_y. The demodulation not only removes the fast oscillation from the signal, but also provides the possibility of observing either of the two components of the precessing magnetisation by an appropriate choice of the local oscillator phase. In prac-

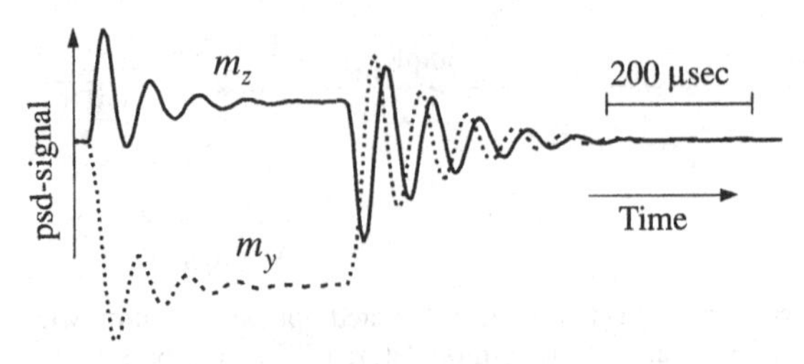

Figure 8.21. Signal of Figure 8.19 after phase-sensitive detection at the modulation frequency.

tice, it is often useful to detect both components simultaneously, by mixing the signal with two phase-shifted local oscillators.

Figure 8.21 shows an example of the signal recorded with phase-sensitive detection. The full line represents the component of the magnetisation that precesses in phase with the modulation, the dashed curve the out-of-phase component. The curves represent the same data as Figure 8.19, but were obtained with the demodulation setup of Figure 8.20. The difference between the modulation frequency and the eigenfrequency of the system, visible as a beat signal in Figure 8.19, appears now as the nutation frequency. The nutation is damped, as in the case of dc pulse excitation, and the system settles into a stationary state. After the end of the pulse, an FID appears; its oscillation frequency is the difference between the Larmor frequency and the modulation frequency.

Applications

As emphasised before, the modulation of the laser intensity has the main effect of reducing the apparent magnetic field. All experiments discussed above, like spin echoes, can also be performed with modulated pump pulses (Tanigawa et al. 1983; Mishina et al. 1987; Suter, Rosatzin and Mlynek 1991c).

In the equations of motion for modulated optical pumping, the reduced pumping efficiency at large magnetic fields appears as an off-resonance effect between the Larmor frequency and the modulation frequency. This is rather evident from the equilibrium magnetisation during the pulse, which is plotted in the left-hand part of Figure 8.22 as a function of the modulation frequency. The circles show the data extracted from time-domain experiments as a function of the modulation frequency. The arrow indicates the modulation frequency at which the time-domain data of Figure 8.22 were measured. The measurements were performed with the laser frequency tuned to the cen-

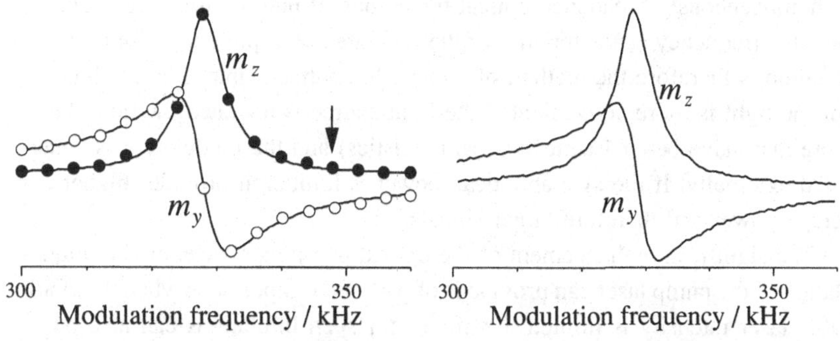

Figure 8.22. Stationary magnetisation during modulated optical pumping as a function of the modulation frequency; m_z and m_y denote the in-phase and out-of-phase components, respectively. The circles in the left-hand side represent the stationary value of the magnetisation measured during a time-resolved experiment and the full line is a least squares fit to a Lorentzian absorption/dispersion line. The right-hand side shows the signal measured in a frequency-domain experiment, using a phase-sensitive detector, while scanning the modulation frequency.

tre of the optical resonance so that no light shift effects occurred. The line through the circles represents the calculated stationary magnetisation. The right-hand part of the figure shows, for comparison, the same data obtained in a frequency-domain experiment, where the modulation frequency was scanned and the equilibrium magnetisation measured through phase-sensitive detection.

8.5.3 Frequency-domain experiments

As Figure 8.22 suggests, optical pumping with modulated light is also useful in frequency-domain experiments where the steady-state signal is measured with a phase-sensitive detector as a function of the modulation frequency (Mlynek et al. 1981b). Similar experiments were performed as early as 1961 with discharge lamps (Bell and Bloom 1961b; Bell and Bloom 1961a).

In the experimental examples presented here, we have assumed that the modulation is sinusoidal. This is particularly attractive for the theoretical analysis and the simplest experimental implementation for cw lasers. If pulsed lasers are available, it may be more convenient to use pulse trains for excitation. Again, the effective sublevel splitting is reduced by the pulse repetition rate or an integer multiple thereof. If the pulses are short enough, very high harmonics of the pulse repetition rate can be used for excitation. Compared to sinusoidal modulation, pulse trains can lead to somewhat higher polarisations and therefore to larger signals. In addition, they can excite systems with inhomogeneously broadened optical transitions. If pulsed lasers are available and the frequency characteristics of the light are not important, pulse train excitation is therefore the method of choice. In contrast, sinusoidal modulation of the light is more convenient if the light source is operated cw (and therefore may have better frequency characteristics) and the modulation is generated externally. If the available peak power is limited, it provides higher average power and therefore larger signals.

The significant enhancement of the optical pumping efficiency that modulation of the pump laser can provide is of particular importance when the available laser intensity is limited. Examples for such limitations can also occur outside the laboratory (Bradley 1992). Most applications, however, have popped up in spectroscopy. As an example, trains of ultrashort pulses from a synchronously pumped dye laser have been used to measure the hyperfine splitting of the Na ground state (1.8 GHz) (Harde et al. 1981; Fukuda et al. 1983b), the Cs ground state (9.2 GHz) (Lehmitz, Kattau and Harde 1986) as well as the excited state of Na (0.2 GHz) (Harde et al. 1981). The accuracy of these experiments can be comparable to rf experiments. In one case (Lehmitz, Kattau and Harde 1986), the coherence was driven and tuned through resonance by the 110th harmonic of the pulse rate (83 MHz). The resulting linewidth was only 30 Hz. Such trains of picosecond pulses can also be generated with semiconductor lasers, which have very low jitter and high stability in the pulse rate. The pulses can be derived directly from a comb generator (Lehmitz, Kattau and Harde 1986; Mishina, Fukuda and Hashi 1988).

8.6 Time-domain spectroscopy

Section 7.3.3 showed that it is possible to obtain spectroscopic information like resonance frequencies and relaxation rates through measurements not only in the frequency domain, but also in the time domain. The two types of data are related through a Fourier transformation. This possibility exists not only in the case of coherent Raman beats discussed in Chapter 7, but also in the case of spin transients in atomic vapours, where Fourier transformation of the time-domain signal provides the magnetic resonance spectrum.

8.6.1 Example

Figure 8.23 shows an example of a time-domain signal, measured in the ground state of sodium, using a transverse magnetic field of 0.7 mT. Clearly, the free induction decay contains more than one frequency component, but a complete

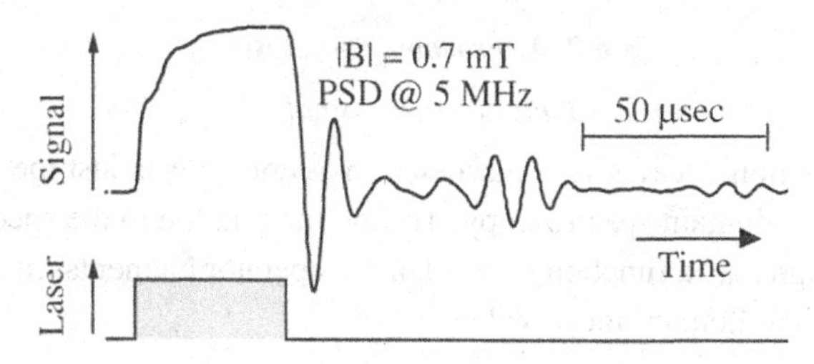

Figure 8.23. Free precession signal of the Na ground state in a field of 0.7 mT.

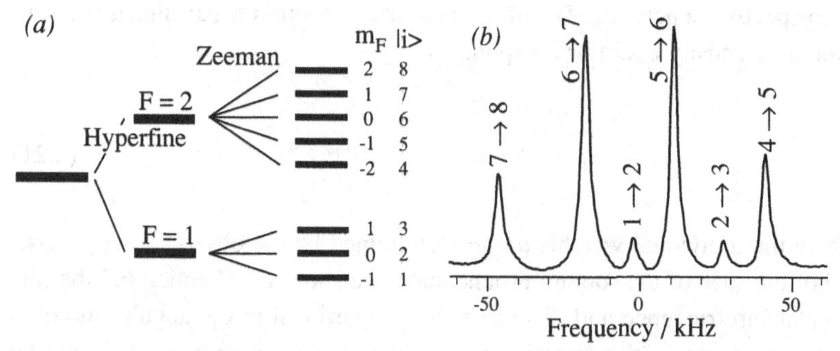

Figure 8.24. **(a)** Level scheme of the Na ground state in a weak magnetic field. The individual levels are labelled with the angular momentum components parallel to the direction of the magnetic field and a running index $|i\rangle$ starting at the lowest energy level. **(b)** Spectrum of the atomic sublevel transitions obtained by Fourier transformation of the free induction decay of Figure 8.23. The numbers give the assignment of the individual resonances to transitions between eigenstates.

analysis is not straightforward from this signal. It is therefore useful to Fourier transform the time domain data. For the data displayed in Figure 8.23, the Fourier transform returns the spectrum shown in Figure 8.24.

Figure 8.24 shows the result of the Fourier transformation on the right-hand side. The separation of the signal into six frequency components is now straightforward. In the spectrum, the six resonance lines are assigned to transitions between neighbouring Zeeman substates, using the numbering scheme defined in the left-hand part of the figure. The four stronger resonances belong to the $F = 2$ multiplet, whereas the two weaker resonances are transitions within the $F = 1$ multiplet. The positions of the individual resonance lines are derived in Section 4.4.3.

8.6.2 Microscopic analysis

Time-domain signal

The free induction decay signal discussed in Section 8.3 is just the signal required for time-domain spectroscopy. To find the relation to the spectrum, we express the signal as a function of the density operator elements. In the eigenbase of the Hamiltonian, they evolve as

$$\rho_{ik}(t) = \rho_{ik}(0)\, \mathrm{e}^{-\mathrm{i}\omega_{ik}t}\, \mathrm{e}^{-\gamma_{ik}t} \tag{8.20}$$

where $\omega_{ik} = \mathscr{E}_i - \mathscr{E}_k$ specifies the Bohr frequency and γ_{ik} the decay rate of the respective coherence. The observable free induction decay signal is a sum over such density operator components:

$$s(t) = \sum_{i\,k} a_{ik}\, \rho_{ik}(0)\, \mathrm{e}^{-(\mathrm{i}\omega_{ik} + \gamma_{ik})t} \tag{8.21}$$

where the individual weights a_{ik} are determined by the observation process.

For the case of the sodium ground state, we showed in Section 6.4 that the circular birefringence and dichroism are proportional to the angular momentum component in the direction of the laser beam. For this evaluation, we choose the quantisation axis along the magnetic field direction and orient the x axis along the laser beam, in contrast to the usual coordinate system, where the two directions are reversed. The observable is now the component F_x of the total angular momentum.

As Figure 8.25 shows, the observable has matrix elements between neighbouring eigenstates of the Hamiltonian. For the Na ground state, we expect a signal

$$\begin{aligned} s(t) = \mathrm{e}^{-\gamma_0 t} \Bigg\{ & \frac{1}{\sqrt{2}} \left[\rho_{12}(0)\, \mathrm{e}^{-\mathrm{i}\omega_{12}t} + \rho_{23}(0)\, \mathrm{e}^{-\mathrm{i}\omega_{23}t}\right] \\ & + \left[\rho_{45}(0)\, \mathrm{e}^{-\mathrm{i}\omega_{45}t} + \rho_{78}(0)\, \mathrm{e}^{-\mathrm{i}\omega_{78}t}\right] \\ & + \sqrt{\frac{3}{2}} \left[\rho_{56}(0)\, \mathrm{e}^{-\mathrm{i}\omega_{56}t} + \rho_{67}(0)\, \mathrm{e}^{-\mathrm{i}\omega_{67}t}\right] \Bigg\} + \mathrm{c.c.} \end{aligned} \tag{8.22}$$

if all coherences decay at the same relaxation rate γ_0. The frequencies ω_{ik} are all close to the Larmor frequency; as discussed in Section 4.4, however, they

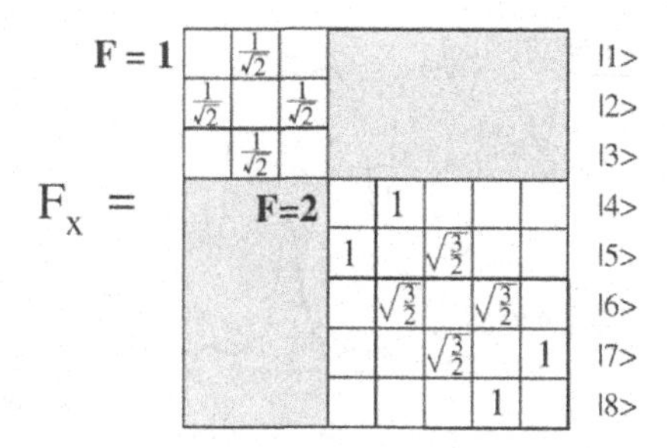

Figure 8.25. Matrix elements of F_x in the two hyperfine multiplets $F = 1, 2$.

can become nonequivalent when the second-order Zeeman interaction and/or the nuclear Zeeman interaction become important.

Fourier transformation

For the data analysis, it is convenient to perform a Fourier transformation of the time-domain data. For the signal given above, this yields the spectrum

$$
\begin{aligned}
s(\omega) = & \frac{1}{\sqrt{2}} [\rho_{12}(0)\, g(\omega - \omega_{12}) + \rho_{23}(0)\, g(\omega - \omega_{23})] \\
& + [\rho_{45}(0)\, g(\omega - \omega_{45}) + \rho_{78}(0)\, g(\omega - \omega_{78})] \\
& + \sqrt{\frac{3}{2}} [\rho_{56}(0)\, g(\omega - \omega_{56}) + \rho_{67}(0)\, g(\omega - \omega_{67})] \qquad (8.23)
\end{aligned}
$$

where $g(\omega - \omega_{rs})$ describes a Lorentzian of width γ_0 centred at frequency ω_{rs}. For low optical pumping rates, the anisotropic part of the initial density operator is close to F_x. We expect that the intensities of the resonance line should have the ratio 1:2:3 for the $F = 1$ and the outer and inner lines of the $F = 2$ transitions. In the case of Figure 8.24, the observed intensities deviate slightly from these approximate values, indicating that the optical pumping has polarised the system sufficiently that other contributions are no longer negligible. In particular, the amplitudes of the transitions in the $F = 1$ multiplet tend to decrease at higher laser intensity, indicating that the optical pumping can completely deplete the levels of the $F = 1$ multiplet. Since the width of the resonance lines is proportional to the decay rate of the coherences, the experiment allows a determination of the relaxation rates as well as of the frequencies.

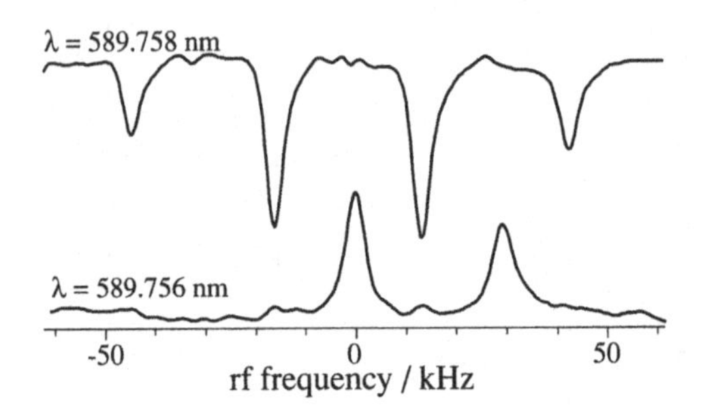

Figure 8.26. Sublevel spectra at two different laser wavelengths, using dispersive detection.

Nuclear spin effects

The separation between the resonance lines of Figure 8.24 shows two effects of the nuclear spin discussed in Section 4.4: the quadratic Zeeman effect and the nuclear Zeeman interaction. Together, they make all six transitions between nearest neighbours nonequivalent. Another effect of the nuclear spin is contained in the susceptibility tensor, even in the simplified form of Equation (6.83). Although both hyperfine multiplets contribute to the signal in proportion to their polarisation, their dependence on the laser frequency differs. If we neglect the excited state hyperfine splitting, their contributions are shifted by the ground-state hyperfine splitting. Even if the optical linewidth is large compared to the hyperfine splitting, this difference allows us to distinguish the two contributions (Suter 1992a). If we tune the laser frequency to the transition from the $F = 1$ ground state and use the dispersive variant of the polarisation-selective detection, the $F = 1$ coherences are no longer visible in the resulting signal.

Figure 8.26 shows two sublevel spectra that were measured with dispersive detection. For the upper trace, the laser wavelength was set to the centre of the transition from the $F = 1$ ground state to the excited state. For the lower trace, the laser was tuned to the $F = 2$ transition. In both cases, only the transitions within one multiplet are visible, since the dispersion and therefore the detection sensitivity vanish for those transitions with which the laser is exactly on resonance.

8.6.3. Possible extensions

In the method discussed so far, the speed of the acoustooptic modulator that generates the pulses limits the time resolution of the experiment, along with the speed of the detection system. If necessary, it is possible to achieve considerably higher time resolution by using a pulsed laser system. The time resolution is then limited only by the length of the optical pulses, which can be as short as a few femtoseconds. In such an experiment, the pulsed laser beam is again split into pump and probe beams. The probe propagates through an optical delay line and through the system. The detection system records the signal stroboscopically (Lange and Mlynek 1978). By varying the delay time through the optical path length, one can sample the evolution of the sublevel coherence (Harde et al. 1981).

With the use of nanosecond pulses, this technique has been applied, e.g., to measure EPR free-induction decay signals in a magnetic field in the ground state of Tm^{2+}:SrF_2 (Kohmoto et al. 1983). In this experiment the Fourier transform of the observed signals gave the EPR spectrum; the origin of the decay was attributed to the superhyperfine interaction between the Tm^{2+} ion and the neighbouring fluorine nuclei. Using mode-locked dye lasers in the picosecond regime, this pump-probe scheme has allowed the study of the hyperfine structure in the *D* lines of Na (Harde et al. 1981) and Cs (Lehmitz and Harde 1986). With subpicosecond pulses, even fine structure beats in Na at 517 GHz could be clearly resolved (Burggraf, Kuckartz and Harde 1986). This experiment is also an example for a measurement of sublevel coherence in an optically excited state, which can be studied with such a pump-probe scheme even if the lifetime of the excited state is very short. Studies of organic dyes using a transmission correlation technique (Rosker, Wise and Tang 1986; Walmsley, Mitsunaga and Tang 1988) have revealed oscillatory behaviour on a femtosecond timescale in the decay of photoexcited dye molecules. It was suggested that this behaviour might be due to a beat phenomenon between coherently excited vibrational levels separated by several THz.

As discussed above, it is also possible to measure higher multipole moments like alignment instead of the magnetisation by using different detection geometries. It is therefore possible to use time-domain spectroscopy for the determination of the relaxation rates of these multipole moments, in particular in ground or near-ground atomic levels. One example is the study of the depolarisation of the $4f^6 6s^2\ {}^7F_1$ near-ground level of samarium by collisions with rare-gas perturbers (Lowe et al. 1987; McLean, Hannaford and Lowe 1990). With an appropriate choice of polarisation and geometry, Zeeman beat signals were obtained whose decay directly yielded the relaxation rates of orientation and alignment. A striking result from those measurements was that a substantial anisotropy in the collisional relaxation in the ${}^7F_1$ level of Sm could be observed.

9
Two-dimensional spectroscopy

Two-dimensional spectroscopy is an extension of conventional spectroscopic methods that characterises resonant systems as a function of two frequency variables. We use this technique to improve on the characterisation of atomic media along the lines discussed in the preceding chapter. On a microscopic level, the most important physical process for two-dimensional spectroscopy is the transfer of coherence between different transitions. We introduce the basics of the technique and discuss a few specific examples to outline its potential.

9.1 Fundamentals

9.1.1 Motivation and principle

Motivation

The preceding chapter showed how light drives the internal dynamics of resonant atomic media and how the measurement of optical anisotropies allows us to monitor these dynamics. The experiments discussed in the preceding chapter, however, can provide only limited information about the system. Most physical systems have more degrees of freedom than we can observe by measurements on transmitted light. As another limitation, we have primarily considered atoms that evolve under their internal Hamiltonian, only weakly perturbed by the probe laser beam. The example of light-induced spin nutation showed that the dynamics of optically pumped atoms differ significantly from those of a free atom. Although it is possible to observe spin nutation for systems with more than two ground-state sublevels, such an experiment suffers from the damping that accompanies optical pumping. The damping drives the system rapidly to an equilibrium, too fast for detailed dynamical observations.

As we will show, two-dimensional (2D) spectroscopy allows the circumvention of both these limitations: It permits a manipulation of the selection rules and allows us to measure arbitrary variables that characterise the microscopic state of the system. In particular, it provides the possibility of observing multipole moments of arbitrary order (Suter and Klepel 1992a). In addition, it allows measurement of the dynamics of radiatively coupled systems with the same precision as that of free atoms (Suter, Klepel and Mlynek 1991b).

In Sections 7.3.3 and 8.6, we showed that time-domain experiments can provide the same information as sublevel spectra obtained by measuring the system response as a function of frequency. For this purpose, one measures the time-domain response of the system, usually in the form of a coherent Raman beat or free induction decay (FID) signal and calculates its Fourier transform to recover the frequency spectrum. As we will show below, this method of time domain spectroscopy is more flexible than the cw method; in particular, we will extend it to the observation of higher order moments not accessible with cw experiments.

Whereas one-dimensional spectroscopy measures the response of the medium to a single-frequency perturbation, two-dimensional spectroscopy analyses the system response to two independent electromagnetic fields that interact with the system. Frequency-domain experiments of this type are known as double resonance experiments. Here we discuss the time-domain version, which is considerably more powerful and flexible. It was first suggested by Jeener (Jeener 1971) and implemented for nuclear magnetic resonance in the group of Richard Ernst (Jeener et al. 1979; Ernst, Bodenhausen and Wokaun 1987). Applications in EPR spectroscopy were also demonstrated (Gorcester and Freed 1988), as well as in rotational spectroscopy (Vogelsanger and Bauder 1990).

Principle

Figure 9.1 summarises the basic idea of the experiment. During the so-called preparation period, an initial laser pulse couples to the electronic transition between a near-degenerate ground state (consisting of the four sublevels i, k, r, s) and an excited state $|e\rangle$. By transverse optical pumping, it excites sublevel coherence in the ground state. The wavy line between substates i, k represents an example of such a coherence. During the evolution period after the end of the preparation pulse, these coherences precess freely for a time t_1 under the influence of the Zeeman interaction. They acquire a phase $\omega_{ik}t_1$ proportional to the Bohr frequency ω_{ik} of the transition in which they evolve and to the duration t_1 of the evolution period. This phase factor contains the spectroscopic information in which we are interested. If the coherence ρ_{ik} contributes to the magnetic dipole moment of the system, this evolution can be observed directly. For higher-order multipole moments, however, this is not possible.

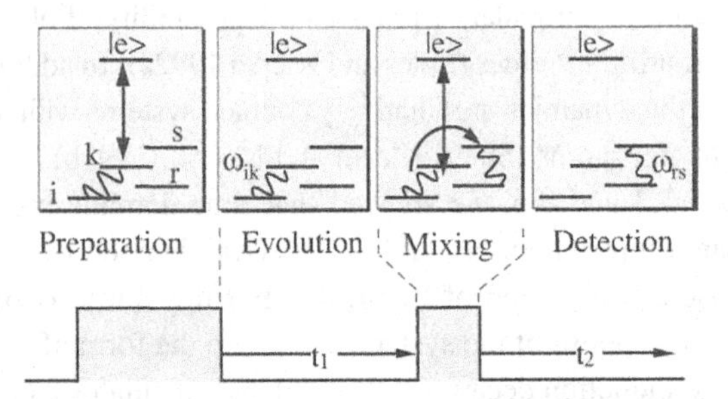

Figure 9.1. Schematic representation of the two-pulse sequence used for two-dimensional spectroscopy and the effect on the atomic system. The names for the different periods are the conventional ones in magnetic resonance spectroscopy.

In the two-dimensional experiment of Figure 9.1, a second laser pulse terminates the evolution period. It may couple to the same electronic transition as the first pulse. It interchanges (mixes) coherences between different transitions and is therefore known as a mixing pulse. In our example, it transfers order from transition *ik* to transition *rs*. This process also transfers the phase information. During the final detection period, the system again precesses freely under the internal Hamiltonian and the order is observed in the form of a free induction decay signal. If the observed signal includes a contribution from the transition *rs*, that contribution is modulated by the phase factor $e^{i\omega_{ik}t_1}$. By repeating the experiment for a number of t_1 values, it is possible to trace out the free precession of the coherence in the sublevel transition ρ_{ik}. With this experimental procedure, it is possible to observe not only multipole moments that change the optical properties of the system, but also those that a laser pulse can convert to directly observable order.

9.1.2 Theoretical analysis

Evolution and mixing

For a demonstration of how this procedure can yield information about "forbidden" transitions or the dynamics of radiatively coupled systems, we write the density operator elements ρ_{ik} after the evolution period as

$$\rho_{ik}(t_1) = \rho_{ik}(0)\ e^{-(i\omega_{ik}+\gamma_{ik})t_1} \tag{9.1}$$

where the indices *i,k* refer to the eigenstates of the Hamiltonian and t_1 is the length of the free precession period. On this basis, the density operator elements are eigenvectors of the evolution.

When the laser couples to the system, this is no longer the case: The interaction with the laser mixes the different density operator elements. We postpone discussion of the physical mechanism to Section 9.2 and summarise here only its overall effect in a mixing matrix η. Its elements $\eta_{rs,ik}$ describe the transfer of coherence from transition ik to transition rs (remember that we refer to populations as coherences ii with equal indices). The density operator for the eight angular momentum states of the sodium ground state has 64 elements, so the exchange matrix η has the dimension 64×64. After the mixing pulse, the matrix element ρ_{rs} depends on the preceding evolution of all other components

$$\rho_{rs}(t_1,0) = \sum_{i\,k} \rho_{ik}(0)\, \mathrm{e}^{-(\mathrm{i}\omega_{ik}+\gamma_{ik})t_1}\, \eta_{rs,ik} \tag{9.2}$$

In particular, it depends on the frequency ω_{ik}, which may not be directly observable. The dependence on the initial coherence $\rho_{ik}(0)$ contains information about the initial state of the density operator, whereas the matrix elements $\eta_{rs,ik}$ summarise the effect of the laser-induced dynamics.

Signal

After the mixing pulse, the system precesses freely again as

$$\rho_{rs}(t_1,t_2) = \sum_{i\,k} \rho_{ik}(0)\, \mathrm{e}^{-(\mathrm{i}\omega_{ik}+\gamma_{ik})t_1}\, \eta_{rs,ik}\, \mathrm{e}^{-(\mathrm{i}\omega_{rs}+\gamma_{rs})t_2} \tag{9.3}$$

where t_2 is the time after the second laser pulse. We write a_{rs} for the weight of the density operator element ρ_{rs} in the observed signal, which becomes

$$s(t_1,t_2) = \sum_{r\,s} a_{rs} \sum_{i\,k} \rho_{ik}(0)\, \mathrm{e}^{-(\mathrm{i}\omega_{ik}+\gamma_{ik})t_1}\, \eta_{rs,ik}\, \mathrm{e}^{-(\mathrm{i}\omega_{rs}+\gamma_{rs})t_2} \tag{9.4}$$

It depends on the two variables t_1 and t_2. Whereas one-dimensional time domain spectroscopy requires the measurement of a signal as a function of a single time variable, two-dimensional spectroscopy starts from a data set that depends on the two time variables t_1 and t_2. This matrix can be obtained by recording the free induction decay after the second pulse for a set of equally spaced t_1 values.

Figure 9.2 shows an example of such a two-dimensional free induction decay (2D FID) containing two frequency components. If the elements of the mixing matrix and the observable $A = (a_{rs})$ are real, the FID signal is

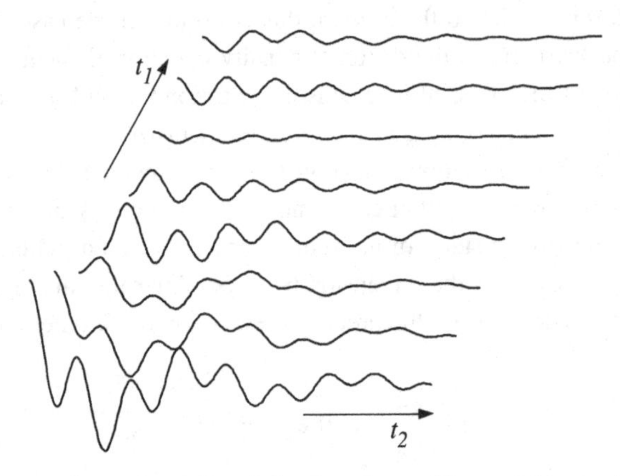

Figure 9.2. Two-dimensional free induction decay. The variable t_2 along the horizontal direction is the detection time, whereas the evolution time t_1 increases towards the rear.

$$
\begin{aligned}
fid(t_1,t_2) = \mathrm{e}^{-\gamma_0(t_1+t_2)} & [\eta_{rs,ik}\cos(\omega_{ik}t_1 + \omega_{rs}t_2) \\
& + \eta_{rs,rs}\cos(\omega_{rs}t_1 + \omega_{rs}t_2) + \eta_{ik,ik}\cos(\omega_{ik}t_1 + \omega_{ik}t_2) \\
& + \eta_{ik,rs}\cos(\omega_{rs}t_1 + \omega_{ik}t_2)]
\end{aligned}
\tag{9.5}
$$

The first term represents a coherence that evolved in transition *ik* during t_1. The mixing pulse transferred it to transition *rs*, where it was observed during t_2. The second term represents coherence that remained in transition *rs*, unaffected by the mixing pulse. The third term remained in transition ik and the fourth was transferred from transition *rs* into *ik*. Each trace in Figure 9.2 represents the signal during the detection period for a given delay t_1 between the two laser pulses, which increases systematically for the traces towards the rear of the figure.

Data processing

To separate these frequency components, we apply a two-dimensional Fourier transformation with respect to the two time variables t_1 and t_2:

$$
f(\omega_1, \omega_2) = \int_{-\infty}^{\infty} dt_1\, \mathrm{e}^{\mathrm{i}\omega_1 t_1} \int_{-\infty}^{\infty} dt_2\, \mathrm{e}^{\mathrm{i}\omega_2 t_2}\, s(t_1,t_2) = \int_{-\infty}^{\infty} dt_1\, \mathrm{e}^{\mathrm{i}\omega_1 t_1}\, sf(t_1,\omega_2) \tag{9.6}
$$

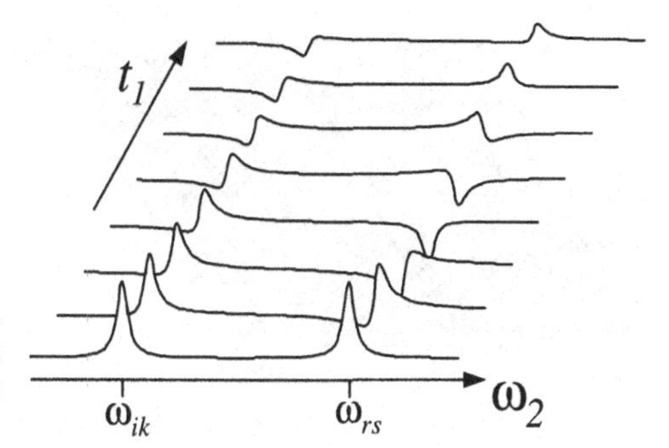

Figure 9.3. Real part of the time-dependent spectrum obtained as the Fourier transform with respect to the second time variable t_2 of the two-dimensional FID.

The inner transform, with respect to t_2, is the same operation that was used in the case of one-dimensional time domain spectroscopy to obtain the spectrum from the FID signal. In the two-dimensional case, this operation is performed on each row of the data matrix. We obtain therefore a sequence of spectra that correspond to different delays t_1

$$sf(t_1,\omega_2) = \mathrm{e}^{-\gamma_0 t_1} \{[\eta_{rs,ik}\, \mathrm{e}^{\mathrm{i}\omega_{ik}t_1} + \eta_{rs,rs}\, \mathrm{e}^{\mathrm{i}\omega_{rs}t_1}]\, g(\omega_2 - \omega_{rs}) + [\eta_{ik,ik}\, \mathrm{e}^{\mathrm{i}\omega_{ik}t_1} + \eta_{ik,rs}\, \mathrm{e}^{\mathrm{i}\omega_{rs}t_1}]\, g(\omega_2 - \omega_{ik})\} \tag{9.7}$$

where $g(\omega_2 - \omega_{\alpha\beta})$ represents a complex Lorentzian centred at $\omega_{\alpha\beta}$.

Figure 9.3 shows the real part of the signal obtained after this first Fourier transform. In the horizontal direction, the signal is now a spectrum, whose frequency variable is ω_2. Each spectrum contains two absorption lines at frequencies ω_{ik} and ω_{rs}. Amplitude and phase of the two absorption lines change with the evolution time t_1. The different behaviour of the two lines indicates that the coherences evolved at different frequencies during t_1.

To obtain a two-dimensional spectrum, we have to perform the second Fourier transform with respect to the second time variable t_1. The four signal components each contribute a resonance line of the type

$$s(\omega_1,\omega_2) = g_2(\omega_1 - \omega_{ik}, \omega_2 - \omega_{rs}) \tag{9.8}$$

to the two-dimensional spectrum; g_2 is a two-dimensional Lorentzian centred at the frequencies $(\omega_1 = \omega_{ik}, \omega_2 = \omega_{rs})$.

The precise form of this resonance line shape depends on the details of the Fourier transform as well as on the relaxation mechanisms, for which we have assumed here an exponential decay. In this case, the resulting lineshape is that of a two-dimensional Lorentzian, as shown in Figure 9.4. As in the one-

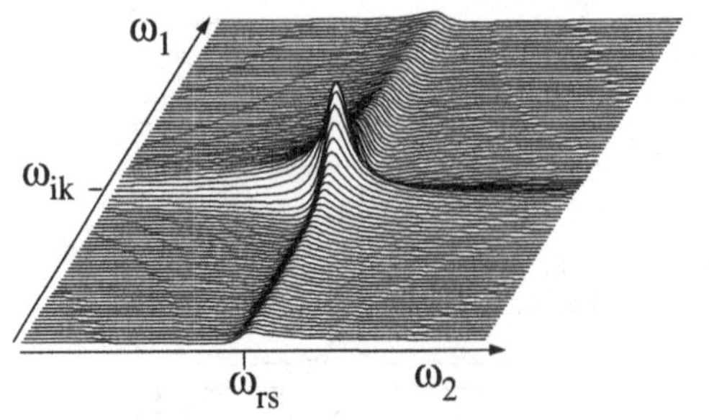

Figure 9.4. Individual resonance line after the second Fourier transform.

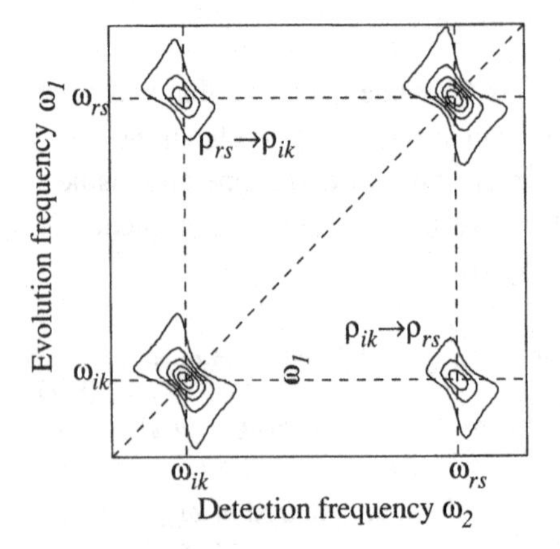

Figure 9.5. Theoretical spectrum indicating exchange of coherence between two sublevel transitions.

dimensional spectra, the width of the resonance lines in the sublevel spectrum is independent of the width of the optical resonance line and of the frequency spectrum of the pump or probe laser. Since the natural width of the transitions is negligibly small, the observed width is due primarily to diffusion of the atoms out of the laser beam, which leads to a lifetime broadening of the resonance lines, and to magnetic field inhomogeneity, which causes an inhomogeneous decay of the signal.

Interpretation of spectra

To provide an idea of the information content of two-dimensional spectra, we consider the example of Figure 9.5. It shows a contour plot representation of a possible 2D spectrum. The system has two transition frequencies, which define four spectral positions in the two-dimensional spectrum. In the example spectrum, each of these positions is occupied by a resonance line. The resonance lines on the diagonal originate from density operator components that have evolved in the same transition before and after the mixing pulse. These so-called diagonal peaks contain the same information as the one-dimensional spectrum. The two off-diagonal or cross peaks at the positions ($\omega_1 = \omega_{ik}, \omega_2 = \omega_{rs}$) and ($\omega_1 = \omega_{rs}, \omega_2 = \omega_{ik}$) indicate that the mixing pulse has exchanged coherence between the two transitions. The density operator components that cause these signals have evolved at different frequencies before and after the pulse. The labels near these peaks indicate the direction of the transfer process. Their amplitudes are directly proportional to the mixing matrix elements $\eta_{\alpha\beta,\nu\mu}$.

9.1.3 Coherence transfer echoes

Two-dimensional spectroscopy can provide more information than one-dimensional experiments whenever a process exchanges order between different sublevel transitions. Only the occurrence of such coherence transfer processes makes this type of two-dimensional spectroscopy meaningful. In the preliminary analysis given above, we did not discuss the nature of these processes, but summarised them in the transfer matrix η. The physical properties that can lead to such processes are diverse, and we discuss an example in Section 9.2. Here, we use coherence transfer echoes (Suter, Rosatzin and Mlynek 1991c) as an example of an experiment that shows such effects and is closely related to two-dimensional spectroscopy, but does not require the analysis of a two-dimensional spectrum. We include it here as a demonstration of the occurrence of these kinds of processes, to make their existence plausible.

In Section 8.4, we demonstrated that the light-shift effect of an off-resonance laser pulse can refocus magnetisation that has dephased in an inhomogeneous magnetic field. Using the simple $J = 1/2$ to $J' = 1/2$ model for the sodium D_1 transition, we calculated the response of the system to the laser pulse sequence and found the occurrence of an echo at time $t = T$ after the second pulse, where T is the pulse separation. This echo is indeed observed experimentally. In the actual experiment, however, the signal does not end after the echo, as one would expect from the analysis presented in Section 8.4: A closer inspection finds a second and even a third echo at times $t = 2T$ and $t = 3T$. This observation is in direct contradiction to the theoretical prediction: There should be no signal after the first echo (Schenzle, Wong and Brewer 1980; Schenzle, Wong and Brewer 1981).

Figure 9.6 shows a typical example of such an echo sequence, together with the excitation laser pulses. The separation between the echoes is equal to the pulse separation time T. These multiple echoes persist under a wide range of

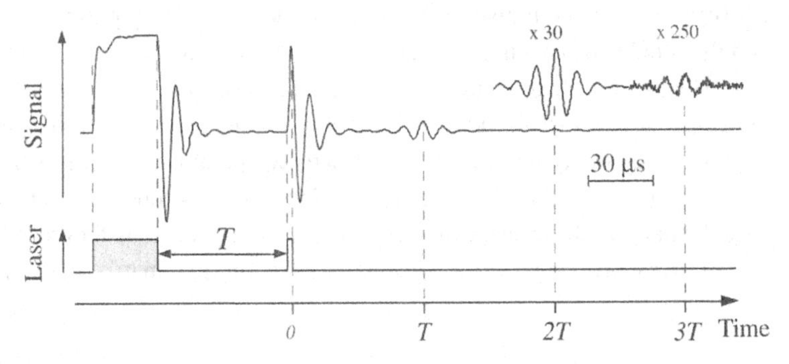

Figure 9.6. Typical experimental signal observed when a sequence of two laser pulses excites sublevel coherences in Na vapour. In this example, the laser pulses had an optical detuning of $\Delta/2\pi = 5.5$ GHz and a power of 80 mW. The length of the refocusing pulse was 1.6 μs, the separation $\tau = 60$ μs and the Larmor frequency $\Omega_L/2\pi = 110$ kHz. The second and third echoes are expanded vertically. The trace at the bottom shows the optical pulse sequence.

experimental conditions. Their dependence on the experimental parameters such as laser intensity, laser detuning, pulse width, and magnetic field strength is similar to that of the first echo. Their presence in the signal is clearly incompatible with the simple echo formation mechanism given above. The resolution of this apparent paradox is that the echo formation mechanism was derived for a ground state with two angular momentum sublevels, thereby neglecting the nuclear spin. Although this is a good approximation for many experiments in pressure-broadened systems, it apparently fails here. A more complete analysis therefore has to take the nuclear spin $I = 3/2$ of Na into account. Its most important effect, in this context, is the increase in the number of available magnetic substates. Although multiple echoes cannot occur in simple spin-1/2 systems, they have been observed in more complicated level systems (Solomon 1958; Maudsley, Wokaun and Ernst 1978). The occurrence of multiple echoes in such systems is possible whenever the refocusing pulse transfers coherences between transitions whose Bohr frequencies have a different dependence on the inhomogeneous interaction, so that the refocusing time differs from the defocusing time.

9.1.4 Possible applications

Two-dimensional spectroscopy allows a direct observation of the creation, redistribution, and decay of sublevel coherences. Such an observational tool appears especially useful whenever the system under study is too complex for

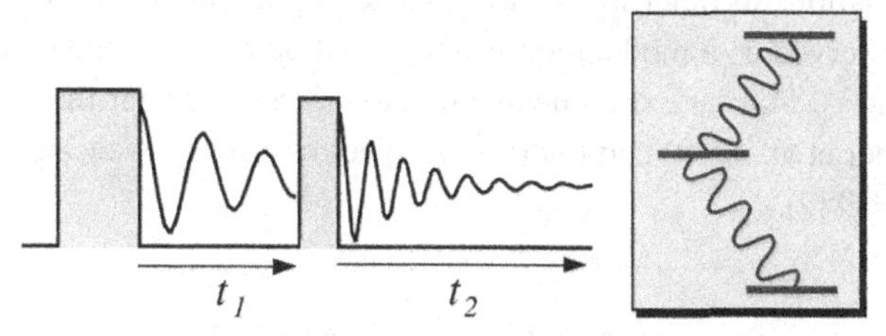

Figure 9.7. Application of the two-pulse experiment to electronic transitions.

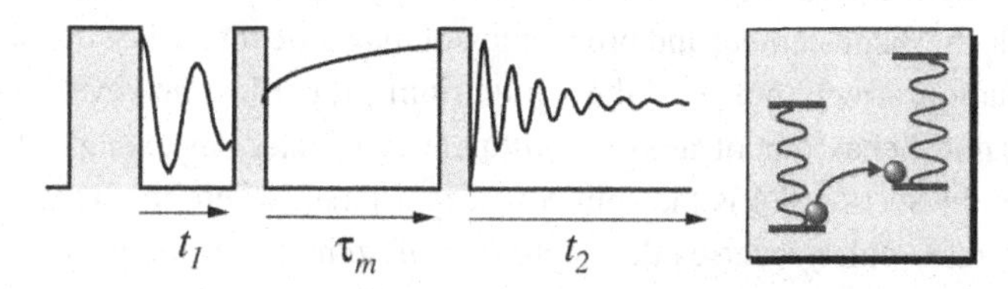

Figure 9.8. Principle of a two-dimensional experiment for spectroscopic investigation of exchange processes.

a theoretical analysis, i.e., when it is too complex for an analytical solution of the equation of motion and driven too far from equilibrium to allow a perturbation treatment. The examples discussed below involve only electronic ground states, where the long lifetimes allow detailed observations. The technique is equally applicable, however, to electronically excited states. It is not even limited to the observation of sublevel transitions, but could also be applied to optical transitions.

Figure 9.7 outlines a possible application to optical transitions. A sequence of two pulses that couple to two connected optical transitions could provide high resolution spectra of transitions between electronically excited states. The two-photon character of the experiment would allow removal of Doppler broadening. This would represent a direct extension to optical transitions of the experiment that we apply here to sublevel transitions.

Another variant, depicted in Figure 9.8, would require a third optical pulse. It would allow this study, e.g., of the dynamics of molecular reactions (Zewail 1988) or fast dynamics in solids (Noll et al. 1990), using pico- or femtosecond pulses. The first two pulses, separated by a time t_1, would establish nonthermal populations in the ground and electronically excited state of one transition, as indicated in the inset at the right. These populations would be labelled with the transition frequency. If molecular dynamics, chemical reactions, or changes in the environment during the mixing period τ_m transfer part of the populations to other levels, as indicated by the arrow, these processes could be observed as a modulation of the signal detected on the second transition during t_2. Similar experiments are known from nuclear magnetic resonance (Jeener et al. 1979) and microwave spectroscopy (Vogelsanger, Bauder and Mäder 1989).

9.2 Coherence transfer

9.2.1 Introduction

Polarised light interacting with atoms not only creates anisotropic states, but also leads to "displacement and broadening of magnetic resonance transitions." The radiation affects not only the equilibrium properties, however, but also the dynamic behaviour of atomic multipole moments. An example that we discussed in Section 8.4 is the spin echo experiment, where the light-shift effect of a laser pulse reverses the dephasing of spins undergoing Larmor precession in an inhomogeneous magnetic field. This experiment demonstrates that light-shift and damping effects modify the precession frequencies and the relaxation rates of the multipole moments. Although displacement and broadening of sublevel transitions can be attributed to changes in the eigenvalues of the effective Hamiltonian, the light-shift–induced spin echo, as an example of dynamical effects, demonstrates that the coupling between light and the atoms also changes the quantisation direction of the Hamiltonian. This reorientation of the quantisation axis of the effective Hamiltonian also lifts the degeneracy of the eigenstates in zero field, which was observed in optical–radio-frequency double-resonance experiments (Dupont-Roc et al. 1967).

If the electronic state has only two angular momentum substates, as we have assumed so far, the dynamical effects are relatively straightforward. They correspond to a virtual magnetic field and a damping effect. When the atomic system becomes more complex, however, the coupling to the radiation field can prepare the atomic system in arbitrary superpositions of the Zeeman substates. The conventional expansion of the resulting atomic state uses irreducible tensor operators as the basis operators (Omont 1977). The multipole moments appear as the expansion coefficients of these operators, as discussed in Section 4.3. In the case of the Na ground state, it is possible to excite moments up to hexadecupoles, which correspond to irreducible tensor operators of rank $k = 4$. These atomic multipole moments evolve under the influence of the atomic Hamiltonian and the coupling to external fields. As the coupling to the radiation field breaks the rotational symmetry of the free atomic Hamiltonian, the multipoles are no longer constants of the motion. The laser can exchange coherence between them, rotate them, create and destroy order in different multipole moments.

If we try to investigate these processes with the techniques discussed up to now, we meet two problems. The first limitation concerns the initial condition: In the single-pulse experiment discussed before, the system is initially always in thermal equilibrium; we have no freedom to choose the initial condition. The second limitation concerns the possible observables: the polarisation-selective detection scheme that we used allows only observation of the magnetic dipole moment and, in some cases, the quadrupole moment. Higher-order multipoles are not directly observable. Two-dimensional spectroscopy removes both limitations. It uses the exchange of order between different multipole moments to observe all multipoles from which order can be transferred to dipole moments (Suter and Klepel 1992b). Light-induced coherence transfer is not only a prerequisite for applying the method, it also represents one of the main study objects (Suter, Klepel and Mlynek 1991c). In this context, the first frequency dimension allows us to specify the initial state of the system, before the interaction with the light, whereas the second frequency dimension analyses the resulting state.

For a theoretical analysis of the static magnetooptic properties of complex multilevel systems, it is often possible to use perturbation methods. These methods lead to a description of the system that is adequate if the optical interaction is small, driving the system not too far from thermal equilibrium. In the situations that we consider, the optical pumping can drive the system arbitrarily far from equilibrium, where a perturbation analysis is not appropriate. The $J = 1/2$ model for the ground state used in Section 5.2 allowed an analytic solution, but this is not the case for the full system in which we discuss coherence transfer processes. In this system, we can only treat certain aspects analytically. In particular, we use the equations of motion derived in Sections 5.2 and 5.4, on optical pumping, which are valid as long as the laser intensity is low enough that optical coherences and excited state populations remain small. Starting from these equations of motion, we will use numerical methods to describe the dynamics of the system.

9.2.2 Example

Spectrum

Before we present a mathematical analysis of light-induced coherence transfer, we describe an experimental example that shows the effect of coherence transfer. The experiment was performed on the ground state of atomic sodium

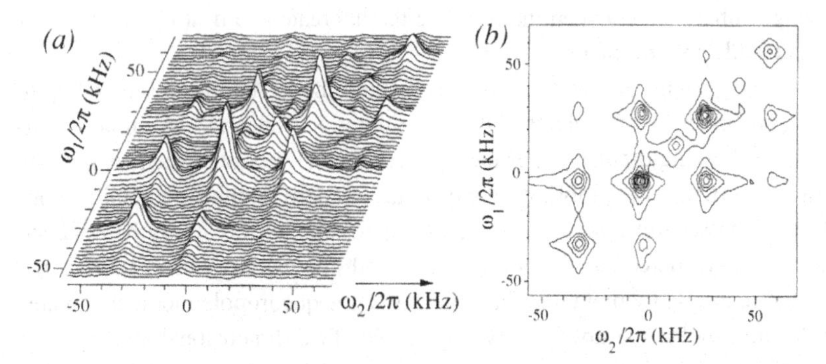

Figure 9.9. Example of a 2D spectrum. **(a)** shows a pseudo-3D representation and **(b)** a contour plot of the same data set. The data were recorded with the laser wavelength tuned 8.6 GHz above the centre of the D_1 resonance line.

while a homogeneous magnetic field of 0.7 mT was applied perpendicular to the direction of the laser beams. At this field strength, the Larmor frequency is close to 5 MHz, and the individual sublevel transitions are easily distinguishable, as discussed in Section 8.6. Excitation and mixing used an amplitude-modulated circularly polarised pump laser beam, and the observed signal was demodulated with the modulation frequency, which was close to 5 MHz.

The driving force behind the coherence transfer process is the light-shift effect which competes with the damping effect of the laser. The damping has a maximum at the centre of the resonance line and decreases rapidly with the resonance detuning. The light-shift effect, in contrast, vanishes at the centre of the optical resonance, has a maximum at a normalised laser detuning $|\overline{\Delta}| = 1$, and falls off less rapidly than the damping at larger detunings, $|\overline{\Delta}| > 1$. It is thus advantageous to use a laser detuning of several optical linewidths for this experiment. For the example presented here, the laser frequency was 8.6 GHz above the centre of the Na D_1 resonance. At the sample cell, the average pump beam intensity during the pulses was $\sim$1 W/cm^2. Under these experimental conditions, excited state effects can be neglected and the observed signal is determined by the ground-state orientation.

The two-pulse experimental scheme of Figure 9.9 was used for the experiment. The evolution time t_1 was incremented from 0 to 283.5 μsec in 64 steps of 4.5 μsec. The resulting data set was Fourier transformed and the absolute value displayed in a pseudo-3D representation (Figure 9.9, left) and as a contour plot (right). In principle, every element $\eta_{rs,ik}$ of the mixing matrix can lead to a resonance line in this spectrum. As can be seen from the spectrum, how-

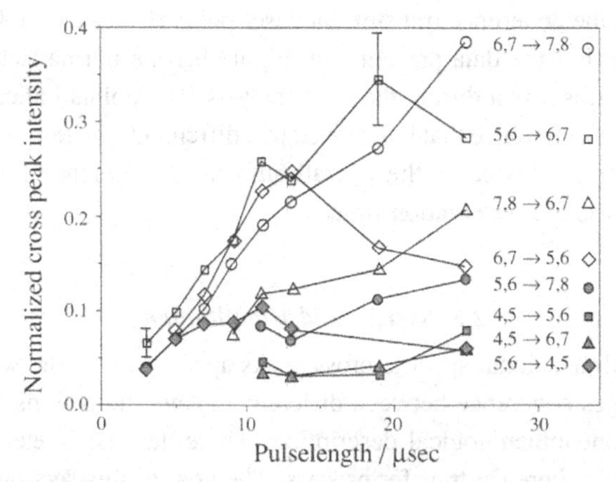

Figure 9.10. Normalised intensities of the various cross peaks as a function of the length of the mixing pulse. The experimental parameters are the same as in Figure 9.9.

ever, the total number of resonance lines is much smaller. This reduction is due to several selection rules, the most important of which is the detection process, which allows only detection of density matrix elements with $|\Delta m_F| = 1$.

The resonance lines along the main diagonal $\omega_1 = \omega_2$, the so-called diagonal peaks, arise from coherences that were not affected by the second pulse; for vanishing length of the second pulse, only these peaks exist. All the remaining peaks indicate that the pulse has transferred coherence from one transition to another. They are usually referred to as cross peaks.

Time dependence

The redistribution of the sublevel coherences by the optical field can be observed by recording a sequence of such spectra with increasing length of the mixing pulse. The transfer of coherence between the different transitions is observable as changes in the intensities of the corresponding resonance lines.

Figure 9.10 shows a partial analysis of these data for the Na system. The different curves represent the transfer of coherence between a representative set of adjacent $|\Delta m_F| = 1$ transitions of the $F = 2$ multiplet. These data clearly show that the exchange process continues with increasing pulse length and that it is quantitatively not a small process. More than a third of the coherence from the transition $6 \leftrightarrow 7$, e.g., has been transferred to the transition $7 \leftrightarrow 8$ after a pulse duration of 25 μsec, and more than half of the coherence has been transferred to other transitions.

Apart from the coherence transfer, the laser pulse also causes a damping of the coherence. For the data presented in Figure 9.10, a normalisation procedure removed this effect during the data analysis. The points indicate the relative amount of coherence that has moved to a different transition. At the same time, the damping decreases the overall amount of coherence present, thus limiting the possible observation time.

9.2.3 System and Hamiltonian

In these two-dimensional spectra, cross peaks appear only if the second laser pulse exchanges coherence between different sublevel transitions. So far, we have used a phenomenological description, where the matrix elements $\eta_{ik,rs}$ summarise the coherence transfer process. The goal of this section is the derivation of a physical model whose outputs are these matrix elements. The system we consider for this purpose is again the electronic ground state of sodium. To keep the notation simple, we further restrict the analysis to the $F = 2$ hyperfine multiplet. The "dark" Hamiltonian for this system includes the coupling between the atomic magnetic moments and the external magnetic fields. For the relatively weak magnetic fields in which we are interested, it is sufficient to use the low-field approximation of the Hamiltonian, which is valid as long as the Zeeman interaction is small compared to the hyperfine splitting. Choosing the z axis parallel to the magnetic field direction, the linear and quadratic terms in the field strength are

$$\mathcal{H}_{\mathrm{Z}} = -\mu_{\mathrm{B}} \, (g_F \, B \, F_z + g_F^{(2)} \, B^2 \, F_z^2) = -\mu_{\mathrm{e}} \, (B \, F_z - \mu_{\mathrm{e}}^{(2)} \, B^2 \, F_z^2) \quad (9.9)$$

as discussed in Section 4.4.3.

Multipole expansion

The Hamiltonian of equation (9.9) is invariant under rotations around the z axis. It is therefore convenient to express the density operator in a representation that transforms irreducibly under rotations around this axis, i.e., to expand it in irreducible tensor operators $T_q^{(k)}$ of rank k and order q (Omont 1977). The ground-state density operator then becomes

$$\rho = \sum_{k\,q} a_{kq} \, T_q^{(k)} \quad (9.10)$$

with a_{kq} representing the expansion coefficients.

The evolution of the density operator under the Zeeman interaction is

$$\rho(t) = \exp(\mathrm{i}\,\mu_\mathrm{e}^{(2)}\,B^2\,t\,F_z^2)\,\exp(\mathrm{i}\,\Omega_\mathrm{L} t\,F_z)\,\rho(0)\,\exp(-\mathrm{i}\,\Omega_\mathrm{L} t\,F_z)\,\exp(-\mathrm{i}\,\mu_\mathrm{e}^{(2)}\,B^2\,t\,F_z^2) \tag{9.11}$$

Here, we have separated the Zeeman effect into a term linear in the field, describing the Larmor precession at frequency $\Omega_\mathrm{L} = \mu_\mathrm{e}\,B$, and a term quadratic in the field. Like every rotation around the symmetry axis, the Larmor precession appears as a phase accumulation of the individual tensorial components:

$$\rho(t) = \exp(\mathrm{i}\,\mu_\mathrm{e}^{(2)}\,B^2\,F_z^2\,t)\left\{\sum_{k\,q} \mathrm{e}^{-iq\Omega_\mathrm{L}t}\,a_{kq}\,T_q^{(k)}\right\}\exp(-\mathrm{i}\,\mu_\mathrm{e}^{(2)}\,B^2\,F_z^2\,t) \tag{9.12}$$

In this form, the linear Zeeman effect appears as the phase accumulation in the braced expression, while the quadratic Zeeman effect is still in operator notation.

Symmetry adapted basis

The quadratic Zeeman effect also conserves the order q of the tensor operators, but not their rank k, indicating that it cannot be described as a rotation in three-dimensional space. We therefore introduce, besides the irreducible tensor operators, another set of operators $A_{q,j}$ that transform irreducibly under rotations around the z axis, as well as under the Hamiltonian evolution. These operators are linear combinations of irreducible tensor operators $T_q^{(k)}$ with the same order q:

$$A_{q,j} = \sum_{k=|q|}^{2F} \alpha_{qjk}\,T_q^{(k)} \tag{9.13}$$

The transformation under rotations around the z axis therefore remains

$$\mathrm{e}^{-\mathrm{i}\phi F_z}\,A_{q,j}\,\mathrm{e}^{\mathrm{i}\phi F_z} = \mathrm{e}^{\mathrm{i}q\phi}\,A_{q,j} \tag{9.14}$$

but the Hamiltonian evolution is now

$$\mathrm{e}^{-\mathrm{i}\mathcal{H}t}\,A_{q,j}\,\mathrm{e}^{\mathrm{i}\mathcal{H}t} = \mathrm{e}^{\mathrm{i}\omega_{q,j}t}\,A_{q,j} \tag{9.15}$$

In the eigenbase of the Hamiltonian, the operators $A_{q,j}$ have for $|q| > 0$ a single nonvanishing matrix element between two states whose magnetic quantum numbers differ by $m_F - m_{F'} = q$. The precession frequency $\omega_{q,j}$ of the operator $A_{q,j}$ is

$$\omega_{q,j} = q\Omega_{\mathrm{L}} + \mu_{\mathrm{e}}^{(2)} B^2 \left(m_F^2 - m'^2_F\right) \tag{9.16}$$

In this operator basis, we expand the density operator as

$$\rho(0) = \sum_{q,j} d_{q,j} A_{q,j} \tag{9.17}$$

and write its time evolution as

$$\rho(t) = \sum_{q,j} \mathrm{e}^{\mathrm{i}\omega_{q,j}t} d_{q,j} A_{q,j} \tag{9.18}$$

which represents the complete time evolution during the evolution and detection periods of the 2D experiment.

9.2.4 Light-induced dynamics

When the pump laser drives the atomic system, optical-pumping, light-shift, and damping effects become important. We discuss these effects in detail in Section 5.4. As a closed solution of the system dynamics in the presence of the laser field does not exist, we recall here only the relevant symmetry properties of the contributions to the dynamics. The light-shift effect, which represents an effective magnetic field, changes the direction around which Larmor precession occurs. This change of the quantisation axis results in a mixing of density operator elements. The rotation caused by the light-shift and the *linear* Zeeman effect mixes spherical tensor operators $T_q^{(k)}$ with the same rank k but different order q. Since the quadratic Zeeman interaction mixes tensors with different rank, the total Hamiltonian mixes all irreducible tensor components. The same conclusion holds for the eigenoperators of the Hamiltonian $A_{q,j}$: The light-induced dynamics mixes all of them.

As discussed in Section 5.2, optical pumping can generate irreducible tensor components of all ranks. The interaction itself has rotational symmetry, thus generating only tensor elements of order zero. In the experiment discussed here, however, the laser beam is perpendicular to the magnetic field direction. In the coordinate system that we use, optical pumping does not conserve any quantum numbers.

Rotating frame

All experiments discussed in the following use modulated optical pumping. We therefore perform the analysis in a coordinate system rotating around the magnetic field direction at the modulation frequency. As shown in Section 5.3, the equations of motion in the rotating frame are the same as in the laboratory frame, if the optical pumping rate and the light shift effect are reduced by a factor that depends on the modulation scheme and is close to 1/2. In addition, the effective damping rate for the populations and for the coherences is somewhat bigger than in the case of a dc pulse.

The Hamiltonian for the rotating frame is obtained with the usual transformation

$$\mathcal{H}^{\mathrm{r}} = U^{-1}\mathcal{H}^{\mathrm{Lab}}U - \mathrm{i}\,\dot{U}^{-1}U \tag{9.19}$$

As we have oriented the z axis of the coordinate system along the magnetic field, the transformation operator is

$$U = \mathrm{e}^{-\mathrm{i}\omega_{\mathrm{m}}t\,F_z} \tag{9.20}$$

In direct analogy to the rotating frame transformation for the two-level system, we find that the precession frequencies for the density operator elements become

$$\omega_{qj}^{(\mathrm{r})} = q(\Omega_{\mathrm{L}} - \omega_{\mathrm{m}}) + \mu_{\mathrm{e}}^{(2)}\,B^2\,(m_F^2 - m'^2_F) \tag{9.21}$$

This reduction by $q\omega_{\mathrm{m}}$ of the effective frequency for tensor components of order q is completely equivalent to a reduction of the magnetic field by $\omega_{\mathrm{m}}/\mu_{\mathrm{e}}$. The factor q does not describe a faster rotation, but only the symmetry of the tensorial component during rotations around the symmetry axis.

Transfer of order

The equations of motion discussed above describe the evolution of the system driven by the laser field. If the system is initially in thermal equilibrium, the resulting dynamics correspond to transverse optical pumping, i.e. to the generation of order in the system. If, on the other hand, the system is already

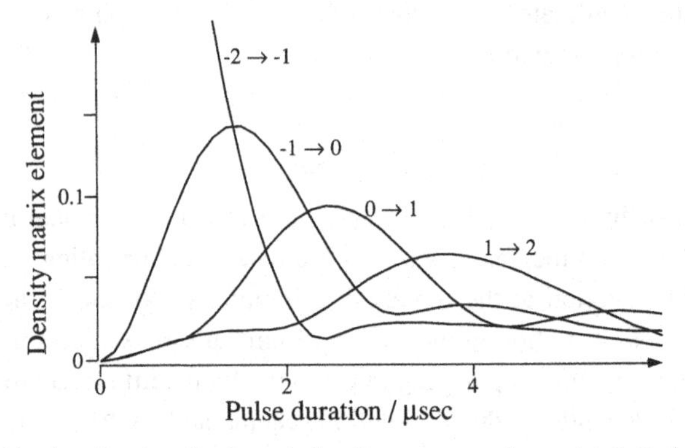

Figure 9.11. Amplitude of selected density operator elements plotted as a function of time during optical pumping. For $t = 0$, the only nonvanishing density operator element is in the $(2,-2 \leftrightarrow 2,-1)$ transition. The curves show the laser-induced transfer of coherence from this density matrix element to other transitions.

in an anisotropic state, the resulting process includes, in addition, an *exchange* of order between different multipoles.

Figure 9.11 demonstrates this process for a system driven by a pulse of circularly polarised light. In this calculation, the system is initially in a coherent superposition of the angular momentum substates $(F, m_F) = (2, -2)$ and $(2, -1)$. The curves in Figure 9.11 represent the amplitude of the density operator elements that correspond to transitions between neighbouring angular momentum substates. The laser pulse apparently causes a transfer of coherence among the different sublevel transitions. The combined effect of light shift and Zeeman interaction causes the oscillatory evolution, whereas the damping effect of the laser field is largely responsible for the overall decay. The different curves were calculated by numerical integration of the equations of motion for the optically driven system derived in Section 5.4.

9.2.5 Signal

Preparation and evolution

As discussed before, the initial laser pulse prepares the system into a state $\rho(0)$ where all (ground-state) density-operator elements are nonzero. The system is then allowed to precess freely under the influence of the magnetic field. Neglecting relaxation, we rewrite the general form of equation (1) in the operator basis $A_{q,j}$

$$\rho^{(\mathrm{r})}(t_1) = \mathrm{e}^{-\mathrm{i}\mathcal{H}^{(\mathrm{r})}t_1}\,\rho_0\,\mathrm{e}^{\mathrm{i}\mathcal{H}^{(\mathrm{r})}t_1} = \sum_{q,j} d_{q,j}\,\mathrm{e}^{\mathrm{i}\omega^{(\mathrm{r})}_{q,j}t_1}\,A_{q,j} \qquad (9.22)$$

In this and the following equations, the Hamiltonian and the density operator are expressed in the rotating coordinate system. For typographical clarity, however, we will now drop the index $^{(r)}$.

The second pump laser pulse concludes the free precession period and redistributes the coherence among the different multipole moments, as shown above. Here, we summarise its effect on the system by the matrix $\eta_{q'j',qj}$

$$\rho(t_1,0) = \sum_{q'\,j'} A_{q',j'} \sum_{q\,j} \eta_{q'j',qj}\, d_{qj}\, e^{i\omega_{q,j}t_1} \tag{9.23}$$

After the second pulse, the system again freely precesses as

$$\rho(t_1,t_2) = \sum_{q'\,j'} e^{i\omega_{q',j'}t_2}\, A_{q',j'} \sum_{q\,j} \eta_{q'j',qj}\, d_{q,j}\, e^{i\omega_{q,j}t_1} \tag{9.24}$$

The experimental setup measures the circular birefringence of the sample and produces a signal proportional to the magnetic dipole moment F_x. If the phase-sensitive detection measures the in-phase and the out-of-phase components, we can distinguish between positive and negative frequencies and selectively observe, e.g., the $T^{(1)}_{-1}$ component. We expand this observable in our operator basis as

$$T^{(1)}_{-1} = \sum_{j=1}^{2F} b_{-1j}\, A_{-1j} \tag{9.25}$$

writing the expansion coefficients as b_{-1j}. With this definition, the demodulated signal is

$$S(t_1,t_2) = \mathrm{Tr}\{T^{(1)}_{-1}\, \rho(t_1,t_2)\} = \sum_{j'} b_{-1,j'}\, e^{i\omega_{1,j'}t_2} \sum_{q\,j} \eta_{1j',qj}\, d_{q,j}\, e^{i\omega_{q,j}t_1} \tag{9.26}$$

where we have used the orthonormality of the basis operators A_{mk}, $\mathrm{tr}\{A_{q,j}^{\dagger}\, A_{q',j'}\} = \delta_{q,q'}\, \delta_{j,j'}$.

Effective observables

This expression lends itself to a reinterpretation if we write it as

$$S(t_1) = \sum_{qj} c_{q,j}\, d_{q,j}\, e^{i\omega_{q,j}t_1} = \mathrm{Tr}\{A_{\mathrm{eff}}\, \rho(t_1)\} \tag{9.27}$$

where the numbers

$$c_{q,j} = \sum_{j'} b_{-1,j'} \, e^{i\omega_{1,j'} t_2} \, \eta_{1j',qj} \tag{9.28}$$

represent the expansion coefficients of the effective observable

$$A_{\text{eff}}(t_2) = \sum_{q,j} c_{q,j} \, A_{-q,j} = \sum_{j'} b_{-1j'} \, e^{i\omega_{1,j'} t_2} \sum_{q,j} \eta_{1j',qj} \, A_{-q,j} \tag{9.29}$$

The operator $A_{\text{eff}}(t_2)$ can be considered as an "effective observable" for the time between the pulses. The expansion coefficients c_{qj} describe the sensitivity with which we can "monitor" the evolution of the multipole moments $A_{q,j}$; they depend on the matrix elements $\eta_{q'j',qj}$ that describe the efficiency of the transfer of sublevel coherence by the mixing pulse. All expansion coefficients c_{qj} may be nonzero and the signal $S(t_1)$ therefore contains contributions from all multipole moments. This method therefore allows observation of all density operator elements, circumventing the selection rules that restrict the values of the multipole order q appearing in the directly observable signal.

The sensitivity with which the individual components contribute to the signal, described by the $c_{q,j}$, depends also on t_2. It is therefore possible to choose a specific observable by measuring the system response at an arbitrary time t_2. In most cases, however, it is more convenient to record the signal as a function of the time t_2 after the second pulse and perform a Fourier transformation to get a sublevel spectrum as a function of the conjugate variable ω_2. This allows a separation of the individual terms contributing to the sum over j' in equation (9.29) if the corresponding resonance frequencies are nondegenerate. The effective observable is then

$$A_{\text{eff}}(\omega_2) = \sum_{q,j} \eta_{1j',qj} \, A_{-q,j} \, g(\omega_2 - \omega_{-1,j'}) \tag{9.30}$$

where $g(\omega_2)$ describes a complex Lorentzian resonance line centered at $\omega_{-1j'}$. With these effective observables, $A_{\text{eff}}(\omega_2)$ or $A_{\text{eff}}(t_2)$, it is possible to observe specific multipole components as a function of the precession time t_1, as shown by equation (9.27). This signal can be Fourier transformed with respect to t_1 to yield a spectrum with resonance lines at positions $\omega_{q,j}$ that have complex amplitudes $c_{q,j} \, d_{q,j}$.

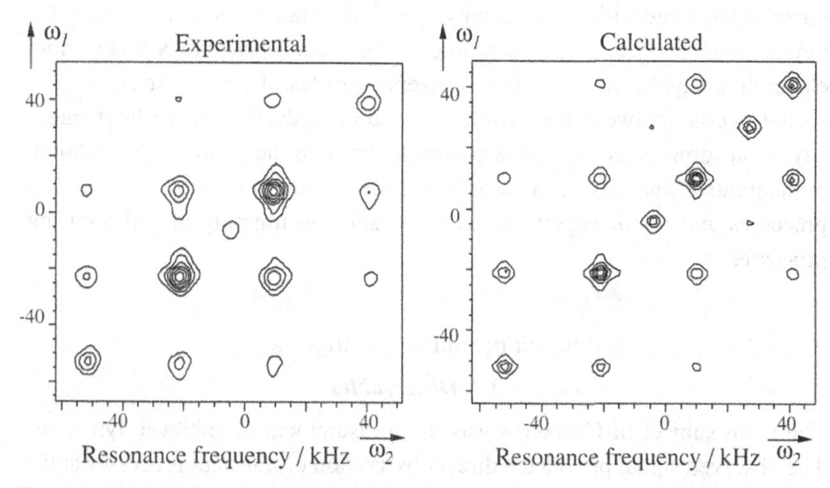

Figure 9.12. Comparison of measured (left) and calculated (right) 2D spectra. Each resonance line shows the transfer of coherence between two different sublevel transitions.

Example

Figure 9.12 shows an example of a resulting spectrum, together with the theoretical data obtained by numerically integrating the equations of motion for the system, starting from thermal equilibrium.

In this experiment, the duration of the second laser pulse was 3 μsec. The horizontal axis, labelled ω_2, corresponds to the precession frequency after the second pulse; the vertical axis, labelled ω_1, represents the precession frequency between the two laser pulses. Without a second laser pulse, the atomic multipoles would precess at the same frequency during the two free evolution periods and all resonance lines would appear on the diagonal, where $\omega_1 = \omega_2$. The mixing of the different multipole moments by the laser pulse, however, transfers coherences from one sublevel transition to another, thereby changing the precession frequency. In the spectrum, the so-called cross peaks with $\omega_1 \neq \omega_2$ give evidence for these processes: Each of these resonance lines corresponds to a different laser-induced coherence transfer process.

The theoretical and experimental data show qualitatively the same behaviour, including the overall timescale. For a more precise comparison between the theoretical and experimental spectra, it is necessary to consider the effect of the spatial dependence of the pump laser intensity. As discussed in Section 5.4.4, this inhomogeneity affects the light shift as well as the damping effect. The calculation takes this effect into account by averaging the result over a range of laser intensities. This is only a semiquantitative approach. Taking the inhomogeneous laser beam fully into account would require a much more elaborate analysis, including the transverse profiles of both laser beams, the spatial overlap between the two beams, and the reduction of the laser intensity by absorption as the beams propagate through the medium. In addition, a quantitative analysis must include a more detailed treatment of diffusion processes and nonlinear effects, such as radiation trapping or self-focusing processes.

9.3 "Forbidden" multipoles

9.3.1 Observables

The main subject of Chapter 8 was the measurement of sublevel dynamics: The observed signal provided a direct observation of transient processes at the microscopic level. The impact of this circumstance depends to a large degree on which variables of the system can be observed in such an experiment. The polarisation-selective detection of transmitted light used in Chapter 8 provided access to the angular momentum component along the laser beam direction. Three laser beams in mutually orthogonal directions can thus observe all three angular momentum components. For the $J = 1/2$ system, these three components characterise the system completely. In this case, the polarisation-selective detection can tell us everything there is to know about the microscopic state of the system.

Density operator components

Most atomic systems consist of more than two angular momentum states. In these systems, optical pumping can excite not only a magnetic dipole moment, but also various higher order moments. Excitation of higher order moments requires higher order processes; its efficiency depends on the laser intensity as well as on the decay rates.

Figure 9.13 shows such a creation of four density operator components. They correspond to tensor components of order $q = 1, 2, 3$ and 4, and are indicated in the density operator near the top of the figure. The curves, which represent their amplitudes, were calculated by numerically integrating the equations of motion for modulated optical pumping in a transverse magnetic field. The parameters used for the calculation correspond to a magnetic field strength $B_0 = 0.7$ mT, a laser intensity of 90 mW/mm^2 and a laser detuning $\Delta/2\pi = 12$ GHz. As Figure 9.13 shows, the element in the transition ($F = 2$, $m_F = -2 \leftrightarrow F = 2$, $m_F = -1$), which belongs to a tensor of order $q = 1$,

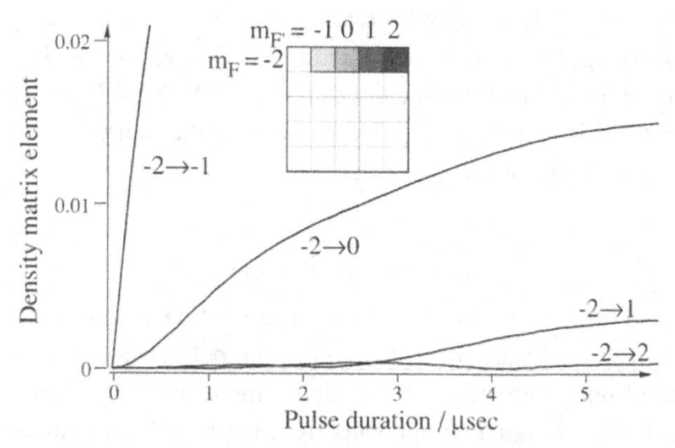

Figure 9.13. Amplitude of selected density operator elements plotted as a function of time during optical pumping. The system is initially in thermal equilibrium and the curves represent the excitation of various atomic multipoles.

starts to grow first and at the highest rate. The process that excites this tensor element is clearly a first-order process. This behaviour is in good agreement with the naive expectation that angular momentum transfer from the photons should first excite those tensor elements that represent magnetic dipole elements. The other tensor elements do not start linearly in time, indicating that they are excited only through multiple absorption processes. In particular the hexadecupole element with $q = 4$ remains small even for long pulse durations. High optical pump rates are required significantly to populate these high order elements.

Not all these multipole moments have an effect on the optical properties of the system. The next higher terms in the multipole expansion, the quadrupole moment, can cause birefringence and dichroism, as discussed in Section 6.4. In excited states, where the multipoles can be monitored through the fluorescence of the medium, the choice of the polarisation of pump light and fluorescence allows selective observation of different components up to rank $k = 2$ (Yodh, Mossberg and Thomas 1986; Vedenin et al. 1989). Using spectroscopic information from experiments of this type, it is possible to study, e.g., the polarisability (Kulina and Rinkleff 1982) and the relaxation (Ghosh et al. 1985; Lowe et al. 1987; McLean, Hannaford and Lowe 1990) of atomic multipoles.

Multipole moments with rank $k > 2$ do not influence the optical properties of an atomic medium and are therefore not directly observable. In systems with total angular momentum $F > 1$, the linear optical properties of a system cannot completely characterise its microscopic state. This situation may be discussed in terms of the density operator of the system. An $F = 2$ system has a 5×5 density operator consisting of 25 independent elements. Since the optical susceptibility of this system can be characterised by a 3×3 tensor with nine elements, it is clearly impossible to observe the complete internal dynamics through linear optical processes.

Effective observables

Time-resolved measurements can extend this limitation somewhat. In the case of the sodium ground state presented above, the polarisation-selective detection measured one component of the dipole moment. The spectral analysis separates it into four signal components from the $F = 2$ multiplet and two of the $F = 1$ multiplet. These six signal components originate from six density operator components. Through the measurement of a single quantity, we can apparently obtain information on six distinct density operator elements. This is possible because the experiment involves several measurements at different times and the density operator elements evolve differently under the internal Hamiltonian. A more detailed analysis shows that the quadratic Zeeman effect of this system converts the dipole moment into higher rank moments at a rate that depends on the frequency difference between the resonance lines in the spectrum. At different times, these density operator elements contribute differently to the dipole moment of the system. This distinct time dependence makes them distinguishable, even if we observe only a single quantity. Many density operator elements remain unobservable, however, as this extension is limited to tensor elements of order $q = \pm 1$. A complete analysis of the internal dynamics of the system therefore requires other techniques.

One such technique, based on two-dimensional spectroscopy, uses the coherence transfer processes introduced above. The effective observables of two-dimensional spectroscopy make it possible to circumvent the usual selection rules and observe all density operator components with a sensitivity linear in the probe-beam intensity. Thus the perturbation of the system by the probe beam can be kept small without sacrificing too much sensitivity. The technique permits the observation of not only the different multipole components in a static manner, but also their evolution as a function of time. Furthermore, it permits the excitation of atomic multipole moments in specific, experimentally controllable, spatial orientations as well as the measurement of their spatial orientation. Comparison of the orientation dependence of the different signal components permits the separation of signal contributions according to the order of the multipole moments.

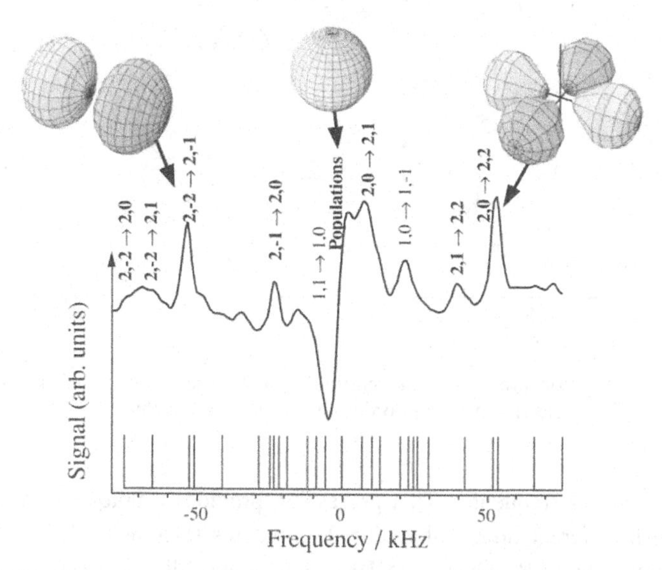

Figure 9.14. Example of an indirectly detected spectrum of the Na ground state showing "forbidden" multipole moments.

Figure 9.14 shows an example of a spectrum that can be obtained with these effective observables. The central part is the experimental spectrum of the sodium ground state, extracted from a two-dimensional spectrum. The stick spectrum at the bottom indicates the possible transition frequencies. Some of the more prominent resonances are assigned to the transitions between individual Zeeman substates. The presence of different orders of tensor elements indicates that the technique is indeed sensitive to higher order multipoles.

9.3.2 Rotations

In a system with so many degrees of freedom, any possibility of distinguishing between different signal components is also important for spectroscopic applications. For a system with axial symmetry, like the atom in a magnetic field, it is often helpful to exploit the rotational symmetry. This procedure uses the definition of the multipole moments, i.e., their different behaviour under rotations.

Larmor precession

The simplest possibility to rotate the atomic multipole moments consists in applying a magnetic field. In the case of linear Zeeman interaction, all multipole moments precess around the direction of the magnetic field at the same

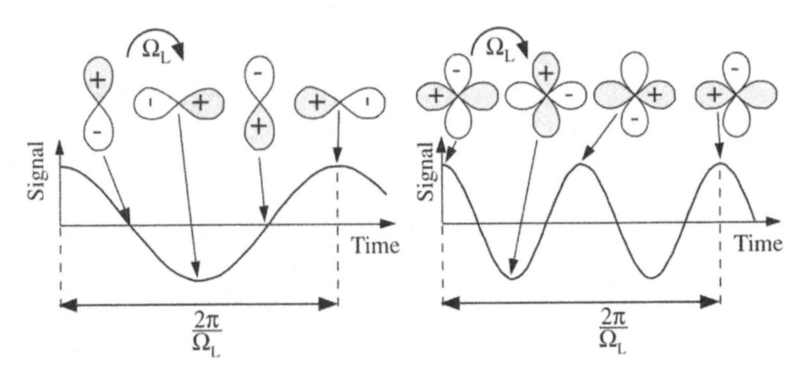

Figure 9.15. Larmor precession of multipoles and observable signal for dipoles (left) and second-order tensor elements (right).

rate, the Larmor frequency. This precession provides a unique signature to distinguish different orders of multipole moments (Dehmelt 1957a; Happer and Mathur 1967b; Pancharatnam 1968; Lange and Mlynek 1978).

As Figure 9.15 shows, this possibility derives directly from the definition of irreducible tensor operators. A tensor component of order q is invariant under rotations by $2\pi/q$. The left-hand side shows the precession of a magnetic dipole moment: The curve represents the observed signal, which is proportional to one component of the dipole moment; the upper part indicates the orientation of the dipole moment at various instances. The time-dependent phase of the signal is a direct representation of the dipole orientation.

The right-hand side shows the same for a second order tensor element. The precession frequency is the same as for the dipole, but a 90° rotation of this multipole moment corresponds to a multiplication with -1, and a 180° rotation turns it into a state indistinguishable from the initial state. The observed signal therefore oscillates at twice the Larmor frequency. Again, the signal phase is a direct measure of the orientation. For a general multipole moment of order q, the signal contribution must be periodic with period $2\pi/(q\ \Omega_L)$, $1/q$ times the Larmor period. Fourier analysis of the signal can separate the time-dependent signal contributions from different tensor components.

Rotation by phase shifts

A rotation of the effective Hamiltonian around the quantisation axis can be implemented by a rotation of the experimental apparatus. Using modulated optical excitation, however, there is a much simpler way to achieve the same goal: It is only necessary to shift the phase of the modulation. To see this, consider the effective Hamiltonian of an optically pumped system in the laboratory frame. According to Section 5.3, it reads

$$\mathcal{H}_{\mathrm{mod}}(\phi) = 2\omega_x \,[1 + \cos(\omega_{\mathrm{m}} t - \phi)]\, F_x - \Omega_{\mathrm{L}}\, F_z \tag{9.31}$$

where ω_{m} is the frequency and ϕ the phase of the modulation, and the z axis is aligned with the magnetic field. In the rotating frame, defined by the unitary transformation $U = \mathrm{e}^{\mathrm{i}\omega_{\mathrm{m}} t F_z}$, this Hamiltonian becomes

$$\mathcal{H}^{\mathrm{r}}(\phi) = \mathrm{e}^{-\mathrm{i}\phi F_z}\, \mathcal{H}^{\mathrm{r}}(0)\, \mathrm{e}^{\mathrm{i}\phi F_z} = -(\Omega_{\mathrm{L}} - \omega_{\mathrm{m}})\, F_z + \omega_x\, (F_x\, cos\phi + F_y \sin\phi) \tag{9.32}$$

where $\mathcal{H}^{\mathrm{r}}(0)$ is the rotating frame Hamiltonian for the reference phase derived in Section 5.3.3. Apparently, the shift of the modulation phase in the laboratory frame corresponds to a (time-independent) rotation around the magnetic field. In the following, we will describe such rotations by an angle ϕ around the z axis as

$$A(\phi) = \mathrm{e}^{-\mathrm{i}\phi F_z}\, A(0)\, \mathrm{e}^{\mathrm{i}\phi F_z} \tag{9.33}$$

To find the effect of the phase shift on the density operator, we calculate the evolution under a rotated Hamiltonian $\mathcal{H}^{\mathrm{r}}(\phi)$. Starting from thermal equilibrium, the system evolves into a state

$$\rho(t;\phi) = \mathrm{e}^{-\mathrm{i}\mathcal{H}^{\mathrm{r}}(\phi)t}\, \rho(0)\, \mathrm{e}^{\mathrm{i}\mathcal{H}^{\mathrm{r}}(\phi)t} = \mathrm{e}^{-\mathrm{i}\phi F_z}\, \rho(t;\phi = 0)\, \mathrm{e}^{\mathrm{i}\phi F_z} \tag{9.34}$$

where the invariance of the thermal equilibrium density operator ρ_{eq} under rotations was used. The final atomic state $\rho(t;\phi)$ is rotated by the same angle ϕ as the Hamiltonian. We can therefore control the orientation of the atomic multipoles experimentally: Shifting the phase of the radio frequency by $\Delta\phi$ rotates the atomic state by an angle $\Delta\phi$. It is straightforward to show that the same holds not only for the Hamiltonian evolution, but also for the optical pumping process.

In principle, it is possible not only to rotate the density operator around the z axis, but also to tilt the orientation of the effective field away from the z axis (Suter and Pearson 1988), as Figure 9.16 illustrates for a specific multipole component. Such a procedure is difficult to implement, however, in a purely optical experiment, in such a way that it changes only the angle θ without affecting the other parameters. We therefore consider only rotations around the z axis, which have the most direct implications. They correspond quan-

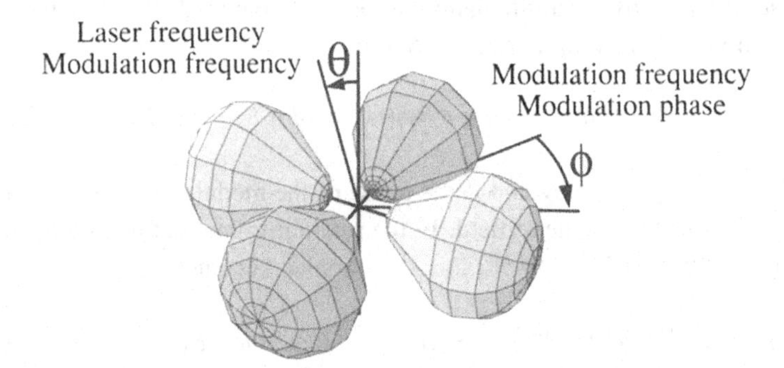

Figure 9.16. Control of the orientation of optically excited multipole moments by different experimental parameters.

tum mechanically to a multiplication of the individual multipole moments by a phase factor proportional to the order of the component. Signal components due to these multipole components are multiplied by the same phase factors. The measurement of the signal phase for different atomic states therefore allows determination of the orientation of the individual components as well as the assignment of signal components to specific multipoles.

Rotated density operator

This rotation affects the density operator not only during the pulse. In the two-pulse experiment, a shift of the phase of the first pulse affects the density operator during the evolution period as

$$\rho(t_1;\phi) = \mathrm{e}^{-\mathrm{i}\phi F_z}\, \rho(t_1;0)\, \mathrm{e}^{\mathrm{i}\phi F_z} = \sum_{qj} \mathrm{e}^{\mathrm{i}q\phi}\, d_{q,j}\, \mathrm{e}^{\mathrm{i}\omega_{q,j}t_1}\, A_{q,j} \qquad (9.35)$$

The density operator components are now labelled with a phase $q\phi$, which is proportional to the tensorial order q. The same factor also appears in the observed signal

$$S(t_1,\phi) = \sum_{qj} c_{q,j}\, d_{q,j}\, \mathrm{e}^{\mathrm{i}q\phi}\, \mathrm{e}^{\mathrm{i}\omega_{q,j}t_1} \qquad (9.36)$$

Signal components associated with a density operator component of order q are phase shifted by $q\phi$ if the phase of the pump modulation is shifted by ϕ. This allows a direct determination of the tensorial order as well as the orien-

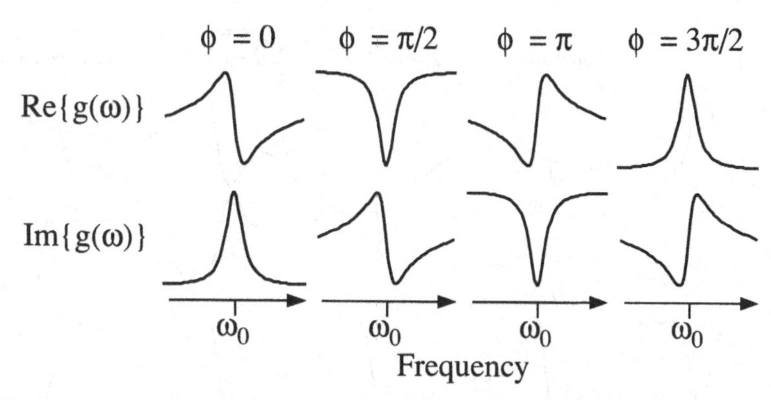

Figure 9.17. Changes in the observed lineshape as a function of signal phase ϕ for a complex Lorentzian lineshape.

tation of this multipole moment, provided this phase factor can be observed experimentally.

Measurement

The signal phase can be most easily measured in sublevel spectra, where the signal phase modifies the lineshape. Figure 9.17 illustrates the expected lineshapes as a function of the signal phase for the case of a Lorentzian lineshape. The upper row shows the real part of the complex lineshape function, the lower row the imaginary part. If the phase changes by 90°, the lineshape of the real part changes from dispersionlike to absorptionlike, and the imaginary part changes from an absorptive to a dispersive shape. A 180° phase shift inverts the signal and intermediate phases lead to mixed phase lines. Their phase can be extracted by lineshape analysis or by multiplication of the observed lineshapes with an additional phase factor to transform the observed line shape into a pure absorption or dispersion line.

Figure 9.18 shows some representative signal components from experimentally observed spectra of the sodium ground state. Each column shows one resonance line associated with a different tensorial order. The assignment of the resonance lines to different sublevel transitions was achieved by comparing the experimentally determined precession frequencies to the calculated transition frequencies. The top row shows the signals for the phase of the pump beam modulation that serves as the reference. The other rows were measured with phase shifts of $\pi/2$, π, and $3\pi/2$, as noted. The four rows correspond to four orientations of the rotating frame density operator along the x, y, $-x$, and $-y$ axes.

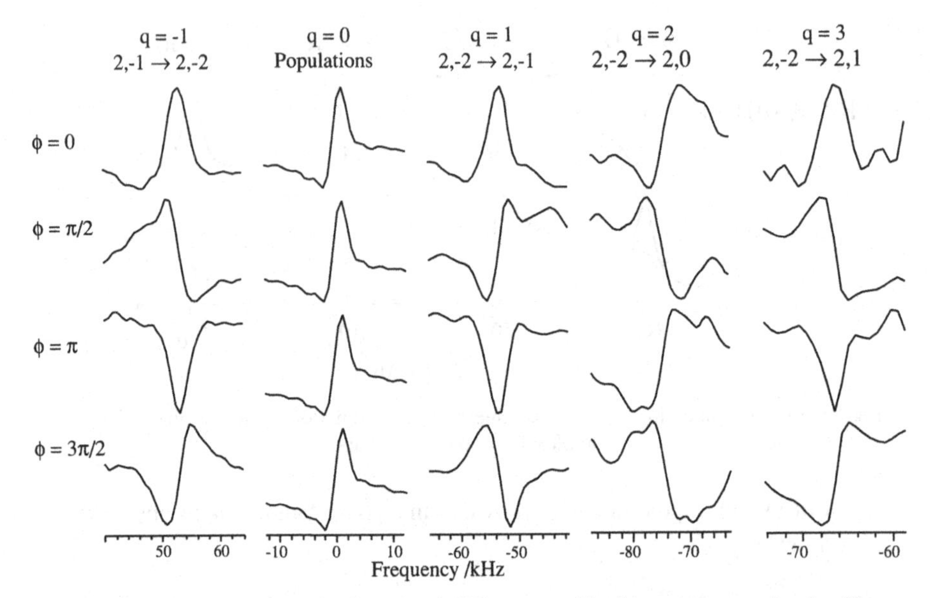

Figure 9.18. Resonance lineshapes of different multipole components in the Na ground state recorded with different modulation phases during the first pulse. Each column represents one resonance line for four different phases. The numbers above each column indicate the assignment to the individual sublevel transitions and the order of the corresponding multipole moment.

As explained above, we expect to see an effect of the phase shift on the different resonances proportional to the tensorial order of the corresponding multipoles. This prediction is checked most easily for the resonance line at the origin of the frequency axis. It originates from the populations of the sublevels, which are invariant under rotations ($q = 0$). The data presented in Figure 9.18 clearly show that the shape of this resonance line is virtually identical in the four experiments, in good agreement with the theoretical expectations.

The behaviour of the line with $q = 2$ is also relatively simple to analyse. The line inverts whenever the modulation phase is shifted by $\pi/2$. The corresponding multipole moments obviously have a periodicity of π. The resonance lines labelled $q = \pm 1$ change the lineshape from approximately absorptionlike to dispersionlike when the modulation phase is shifted by $\pi/2$. The periodicity of this signal is 2π, as expected for a first-order tensor element. The resonance line associated with the third order tensor appears to have the same behaviour as the $q = -1$ term. Theoretically, we expect that a shift of the modulation phase by $\pi/2$ should change the phase of this resonance

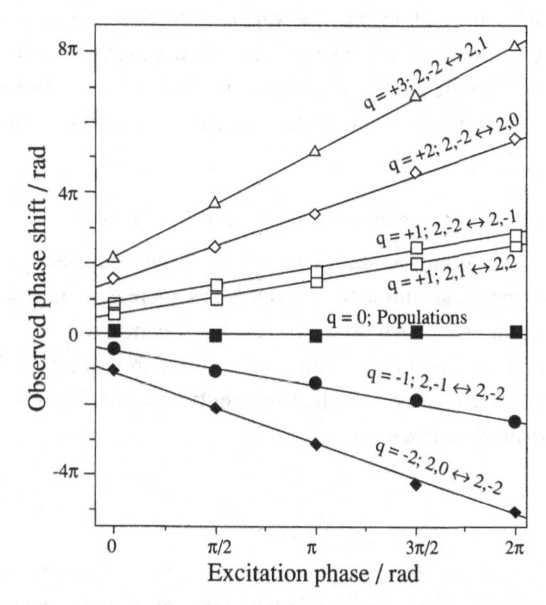

Figure 9.19. Phase of the resonance lines as a function of the excitation phase. The solid line represents the theoretical behaviour; the circles, squares, and triangles show the experimental values.

line by $3\pi/2$; however, such a phase change is indistinguishable from the $-\pi/2$ phase shift observed for the $q = -1$ term. To distinguish these two tensor elements, shifts of the modulation phase by less than $\pi/2$ are needed. More generally, distinguishing tensors of order q, q' requires rotation angles smaller than $2\pi/(q - q')$.

Comparison of multipole orders

Figure 9.19 analyses the phase variation more systematically by comparing the expected (straight lines) and measured (squares, circles, and triangles) phases for a representative set of multipole components as a function of the modulation phase. Four different orientations of the atomic system were prepared by modulated optical pumping. For each orientation, the phases of the resonance lines were measured. When necessary, the measured phase was incremented by multiples of 2π to bring it from the range $[0, 2\pi]$ to the theoretically expected range $[2\pi n, 2\pi(n + 1)]$. The uncertainties in the measured data are highest for the weak resonance lines ($q = 2, 3$), but remain smaller than $\sim\pi/10$. The offset of the observed values, i.e., the phase at the reference orientation, is arbitrary and depends on various experimental parameters. The slope of the lines, however, demonstrates that the comparison of signal phases for atomic states with different orientations allows a determination of the tensorial order of the multipole components that cause these signal contributions.

9.3.3 Separation of multipole orders

In a spectrum of an atomic system with nonvanishing nuclear spin, like that of Figure 9.14, the large number of resonances increases the likelihood of overlap between the resonance lines. In crowded spectra, overlap between different resonance lines can make it impossible to separate the signal contributions and permit an assignment to the sublevel transitions. In those cases, additional means for characterising signal contributions are often helpful.

Procedure

The most direct method for this purpose uses the rotations discussed here. Consider a sequence of experiments, in which the modulation phase takes the values $\phi = p\ \pi/2$, with p integer. The individual experiments prepare the system into states that are rotated by multiples of $\pi/2$. As we have seen, this rotation changes the phase of the individual multipoles by a factor $e^{iq\phi}$. This different behaviour allows separation of the different orders by Fourier analysis: We calculate linear combinations, using weighting coefficients c_{qp} that depend on the order q of the tensor component we want to extract, and on the excitation phase:

$$c_{qp} = e^{i\,q\,p\,\pi/2} \tag{9.37}$$

Figure 9.20 shows the spectra that can be obtained with this procedure. The top trace represents a single experiment, which includes contributions from all possible multipole moments. The second trace shows the orientation-independent signal component, obtained by adding the four experimental spectra with equal weights, $c_{qp} = 1$. It contains the signal components originating from $q = 0$ multipole components, i.e., from the population differences. They are invariant under rotations like the Larmor precession and therefore appear at the origin of the frequency axis, $\omega = 0$. The $q = \pm 4$ components should also appear in this spectrum, since a rotation by multiples of $\pi/2$ brings them into a state indistinguishable from its initial orientation. The intensity of these signal contributions is too small, however, to be seen in this spectrum. The third trace contains the $q = 1, -3$ coherences; the strongest signals, due to

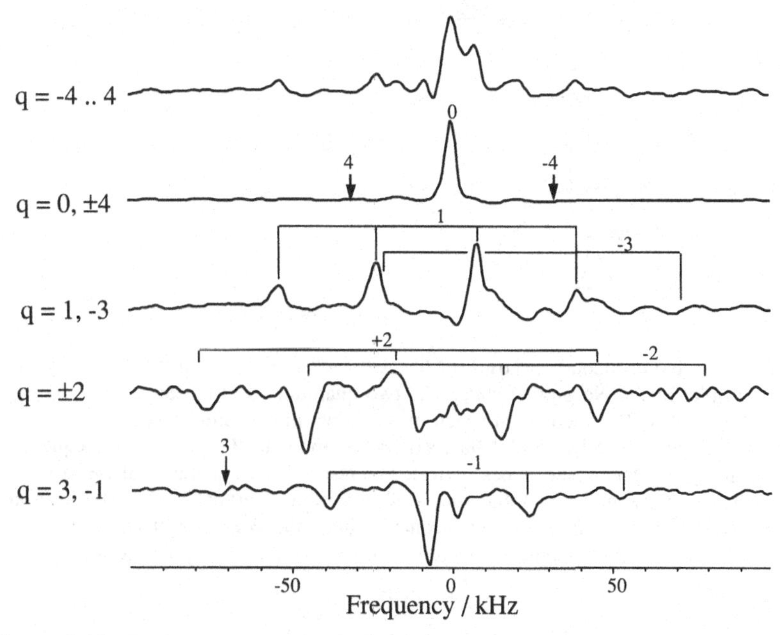

Figure 9.20. Sublevel spectra for individual multipole orders obtained by linear combination of four spectra. The four experiments used modulation phases of 0, $\pi/2$, π, and $3\pi/2$.

$q = 1$ components, are those that appear in the normal one-dimensional experiment. The fourth trace contains the $q = \pm 2$ subspectrum with three transitions from both tensorial orders. By comparing this subspectrum with the unedited spectrum at the top of the figure, we see clearly that this separation of subspectra makes it possible to observe the weak resonance lines associated with $q = \pm 2$ operators that are just barely visible in the unedited spectrum.

Two-quantum transitions

The individual spectra represented in Figure 9.20 are cross sections through a two-dimensional spectrum. Figure 9.21 shows a complete 2D spectrum in contour plot representation. The data presented in the figure are a linear combination of four separately recorded spectra, using modulation phases (0, $\pi/2$, π and $3\pi/2$) and the coefficients (1, -1, 1, -1) for the linear combination.

The frequency variable, ω_1, drawn along the vertical axis, measures the precession frequency during the evolution period, when we can observe all mul-

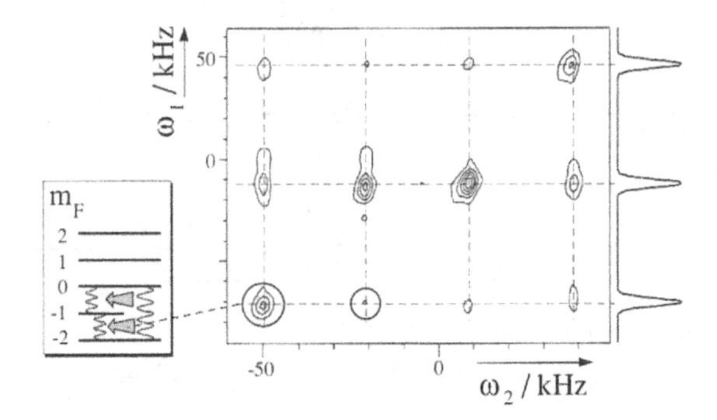

Figure 9.21. Experimental spectrum of the $\Delta m_F = 2$ transitions in the $F = 2$ hyperfine multiplet of the Na ground state. The two-quantum spectrum appears along the vertical (ω_1) axis. The theoretical spectrum is shown in one-dimensional form parallel to the ω_1 axis. The horizontal (ω_2) axis corresponds to the precession frequency during detection. The dashed lines indicate the resonance frequencies of the $\Delta m_F = 2$ transitions (horizontal lines) and $\Delta m_F = 1$ transitions (vertical lines). The inset at the left shows for the two marked resonance lines the coherence transfer process that turns the two-quantum coherence into an observable dipole moment.

tipole orders. For this particular linear combination, though, we expect to find only transitions with $\Delta m_F = \pm 2$. The theoretical two-quantum spectrum of the $F = 2$ multiplet appears in one-dimensional form on the right-hand side of Figure 9.21 and the horizontal dashed lines indicate the positions of the corresponding resonance lines.

The frequency variable ω_2, drawn along the horizontal axis, describes the precession frequency during the detection period. For this period, the selection rule $\Delta m_F = \pm 1$ remains valid and we expect to find the usual one-dimensional spectrum in this direction. The vertical dashed lines indicate the positions of these transitions. Each of the twelve resonance lines in the two-dimensional spectrum indicates the transfer of coherence from a two-quantum ($\Delta m_F = 2$) transition into a single quantum ($\Delta m_F = 1$) transition. The inset at the left shows for the two resonance lines marked with a circle the underlying coherence transfer process. Both signal contributions arise as double quantum coherence in the transition $m_F = 0 \leftrightarrow m_F = -2$. The mixing pulse transfers them into single quantum transitions. The component at the far left ends up in the transition $m_F = -1 \leftrightarrow m_F = -2$, whereas the second component is transferred into the transition $m_F = -0 \leftrightarrow m_F = -1$.

In such an experiment, where two-quantum transitions along the vertical are associated with single quantum transitions along the horizontal dimension, the spectrum is clearly asymmetric, in contrast to the spectra shown in Figure 9.12, where both dimensions correspond to single quantum ($\Delta m_F = 1$) transitions. This asymmetry characterises not only the positions of the lines; their shapes are asymmetric as well. Diffusion causes a similar line broadening in both dimensions, but magnetic field inhomogeneity affects the two-quantum transitions twice as strongly as the single quantum transition (as discussed in the following section). Accordingly, the resonance lines tend to be wider in the vertical dimension. Strong magnetic field inhomogeneity can cause significant distortion of the resonance lines, resulting in line shapes elliptically elongated along the axis $2\omega_1 = \omega_2$.

9.3.4 Coherence transfer echoes

Coherence transfer echoes were briefly mentioned in the introduction as an example of a phenomenon that depends on laser induced coherence transfer. It is not a two-dimensional experiment, although the experimental scheme is the same: A sequence of two pump laser pulses excites the system, and the response after the second laser pulse is observed (Suter, Rosatzin and Mlynek 1991c). The physical processes that cause the coherence transfer echoes are the same as those in the two-dimensional experiment. The difference is mainly the presence of an inhomogeneous magnetic field and the different analysis of the data. We include here a discussion of the experiment, using the formalism developed for the two-dimensional experiment. In addition, we discuss an experimental verification of the proposed mechanism for the formation of the multiple echoes.

Mechanism

Since the experimental scheme is the same as for two-dimensional spectroscopy, we can use the formalism developed above for the description of coherence transfer processes. Starting from equation (9.22), describe the system before the second laser pulse as

$$\rho(T) = \sum_{q\,j} d_{q,j}\, e^{i\omega_{q,j}T} A_{qj} \tag{9.38}$$

where T is the separation of the two pulses and relaxation mechanisms are neglected. The system evolves in a magnetic field $\vec{B} = (0, 0, B)$, and we assume that the field strength is low enough that only the linear Zeeman effect is relevant. The precession frequency of the individual tensor components is $\omega_{q,j} = q\Omega_L$, where Ω_L is the Larmor frequency.

As described before, the second pulse changes the evolution of the system through light shift and optical pumping and we summarise the coherence transfer process by the matrix η. After the pulse, the density operator is

$$\rho(T+) = \sum_{q'\,j'} \sum_{q\,j} \eta_{q'j',qj}\, d_{q,j}\, \mathrm{e}^{-\mathrm{i}q\Omega_{\mathrm{L}}T}\, A_{q',j'} \tag{9.39}$$

The system precesses freely again, so that at a time t after the second pulse, the density operator becomes

$$\rho(T;t) = \sum_{q'\,j'} \sum_{q\,j} \eta_{q'j',qj}\, d_{q,j}\, \mathrm{e}^{-\mathrm{i}\Omega_{\mathrm{L}}(Tq+tq')}\, A_{q',j'} \tag{9.40}$$

The effect of the inhomogeneous magnetic field is a variation of the magnetic field strength and therefore of the phase factors in equation (9.40). If the inhomogeneity $\Delta\Omega_{\mathrm{L}}$ is large enough ($\Delta\Omega_{\mathrm{L}} > 1/T$), the observable signal vanishes unless the phase factor is identically zero. Echoes may therefore occur at $t = -T\, q/q'$. If we observe the magnetic dipole moment, the order of the observable is $q' = \pm 1$ and echoes should appear at $t = n\,T$ ($n = 0, \ldots, 2F$), i.e., at multiples of the pulse separation time T after the second pulse. The first echo should appear during the second pulse, whereas the last echo should occur at $2F$ times the pulse separation.

Interpretation of results

Using this mechanism, we may interpret the experimental results of Figure 9.6 as follows: The simultaneous action of the laser field and the magnetic field during the first laser pulse creates atomic multipole moments of order q with $-4 \leq q \leq 4$. After the end of the pulse, the coherences precess in the transverse magnetic field, each at its characteristic precession frequency $q\Omega_{\mathrm{L}}$. The time evolution of the one-quantum coherence leads to a free induction decay signal. The second pulse causes a redistribution of these coherences, which subsequently evolve with a different frequency. The first "echo," for $n = 0$ corresponds to a conversion of population ($q = 0$), which was not affected by the inhomogeneous magnetic field, into observable coherence ($q' = \pm 1$). It produces an observable signal immediately after the pulse and is therefore observed as a second free induction decay.

In an inhomogeneous magnetic field, multipole components dephase at rates proportional to their order q, as shown in the lower part of Figure 9.22. The

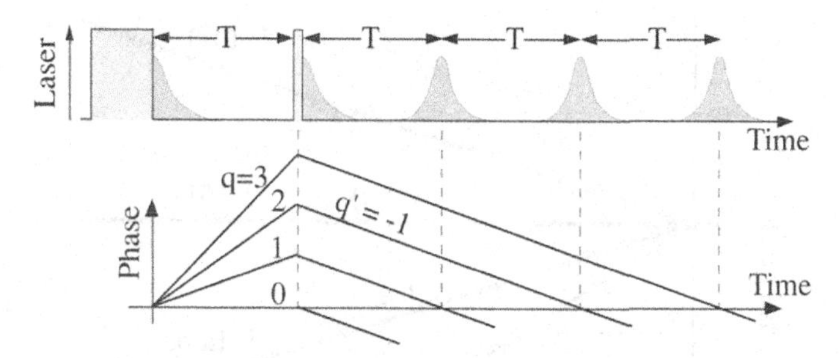

Figure 9.22. Mechanism of coherence transfer echoes.

transfer into visible coherence of order $q' = -1$ causes the phase to decrease at the same rate for all components. The time at which they form echoes is thus a multiple of the tensorial order during the defocusing period. The echoes corresponding to $q = 0, \ldots, 3$ are easily visible in the experimental data of Figure 9.6; the echo corresponding to $q = 4$, which is theoretically possible in the Na ground state, is too weak to be seen.

Orientation dependence

According to this mechanism, the echo at time $t = -qT$ should "remember" any characteristics of the corresponding multipoles from the time between the two pulses. This prediction suggests the following experiment to confirm the proposed mechanism: Using modulated excitation, it is possible to multiply tensor components of order q with a phase factor $e^{iq\phi}$, where ϕ is the phase of the modulation. The initial density operator is then

$$\rho(0;\phi) = \sum_{qj} d_{q,j}\, e^{iq\phi}\, A_{q,j} \tag{9.41}$$

As a scalar, this phase factor is not affected by the subsequent evolution, and can be used to track the coherence through the experiment. After the refocusing pulse, the density operator is

$$\rho(T;t;\phi) = \sum_{q'j'} \sum_{qj} \eta_{q'j',qj}\, d_{q,j}\, e^{-i\Omega_L(Tq+tq')}\, e^{iq\phi}\, A_{q',j} \tag{9.42}$$

The phase factor therefore depends on the tensorial order q during the evolution time, not during the detection time. By systematic variation of the exci-

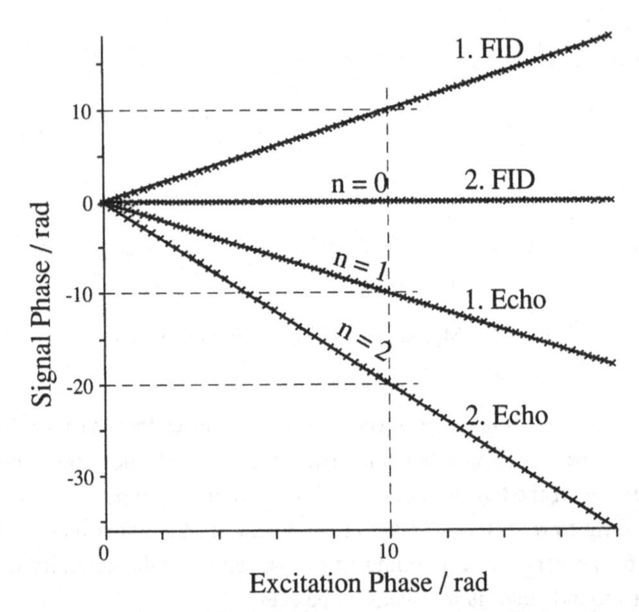

Figure 9.23. Phases of the various signal components vs. excitation phase. Crosses represent experimental points; the full lines show the predicted behaviour. The dashed lines help to compare the slope of the different curves.

tation phase ϕ, it is possible to separate the different tensor components of the coherence, e.g., by Fourier analysis.

Figure 9.23 shows the observed phases from such an experiment, using modulated excitation and systematically incrementing the modulation phase during the first laser pulse. The modulation phase during the second pulse was constant between experiments. The data labelled "1.FID" and "2.FID" refer to the free induction decay following the excitation and refocusing pulse. The first and second echoes occurred at times T, $2T$ after the refocusing pulse. Figure 9.23 shows that the phase of the first FID was equal to the modulation phase, confirming the rotation of the density operator prepared by the modulated pulse. The second FID signal did not depend on the modulation phase since the magnetisation that caused this signal was created during the second pulse. The first echo is a normal Hahn echo and its phase is the inverse of the excitation phase. The magnetisation that appears during the second echo was created as double quantum coherence that acquires twice the phase of single quantum coherence. The second echo therefore appears with twice the phase of the first echo. These experimental results corroborate the hypothesis that the observed multiple echoes are due to a laser-induced coherence transfer.

This mechanism for the observed multiple echoes agrees with the extended form of the "theorem on coherent transients" by Schenzle and colleagues (Schenzle, Wong and Brewer 1981). It states that for this mechanism, where the defocusing is faster than the refocusing process, the signal should end at $4T$. For an infinitely broadened spin transition, this condition is obviously fulfilled. There is a certain analogy between these coherence transfer echoes and the three-level echoes discussed in Section 3.3.3, but the physical mechanism driving the processes is radically different. The main difference is that in the case of the "three-level echoes," the coherences are created and observed in optical transitions directly coupled to the radiation field. In the coherence transfer echo experiment, in contrast, the dephasing and rephasing of the coherences occur in sublevel transitions that do not interact with the laser beam. The refocusing process is initiated by higher order effects, in particular the light-shift effect.

10

Nonlinear dynamics

In all preceding chapters, we have tried to separate the effect of the light on the atoms from the reverse effect, the modification of the light by the atomic medium. In most cases this allows an explicit solution of the equations of motion and provides stable stationary states. In recent years, another approach to the interaction between light and matter has appeared, which emphasises strong coupling between the two. It uses the interactions as a tool to study general aspects of nonlinear dynamics. This chapter provides an introduction to this field, without trying to give a complete summary.

10.1 Overview

10.1.1 Resonant vapours as optically nonlinear media

Mutual interactions

So far, we have considered the effect of the laser on the atomic medium separately from the measurement of microscopic dynamics by the polarisation selective detection of transmitted light. In this approximation, the pump laser drives the atoms without suffering significant attenuation or polarisation changes. Conversely, the probe beam, which monitors the optical properties of the atomic medium, changes the microscopic state of the atoms only infinitesimally. This approach guarantees, e.g., that the response of the medium to the probe beam is linear. As stressed before, this assumption of one-sided interactions is always an approximation, since the conservation of energy and angular momentum make it impossible to change either of the two partial systems without compensating changes in the other part. Nevertheless, it is always possible to come arbitrarily close to this ideal situation by adjusting the light intensity and/or the particle density.

Figure 10.1 illustrates these conditions graphically: At low intensity and low particle density, the two partial systems influence each other only weakly. Ideal conditions for optical detection of microscopic properties include a par-

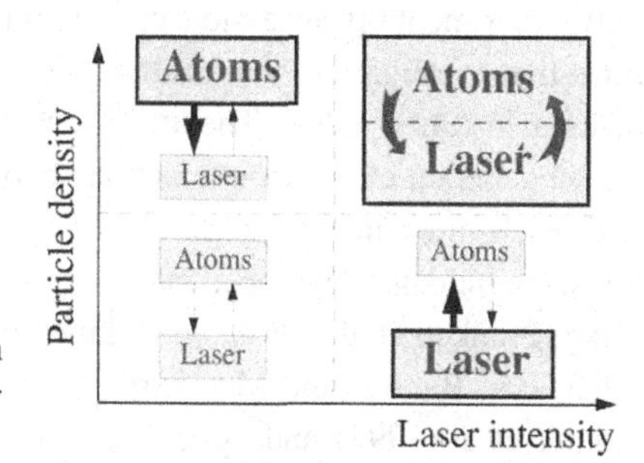

Figure 10.1. Different types of interaction as a function of laser intensity and particle density.

ticle density high enough that its effect on the laser beam is easily observable and a laser intensity low enough that the atoms absorb on average less than one photon during the lifetime of the state that we wish to observe. For ground-state atoms in a vapour cell, confined by buffer gas, this lifetime is of the order of one msec. Under these conditions, the atoms absorb approximately one photon per msec if the laser intensity is 15 μW/mm^2 and its frequency is at the optical resonance. A laser beam whose intensity is below this limit is a good approximation of an ideal probe. Conversely, the laser beam is not significantly attenuated if the number of atoms it encounters during its passage through the sample is less than the ratio A/σ of beam cross section A to the absorption cross section σ of the atoms. Under the conditions specified above, the absorption cross section for an optically allowed transition in an isotropic medium is some $2 \cdot 10^{-16}$ m^2. For a cell length of 4 cm, a particle density of $6.7 \cdot 10^6$/mm^3 causes an attenuation of 10%.

If the experimental conditions are outside these limits, the system starts to show deviations from the idealised behaviour discussed above. If both conditions are violated, corresponding to the upper right corner in Figure 10.1, the behaviour may change qualitatively: The effect of the medium on the light, as well as the effect of the light on the medium, starts to becomes nonlinear. One of the most interesting consequences of such behaviour is that it may then no longer be possible to separate the dynamics of the atomic medium from that of the light. The evolution of the combined system differs qualitatively from that of the individual subsystems. This change of behaviour corresponds to a phase transition associated with spontaneous symmetry breaking of spatial or temporal coordinates. We discuss an example for both cases below.

Nonlinear optics

A realistic description of the interaction between light and atoms must therefore include a dependence on the laser intensity. The linear case discussed up to now thus appears as the low-intensity limit. This limit was traditionally the only experimentally accessible case until lasers became available. In the meantime, the investigation of optically nonlinear processes has matured into a field of its own (Shen 1984). The attraction of the field is that optically nonlinear media generate effective interactions between photons, which do not interact with each other in free space. Apart from the intrinsic interest in these systems, the field has developed many important applications, like frequency doubling (Franken et al. 1961), optical memories (Carlson, Babbitt and Mossberg 1983; Bai, Babbitt and Mossberg 1986; Mitsunaga, Yano and Uesugi 1991; Kohler et al. 1993) and optical computers (Jenkins et al. 1984; Prise et al. 1991; Ingold et al. 1992; Wild et al. 1992a). Most investigations in the field of nonlinear optics use dielectric solids as the nonlinear medium, but several ground breaking experiments have used resonant atomic vapours.

Nonlinear optics has grown into a large field and it would be hopeless to attempt a survey in this limited space. It seems important, however, to provide at least an idea of what kinds of experiment resonant atomic vapours are useful and how their internal level structure modifies the behaviour of the systems. Optically nonlinear effects include homogeneous changes of the susceptibility, which affect amplitude and phase of the transmitted light and therefore the intensity as well as the polarisation. This aspect of the interaction is not qualitatively different from the linear interactions. If the inhomogeneity of the optical properties becomes relevant, intensity-dependent changes of the refractive index profile lead to lensing effects that change the laser beam cross section as it propagates through the medium.

In previous sections, we discussed some simple examples of nonlinear optical effects. In Section 2.2.4, we calculated the optical polarisation of a two-level system as

$$(s_x,s_y,s_z)_\infty = \frac{1}{1 + \delta^2 + \kappa^2\Gamma_2/\Gamma_1}\,(\delta\,\kappa,\ \kappa,\ 1 + \delta^2) \qquad (10.1)$$

where the dimensionless Rabi frequency κ is proportional to the laser field amplitude. At low intensity, the transverse components (s_x, s_y), which cause absorption and dispersion, are proportional to the field amplitude, whereas the excited state population appears quadratically in the field amplitude. At higher intensity, their increase becomes slower, as Figure 10.2 shows.

When the coupling strength κ approaches unity, the increase of the transverse components with the field strength becomes less than linear and eventually turns into a decrease. This behaviour is known as saturation. For two-level atoms, saturation occurs at the intensity at which the atom absorbs one photon during the radiative lifetime of the excited state. In multilevel atoms, under conditions of

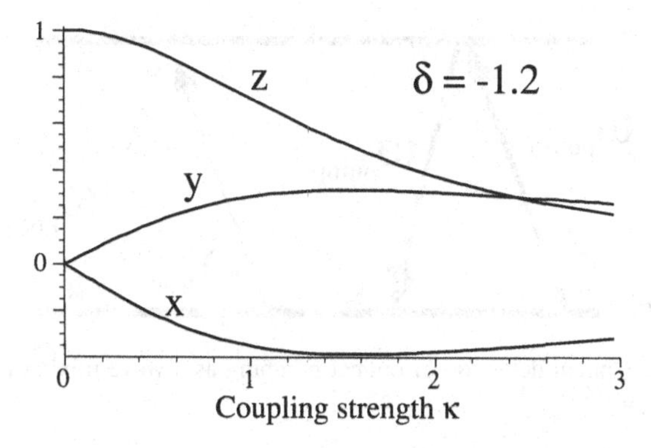

Figure 10.2. Variation of the density operator components with the laser field strength.

optical pumping, the relevant lifetime is that of the ground-state polarisation, which can be more than five orders of magnitude longer. The saturation intensity decreases correspondingly, while nonlinear effects increase.

10.1.2 Wave mixing

As long as the response of the medium is linear, different waves interacting with the medium are independent of each other. When the total intensity becomes too high, the atoms generate an interaction between the different waves. In the language of nonlinear optics, this effect is known as wave mixing. The conventional description (Shen 1984) expands the response of the medium in a power series

$$P_\alpha = \sum_{n=1}^{\infty} P_\alpha^{(n)} \qquad (10.2)$$

where α runs over the coordinates (x, y, z). The nth-order polarisations depend on n field components. The first terms of the series are

$$\begin{aligned} P_\alpha^{(1)}(\omega) &= \chi_{\alpha\beta}^{(1)} E_\beta(\omega) \\ P_\alpha^{(2)}(\omega = \omega_1 + \omega_2) &= \chi_{\alpha\beta\gamma}^{(2)}(\omega = \omega_1 + \omega_2)\, E_\beta(\omega_1) E_\gamma(\omega_2) \end{aligned} \qquad (10.3)$$

The susceptibilities are (reducible) nth-rank tensors and depend on all involved frequencies. They form the source of an additional wave with fre-

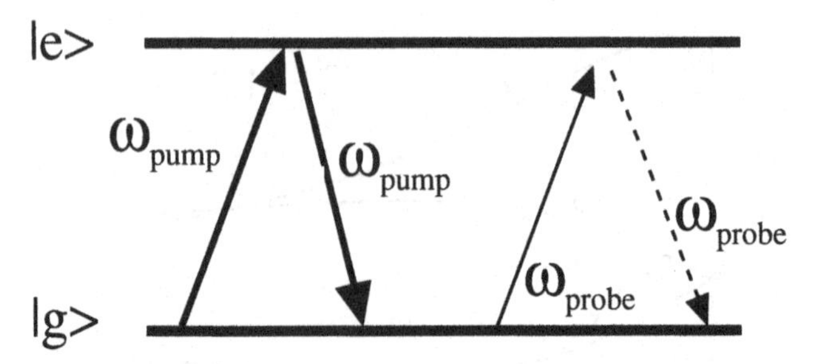

Figure 10.3. Optical detection of optical pumping as a wave-mixing process.

quency ω. An nth-order susceptibility thus mixes n incoming waves with one outgoing wave. The corresponding process is known as $(n + 1)$ wave mixing. The incoming waves do not, however, have to be distinct waves. In many cases, like harmonic generation or Kerr effect, they all represent the same laser beam. Conservation of energy and momentum require that the frequency and wavevector of the outgoing wave be an algebraic sum of the contributions from each wave.

Optical detection of optical pumping can also be cast into the language of nonlinear optics. In lowest order, it appears as a third-order effect (four-wave mixing), where two incident photons originate from the pump laser beam and the third from the probe laser beam, as shown in Figure 10.3.

The optical pumping, which creates a ground-state orientation, appears in this picture as a Raman process. The second Raman process is initiated by the probe laser, which scatters off the coherent excitation of the medium, as discussed in Section 6.3.3. The outgoing photon, dashed in Figure 10.3, has the same frequency as the probe photon if the optical pumping creates the order between degenerate states; otherwise it is frequency shifted by the difference between the two pump photons.

An important step in the evolution of nonlinear optics was the first demonstration of frequency doubling in quartz (Franken et al. 1961). In atomic vapours, frequency doubling is possible only if static fields reduce the symmetry of the medium (Hakuta, Marmet and Stoicheff 1991; Hakuta, Marmet and Stoicheff 1992). This scheme allows also exact phase matching, while keeping absorption low. Most wave mixing experiments have used odd-order nonlinear susceptibilities, which do not require external fields. Third harmonic generation, e.g., was demonstrated early in rubidium vapour in the presence of Xenon buffer gas (Bloom et al. 1975b; Bloom, Young and Harris 1975). Fifth and seventh harmonic generation can provide coherent light in the deep

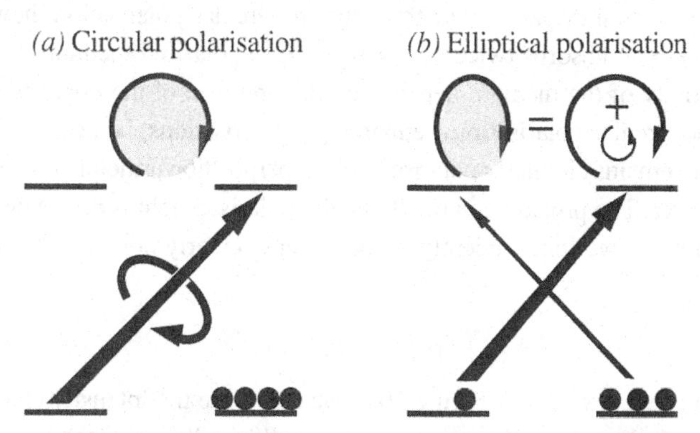

Figure 10.4. Optical pumping with circularly polarised light (left) and elliptically polarised light (right). Although the absorption of purely circularly polarised light can be arbitrarily small, the presence of the opposite polarisation increases the absorption of both components.

UV and soft X-ray region (Reintjes et al. 1976; Reintjes et al. 1977). Other four-wave mixing experiments (Bloch and Ducloy 1981; Ducloy and Bloch 1994) have demonstrated phase conjugation (Bloom, Liao and Economou 1978; Steel and Lind 1981; Carlson, Babbitt and Mossberg 1983; Köster, Mlynek and Lange 1985; Pinard, Grancelement and Grynberg 1986; Oria et al. 1989; Maki, Davis, Boyd et al. 1992) and squeezing of light (Slusher et al. 1985; DeOliveira, Dalton and Knight 1987; Maeda, Kumar and Shapiro 1987; Hilico et al. 1992a; Hilico et al. 1992b; Kimble 1992; Reynaud et al. 1992; Olsen, Gheri and Walls 1994). Since these applications require high nonlinearities at low power, it is not surprising that optical pumping appears to increase significantly the efficiency of squeezing (Kupriyanov, Sokolov and Subbotin 1988b; Kupriyanov and Sokolov 1988a).

10.1.3 Coupled absorption

Principle

When the laser intensity is high enough to modify significantly the state of the atomic system, the different polarisations of the light can no longer propagate independently through the medium. Absorption as well as dispersion of one beam also depends on the properties of the second beam. This can be seen most easily in the case of optical pumping.

If the polarisation of the light is circular, as in the left-hand side of Figure 10.4, it is possible to bring all atoms into the nonabsorbing state and the medium becomes transparent. For the opposite circular polarisation, however, the same medium absorbs twice as much as the unpolarised medium. Overall, the absorptivity of the medium depends on the presence of the opposite polarisation. The circular polarisations remain eigenpolarisations, i.e., circularly polarised light remains circularly polarised. The superposition principle is no longer valid, however. The propagation of elliptically polarised light is not equal to the superposition of two independently propagating circularly polarised beams.

Propagation

For the four-level system of Figure 10.4, we can write the intensity change of the two circularly polarised components of an elliptically polarised laser beam as

$$\frac{\partial A_+}{\partial z} = -2\,\alpha_0\,A_+\,\rho_{11} \qquad \frac{\partial A_-}{\partial z} = -\,2\,\alpha_0\,A_-\,\rho_{22} \tag{10.4}$$

where the absorption coefficient α_0 of the unpolarised medium depends on the absorption cross section of the atoms and the particle density. $A_\pm$ is the intensity of $\sigma_\pm$ light. This expression of the absorption remains valid as long as the excited state population remains small.

If the laser intensity is high enough that we can neglect losses due to diffusion, the populations are

$$\rho_{11} = \frac{A_-}{A_+ + A_-} \qquad \rho_{22} = \frac{A_+}{A_+ + A_-} \tag{10.5}$$

This allows us to eliminate the populations from the equation of motion. The resulting equation for the intensities alone becomes

$$\frac{\partial A_+}{\partial z} = -2\,\alpha_0 \frac{A_+ A_-}{A_+ + A_-} = \frac{\partial A_-}{\partial z} \tag{10.6}$$

The intensity change is the same for both components, and it vanishes if either of the two components vanishes. In this medium, photons of opposite circular polarisation behave like antiparticles that can annihilate each other. This mutual annihilation continues until the polarisation of the light becomes circular and only one type of photon remains.

If the light is linearly polarised, its polarisation remains linear, as the number of absorbed photons of both circular polarisations is the same. In the case

Figure 10.5. Pseudo-potential for the evolution of the polarisation in an optically pumped medium. The stable fixed points are the circular polarisation; linear polarisation is an unstable fixed point. The polarisation represented by the ball is initiallyclose to a linear polarisation, but evolves towards the circular polarisation state.

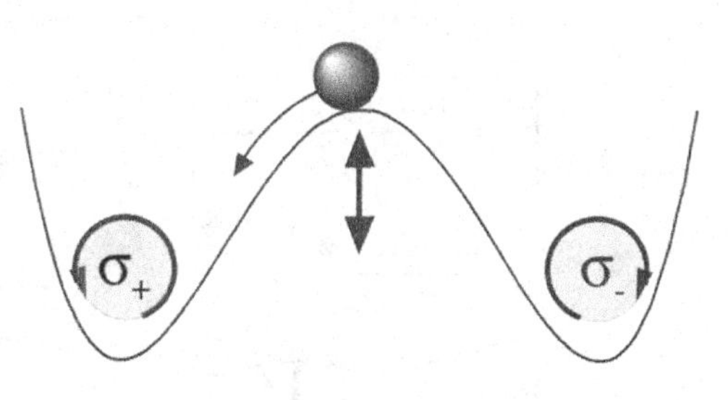

of elliptically polarised light, however, where one circular polarisation component is stronger than the other, the weaker component decays to zero over an absorption length shorter than for linearly polarised light by a factor of two, as most of the population is in that ground state to which the weaker polarisation couples. The system thus has a tendency to evolve towards circular polarisations by eliminating linearly polarised components of the light.

Qualitatively, we may represent the evolution of the system by the pseudo-potential in Figure 10.5: Linear polarisation is an unstable fixed point, whereas the two circular polarisations are stable fixed points. The system evolves towards the circular polarisations unless it is initially exactly linearly polarised. The true evolution is not oscillatory, as this potential might suggest, but the decay is irreversible. This interaction between opposite polarisations is responsible for many optically nonlinear processes like polarisation instabilities (Gauthier, Malcuit and Boyd 1988).

10.2 Nonlinear propagation: self-focusing

10.2.1 Light-induced waveguides

Optical pumping

In optically nonlinear media, the refractive index depends on the optical intensity. In atomic vapours, the strongest nonlinearity is due to optical pumping, which reduces the number of atoms interacting with the pump beam while increasing the number that interacts with the opposite polarisation. As the optical susceptibility of a medium is proportional to the number of interacting particles, optical pumping reduces the difference between the refractive index and one.

In the optically pumped medium, the number of atoms interacting with the laser is reduced, as shown in the inset of Figure 10.6. This reduction reflects itself directly in the amplitude of the dispersion curve. The refractive index of the optically pumped curve is thus always closer to unity than in the lin-

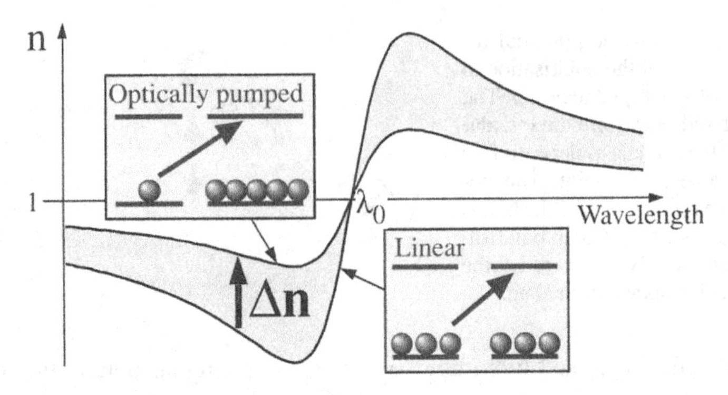

Figure 10.6. Refractive index change Δn by optical pumping as a function of the laser wavelength.

ear regime. In nonlinear optics, the intensity dependence of the refractive index is known as the Kerr effect. The modification can be positive or negative, depending on the medium. In the case of resonant vapours, the change is positive (i.e., the refractive index increases with intensity) for blue laser detuning and negative for red laser detuning, as Figure 10.6 shows.

For purely circular light propagating through our model atomic medium, the dependence of the refractive index on the laser intensity is

$$n = n_0 + (1 - n_0)\frac{P_+}{P_+ + \gamma_0} \tag{10.7}$$

where the optical pumping rate P_+ is proportional to the intensity and n_0 is the refractive index of the linear medium. For high enough pumping rate, the refractive index tends towards unity.

Beam and index profile

Laser beams always have a finite extent and usually an approximately Gaussian intensity distribution across the beam, with a maximum at the centre. In an optically nonlinear medium, the refractive index shows an intensity-dependent variation across the beam. At the centre of the beam, where the intensity is highest, it reaches values close to unity and tends towards the linear value far from the beam axis.

Figure 10.7 represents the situation graphically. In the left-hand part, it shows the increase (for a blue-detuned laser) or decrease (for a red-detuned laser) of the refractive index towards unity as the laser intensity increases. The

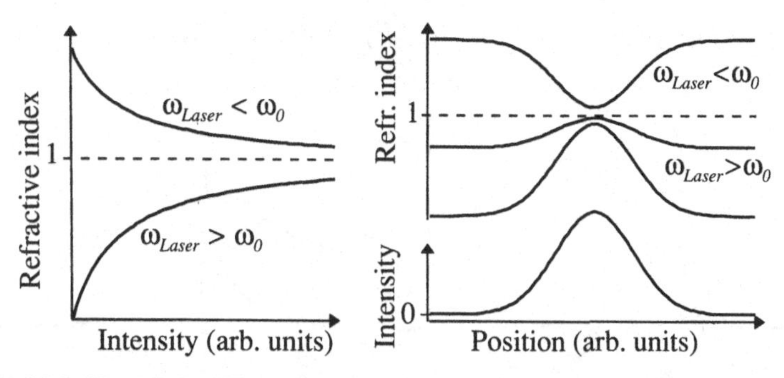

Figure 10.7. The left-hand side shows how the refractive index tends towards unity for high enough intensity. In a typical Gaussian beam profile shown in the bottom trace at the right, this results in an index profile that has qualitatively the same shape as the beam profile itself. The different cases shown correspond to different laser frequencies.

right-hand part shows the spatial variation across the laser beam for a Gaussian beam profile. At the centre of the beam, where the intensity is highest, the refractive index is close to unity. If the laser detuning is towards the blue, the linear refractive index is smaller than one and the index is therefore highest at the centre, as shown by the two curves below the dashed line. For red detuning, the refractive index at the centre of the beam decreases.

Diffusion

So far, we assumed that the response of the medium to the laser is purely local, i.e., it depends only on the laser intensity at the position for which we evaluate the index. In gaseous media, we also have to take the effect of atomic motion into account. If the mean free path of the atoms is small compared to the diameter of the laser beam, we can describe the motion as a diffusion process. The atomic motion preserves the state of the atom and therefore averages the refractive index. Assuming that the system is cylindrically symmetric, we can describe the refractive index change by the equation

$$\frac{\partial}{\partial t}\Delta n(r,t) = \alpha A^2[1 - n_0 - \Delta n(r,t)] - \gamma_0\,\Delta n(r,t) + D\left[\frac{\partial}{\partial r^2} + \frac{1}{r}\frac{\partial}{\partial r}\right]\Delta n(r,t) \tag{10.8}$$

where $\Delta n = n - n_0$ represents the light-induced deviation of the refractive index n from the linear index n_0. D is the diffusion constant and γ_0 the depo-

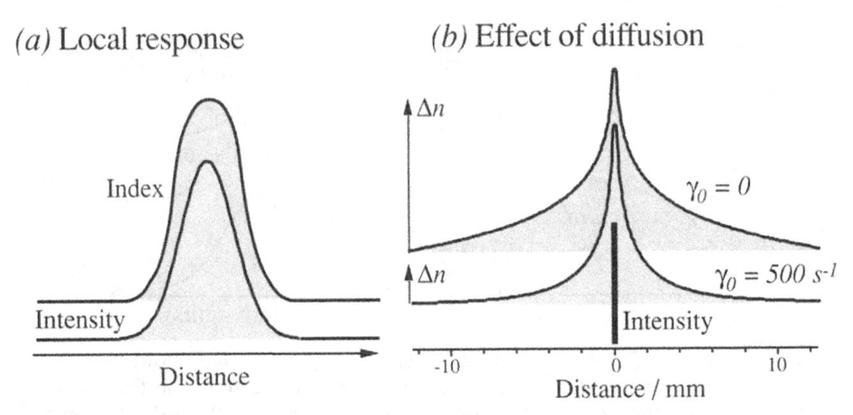

Figure 10.8. Local (**a**) and nonlocal (**b**) response of an optically nonlinear medium. The left-hand side shows schematically the case of a solid or an atomic gas with a short lifetime of the nonlinearity. The right-hand side represents the effect of diffusion for two different relaxation rates.

larisation rate of the ground-state magnetisation, which may include collisional depolarisation and magnetic fields. The proportionality constant α between the optical pumping rate and the laser intensity depends on the transition strength and the laser frequency. Through the diffusion term, the index now depends not only on the local field amplitude, but also on a spatially averaged intensity.

Figure 10.8 illustrates this effect. The left-hand part of the figure shows the connection between the laser intensity and the refractive index for the case of a local response of the medium. The nonlinear index change extends over the same spatial region as the laser beam. The right-hand part of the figure shows the modification of the index profile by atomic diffusion. In this case, the index change extends well beyond the laser beam diameter. To find a simple analytical expression for the resulting profile, which is the stationary solution of equation (10.8), we make the following assumptions:

(1) The laser beam has a cylindrical intensity profile with radius r_0 and propagates along the axis of a cylindrical cell of radius R.
(2) The relaxation vanishes inside the sample volume, but is instant and complete on the surface of the sample cell (Tanaka et al. 1990), i.e., $\gamma_0 = 0$, $\Delta n(R, t) = 0$.
(3) The optical pump rate is sufficiently large that the index change Δn within the laser beam reaches the maximum possible value $\Delta n_{\max} = 1 - n_0$.

In this case, the resulting index profile has a logarithmic dependence on the distance from the laser beam

$$\Delta n(r) = \Delta n_{\max} \frac{\ln(r/R)}{\ln(r_0/R)} \tag{10.9}$$

and is $\Delta n_{\max}$ inside the beam. The assumptions made here are sufficiently close to actual experimental conditions that the index profile is a good approximation to an experimental one.

Relaxation effects

A nonvanishing relaxation rate γ_0 reduces the effect of diffusion. The second curve in the right-hand part of Figure 10.8 illustrates this for a relaxation rate $\gamma_0 = 500$ sec^{-1}; the resulting index profile has become visibly narrower compared to the case without relaxation. Both index profiles reach zero at the walls of the sample cell and the same maximum at the centre. The physical mechanism causing these relaxation processes can include collisions with buffer gas atoms (Dehmelt 1957b). Under typical experimental conditions used for self-focusing experiments, however, other effects are considerably more effective. In particular, the high number densities required for effective self-focusing lead to significant radiation trapping effects (Kenty 1932; Zemansky 1932; Holstein 1947; Holstein 1951; Bicchi et al. 1994), which contribute to the relaxation rate (Tanaka et al. 1990; Lange et al. 1992).

The effect of diffusion therefore depends on the relative size of the diffusion rate D and the relaxation rate γ_0. A good estimate of the cross-sectional area F of the induced waveguide is $F = D/\gamma_0$. For the example shown in Figure 10.8, the diffusion constant was $D = 5 \cdot 10^{-3}$ m^2 sec^{-1}, and the relaxation rate $\gamma_0 = 500$ sec^{-1}. With these values, the above estimate yields an effective cross section $F = 0.1$ cm^2, corresponding to a radius of 1.8 mm, in good agreement with the calculated profile. The radius of the induced waveguide can therefore become much larger than the radius of the laser beam, which was 0.1 mm in the calculation. Since this widening of the induced waveguide depends on the relaxation rate γ_0, it is effective only for optical nonlinearities due to ground-state polarisation, where the lifetime is long enough. If the optical nonlinearity is due to saturation of an optical transition (Grischkowsky 1970; Bjorkholm and Ashkin 1974), the decay rate γ_0 of the refractive index profile is the relaxation rate of the excited state population. For a strong optical transition, the range of the medium response is then limited to a few micrometers, some three orders of magnitude less than in the case considered here.

10.2.2 Self-focusing

General

A laser beam propagating through a medium with an inhomogeneous refractive index can no longer be approximated as a plane wave or a freely propagating Gaussian beam. A region of high refractive index acts like a waveguide or an optical fibre, which prevents diffraction of the light and confines it to the high-index region. This is just the situation that a blue-detuned laser finds in an optically pumped medium, as shown in Figure 10.7. The difference from a normal waveguide is that in this case the waveguide was written by the laser beam itself. Instead of a waveguide, the refractive index profile may also be compared to a focusing lens. This interaction of the laser with the medium is therefore known as self-focusing (Askar'yan 1962; Hercher 1964; Shen 1984). It occurs if the total index of refraction increases with the laser intensity. In such a medium, laser beams with Gaussian intensity profiles induce an index profile that counteracts the normal effect of diffraction. When the two effects cancel each other, the beam can propagate without diffraction. This situation is known as self-trapping (Chiao, Garmire and Townes 1964; Bjorkholm and Ashkin 1974).

Since both the Kerr effect leading to self-focusing and the opposing tendency of light to diffract are indirectly proportional to the cross-sectional area of the beam, the self-trapped state is unstable in ideal three-dimensional Kerr media (Derrick 1964; Shen 1984). If the laser intensity is below a critical intensity that depends on the nonlinearity of the medium, the beam diffracts as usual. Above the critical intensity, the beam self-focuses to an area of the dimension of the optical wavelength (Marburger and Dawes 1968), unless damage to the material stops the process (Hercher 1964). At the critical intensity, the beam can propagate without diffracting, but arbitrarily small perturbations cause it to fall into either of the two other regimes. In two-dimensional systems, such as planar waveguides, self-trapped beams can be stable and lead to nondiffracting beams (Maneuf, Desailly and Froehly 1988; Maneuf and Reynaud 1988a; Aitchison et al. 1990). In analogy to other stable, localised solutions of nonlinear wave equations, such self-trapped beams have been called "spatial solitary waves" or "transverse solitons."

In three-dimensional media, the collapse of the laser beam may come to a stop if the index of refraction does not increase indefinitely with intensity, but saturates before the material suffers laser-induced damage (Shen 1984). In such a medium, laser beams can become localised in the two transverse dimensions. Their diameter is not constant, but shows characteristic oscillations that reflect the competing effects of self-focusing and diffraction (Shen 1984;

Snyder et al. 1991; Karlsson 1992). This saturation effect has also been proposed as the relevant mechanism for the formation of the small-scale filaments that tend to form in many self-focusing processes after the collapse of the beam (Gustafson et al. 1968).

Self-focusing in atomic media

For the observation of these effects, resonant atomic media offer several attractive properties. Due to the resonant enhancement of the interaction, nonlinear effects appear at relatively low intensities. In addition, the nonlinear refractive index change can easily be saturated, even with a low power ($\leq$100 mW) cw laser and the nonlinear optical properties are of the same order of magnitude as the linear effect. Both the linear and nonlinear contributions can be adjusted by tuning the laser frequency close to an optical transition frequency. Atomic vapours, in particular, have the additional advantage of being essentially immune to damage and allow large variations of the refractive index by changing the particle density through the temperature. In multilevel systems, these effects are even more pronounced. The basic reason is that multilevel systems can combine high transition strengths with long lifetimes – the ideal combination for large nonlinearities. Using optical pumping, it is possible to saturate a transition with intensities of the order of a few μW/mm^2.

Early demonstrations of this type of experiment were reported by Grischkowsky (Grischkowsky 1970) and by Bjorkholm and Ashkin (Bjorkholm and Ashkin 1974), who used cw lasers to achieve not only self-focusing, but also self-trapping. Subsequent work by other groups improved the understanding of these experiments, using theoretical models that treat the atomic medium as an ensemble of two-level atoms (Wagner, Haus and Marburger 1968; Boshier and Sandle 1982; LeBerre, Ressayre and Tallet 1985; McCord, Ballagh and Cooper 1988). Yabuzaki and colleagues (Yabuzaki, Hori and Kitano 1982) suggested that optical pumping with circularly polarised light should cause self-focusing at even lower power.

Wave equation

For a theoretical description of this process, we start from the usual nonlinear wave equation in the paraxial approximation, using cylindrical coordinates (r, ϕ, z)

$$k \frac{\partial}{\partial z} A^2 = -\nabla_\perp \cdot (A^2 \, \nabla_\perp \phi) \tag{10.10}$$

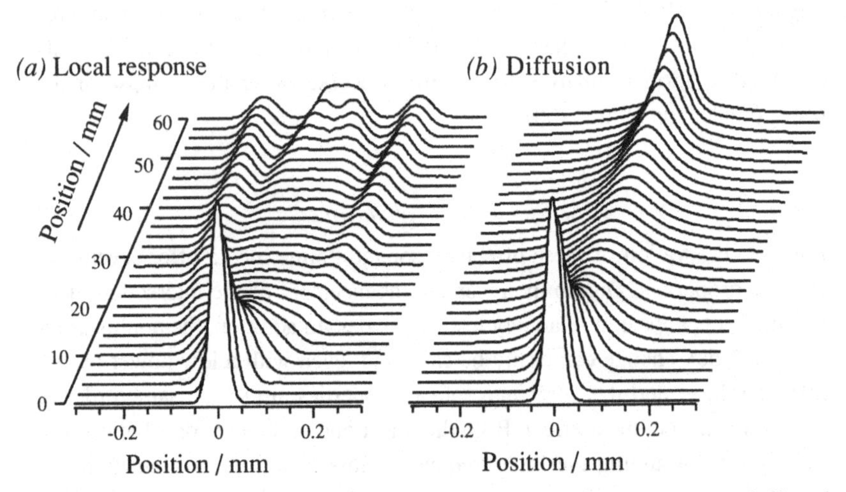

Figure 10.9. Propagation of a laser beam in a medium with local response (left) and a vapour with diffusive atomic motion (right). The absolute value of the laser beam amplitude is plotted as a function of the distance from the centre of the laser beam as it propagates into the medium.

$$\frac{\partial}{\partial z}\,\phi = -\frac{1}{2k}\,(\nabla_\perp\phi)^2 + \frac{k}{2}\left(\frac{\nabla_\perp^2 A}{k^2 A} + 2\,\frac{\Delta n}{n_0}\right) \tag{10.11}$$

where $A\,\mathrm{e}^{\mathrm{i}\phi}$ represents the slowly varying part of the complex field amplitude, k the wavevector, n_0 the linear refractive index, and $\nabla_\perp$ refers to the transverse coordinates r, θ. The variation of the refractive index change Δn with the laser intensity was discussed in Section 10.2.1.

This wave equation can be integrated numerically. Figure 10.9 shows the result of such an integration, using the parameters $k = 10.7\ \mu\mathrm{m}^{-1}$ (Na D_1 transition) and $n_0 = 1.7 \cdot 10^{-5}$. The incident laser beam amplitude was a plane wave with Gaussian cross section and a beam diameter of 36 μm (FWHM). The maximum pump rate was $3 \cdot 10^7\ \mathrm{sec}^{-1}$, corresponding to an intensity of 20 W mm^{-2} at a normalised laser detuning of $\overline{\Delta} = 5$. The total beam power was 26 mW. The relaxation rate γ_0 was set to $10^4\ \mathrm{sec}^{-1}$, corresponding to a diffusion-induced width of the waveguide of $\sqrt{D/\gamma_0} = 400\ \mu$m, an order of magnitude larger than the laser beam diameter.

Figure 10.9 shows the effect of a nonlocal medium response on the self-focusing process. The absolute value of the laser amplitude is plotted as a function of the distance from the centre of the laser beam. Successive traces correspond to cross sections further into the medium. The data in the left-hand column illustrate the case of purely local response of the system, corresponding

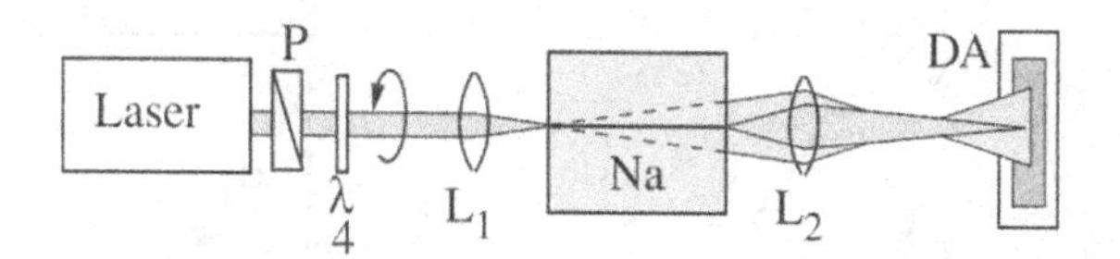

Figure 10.10. Setup for the observation of self-focusing by optical pumping. P = polariser, $\lambda/4$ = retardation plate, L = lens, DA = photodiode array. The dark-shaded region shows the beam path with self-trapping, the light-shaded area enclosed by dashed lines corresponds to the free beam.

to a vanishing diffusion constant. In this case, the initially Gaussian shape of the laser beam is unstable; it separates into a ring structure whose diameter increases as the beam propagates further into the medium. Similar structures are well known from self-focusing experiments in dielectric solids (Shen 1984). The data shown in the right-hand column were calculated with identical parameters, except that the diffusion coefficient was set to the experimentally determined value of $D = 5000\ \mu\text{m}^2/\mu\text{sec}$. In this case, the laser beam remains stable, apart from a small oscillatory motion of the beam diameter. This qualitatively different behaviour is due to the nonlocal response of the atomic medium, which stabilises the transverse beam profile.

This stabilisation effect can be understood qualitatively by considering that diffusion is an efficient decay process for small structures, but relatively inefficient for larger structures. As a result, the optical pumping process induces a waveguide that is smoother than it would be without atomic motion, thereby preventing the beam from breaking apart. In addition, the width of the waveguide can be larger than the width of the laser beam, resulting in better confinement of the beam. The existence of these stable solutions is not a contradiction of Derrick's theorem (Derrick 1964), although these two-dimensional structures are stable against the scaling perturbations considered by Derrick. Not only are the solutions time-dependent, but the conditions considered here do not agree with the assumptions used by Derrick, who assumed that the refractive index depends only on the local intensity.

10.2.3 Experimental observation

An experimental observation of these effects is possible with a setup like that shown schematically in Figure 10.10. The laser beam is circularly polarised and focused on the entrance window of the Na cell. The second lens images the exit window of the Na cell onto a linear photo diode array, thereby enlarging the linear dimensions by a factor of four. The dark-shaded area in Fig-

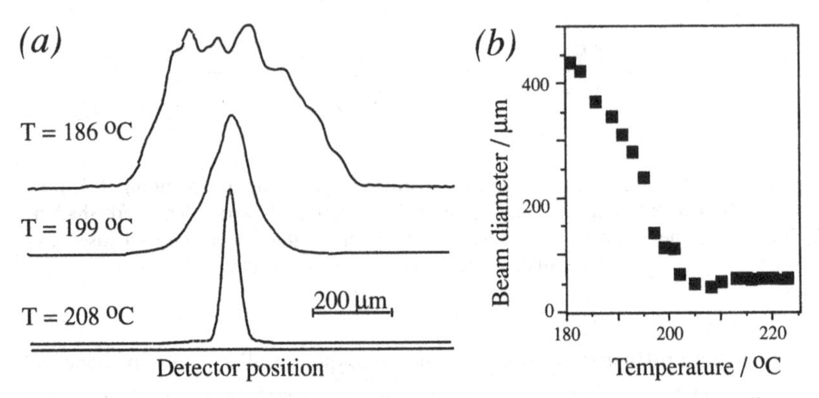

Figure 10.11. (**a**) Beam profiles for three different temperatures corresponding to different (nonlinear) optical path lengths, the laser is blue-detuned from the Na D_1 transition ($\Delta = -10$ GHz). (**b**) Beam diameter (FWHM) at the exit window as a function of temperature.

ure 10.10 corresponds to the path of the self-focused beam, whereas the light-shaded area shows the propagation of the beam in a linear medium.

Particle density

Instead of measuring the beam profile as a function of the cell length, as in the theoretical results of Figure 10.10, it is experimentally simpler to measure the beam profile behind a cell of constant dimensions, but variable effective optical path length. This variation can be achieved by raising the cell temperature and thereby the particle density.

In this way, the onset of self-focusing can easily be observed, as shown in Figure 10.11, where the observed beam profile is plotted for different cell temperatures. The curves in the left-hand column show the observed beam profiles at three different temperatures; the right-hand side summarises the measured beam widths (FWHM) over a range of temperatures. For these measurements, the laser frequency was kept constant, at $-\Delta = 10$ GHz above the atomic resonance frequency, and the external magnetic field was set to zero. A 45 mW laser beam was focused onto the entrance window, resulting in an intensity of 7 kW cm^{-2}, an order of magnitude above the observed critical intensity. The beam diameters were determined from the digitised data and the geometrical enlargement by the imaging system was taken into account to calculate the beam diameter at the exit window plotted in the figure. At the lowest temperature, $T = 186$ °C, the propagation is linear and the laser beam emerges with a diameter of several hundred μm. At higher temperatures, the increasing particle density increases the nonlinear effects and self-

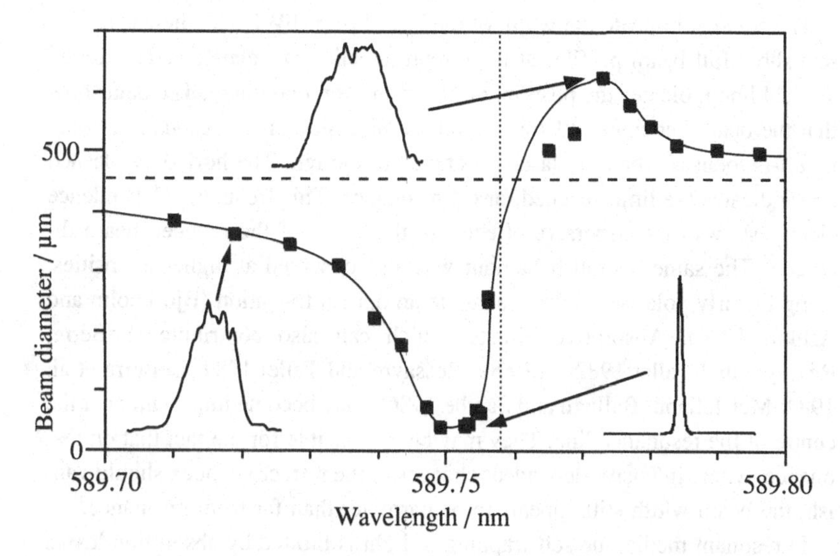

Figure 10.12. Beam diameter (FWHM) of the laser beam at the exit window as a function of the laser wavelength (black squares). The insets show three different beam profiles for the wavelengths indicated by the arrows. The line through the squares is intended as a guide to the eye and does not represent a theoretical curve.

focusing starts to decrease the beam diameter. Above 205 °C, a further temperature increase does not cause an additional decrease of the beam diameter, indicating saturation of the self-focusing process and a stable, self-trapped beam. The shallow minimum at 205 °C may indicate a soliton oscillation.

Wavelength dependence

The calculation assumed that an increase in the laser intensity would raise the refractive index, which should occur for a blue-detuned laser. As discussed above, the effect of the optical pumping process on the index of refraction is to bring it closer to one. Since the index of refraction of a thin resonant medium is smaller than one for blue-detuned light, and larger than one on the red-detuned side, optical pumping induces a positive Δn at shorter wavelengths and a negative Δn for longer wavelengths. We therefore expect that a red-detuned laser should decrease the refractive index, inducing an index profile that would defocus the laser rather than focus it. This prediction can easily be checked by measuring the laser frequency dependence of the beam diameter at the exit window. Figure 10.12 summarises the results; it shows the measured beam profiles at the exit window of the Na cell for different laser frequencies.

The squares indicate the width of the laser beam (FWHM) whereas the insets show full beam profiles at three representative wavelengths. The dotted vertical line indicates the position of the absorption maximum. The data show that the beam becomes self-trapped on the blue side of the resonance, while it self-defocuses when the laser is detuned to the red. The horizontal dashed line indicates the limit reached far off resonance. This frequency dependence clearly shows that dispersive effects are the cause of the reduced beam diameter. The same overall behaviour was also observed at higher intensities, using linearly polarised light to saturate an optical transition (Bjorkholm and Ashkin 1974). Absorptive effects, which can also contribute (LeBerre, Ressayre and Tallet 1982; LeBerre, Ressayre and Tallet 1984; LeBerre et al. 1984; McClelland, Ballagh and Sandle 1986), may become important near the centre of the resonance line. They may be responsible for the fact that on resonance, where intensity-dependent changes of the refractive index should vanish, the beam width still appears to be narrower than far from resonance.

In resonant media, the self-trapping of light is limited by absorption losses not explicitly considered here. Due to the different dependence of absorption losses (quadratic) and dispersive effects (linear) on the optical detuning, these losses can be reduced by increasing the laser detuning. They cannot be eliminated completely, since they are part of the optical pumping process that generates the nonlinearity; however, the beam travelling inside the self-induced waveguide experiences a considerably reduced absorption, compared to the case of linear propagation.

10.2.4 Other structures

As the theoretical analysis showed (see, e.g., Figure 10.9), a laser beam travelling through a self-focusing medium does not always form a stable self-trapped beam, but may also break apart into several partial beams, which may be ring structures showing axial symmetry (Grischkowsky 1970; LeBerre et al. 1984b; Suter and Blasberg 1993), but also separate into several approximately Gaussian beams (Grantham et al. 1991). In this case, the resulting pattern does not show axial symmetry. Using the experimental parameters, like intensity and detuning, as control parameters, it is possible to move from one regime to the other. This breaking apart of the laser beam can also be controlled by introducing experimentally controlled asymmetries into the beam. Grantham and colleagues introduced aberrations, like astigmatism, into the beam to generate "kaleidoscopic" spatial instabilities (Grantham et al. 1991).

Such observations show that the stabilisation of the self-trapped beam by the saturation process and the nonlinear response of the medium is effective

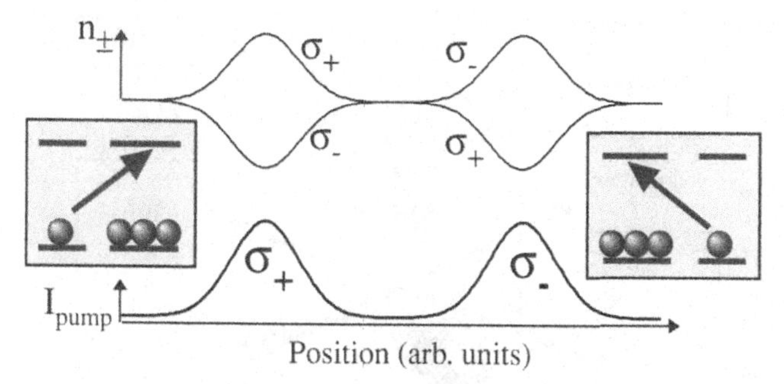

Figure 10.13. Modification of the refractive index profile by optical pumping with blue-detuned light. Two laser beams of opposite circular polarisation modify the refractive index by optical pumping, as shown in the insets. The lower curve shows the intensity distribution of the laser; the upper curves indicate the refractive index profile for both polarisations.

only in a limited parameter range. Within this parameter range, however, self-trapped beams can be quite stable. If two self-trapped beams of opposite circular polarisation converge at a small angle under conditions of self-focusing, they may even bounce off each other (Tam and Happer 1977; Holzner et al. 1992). To understand this behaviour, we have to consider that the optical pumping of one laser beam also modifies the refractive index profile for the opposite polarisation.

As Figure 10.13 shows, blue-detuned circularly polarised light increases the refractive index for itself at the centre of the beam, as discussed above. At the same time, it increases the number of atoms interacting with the opposite polarisation and therefore creates an opposite variation of the index profile for the other polarisation. As in the case of total internal reflection, the laser beam does not easily penetrate into the region of lower refractive index. As a result, the two beams repel and bounce off each other. We can summarise the self-trapping and beam bouncing in a simple qualitative model: The medium introduces an effective interaction between photons that is attractive between equal polarisations and repulsive for opposite polarisations.

The modification of the refractive index by the laser intensity affects not only the transverse profile of the laser beam, but in the case of pulsed lasers also the spectrum and the shape. During a short laser pulse, the refractive index can change rapidly enough that the resulting phase modulation changes the frequency of the propagating light significantly (Gustafson et al. 1969). This effect is known as self-phase modulation (Wong and Shen 1972). It causes spectral broadening of pulses propagating through nonlinear optical media, but can also be used to stabilise the pulses. If the self-phase modulation com-

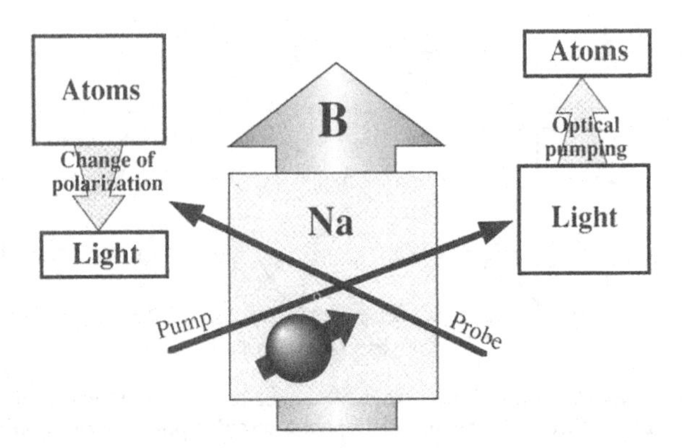

Figure 10.14. A pump-probe experiment approaches the limit of two one-sided interactions.

bines with group velocity dispersion, it is possible to form soliton pulses (Drummond et al. 1993) which have a constant length, while linear pulses travelling through dispersive media increase in length. The self-induced change in the refractive index also modifies the temporal shape of the pulse. This effect is known as self-steepening (Grischkowsky, Courtens and Armstrong 1973).

10.3 Temporal instabilities

10.3.1 Feedback

Coupled beams

Resonant optical media coupled to laser beams may exhibit not only spatial pattern formation, but also temporal instabilities like phase transitions leading to self-pulsing (Pauliat, Ingold and Günter 1989) or chaos (Ikeda, Daido and Akimoto 1980). In most cases, these systems include interactions with multiple laser beams or optical resonators. Since a detailed analysis or even a complete summary of these effects would require much more space than a section of this book, we discuss only one of the simplest models and hint at the connections to other examples. The specific example discussed here exhibits many of the features that occur in other systems. The relevant equations of motion are simple enough that they permit a discussion of these effects using analytical solutions to the equation of motion. For this purpose, we start with the usual pump-probe scheme, where each of two laser beams realises a one-sided interaction.

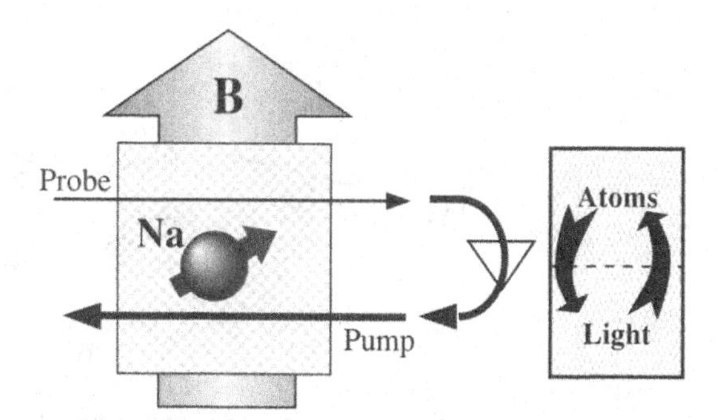

Figure 10.15. A coupling between pump and probe laser beam turns atoms and laser light into a strongly coupled system.

In an idealised pump-probe experiment, as represented in Figure 10.14, the strong pump laser beam modifies the atomic medium without being affected significantly, whereas the intensity of the probe laser beam is too low for a significant modification of the atomic system. As long as the two laser beams are independent, the interaction sequence is strictly one-sided, from the pump laser beam to the atomic medium to the probe laser beam. This is the situation discussed in all preceding sections. Here, we introduce a small modification that changes qualitatively the nature of the system behaviour: We lock the polarisation of the two laser beams to each other.

As Figure 10.15 shows, this coupling changes the one-sided interaction into a closed loop. The effect that the medium has on the probe beam changes the polarisation of the pump beam, which modifies the medium. As a result, the system can exhibit self-organisation even at relatively low laser intensities and can be controlled through a number of easily accessible experimental parameters. Depending on these parameters, the system can fall into stable limit cycles, bi- and multistability, or chaos. The variables that participate in this nonlinear evolution involve variables of the atomic system like the angular momentum as well as the polarisation of the laser beams. The gain of the feedback loop that locks the two polarisations acts as a control parameter, which can cause a phase transition of the combined system between a stationary isotropic and the anisotropic oscillatory behaviour.

Taking the point of view of the atomic medium, the two coupled laser beams introduce a coupling between the atomic spins. If this coupling is stronger than the dissipative forces, the system develops long-range order, in close analogy to a ferromagnet. In an external magnetic field, the resulting magnetisation undergoes Larmor precession, which can be observed as an oscillation. The laser beams stabilise this state as optical pumping replaces the or-

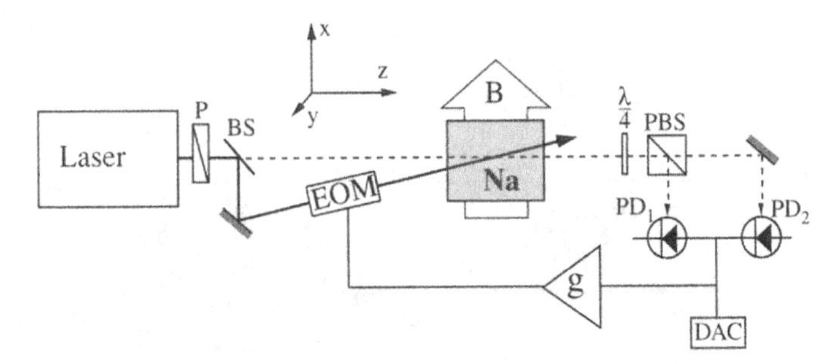

Figure 10.16. Schematic representation of the experimental setup. P = polariser, BS = beam splitter, EOM = electrooptic modulator, B = magnetic field, $\lambda/4$ = retardation plate, PBS = polarising beam splitter, $PD_{1,2}$ = photodiodes, DAC = D/A converter.

der lost by dissipation. This input of light, a feature that is typical for systems exhibiting temporal self-organisation, makes it an open system.

Experimental realisation

Figure 10.16 shows the setup used for the experimental implementation of this scheme. A pump laser beam (full line) excites population differences in the sodium vapour. A second laser beam (dashed line) probes this population difference and the polarisation-selective detector shown in the figure returns a signal directly proportional to the magnetic dipole moment in the direction of the laser beam. This detector signal is digitised and transferred to a computer for data analysis and storage. Simultaneously, it controls the polarisation of the pump laser beam by an electrooptic modulator (EOM) whose fast axis is rotated 45° from the (linear) polarisation of the incident pump beam. After passing through the EOM, the pump beam polarisation is elliptical, with the ellipticity depending on the modulator voltage; it becomes circular when the amplified signal equals the quarter-wave voltage of the EOM. In the experiments discussed in the following, the small signal absorption was ~60%. The pump beam had an intensity of 200 μW/mm^2 and the probe beam 50 μW/mm^2. The transverse magnetic field was $|B_x|$ = 14.3 μT, corresponding to a Larmor frequency of 100 kHz.

10.3.2 Evolution

Equation of motion

For the mathematical analysis of the system, we use the $J=1/2 \leftrightarrow J'=1/2$ model and choose a coordinate system whose z axis is parallel to the laser beams and whose x axis points along the magnetic field. We expand the ground-state density operator in the angular momentum operators $J_{x,y,z}$ and write the expansion coefficients as the three components of the magnetisation vector $\vec{m} = (m_x, m_y, m_z)$. As discussed in Section 5.3, on modulated optical pumping, the optical pumping rate is proportional to the sine of the EOM voltage. Since the overall intensity does not change during the modulation cycle, we have

$$P_+ + P_- = P_0 \qquad P_+ - P_- = P_0 \sin(g\, m_z) \tag{10.12}$$

where P_0 is the pumping rate of a circularly polarised beam. The dimensionless gain factor g, which describes the coupling between the two laser beam polarisations, includes the sensitivity of the detector, the gain of the electronic feedback loop, and the electrooptic coefficient of the modulator.

Neglecting light shift effects, we write the equation of motion for the atomic system as

$$\begin{aligned} \dot{m}_y &= \Omega_L\, m_z - (\gamma_0 + P_+ + P_-)\, m_y \\ \dot{m}_z &= P_+ - P_- - \Omega_L\, m_y - (\gamma_0 + P_+ + P_-)\, m_z \end{aligned} \tag{10.13}$$

where γ_0 describes the loss of magnetisation due to diffusion. We do not include the x component, which is not coupled to the other two variables. Using equation (10.12), we eliminate the optical pumping rates $P_\pm$ to obtain an equation of motion for the atomic medium alone,

$$\begin{aligned} \dot{m}_y &= \Omega_L\, m_z - (\gamma_0 + P_0)\, m_y \\ \dot{m}_z &= -\Omega_L\, m_y - (\gamma_0 + P_0)\, m_z + P_0 \sin(g\, m_z) \end{aligned} \tag{10.14}$$

The optical pumping term $P_0 \sin(g\, m_z)$, which increases the polarisation along the laser beam, is proportional to the polarisation already present. It introduces a positive feedback into the system.

Solutions

These equations do not have a simple analytic solution. To analyse the behaviour of the system, we first discuss the situation close to the thermal equilibrium, where the sublevel polarisation is small, $m_z\, g << 1$. The equations of motion then reduce to

$$\dot{m}_y = \Omega_L\, m_z - (\gamma_0 + P_0)\, m_y$$
$$\dot{m}_z = -\Omega_L\, m_y + (P_0(g - 1) - \gamma_0)\, m_z \qquad (10.15)$$

An additional simplification is possible by restricting attention to the case of small gain, $gP_0 << \Omega_L$, where the system does not change significantly over one Larmor period. This situation is closely related to the case of transverse optical pumping with a modulated pump beam. We make the ansatz

$$m_y(t) = m_0 \sin(\Omega_L t - \phi)\, e^{kt} \qquad m_z(t) = m_0 \cos(\Omega_L t - \phi)\, e^{kt} \qquad (10.16)$$

where the starting amplitude m_0 and the phase ϕ depend on the initial conditions.

Inserting this ansatz into the equation of motion, we find a growth rate $k = (P_0(g/2 - 1) - \gamma_0)$. Its sign determines whether the magnetisation in the system decays to zero or grows exponentially. In the case of small or vanishing gain ($0 \leq g < 2$), the equilibrium position of the system is isotropic. A magnetisation decays with a damping rate that is higher than for a free atom by $P_0(1 - g/2)$, due to the damping effect of the partially polarised pump beam. For $g = 2$, this additional damping effect by the pump beam is cancelled exactly, and in the region $2 < g < 2(1 + \gamma_0/P_0)$, the FID decays with a time constant longer than during free precession. For $g = 2(1 + \gamma_0/P_0)$ the time constant diverges. When the coupling strength g exceeds the critical value, the isotropic state of the system becomes unstable and the magnetisation starts to grow from small fluctuations at the rate k.

This behaviour closely resembles that of a ferromagnet. The feedback between the two laser beams introduces an effective coupling between the spins, which favours parallel orientation. As long as the losses described by $P_0 + \gamma_0$ exceed the coupling strength $P_0\, g/2$, the system remains isotropic. However, at the critical "temperature," it spontaneously breaks its symmetry, falling into an anisotropic state. In contrast to a true ferromagnet, the couplings introduced by the laser in this system extend over macroscopic distances. The coupling between two spins does not depend on their position, as long as both spins interact with both laser beams. In addition, the magnetisation that forms in the sample does not have a fixed orientation, but precesses around the magnetic field at the Larmor frequency. The pseudo-potential of Figure 10.17 suggests that the system can choose between two different directions in which the magnetisation points. More precisely, this corresponds to an axial symmetry around the magnetic field direction: All possible phases ϕ are equivalent.

Figure 10.18 shows an example of this behaviour. The system was initially

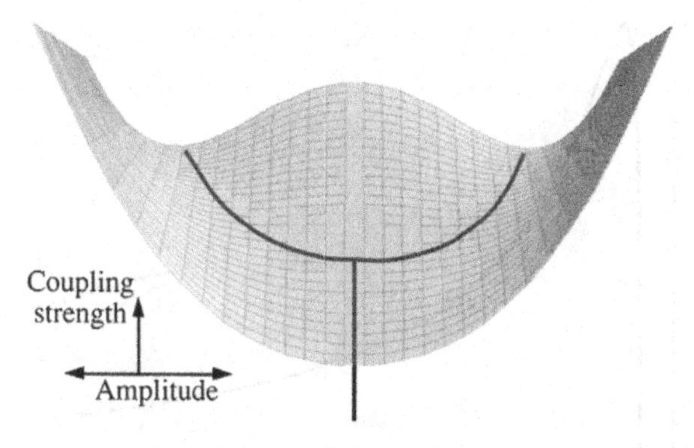

Figure 10.17. Pseudo-potential determining the state of the atomic system as a function of anisotropy (horizontal axis) and coupling strength (into paper).

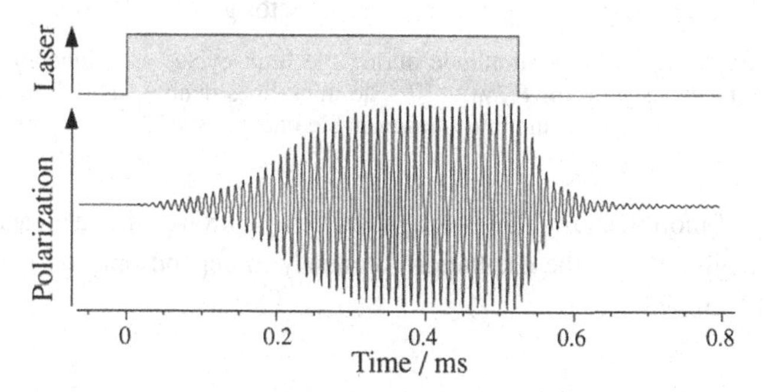

Figure 10.18. Polarisation oscillation observed when the feedback loop is closed by switching the pump laser on. The upper trace shows the pump laser intensity, the lower trace the signal of the polarisation-selective detector.

in thermal equilibrium until the feedback loop was closed at time $t = 0$ by switching the pump laser beam on with an acoustooptic modulator (not shown in Figure 10.18). As displayed in the figure, the polarisation of the two laser beams starts to oscillate soon after the pump laser beam begins to drive the system. The oscillations grow out of the noise with a delay that depends on the loop gain. The phase varies randomly between experiments, as expected for a random initial condition. The growth of the oscillation amplitude is initially roughly exponential, until the system saturates and reaches a stable limit cycle.

When the pump laser is switched off, the magnetisation built up during the pulse is observed as a free induction decay. Comparing the oscillation period of the FID with that of the driven magnetisation shows that the driven system does indeed oscillate at the Larmor frequency. A closer examination of the

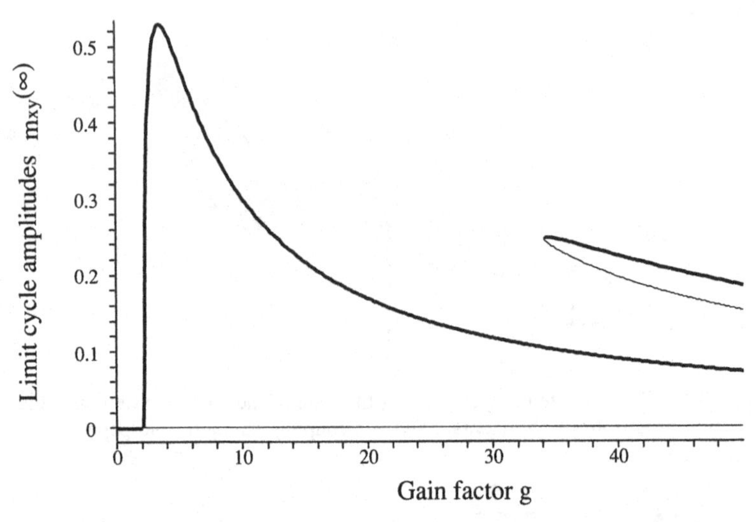

Figure 10.19. Magnetisation amplitude during the limit cycles as a function of the gain in the feedback loop for $P_0/\gamma_0 = 10$. The thick lines denote stable limit cycles, the thin lines unstable ones.

equations of motion shows that propagation delays through the feedback loop must be negligible for the oscillation frequency to depend only on the magnetic field strength.

10.3.3 Limit cycles

As Figure 10.18 shows, the oscillation amplitude does not continue to grow exponentially but saturates and reaches a limit cycle. The system deviates from the exponential growth when the approximation of low polarisation becomes invalid. To find the amplitude of the stationary limit cycle, we assume that the magnetisation precesses at the Larmor frequency Ω_L and analyse its long-term behaviour, using the assumption that the amplitude $m_{yz} = \sqrt{m_y^2 + m_z^2}$ changes only slightly during a Larmor cycle, which can be formulated as $k << \Omega_L$. We then obtain an equation of motion for the amplitude

$$\dot{m}_{yz} = -(\gamma_0 + P_0)\, m_{yz} + P_0\, J_1\, (g\, m_{yz}) \tag{10.17}$$

where J_1 represents the first-order Bessel function. The first term describes the losses of the system; the second term represents the effect of optical pumping. A stable limit cycle results when the Bessel function exactly cancels the losses of the first term.

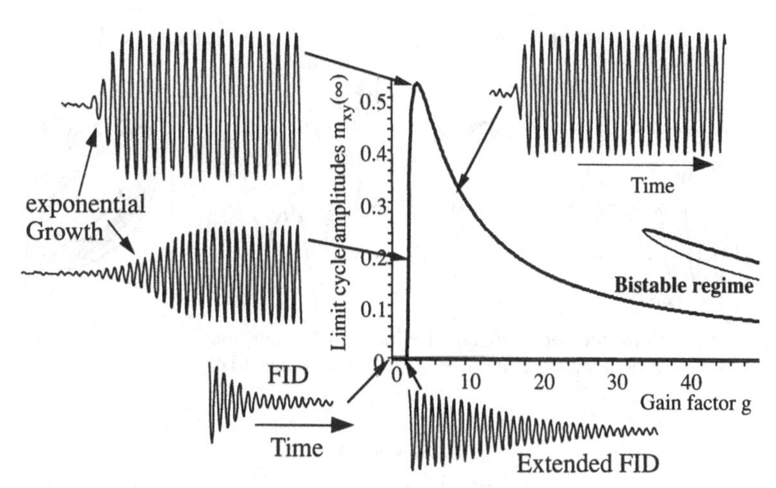

Figure 10.20. Summary of system behaviour as a function of the gain factor acting as a control parameter.

Figure 10.19 shows the stationary solutions of equation (10.17), which represent the amplitudes of the precessing magnetisation during the stable limit cycle as a function of the loop gain g. The amplitude reaches a maximum at a gain factor $g \sim 3.5$. Increasing the gain further lowers the amplitude because the increase in the voltage across the EOM moves the intensity to higher harmonics of the Larmor frequency. As discussed in Section 5.3, these higher harmonics do not contribute to the buildup of magnetisation. Multiple solutions are possible if the gain increases further; these additional solutions reflect the oscillatory nature of the Bessel function. The solutions corresponding to stable limit cycles are drawn as bold lines in the figure; those corresponding to unstable trajectories are represented as thin lines.

Figure 10.20 summarises the behaviour of the system as a function of the loop gain. The oscillatory curves represent experimentally observed signals, superimposed on the theoretical curve of Figure 10.19. For low enough gain, the equilibrium position of the system is isotropic. If an initial magnetisation is present, it decays while undergoing Larmor precession. This is represented by the free induction decay at the lower left. For increasing gain, the feedback partly cancels the losses of the system and a free induction decay with extended decay time is observed. Above the critical coupling strength, the magnetisation forms spontaneously and grows exponentially, until it reaches a stable limit cycle. Further increase of the loop gain primarily increases the growth rate k, while the amplitude of the magnetisation reaches a maximum and finally decreases.

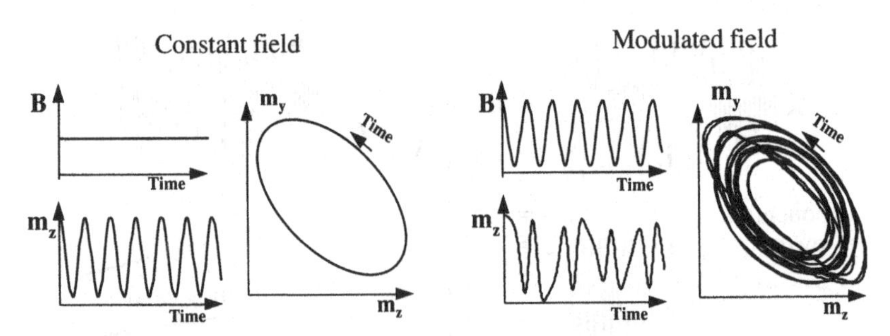

Figure 10.21. Behaviour of the magnetisation in a constant magnetic field (left) and in a field whose amplitude is modulated sinusoidally (right).

10.3.4 Chaos

Field modulation

Since the magnetisation precesses only in the *yz* plane perpendicular to the magnetic field, the effective dimension of the system is two. Such systems always show regular behaviour like the stable limit cycles discussed above. A well-known way to change this behaviour is to modulate a system parameter, e.g., the magnetic field.

Figure 10.21 compares the evolution of the system in a static (left) and modulated (right) magnetic field, calculated by numerical integration of the equations of motion. The upper trace shows the magnetic field as a function of time. The lower trace indicates the evolution of the magnetisation component m_z, which we used as the signal in the experiments discussed above. In a constant magnetic field, it shows a sinusoidal modulation at the Larmor frequency, whereas the evolution in a modulated field is aperiodic. This is even more visible in the two-dimensional representation of the magnetisation trajectory in the *yz* plane. In the constant field, the magnetisation vector follows a closed trajectory, whereas the trajectory in the right-hand part of the figure carries the signature of a strange attractor.

A closer examination shows that the modulation of the magnetic field drives the system into a chaotic regime, characterised by magnetisation trajectories that diverge exponentially in time. Although the magnetisation still precesses in the *yz* plane, the effective dimension of the system is now three, as the equations of motion are time-dependent. For small modulation amplitudes, the deviations from the regular motion are small, but they increase with the amplitude. The fractal dimension of the trajectory is of the order of 1.8.

Figure 10.22 gives a survey of the behaviour of the system as a function of the modulation amplitude in the form of a Poincaré section. For each ampli-

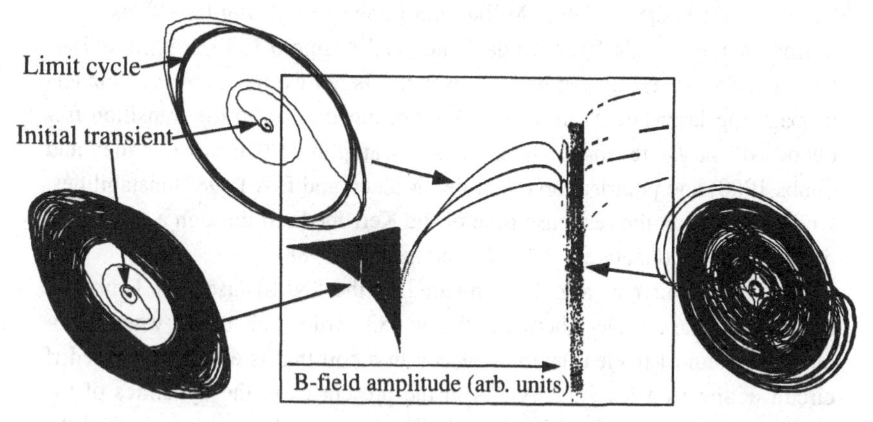

Figure 10.22. Poincaré section of magnetisation trajectory.

tude, the dots represent the length of the magnetisation vector when it crosses the *xz* plane. For this plot, the equations of motion were integrated numerically, starting at a time where the initial transients had died out. The three insets show the precessing magnetisation in the *yz* plane for three different modulation amplitudes. At low modulation, the evolution appears to be chaotic. For higher amplitudes, the system falls into a period-three orbit, as shown by the second magnetisation trajectory. The third trajectory shows another chaotic region within the periodic regime.

Related systems

Optical bistability (Gibbs 1985) and tristability (Kitano, Yabuzaki and Ogawa 1981a; Kitano, Yabuzaki and Ogawa 1981b) was observed early in optical resonators that contained a Kerr material (Szoke et al. 1969; Gibbs, McCall and Venkatesan 1976). The nonlinearity is in this case the Kerr effect, which makes the optical length of the cavity intensity dependent. On the other hand, the power circulating in the optical cavity depends on the optical length. Depending on the cavity length, there may be several modes in which the system can operate.

Field modulation is not the only way to increase the effective dimension of the system. The light shift effect offers an alternative possibility: It changes the direction of the effective magnetic field, thus coupling the third component of the magnetisation to the system. Several groups have observed self-pulsing and chaos in optical resonators that contained resonant atomic vapours and were placed in a transverse magnetic field (Mitschke et al. 1986; Boden,

Dämmig and Mitschke 1992; Möller and Lange 1992). Similar effects occur in ring cavities (Ikeda 1979; Ikeda, Daido and Akimoto 1980; Hamilton, Ballagh and Sandle 1982). An optical resonator is not even necessary. Counter-propagating laser beams in atomic sodium, tuned close to the transition frequency of the D_1 resonance line, can show amplitude (Khitrova, Valley and Gibbs 1988) and polarisation (Gauthier, Malcuit and Boyd 1988) instabilities, which depend on the response time of the Kerr medium through a four-wave mixing process (Gaeta et al. 1987; Narum et al. 1988).

This behaviour can also be compared to the "NMR-laser" of Brun and coworkers (Marxer, Derighetti and Brun 1983; Brun et al. 1985), where a polarised system of nuclear spins is placed in a coil that is a part of a tuned rf circuit acting as a laser cavity. As in the present case, the dynamics of the NMR laser are determined by the coupling between the spin system and the magnetic field.

11
Mechanical effects of light

In the preceding chapters, we discussed the manipulation and observation of atomic variables associated with the internal degrees of freedom. A complete separation between internal and external degrees of freedom, like position and momentum, is not always possible, however. It therefore appears appropriate to include a discussion of the mechanical effects of light, which have received increasing attention in recent years. The transfer of momentum between photons and atoms, first observed more than sixty years ago, had to wait for the wide availability of lasers to become a useful experimental tool. In particular the cooling of atoms and atomic ions to very low temperatures has found numerous applications.

11.1 Light-induced forces

That light exerts mechanical forces on massive particles like atoms may appear surprising. To motivate the existence of such an effect two different approaches are possible. The first approach considers the light as a collection of photons that carry, apart from energy and angular momentum, linear momentum as well. Photons interacting with atoms can therefore change the momentum of the atoms. The second approach considers light as a wave, i.e., an inhomogeneous electromagnetic field interacting with the atomic dipole moment. Both approaches provide a possible description for the numerous phenomena that can occur in this context, but in many situations, one of them turns out to be more intuitive or more useful for calculations than the other.

Apart from the purely light-induced forces, there are also light-mediated forces that not only depend on the interaction between atoms and light, but also include other interactions. An important example is light-induced drift (Gel'mukhanov and Shalagan 1980; Werij et al. 1984; Mariotti et al. 1988; Gozzini et al. 1989; Meer et al. 1993). In these experiments, a laser beam controls the diffusive motion of atoms in a buffer gas. Depending on the laser

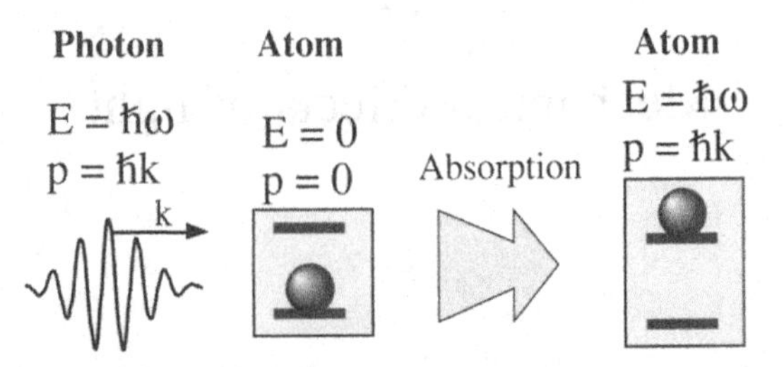

Figure 11.1. Conservation of energy and linear momentum: The atom must absorb not only the energy, but also the momentum of the photon. Left: photon with atom at rest. Right: the atom has absorbed the photon and acquired its momentum and energy.

frequency, it excites atoms with either positive or negative velocity along the laser beam direction into the excited state. If the diffusion cross sections of ground and excited states differ, this will result in a net transfer of atoms parallel or antiparallel to the laser beam direction. Here, we concentrate on free atoms and forces that arise only from the interaction of light with atoms.

11.1.1 Momentum conservation

Single photon absorption

From Maxwell's description of electromagnetic waves, it was known that light carries momentum, but it was Einstein's concept of a photon that turned this knowledge into a powerful idea (Einstein 1917). The consequences become easily visible if we consider the absorption of a photon by an atom.

Figure 11.1 summarises the process. Before the absorption process, the photon, represented here as a wave packet, has an energy $\hbar\omega$ and a momentum $\hbar k$, while the atom is in the electronic ground state at rest. After the absorption processes, the photon is no longer present. Since momentum is a conserved quantity, the photon momentum of $\hbar k$ cannot vanish during the absorption process, but must be absorbed by another part of the system. The only reservoir that can absorb the photon momentum in this system is the atomic velocity. In the final state, the atom must therefore be in an electronically excited state to accommodate the energy of the photon, and its velocity must have changed to accommodate the photon momentum. During the reverse process, the emission of a photon, the atom must change its velocity by another photon momentum, again conserving the total momentum. This velocity change of the atom is known as the photon recoil.

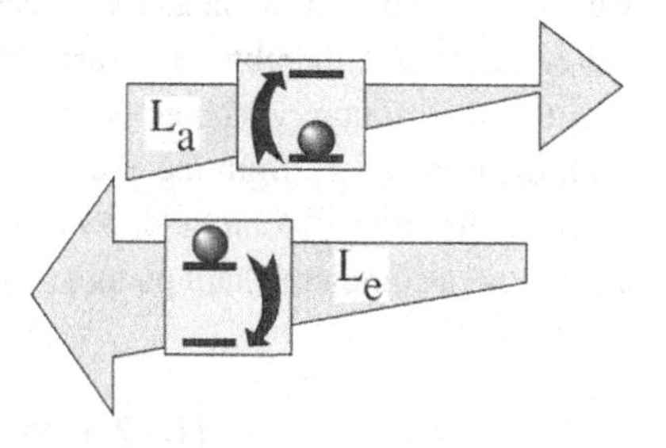

Figure 11.2. Stimulated absorption and stimulated emission by counterpropagating lasers. Beam L_a, which stimulates the absorption process, loses energy to the counterpropagating beam L_e, which stimulates the emission process.

Repeated events

The momentum change accompanied by the absorption or emission of a single photon has a relatively small effect on the momentum of an atom moving at thermal velocities. The importance of laser-induced forces relies on the possibility of repeating many events of this type in a short time. The photons absorbed by the atom are always supplied by a laser beam. The emission of a photon can occur spontaneously or it can be induced by the laser field. Since spontaneous emission is isotropic, the recoil associated with it averages to zero over many events. In experiments involving no induced emission, the net force on the atom results only from the absorption events. This type of force is known as the spontaneous scattering force.

If the emission process is induced by a laser field, the direction of propagation of the laser breaks the symmetry and selects a preferred direction in space. It is then possible also to achieve a net momentum transfer for the emission process. Since the net force induced by the laser results from the sum of the contributions of absorption and emission, the emission process must differ somehow from the absorption process. If the stimulated emission occurred under the same condition as the absorption process, the atom would return to its initial state and the net momentum transfer would vanish. An ideal situation occurs when the laser beam L_e that induces the emission process propagates in a direction opposite to L_a, which induces the absorption process. Every absorption–emission cycle transfers one photon from beam L_a into beam L_e, changing the momentum of the photons by $2\hbar k$, as shown in Figure 11.2.

Such an asymmetry between absorption and emission moves photons from one laser beam to the counterpropagating beam (Söding et al. 1992), thus changing the photon momentum and simultaneously the atomic momentum. Such an arrangement yields the largest momentum transfer per cycle, but it is not the only possibility. Other arrangements under discussion rely on spatial variations of the laser intensity, like tightly focused beams or evanescent waves, to achieve an imbalance between the directions of induced absorption and emission and thereby a net force on the atom.

In this description, which emphasises the particle properties of the radiation field, the force arises from the exchange of discrete packets of momentum between atoms and radiation field. As usual in quantum mechanics, it is also possible to use a complementary picture that describes the radiation field as a wave.

11.1.2 Optical potential

In this case, the force on the atom arises from the variation of the interaction energy between the atomic dipole moment and the electric field of the laser beam. As usual, the resulting force is the gradient of the potential energy, i.e., of the interaction energy. In the simplest case of a monochromatic plane wave, the only variation arises from the phase variation along the direction of propagation, which is proportional to the optical wavevector k. Under these conditions, the force acting on a two-level atom is proportional to this gradient, to the population of the excited state, and to the spontaneous emission rate (Stenholm 1986). This is the spontaneous scattering force, which corresponds in the particle picture to the absorption of photons, combined with spontaneous emission. In a monochromatic plane wave, it is the only force acting on a two-state atom.

In an inhomogeneous field, the gradient of the interaction potential also includes variations of the amplitude (Gordon and Ashkin 1980). The force corresponding to this part of the gradient is proportional to that component of the induced atomic dipole moment that is in phase with the exciting field. This contribution to the light-induced forces is known as dipole or gradient force. In the particle picture, it corresponds to cycles of induced absorption and emission and therefore behaves qualitatively differently from the spontaneous scattering force. Spontaneous emission nevertheless has an effect on this force since it determines the equilibrium state of the atomic system, as discussed in Chapter 2. The momentum transfer that mediates the force, however, occurs through induced processes. Correspondingly, the resulting force is conservative: The atom moves with very little damping if the experimental conditions are chosen such that dipole forces dominate.

As in the case of the internal dynamics considered in the preceding sections, the light-induced forces can be derived in the semiclassical approximation (Stenholm 1986). This approach yields the correct average forces, but it cannot describe the more subtle effects associated with the quantisation of the radiation field, in particular the heating effects associated with quantum fluctuations (Gordon and Ashkin 1980).

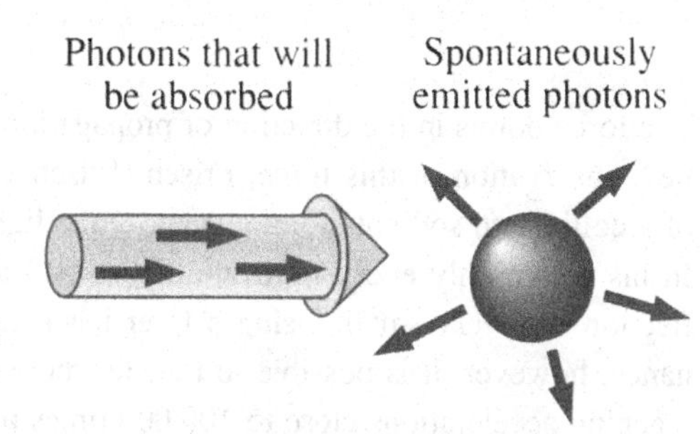

Figure 11.3. Directed absorption vs. isotropic spontaneous emission of photons.

11.2 Spontaneous forces

11.2.1 Scattering force

Origin

The first observation of the photon recoil was reported as early as 1933 (Frisch 1933). The conventional light sources that were available at that time, however, had only a small effect on the atoms. The momentum of a single visible photon ($\lambda = 589$ nm) is $1.125 \cdot 10^{-27}$ mkg/sec, more than four orders of magnitude less than the thermal momentum of an Na atom. The recoil velocity $\hbar k/M$ due to a single photon is thus rather small (3 cm/sec for sodium, corresponding to a Doppler frequency shift of 50 kHz), and the number of photons that an atom absorbs from a thermal light source is also quite small. This situation changed dramatically with the introduction of lasers. As a resonant laser beam can easily saturate an optical transition, an atom can absorb up to $6 \cdot 10^7$ photons per second. Obviously, the atoms must emit photons at the same rate. During the emission process, each photon also carries away one quantum of linear momentum.

As Figure 11.3 shows, however, the momentum carried by the spontaneously emitted photons is, in contrast to the absorbed momentum, not oriented. Its average vanishes and it has no directional effect on the motion of the atom. All the absorbed momentum, on the other hand, is directed along the laser beam. The contributions of the absorbed photon momenta therefore add coherently.

Average force

The momentum that an atom receives from the laser beam per unit time is $\hbar kr$, where r stands for the rate at which the atom absorbs and reemits photons. Using Newton's second law, this momentum transfer is usually described as a force $\vec{F}$ acting on the atom

$$\vec{F} = \hbar \vec{k}\, r \tag{11.1}$$

The force points in the direction of propagation of the laser beam. In the earliest observation of this force, Frisch (Frisch 1933) measured the deflection of a collimated sodium atomic beam by the light pressure of a sodium lamp. In his setup, only every third atom absorbed a photon, so the resulting deflection was very small. Using a laser beam tuned close to an optical resonance, however, it is possible to transfer more than 10^7 photons per second, reaching accelerations close to 100,000 times that of the earth's gravitational field.

The number of absorbed photons depends not only on the intensity of the light, but also on the frequency detuning between the laser and the atomic resonance frequency. In the laboratory frame, the resonance frequency depends on the atomic velocity through the Doppler shift $\vec{k} \cdot \vec{v}$. In a single dimension, the scattering rate for a free atom is, according to equation (2.59),

$$r = \Gamma_1 \frac{\omega_x^2}{\Gamma_1^2 + 4(\Delta\omega_0 + kv)^2} \tag{11.2}$$

where $\omega_x = \mu_e E/\hbar$ is the optical Rabi frequency, Γ_1 the decay rate, k the wave vector, and v the atomic velocity. We have simplified the formula to the case of small saturation. The laser detuning $\Delta\omega_0$ is the difference $\omega_0 - \omega_L$ between the resonance frequency ω_0 of the atom at rest and the laser frequency ω_L.

Inserting the scattering rate into equation (11.1), we obtain the velocity dependence of the spontaneous scattering force

$$F = \hbar k\, \Gamma_1 \frac{\omega_x^2}{\Gamma_1^2 + 4(\Delta\omega_0 + kv)^2} \tag{11.3}$$

It reaches a maximum for atoms with velocity $v = -\Delta\omega_0/k$, which Doppler shifts the atom into resonance, and decreases with a half width of $v_{1/2} = \Gamma_1/k$.

Frequency chirp

This force can accelerate or decelerate atoms at a high rate (Ertmer et al. 1985; Prodan et al. 1985). One consequence of the velocity change, however, is that the Doppler shift associated with this velocity change soon moves the atom out of resonance with the laser beam. For sodium, every scattered photon shifts the resonance frequency by 50 kHz. Correspondingly, one hundred photons shift the resonance frequency by the homogeneous width of the optical reso-

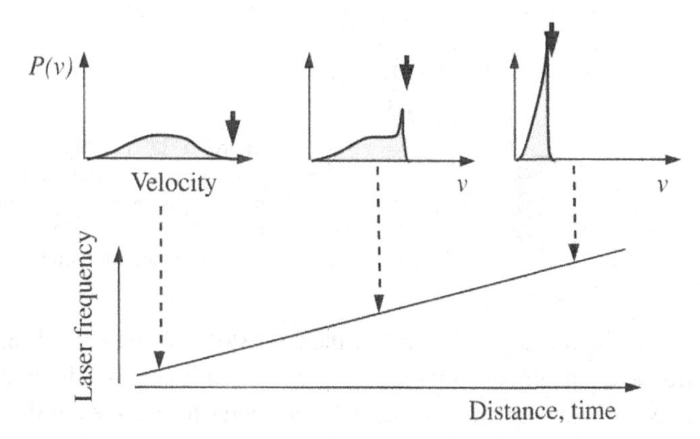

Figure 11.4. Effect of a frequency chirp on the velocity distribution of atoms in a beam.

nance line. More generally, the frequency shift exceeds the homogeneous linewidth after the absorption and reemission of

$$n_\Gamma = \Gamma_2 \frac{M}{\hbar k^2} \tag{11.4}$$

photons, where M is the atomic mass. The atoms are then no longer resonant with the laser and the force acting on them drops significantly.

Since the Doppler shift of atoms moving at thermal velocities is significantly larger than the homogeneous linewidth over which atoms are resonant, a large change of the atomic velocity requires some means by which to adjust the laser detuning to the changing Doppler shift. A typical application of light-induced forces that depends on this possibility is the slowing and stopping of atomic beams (Ertmer et al. 1985). One possible solution is to sweep the laser frequency through the spectrum, following the resonance frequency of the atoms as their velocity decreases (Balykin, Letokhov and Mishin 1980a; Balykin, Letokhov and Mishin 1980b). This was one of the earliest methods used to decelerate an atomic beam. It requires a pulsed atomic beam to allow the laser frequency to be resonant with a single packet of atoms while they are decelerated.

Figure 11.4 shows the effect of such a frequency chirp when the laser decelerates an atomic beam. The curves at the top of the figure represent the velocity distribution at sequential locations. At the left-hand part of the figure, where the velocity distribution is thermal, the laser frequency is resonant with those atoms travelling at the highest velocity towards the laser. As these atoms

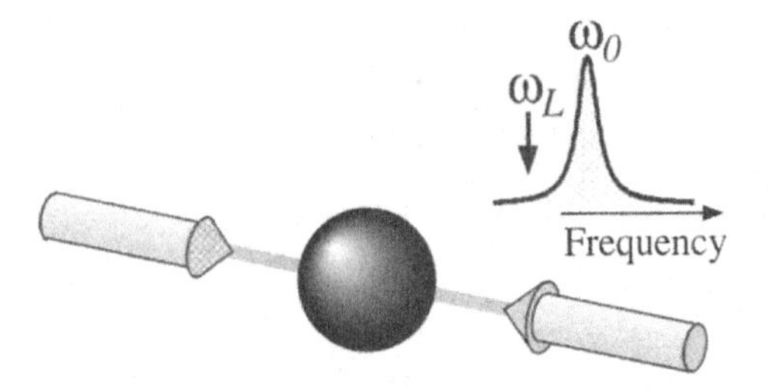

Figure 11.5. Principle of laser cooling: Two counterpropagating laser beams are slightly detuned to the red wing of the atomic resonance line.

travel to the right, the opposing force of the laser slows them down. Their resonance frequency therefore shifts towards shorter wavelengths. The laser frequency must therefore be swept towards higher frequencies to retain the slowing effect on the atoms. In this way, it is possible to reduce the atomic velocity until the atoms come to a stop or even reverse their motion (Ertmer et al. 1985).

11.2.2 Doppler cooling

Principle

One possible application of light-induced forces is the reduction of atomic velocities to zero relative to the laboratory. Although stopping an atomic beam is an important step towards this goal, it has its limitations: Slowing the velocity in one direction broadens the velocity distribution perpendicular to the laser beam to some extent. In addition, zero velocity is not preferred over other values – depending on the laser frequency, the atoms can end up moving in the original direction or backwards. The ideal goal of reducing all velocity components to zero requires a setup that prefers this particular velocity group. One possible selection mechanism, known from Doppler-free spectroscopy, uses two counterpropagating laser beams. In such a laser field, only those atoms that are at rest relative to the laser see a monochromatic field. Atoms with a nonvanishing velocity component in the direction of the two lasers see the two frequencies Doppler shifted in opposite directions.

Figure 11.5 shows schematically how two laser beams can reduce the atomic velocity distribution along their axis. The two laser beams propagate in opposite directions. Their frequencies are detuned slightly to the red wing of the atomic resonance line. For an atom at rest, the two counterpropagating laser beams exert equal but opposite forces so their effects cancel. If the atom is moving towards one of the sources, however, it sees the corresponding light blue-shifted and therefore closer to resonance. It absorbs more photons from this laser beam than from the copropagating one and feels a net force that increases with its velocity. Over a certain range, the resulting force is proportional to the velocity, with a negative proportionality constant. Like a very

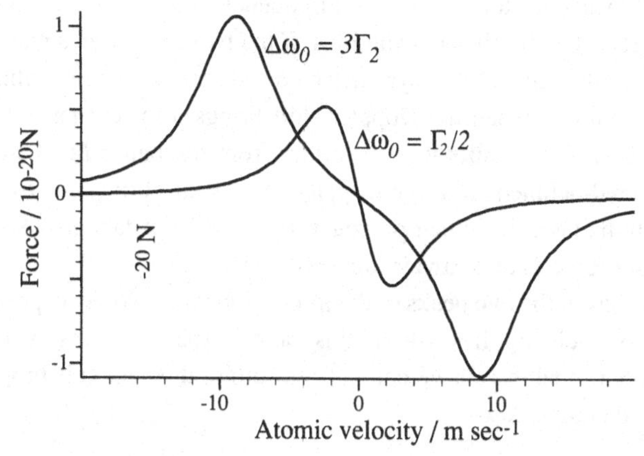

Figure 11.6. Velocity dependence of the force due to two counterpropagating beams with different detunings $\Delta\omega_0$. The optical Rabi frequency is $\omega_x = 3 \cdot 10^7\ \mathrm{sec}^{-1}$.

viscous medium, the light slows the atoms down efficiently. This setup is therefore known as "optical molasses."

Force

To calculate the force on the atom, we assume that the absorption of photons from the two counterpropagating laser beams is independent, one from the other, which is a good approximation for intensities well below saturation. The force is then the combined effect of the two individual laser beams, which contribute

$$F_{\pm} = \pm\, \hbar k\, r_{\pm} \tag{11.5}$$

where the rate

$$r_{\pm} = \frac{\Gamma_1\, \omega_x^2}{\Gamma_1^2 + 4(\Delta\omega_0 \pm kv)^2} \tag{11.6}$$

refers to the number of photons scattered from the two beams, valid for low intensities. The two contributions combine to a total force

$$F = \hbar k\, \Gamma_1\, \omega_x^2 \left(\frac{1}{\Gamma_1^2 + 4(\Delta\omega_0 + kv)^2} - \frac{1}{\Gamma_1^2 + 4(\Delta\omega_0 - kv)^2} \right) \tag{11.7}$$

Figure 11.6 shows this force as a function of atomic velocity for two different laser detunings $\Delta\omega_0$. In both cases, the laser is detuned towards longer wavelengths from the atomic resonance frequency of the motionless atom. For an atom at rest ($v = 0$), the contributions of the two laser beams exactly cancel, as we would expect from symmetry considerations. The resulting force reaches a maximum when the Doppler shift brings it to resonance with one of the two beams, but shifts it even further from resonance for the opposite beam. For a red-detuned laser, the Doppler shift always brings it closer to resonance with the laser beam that propagates against its instantaneous velocity. The resulting force is thus always opposed to its velocity.

The positions of the two peaks of the force define the velocity capture range. Atoms whose velocity lies within this range experience a slowing force roughly proportional to their velocity. The width of this region is proportional to the laser detuning $\Delta\omega_0$

$$|v_{\text{capture}}| \leq |\Delta\omega_0/k| \tag{11.8}$$

Figure 11.6 shows how a larger laser detuning increases the capture range. This increased capture range occurs at the expense of a smaller friction force near zero velocity, as the reduced slope of the curve in Figure 11.6 shows.

Molasses

Within the capture range, the atoms experience a friction force roughly proportional to their velocity

$$F_{\text{om}} = -\eta\, v \tag{11.9}$$

with η representing a friction coefficient. In this regime, the atomic motion obeys Stokes's law. The two counterpropagating beams create the effective viscous medium known as optical molasses. We determine the friction coefficient by expanding the expression (11.7) for small velocities,

$$\eta = \frac{\partial F}{\partial v}\bigg|\, v = 0 = 16\, \Delta\omega_0\, \Gamma_1\, \hbar k^2 \,\frac{\omega_x^2}{(\Gamma_1^2 + 4\Delta\omega_0{}^2)^2} \tag{11.10}$$

This expression may be simplified somewhat by using the photon scattering rate for atoms at rest

$$r_0 = \frac{\Gamma_1\, \omega_x^2}{\Gamma_1^2 + 4\Delta\omega_0^2} \tag{11.11}$$

Figure 11.7. Atomic momentum before and after an absorption-emission cycle. p_0 and p are the atomic momenta before and after the cycle, p_{abs} is the absorbed momentum, and p_{em} is the change by the spontaneous emision.

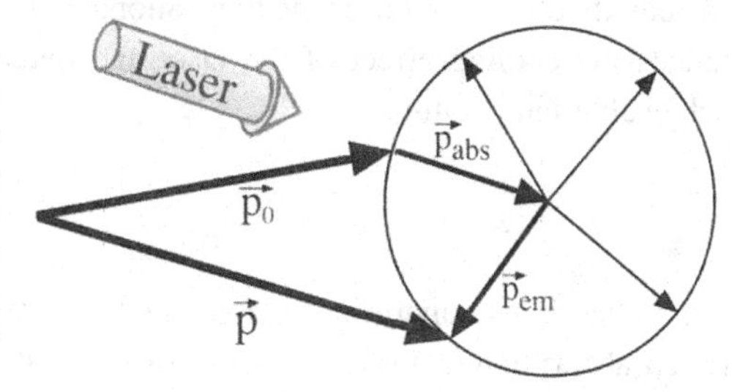

The viscosity then becomes

$$\eta = \frac{16\,\Delta\omega_0\, r_0\, \hbar k^2}{\Gamma_1^2 + 4\Delta\omega_0{}^2} \tag{11.12}$$

The viscosity has a dispersionlike dependence on the laser detuning. It reaches a maximum when the laser is tuned to the red wing of the absorption line and changes sign at the origin. This optically induced viscosity decreases the atomic velocity very effectively.

11.2.3 Velocity diffusion

Phenomenology

Within this simple model, the atomic velocity should decrease exponentially to zero. Simply by waiting long enough, it should be possible to reach arbitrarily low temperatures. Since the cooling effect is proportional to the temperature, however, the system eventually reaches an equilibrium at that temperature where the velocity-dependent cooling force is balanced by heating effects that are often independent of temperature. Although many heating processes, like collisions with background atoms, can be reduced in principle, there is one contribution directly associated with the cooling process itself and therefore difficult to eliminate. It arises from the granular nature of the absorption process and is known as velocity diffusion. (See Figure 11.7.)

As discussed above, spontaneous emission has no effect on the average velocity; it does, however, change the spread of the atomic velocities. After the emission of a single photon, the atomic momentum may be anywhere on a sphere whose centre is the atomic momentum before the spontaneous emission and whose radius is the momentum transferred by a single photon. A sequence of spontaneous emissions corresponds to a random walk in momentum space, which broadens the atomic velocity distribution. This effect, known as velocity diffusion, raises the translational temperature of the atoms. It counteracts the cooling effect of the laser and determines a limit on the lowest achievable temperature.

Dynamics

For a quantitative analysis, we compare the atomic momentum before and after an absorption–emission cycle for a single photon. Writing $\vec{p}_0$ for the initial atomic momentum, we obtain the final momentum as (Einstein 1917)

$$\vec{p} = \vec{p}_0 + \hbar\vec{k}_{\rm L} - \hbar\vec{k}_{\rm se} \tag{11.13}$$

where the wavevector $\vec{k}_{\rm L}$ refers to the laser beam and $\vec{k}_{\rm se}$ to the spontaneously emitted photon. The second term, which describes the recoil during the absorption of the photon, points in the direction of the laser beam. Although we do not include the average effect of the laser here, the spacing between the individual absorption events varies stochastically, forcing us also to include a stochastic term in the equations of motion. The third term in equation (11.13), which describes the momentum change by the emitted photon, is isotropic.

To find the effect of the velocity diffusion on the kinetic energy, we calculate the average change in translational energy per absorption–emission cycle

$$\Delta E_{\rm t} = \frac{1}{2m}\langle \Delta p^2 \rangle = \frac{1}{2m}\langle (\vec{p}_0 + \hbar\vec{k}_{\rm L} - \hbar\vec{k}_{\rm se})^2 - \vec{p}_0^{\,2} \rangle$$

$$= \frac{1}{2m}\hbar[2\langle \vec{p}_0\vec{k}_{\rm L}\rangle - 2\langle \vec{p}_0\vec{k}_{\rm se}\rangle + \hbar\vec{k}_{\rm L}^2 + \hbar\vec{k}_{\rm se}^2 - 2\hbar\vec{k}_{\rm L}\vec{k}_{\rm se}] \tag{11.14}$$

The cross terms between the three contributions to the final momentum all vanish, since the three vectors $\vec{p}_0$, $\vec{k}_{\rm L}$, and $\vec{k}_{\rm se}$ are all uncorrelated with each other. The remaining terms yield

$$\Delta E_{\rm t} = \frac{(\hbar k)^2}{m} \tag{11.15}$$

For every scattered photon, the translational atomic energy increases on average by twice the photon recoil energy, since absorption and emission contribute identical amounts.

The number of absorption–emission events required to slow down the atom over the capture range defined by the optical linewidth is large enough that we can treat the sequence of absorption–emission events as a continuous process. If the atom scatters photons from the two laser beams at a total rate $r_+ + r_-$, its translational energy increases due to velocity diffusion as

$$\frac{dE_t}{dt} = \frac{r_+ + r_-}{m} (\hbar k)^2 \tag{11.16}$$

This is the most important heating process that counteracts the cooling process of the optical molasses.

11.2.4 Doppler limit

Translational temperature

Boltzmann's law allows us to rewrite the translational energy as a temperature. For a single degree of freedom, the relation is

$$E_t = \frac{1}{2} k_B T \tag{11.17}$$

Correspondingly, the translational temperature of an atomic gas with a single degree of freedom may be defined as

$$T_t = 2 = \frac{\langle E_t \rangle}{k_B} = \frac{m}{k_B} \langle v^2 \rangle \tag{11.18}$$

This definition of the translational temperature implies that the atomic velocity has a Maxwellian distribution, but it is conventionally applied to a much wider range of distributions. Quotations of temperatures in most cases imply nothing about the actual velocity distribution, but are used as a parametrisation of the mean kinetic energy.

The friction of the optical molasses and the velocity diffusion process change the kinetic energy by

$$\frac{dT_t}{dt} = \frac{m}{k_B} \frac{d\langle v^2 \rangle}{dt} = \frac{2}{k_B} \left\{ \frac{r_+ + r_-}{m} (\hbar k)^2 - \eta v^2 \right\} \tag{11.19}$$

The effect of velocity diffusion becomes relevant only when the Doppler shifts are well below the homogeneous linewidth. We may therefore simplify the expression by approximating the sum of the two scattering rates by twice the scattering rate at zero velocity, putting $r_+ + r_- = 2\,r_0$. Using equation (11.18) to express the velocity by the temperature and using equation (11.12) for the viscosity, we obtain

$$\frac{\mathrm{d}T_\mathrm{t}}{\mathrm{d}t} = \frac{4r_0}{k_\mathrm{B}m}(\hbar k)^2 - \frac{16\,\Delta\omega_0\,r_0\,\hbar\,k^2}{\Gamma_1^2 + 4\Delta\omega_0^2}\,\frac{2}{m}\,T_\mathrm{t} \tag{11.20}$$

The first, temperature-independent term on the right-hand side represents the heating due to velocity diffusion; the second term, which is proportional to the temperature, represents the cooling effect of the optical molasses. The different temperature dependence causes the system to reach an equilibrium temperature at which the two terms cancel.

Temperature limit

The stationary situation is reached when

$$\frac{\partial T_\mathrm{t}}{\partial t} = 0 \Rightarrow (\Gamma_1^2 + 4\Delta\omega_0^2)\,\hbar = 8\,\Delta\omega_0\,k_\mathrm{B}T_\mathrm{t} \tag{11.21}$$

This corresponds to a temperature

$$T_\mathrm{t} = \hbar\,\frac{\Gamma_1^2 + 4\Delta\omega_0^2}{8\,\Delta\omega_0\,k_\mathrm{B}} \tag{11.22}$$

The final temperature therefore depends on the laser frequency, not on the laser intensity. This is only true as long as the intensity remains well below the saturation intensity, as we have assumed here. The temperature can be minimised by setting the laser frequency detuning to half the homogeneous linewidth, $\Delta\omega_0 = \Gamma_2 = \Gamma_1/2$. At this detuning, the temperature becomes

$$T_\mathrm{D} = \frac{\hbar\,\Gamma_1}{2\,k_\mathrm{B}} \tag{11.23}$$

which is known as the Doppler limit. For the Na D lines, the corresponding temperature is 240 μK. Here, we have assumed that laser cooling operates in only one dimension. For laser cooling in more than one dimension, the equilibrium temperature changes by factors close to unity.

Historical

This method for cooling atoms with laser light was suggested by Hänsch and Schawlow (Hänsch and Schawlow 1975) for cooling of atomic vapours and independently by Wineland and Dehmelt (Wineland and Dehmelt 1975) for cooling of trapped ions. First results for cooling of trapped ions were published in 1978 (Neuhauser et al. 1978; Wineland, Drullinger and Walls 1978). Balykin and colleagues (Balykin, Letokhov and Mishin 1980a; Balykin, Letokhov and Mishin 1980b) reported the successful deceleration of atoms in an atomic beam using a frequency chirp of the laser; a group from NBS (Phillips and Metcalf 1982) used a different technique for the same purpose. The Russian group (Balykin et al. 1985; Balykin, Letokhov and Sidorov 1986) achieved transverse cooling of atomic beams in one and two dimensions. Steve Chu and colleagues (Chu et al. 1985) from AT&T Bell Labs reported three-dimensional results. These first experiments found temperatures that were, within experimental error, equal to the Doppler temperature.

Soon afterwards, even lower temperatures were reported: 40 μK in sodium (Lett et al. 1988) and 2.5 μK in cesium (Salomon et al. 1990), both corresponding to an average atomic momentum of only 3.5 photon momenta. These results indicated that there were other, more important, mechanisms which were more efficient than the Doppler cooling considered here. These deviations prompted further theoretical work to remove these difficulties from the theoretical model (Dalibard and Cohen-Tannoudji 1989; Ungar et al. 1989; Shang et al. 1991). It turned out that the additional mechanisms all depended on the details of the atomic level structure and would not occur in two-level atoms. We discuss some of these mechanisms in Section 11.4.

11.3 Stimulated forces

11.3.1 Gradient force

Interaction

The gradient or electric dipole force is another manifestation of the mechanical effect of light on the atoms whose properties differ significantly from those of the spontaneous scattering force. Its basic characteristics can be derived directly from a classical description of the field. Classically, the interaction energy between the atom and the field is

$$E_{\text{pot}} = -\vec{d} \cdot \vec{E} \tag{11.24}$$

For atoms, which have no permanent electric dipole moment, the dipole is induced by the field and the energy becomes

$$E_{\text{pot}} = -\frac{\alpha}{2} E^2 \tag{11.25}$$

where α is the polarisability of the atom. Using equation (2.54) to express the in-phase component of the induced dipole moment, we obtain

$$E_{\text{pot}} = -\frac{\mu_e E}{2} \frac{4\,\Delta\omega_0\,\omega_x}{\Gamma_1^2 + 4\Delta\omega_0^2 + 2\omega_x^2} \tag{11.26}$$

According to this equation, the potential energy of an atom in an intense laser field can be positive or negative, depending on the laser detuning $\Delta\omega_0$. This effect can be understood classically: For a red-detuned laser ($\Delta\omega_0 > 0$), the induced dipole moment is in phase with the driving field and the atom can lower its energy by moving to the high field region. For a blue-detuned laser, the induced dipole lags behind the field by more than $\pi/2$ and the interaction energy changes sign.

If we are interested in studying the gradient force alone, it is convenient to use a large laser detuning so that the first and third terms in the denominator of equation (11.26) can be neglected.

$$E_{\text{pot}} = -\frac{\mu_e E}{2} \frac{\omega_x}{\Delta\omega_0} \tag{11.27}$$

The force acting on the atom is proportional to the gradient of the energy. In the one-dimensional case, the force is (Minogin and Serimaa 1979; Dalibard and Cohen-Tannoudji 1985)

$$F = -\vec{\nabla} E_{\text{pot}} = \hbar \frac{\omega_x^2}{\Delta\omega_0} \frac{1}{E} \frac{\partial E}{\partial x} \tag{11.28}$$

Photon picture

Instead of this semiclassical analysis of the gradient force, it is also possible to derive the effect in a quantum mechanical picture. In particular, the dressed state analysis of Dalibard and Cohen-Tannoudji (Dalibard and Cohen-Tannoudji 1985) has proved very successful. On a semiquantitative level, we can also understand the force as arising from stimulated absorption and emission

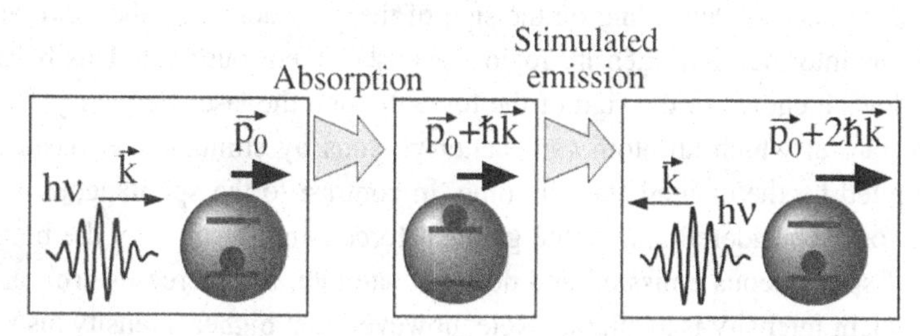

Figure 11.8. Gradient force in a standing wave as an absorption–stimulated-emission sequence.

events. In a standing wave, the atom may absorb a photon from one of the two running waves and reemit it into the other wave, as shown in Figure 11.8.

Both stimulated processes transfer the momentum of one photon to the atom. Together, they change the atomic momentum in units of $2\,\hbar\,\vec{k}$ or multiples thereof (Gould, Ruff and Pritchard 1986). The rate at which photons are absorbed and reemitted is

$$r_{\rm d} = \frac{\omega_x^2}{4\,\Delta\omega_0} \tag{11.29}$$

if the field strength varies as

$$\frac{\partial E}{\partial x} = kE_0 \tag{11.30}$$

In a standing wave, the gradient changes sign every quarter wavelength. The stimulated force changes its sign with the gradient. In a periodic potential like a standing wave, the stimulated momentum transfer therefore becomes inefficient and averages out over macroscopic distances.

Properties

Since the optical Rabi frequency ω_x is proportional to the electric field amplitude, the gradient force is proportional to the field strength, the field gradient, and the laser detuning. It vanishes in a homogeneous field, but can be quite strong in an inhomogeneous field. The largest field gradients appear when the light intensity changes from zero to its maximum over a distance comparable to an optical wavelength. The most common geometries that allow strong forces over short distances include standing waves, tightly focused laser beams, and evanescent waves. Depending on the sign of the laser detuning, the field pulls the atoms into the high-intensity region or pushes them out of it. This behaviour allows a choice of the sign of the force through the laser frequency.

The rate at which an atom can scatter photons by stimulated emission is not limited by the excited state lifetime. In contrast to the spontaneous scattering force considered above, the gradient force is not limited by the bottleneck of spontaneous emission and does not saturate, but increases arbitrarily if sufficient intensity is available. Note, however, that higher intensity also requires higher detuning to prevent the transition from saturating and to reduce unwanted spontaneous emission. The absence of spontaneous processes, which destroy the phase coherence of the atomic state, is important for experiments where the phase of the atomic system must be conserved. Since the increase of the dipole moment near an atomic resonance is important for making this force strong enough, spontaneous emission processes can never be suppressed completely, but it is often possible to reduce them sufficiently by combining sufficiently large detuning of the laser with high intensity.

11.3.2 Applications

Optical traps

Apart from the possibility of increasing the gradient force to arbitrarily high values, it has the interesting property that it depends much less on the details of the level structure than the spontaneous scattering force. Good examples that demonstrate the application of the gradient force to systems whose optical properties are not known in detail include early applications of the gradient force for levitating macroscopic particles in a focused laser beam (Ashkin 1970b; Ashkin 1970a; Ashkin and Dziedzik 1971). The possibility of trapping even living organisms in the focus of a laser beam (Ashkin, Dziedzic and Yamane 1987; Ashkin and Dziedziec 1987) has bean summarised by calling the laser beams "optical tweezers." They have found various applications in biology, e.g. for manipulating individual DNA molecules (Berns et al. 1989; Perkins et al. 1994; Perkins, Smith and Chu 1994). The insensitivity to the details of the level structure also allows application of the gradient force to molecules (Berns et al. 1989). The problem that the gradient force averages to zero over macroscopic distances requires the use of laser fields with spatial variations comparable to the optical wavelength.

Early experiments with the dipole force used a focused laser beam to create a high-intensity region into which the atom is pulled by the gradient force F_d, as indicated in Figure 11.9. The spontaneous scattering force F_{sp} simultaneously pushes it in the direction of the laser beam. Which of the forces dom-

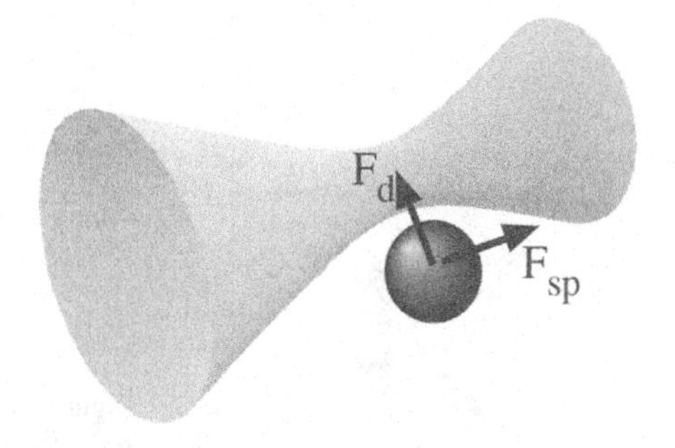

Figure 11.9. Focusing of a laser beam creates a small high-field region. If the laser is red-detuned, the gradient force, which is indicated by F_d, pulls the atom into the high-field region. The spontaneous force F_{sp} pushes it in the direction of the beam.

inates depends on the detuning and intensity of the laser. It was proposed early that such an arrangement should allow trapping of neutral atoms (Ashkin 1978; Ashkin and Gordon 1979), but the depth of the trap was too low for efficient confinement. In addition, the trapping laser heats the atoms through velocity diffusion (Gordon and Ashkin 1980). Traps for atomic ions had already been built (Fischer 1959; Paul 1990), using magnetic (Penning 1936) and electric (Paul 1990) fields to create an effective potential that keeps the ions in the trap. In the trap the ions could be cooled optically (Neuhauser et al. 1978; Wineland, Drullinger and Walls 1978).

Trapping of neutral atoms with optical fields had to await laser cooling to provide atoms with low enough kinetic energy. In laser-optical traps (Chu et al. 1986), the cooling lasers decelerate the atoms to a low velocity and an electromagnetic potential pulls the atoms towards the centre of the trap. Trapping without cooling is difficult if not impossible: The trapping force is a conservative potential and atoms falling into the trap acquire sufficient energy to leave it again, unless some dissipation mechanism is available that removes this energy. In most cases, this mechanism is the spontaneous scattering force or another cooling mechanism discussed below. The first trap for neutral atoms was reported by a group at AT&T Bell Labs (Chu et al. 1986). They used a pulsed atomic beam to provide the atoms and a chirped laser beam to decelerate them. Three orthogonal pairs of counterpropagating laser beams created optical molasses that cooled the atoms to low enough temperatures that another laser beam could trap them in its focus.

Mirrors for atoms

Another setup that achieves large gradients of the laser intensity uses an evanescent wave. If an optical field is incident on an interface between two dielectrics from the optically denser side and its angle of incidence is larger

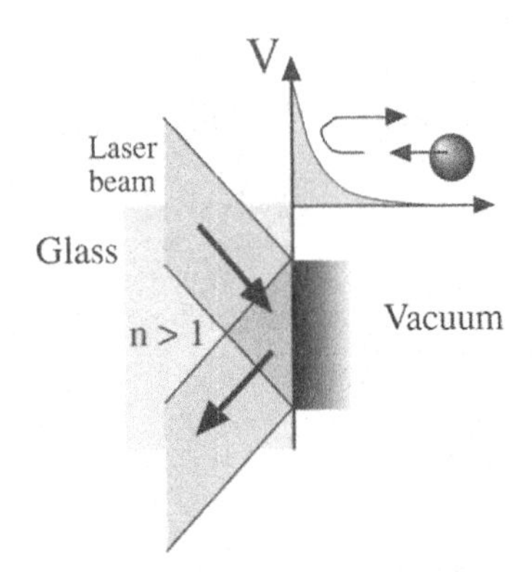

Figure 11.10. The evanescent field of a totally reflected laser beam creates an effective potential for an atom. For a blue-detuned laser beam, the atom is reflected.

than the critical angle, it is totally reflected at the interface. The field does not stop suddenly at the interface, however, but drops off exponentially as a function of the distance from the interface, as shown in Figure 11.10. The decay distance is of the order of an optical wavelength.

The fast decay of the laser field provides a large intensity gradient and therefore a strong dipole force. The possibility of using this evanescent wave as a mirror for atoms was first suggested by Cook and Hill (Cook and Hill 1982). They calculated that a field of 1 W cm^{-2} should be sufficient to reflect atoms whose velocity component perpendicular to the wall was a few metres per second. If the kinetic energy of the atoms is higher than the electromagnetic potential of the wave, it penetrates through the potential barrier and interacts directly with the wall. A group at the Russian Institute of Spectroscopy (Balykin et al. 1987; Balykin et al. 1988) implemented this proposal. They directed an atomic beam to the evanescent wave at a small angle and observed its deflection when the light was turned on. More recently, cold cesium atoms were observed to bounce off an evanescent field in a gravitational cavity (Aminoff et al. 1993).

Beam splitters

Standing light waves form the third arrangement that allows large intensity gradients. At high enough intensity, it is possible to confine a well collimated atomic beam to the nodes or the antinodes of the field, depending on the laser detuning (Salomon et al. 1987). Under suitable conditions, this potential variation can also act like a grating on a collimated atomic beam, splitting it into several partial beams whose momenta differ by multiples of $2\hbar\vec{k}$ from that of the incident beam, where $\vec{k}$ is the wavevector of the laser field perpendicular to the incident atomic beam. (See Figure 11.11.)

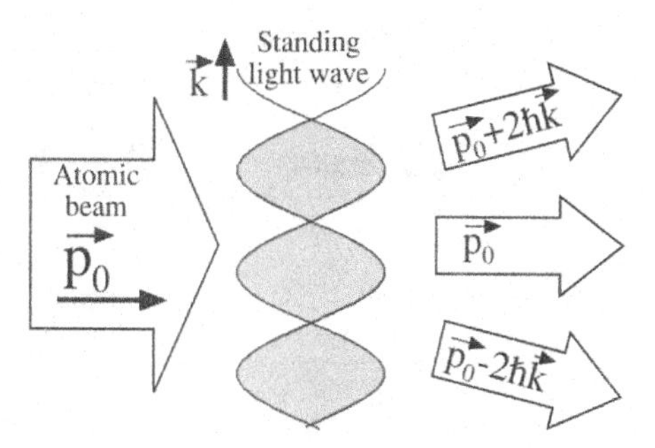

Figure 11.11. A standing laser beam acting as a beam splitter.

Since the gradient force averages to zero over macroscopic distances, this momentum transfer is only possible if the deflection of the partial beams within the standing wave laser beam is small compared to an optical wavelength. This can also be understood in terms of momentum conservation: The momentum transfer cannot change the absolute value of the atomic momentum, only its direction (Martin et al. 1988). The momentum transferred from the laser field must therefore be perpendicular to that of the incident atoms and the laser field must contain a wavevector component perpendicular to the atomic velocity vector. This requirement is fulfilled in a focused laser beam, where the distribution of photon momenta is wide enough. The light can then split the atomic beam into a large number of partial beams (Gould, Ruff and Pritchard 1986). If, on the other hand, the laser beam approximates a plane wave, it can only transfer momentum to atoms travelling in a very narrow region around the matching condition. In this case, momentum conservation restricts the number of deflected beams to one, in close analogy to Bragg scattering of X-rays from a crystal (Martin et al. 1988).

Lenses

Focusing of an atomic beam becomes possible if the interaction between the electric field and the atomic dipole moment creates an effective potential that has a parabolic spatial dependence.

Figure 11.12 represents three possibilities of implementing a potential that is approximately quadratic in the region of the atomic beam. The first implementation of this idea by Bjorkholm and colleagues (Bjorkholm et al. 1978; Bjorkholm et al. 1980) used the Gaussian profile of a laser beam copropagating and superimposed on the atomic beam, as shown in the left-hand side of Figure 11.12. The laser frequency is red-detuned in this case, pulling the atoms towards the centre of the beam. The situation is reversed when a TEM_{01}

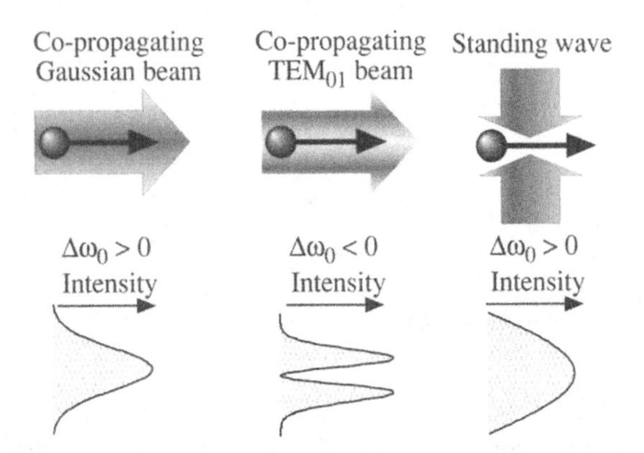

Figure 11.12. Three schemes for the construction of lenses for atomic beams.

beam is used instead, which has a node at the centre of the beam, as shown in the second column of the figure. In this case, the laser frequency must be detuned to the blue to push the atoms towards the low-intensity central part of the beam. Instead of a copropagating beam, it is also possible to use a standing wave field perpendicular to the atomic beam (Sleator et al. 1992), as shown in the right-hand side of the figure. The atoms must then pass through the antinode of the field to be focused in one dimension. As a first step towards technological applications, this lensing effect has allowed control of the deposition of chromium atoms on a silicon substrate (McClelland et al. 1993).

11.3.3 Rectified dipole force

Stimulated forces like the gradient or dipole force have several important advantages over the spontaneous scattering force: They do not saturate with increasing intensity of the laser field and can therefore be made arbitrarily strong. In addition, the experimental conditions can be chosen such that spontaneous emission of photons is minimised while the stimulated forces are still large. This is important in the context of experiments that rely on atomic coherence, which is destroyed by spontaneous emission. Since the stimulated forces rely on the spatial variation of the interaction energy, however, they vary in space and their average over distances larger than an optical wavelength vanishes.

This problem can be solved by "rectification" of the dipole force (Kazantsev and Krasnov 1989; Grimm et al. 1990; Ovchinnikov et al. 1994). In a normal standing wave, the dipole force changes sign every quarter wavelength. The rectification reverses this sign change, thus retaining a nonvanishing spatial average over macroscopic distances.

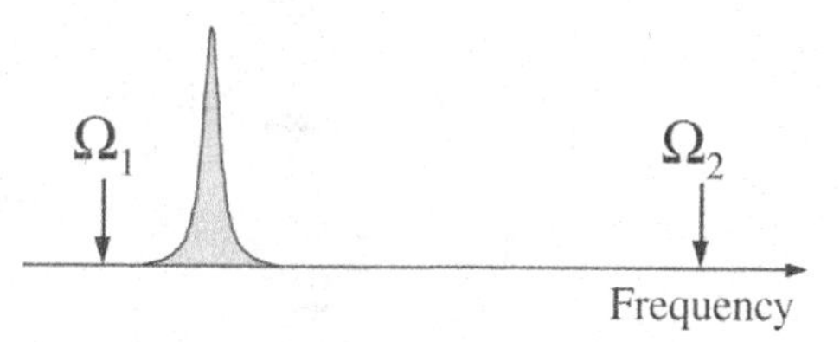

Figure 11.13. Bichromatic irradiation for rectified dipole force.

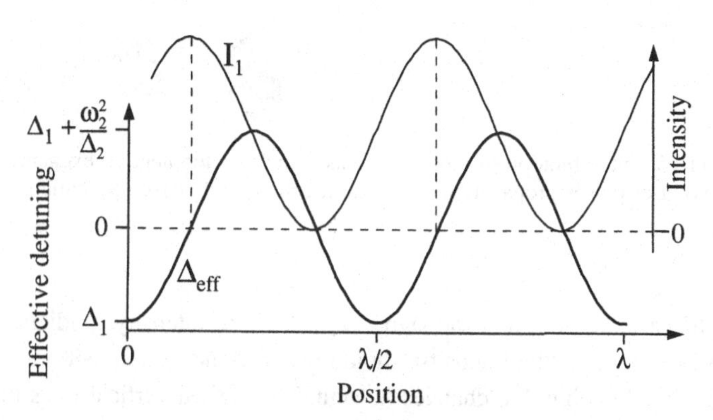

Figure 11.14. Spatial variation of the effective detuning Δ_{eff} and the intensity of the laser field component at Ω_1.

Frequency-domain interpretation

One possibility uses a bichromatic standing wave formed by two counter-propagating laser beams.

Figure 11.13 shows the frequencies of the two laser fields. Both are detuned from the atomic resonance frequency. The field at frequency Ω_1 generates the actual force, whereas the second field, which is considerably further from resonance, causes a light shift of the atomic resonance line. As discussed in Section 3.5.4, this light shift is for large detunings equal to the square of the optical Rabi frequency ω_2 divided by the resonance detuning Δ_2

$$\delta_2 = \frac{\omega_2^2}{4\Delta_2} \tag{11.31}$$

In the standing light wave, the intensity and the Rabi frequency are modulated with a period of half a wavelength. The resonance frequency of the atom, dressed with the laser field at Ω_2, varies correspondingly with this spatial period.

Figure 11.14 shows how the effective detuning $\Delta_{eff} = \Delta_1 + \delta_2$ varies over one wavelength. To optimise the rectification effect, the variation should be

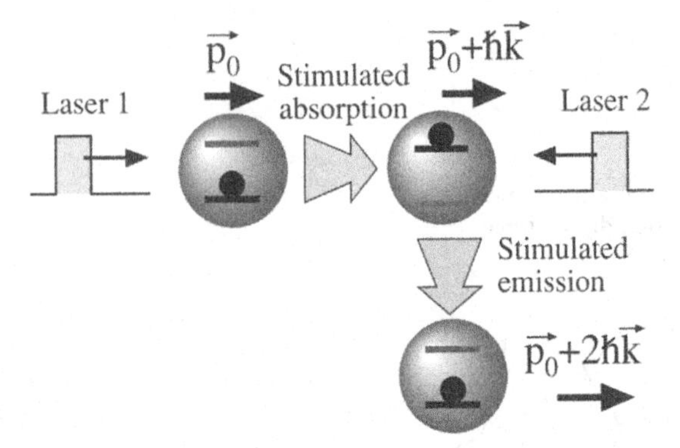

Figure 11.15. Interpretation of dipole force as stimulated momentum exchange. The two laser pulses propagate in opposite directions and arrive sequentially.

symmetric around zero and the sign reversal of the intensity gradient of I_1, i.e., an extremum of the intensity, should occur at the same positions where the effective detuning Δ_{eff} changes its sign. The dashed vertical lines in Figure 11.14 mark the positions of this sign change. As the figure shows, the nodes of the two standing waves must be shifted against each other by a quarter of a spatial period. Under these conditions, the two sign changes cancel, thus "rectifying" the dipole force.

Time-domain interpretation

As discussed above, it is possible to interpret the dipole or gradient force as a sequence of stimulated absorption–emission events. Rectification of this process would require that the atom always absorb light from beam one and emit photons into beam two, as indicated in Figure 11.15.

Conceptually, this could be accomplished by irradiating the atom with a π pulse in forward direction, thus bringing it into the excited state and transferring one quantum of momentum from the photon. A subsequent π pulse from the backward beam would stimulate the emission of a photon, thus adding a quantum of momentum in the same direction (Voitsekhovich et al. 1989; Voitsekhovich et al. 1991). In contrast to this simplified picture, the true laser beams do not consist of π pulses, but have instead a sinusoidally modulated intensity, which may be interpreted as an experimentally convenient implementation of π pulses.

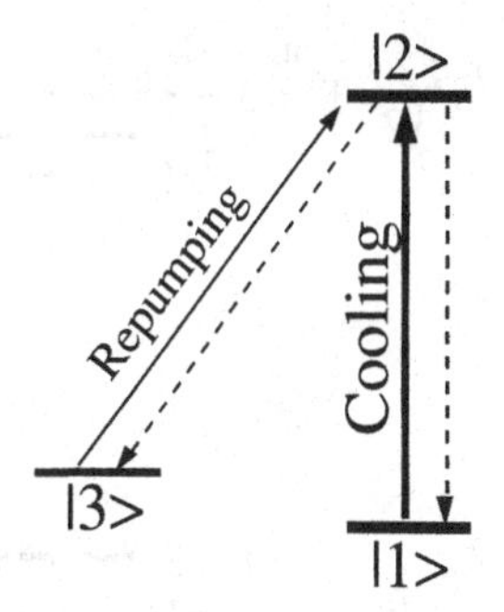

Figure 11.16. Cooling and repumping transitions in multilevel systems.

11.4 Forces on multilevel atoms

11.4.1 Multilevel effects

So far, we have treated the atoms as two-level systems. This is a convenient model to illustrate the basic principles, but actual experiments always involve more complicated atoms. When multilevel atoms interact with laser beams, the same effects discussed for two-level atoms cause mechanical forces on the atoms. The internal degrees of freedom of these atoms, however, change the interaction with the radiation field qualitatively. Although the extended level structure provides some difficulties, it also provides more flexibility. As an example, it is possible to cool atoms to much lower temperatures than in two-level atoms.

Optical pumping

One effect that occurs only in multilevel atoms is that spontaneous emission may bring them into a state different from their initial state. This represents a difficulty in the case of laser cooling, where the atoms must scatter a large number of photons before they accumulate enough momentum from the laser field. Even if an atom cycles several thousand times through the cooling transitions before it "escapes," the momentum it has acquired until then is not enough sufficiently to decelerate a thermal atom. Since most atoms have possible escape routes, it is usually necessary to use a second laser field for pumping them back into the cooling transition.

Figure 11.16 illustrates the principle: The cooling laser operates on the transition $|1\rangle \rightarrow |2\rangle$. If an atom falls into the long-lived level $|3\rangle$, whose energy differs from that of level $|1\rangle$, it is no longer resonant with the cooling laser.

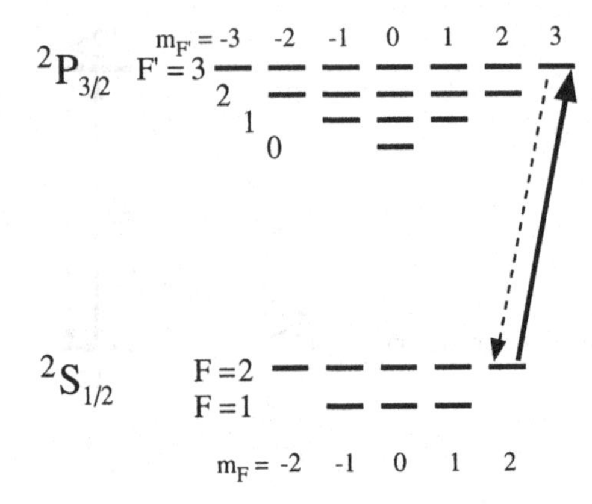

Figure 11.17. Example of a cycling transition in the D_2 transition of sodium.

A second laser field is required to excite the escaped atoms to level $|2\rangle$, whence they may participate again in the cooling process. A typical example is the hyperfine structure of the alkali. The cooling process uses the transition from one of the hyperfine levels to the $^2P_{3/2}$ excited state, whence atoms may decay to either of two ground-state multiplets.

The repumping problem can be reduced significantly if the cooling laser couples to a transition between a ground state $|g\rangle$ and an excited state $|e\rangle$ such that the only decay path from the excited state $|e\rangle$ leads back to $|g\rangle$. The two levels then form a closed two-level system.

Figure 11.17 shows an example of such a cycling transition. The level scheme represents the D_2 transition of atomic sodium. Circularly polarised light couples the $|\text{g};F = 2, m_F = 2\rangle$ state only to the $|\text{e};F = 3, m_F = 3\rangle$ state. The selection rule $\Delta m_F = 0, \pm 1$ for the spontaneous emission process forces the excited state to decay to the initial state. Optical pumping can transfer the population to this ground-state sublevel, thus forcing all atoms into this two-level system. The two levels thus form a good approximation of a two-level system.

The approximation is not perfect, however. Although the hyperfine splitting of the excited state is several times larger than the optical linewidth, the atom may still make a transition to one of the other excited state sublevels, from which it can decay to the second hyperfine multiplet. If this process occurs before the atom has scattered the large number of photons required to stop it, a repumping laser is required.

Magnetic dipole force

Multilevel atoms have a permanent magnetic dipole moment, which allows use of static magnetic fields for applying forces to the atoms. In close anal-

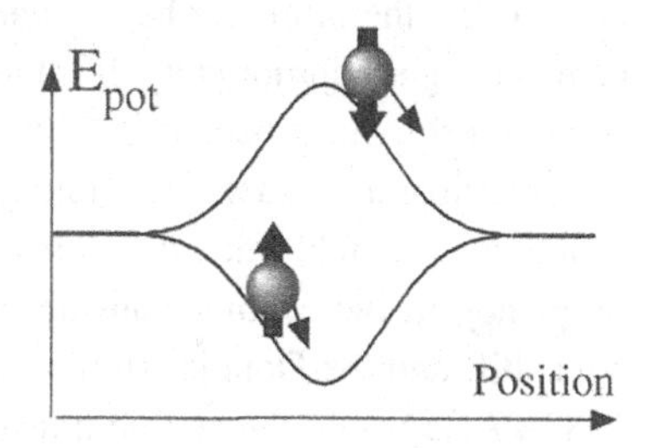

Figure 11.18. In an inhomogeneous magnetic field, the force on the atoms depends on its spin state. The atom may be pushed out of the high field region or drawn into it.

ogy to an inhomogeneous laser field that pushes the atoms towards the potential energy minimum, an inhomogeneous magnetic field pushes them towards the location of the minimum or maximum field strength, depending on the spin orientation.

Figure 11.18 illustrates this behaviour for spins with opposite spin orientation. The magnetic field strength has a maximum at the centre of the figure, generating extrema in the potential for the two opposite spin states. The spin-down particle is pushed away from this maximum, while the spin-up particle is pulled into it. This effect is known from the Stern–Gerlach experiment (Gerlach and Stern 1924), where it separates the spin states. Magnetic dipole forces are also used for storing cold neutrons in magnetic storage rings (Kügler, Paul and Trinks 1978).

The possibility of using this force for trapping neutral atoms was suggested by Heer (Heer 1960; Heer 1963). The depth of such a trap is relatively low. For an atom with a magnetic moment of one Bohr magneton, a field strength of two Tesla is required to confine atoms at a temperature of one Kelvin. The implementation therefore had to await laser cooling to prepare sufficiently cold atoms (Migdall et al. 1985; Bagnato et al. 1987; Cornell, Monroe and Wieman 1991). Evaporative cooling in such a trap lowered the temperature further until Bose–Einstein condensation occurred (Cornell et al. 1995).

Zeeman tuning

When spontaneous scattering forces are used to decelerate atoms in an atomic beam from their thermal velocity to rest, the laser field must remain resonant with the atoms throughout the whole process, while the Doppler shift of the atom decreases by several hundred MHz. With conventional lasers, this is possible only if either the laser frequency or the atomic resonance frequency is adjusted in such a way that the difference between them remains constant. If the laser frequency is swept, as in the case of chirp cooling (Balykin, Letokhov and Mishin 1980a; Balykin, Letokhov and Mishin 1980b), discussed in Section 11.2.1, the laser can be resonant only with a group of atoms whose velocity changes appropriately. It is then not possible to operate the atomic beam continuously, but it becomes necessary to switch to pulsed operation.

A possible alternative uses tuning of the atomic energies: A space-dependent magnetic field shifts the atomic energies in such a way that the resonance frequency of the cycling transition remains constant as the sum of the Zeeman shift and the Doppler shift compensate each other (Phillips and Metcalf 1982). As a result, the atomic transition frequency remains constant throughout the deceleration process and the atomic beam can be operated continuously. This scheme works only for a particular atomic state. This is also required in laser cooling experiments, however, which must prepare the atoms in a specific magnetic substate that couples to a cycling transition, as explained above. In most experimental implementations, a tapered solenoid generates the variation of the magnetic field strength.

11.4.2 Magnetooptic traps

Principle

For many applications, cooling of atoms is not sufficient. In addition, it is necessary to trap them, i.e., to localise them at a position in space. This requires, besides the velocity-dependent force, a position-dependent force that prevents the atoms from escaping the trapping region. The Zeeman tuning mechanism explained above provides the most popular implementation of such a scheme (Raab et al. 1987). It uses inhomogeneous magnetic fields to shift the energy levels of the atoms in such a way that whenever they move towards the edge of the trapping region, they get closer to resonance with that laser beam that pushes them back towards the centre of the trap.

Figure 11.19 illustrates the principle for a one-dimensional arrangement. Two counterpropagating laser beams with opposite circular polarisation drive two transitions of a $J = 0 \leftrightarrow J' = 1$ transition. The frequency of both beams is red-detuned from the atomic resonance frequency. The magnetic field shifts the $m_{J'} = \pm 1$ states in opposite directions. If the atom moves towards the left, the Zeeman effect brings the σ_- transition closer to resonance and moves the σ_+ transition out of resonance, as indicated by the inset on the left-hand side. The atom therefore absorbs more photons from the laser beam propagating towards the centre of the trap and the associated spontaneous scattering force pushes the atom back towards the centre of the trap. The same effect acts in the opposite direction if the atoms move out of the trap towards the right, as the inset at the right of the figure illustrates.

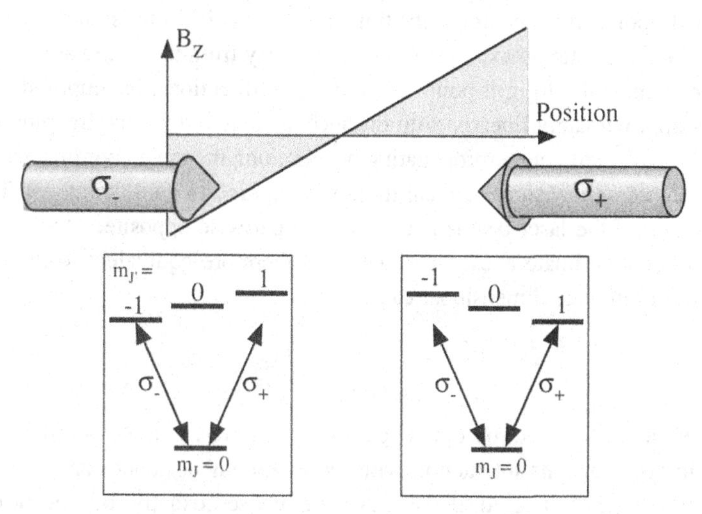

Figure 11.19. Use of Zeeman tuning for the implementation of a space-dependent force. The field is parallel to the two counterpropagating laser beams and varies through zero in the centre of the setup. The two insets show the resulting level schemes at the left- and right-hand parts of the setup for a $J=0 \leftrightarrow J'=1$ transition.

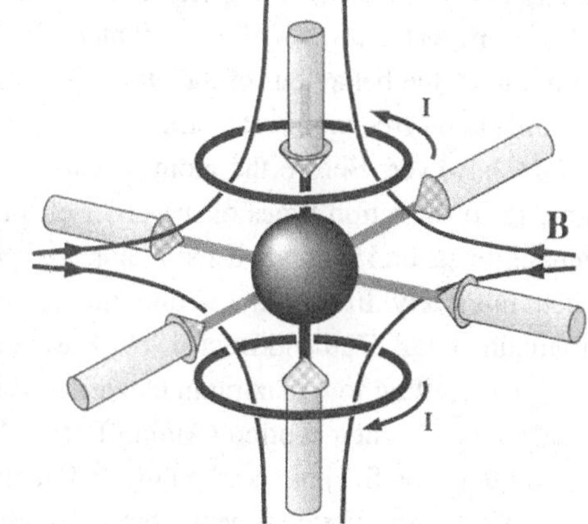

Figure 11.20. Arrangement of laser beams and magnetic field for a magnetooptic trap. The three pairs of counterpropagating laser beams are mutually perpendicularand circularly polarised. The two loops in anti-Helmholtz arrangement form a quadrupole magnetic field.

Three-dimensional arrangement

This idea can be extended to three dimensions by superimposing a quadrupole magnetic field over three pairs of circularly polarised counterpropagating laser beams, as shown in Figure 11.20.

The two field coils are arranged in an anti-Helmholtz configuration carrying a current in opposite direction. The resulting field has the characteristics of a quadrupole. At the centre of the trap, the field vanishes for symmetry reasons. Along the rotation axis, the field points away from the centre and in the plane perpendicular to it, it points inward. In all directions, its amplitude increases approximately linearly with distance, as required for the trapping effect. Three pairs of counterpropagating beams along the three coordinate axes form the three-dimensional optical molasses required to cool the atoms. The polarisation of the laser beams is circular and pairwise opposite. Along each of the three coordinate axes, the situation is therefore equivalent to the one discussed in the one-dimensional case.

Applications

The ideal atom for spectroscopic experiments is at rest at a well-defined position in space and its interaction with the environment under complete experimental control. This ideal will never be realised exactly, but recent experiments are approaching this goal quite closely. Laser cooling has reduced atomic velocity and therefore the Doppler shift to values negligible compared to the natural linewidth. This has made it possible to localise the atoms in various kinds of traps, keeping them available for spectroscopic studies for a virtually unlimited time. In experiments performed on earth, this is only possible as long as the trapping fields counteract the gravitational field. Under these conditions, the behaviour of the atoms is distinct from that of free atoms. Experiments on free atoms can still be performed by switching off the trapping fields, however. Before the atoms escape from the trap, they are still available for observation times of the order of milliseconds. This time can be extended to several seconds if the atoms are accelerated upwards before letting them fall freely. In the gravitational field, the atoms will then form an "atomic fountain" (Hall, Zhu and Buch 1989; Kasevich and Chu 1991; Kasevich et al. 1991), travelling to a maximum elevation while gravity decelerates and eventually reverses their motion (Aminoff et al. 1993).

Although the field of laser cooling and trapping is still young, there is already a growing body of experiments that use these newly developed techniques to perform measurements on atomic systems with unsurpassed precision. Most applications use cooling and trapping to confine the atoms to well-defined spatial locations and eliminate Doppler broadening while simultaneously extending the effective lifetime, i.e., the time that the atoms spend inside the experimental area. This possibility should be of vital importance in future time standards, where the time that the atoms spend inside the measurement area limits the precision of the measurement (Hall, Zhu and Buch 1989). Another example is a precision measurement of the gravitational field (Kasevich and Chu 1991).

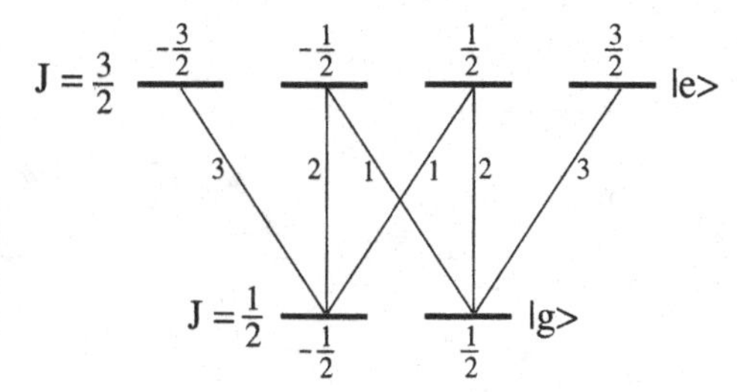

Figure 11.21. Level scheme considered for Sisyphus cooling. The numbers close to the transitions indicate relative transition strengths.

11.4.3 Sisyphus cooling

Soon after the implementation of laser cooling for neutral atoms, several mechanisms were discovered that allowed multilevel atoms to be cooled to temperatures below the Doppler limit. We summarise here one of the relevant mechanisms to give an idea how it is possible to break the Doppler limit.

System

Figure 11.21 shows a possible energy level scheme for the analysis of Sisyphus cooling. It consists of an electronic ground state with angular momentum $J = 1/2$ and an excited state with angular momentum $J = 3/2$. The numbers near the transitions indicate relative transition strengths, i.e., the squares of the dipole matrix elements. The scheme discussed here uses two counter-propagating beams with orthogonal linear polarisations, as indicated in the upper half of Figure 11.22.

As the upper half of Figure 11.22 shows, the two laser beams interfere to form a field whose polarisation changes along the laser beam. When the two laser fields are in phase, their superimposed light is linearly polarised at $\pm 45^\circ$ from the polarisation of the individual beams. Over a distance of $\lambda/8$, the relative phase between the two beams changes by $\pi/2$, making the light circularly polarised. The atoms that interact with the superposition of the two beams experience fast variations of the interaction.

Mechanism

To understand how Sisyphus cooling works in this system, we need to consider the variation of the optical pumping and the light shift effect along the laser beam. As discussed in Section 3.5.4, a laser that couples to a transition between a ground-state sublevel and a sublevel of the excited state causes a light shift proportional to the square of the electric dipole matrix element for

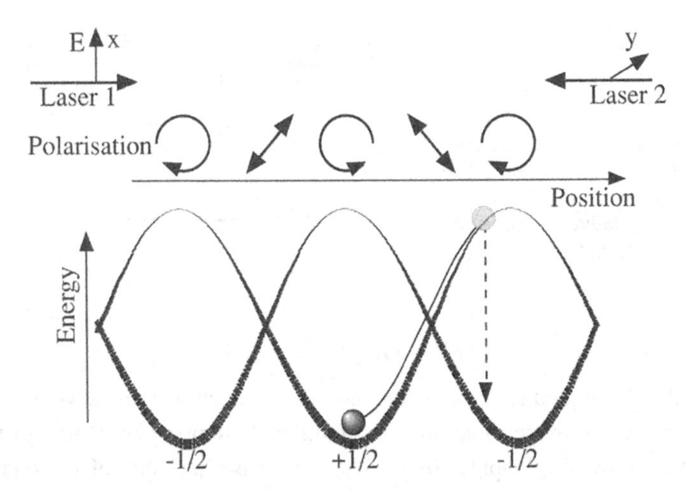

Figure 11.22. Principle of Sisyphus cooling in the level scheme of Figure 11.21: The upper half indicates the variation of the polarisation along the laser beam if the polarisations of the two counterpropagating beams are orthogonal. The lower half of the figure indicates the light-shifted energies of the two ground-state sublevels.

that transition. In the level system of Figure 11.21, we find that σ_+ light couples more strongly to the $m_J = +1/2$ ground-state sublevel than the $-1/2$ sublevel. For off-resonance irradiation, both levels are frequency shifted, but the effect is three times stronger for the $+1/2$ state. Conversely, σ_- light shifts the $-1/2$ level more strongly. The lower half of Figure 11.22 shows the energies of these states as a function of the position along the laser beam.

Simultaneously, the optical pumping changes the equilibrium populations of the sublevels. Figure 11.22 shows the populations of the two states along the direction of the laser beam as the thickness of the two lines that represent the two ground states. Apparently, the combination of light shift and optical pumping always enhances the population of the state with the lowest energy.

The trajectory indicated in the level scheme of Figure 11.22 shows one example as to how the motion of the atom in this potential causes a reduction of the kinetic energy. The atom symbolised by the ball starts in the $|g;1/2\rangle$ level, at the position where the energy of this state goes through its minimum. The atomic momentum moves it towards the right. Since the energy of the atomic state increases along the trajectory, the atom loses momentum. After a distance $\lambda/4$, the sense of the laser field polarisation has inverted and the light shift now favours the other state. At this position, the probability for an optical pumping process into the opposite state is highest. The spontaneous emission brings the atom into a state with lower internal energy, without changing its kinetic energy. The motion between the two potential minima has thus reduced the kinetic energy of the atom by the amount of the light shift effect.

The most efficient cooling is obtained if the atoms travel, on average, half a period of the optical potential before being optically pumped. Higher pump rates increase the heating by velocity diffusion without enhancing the cooling effect, whereas lower pump rates make the cooling process too slow.

Recoil limit

As the atoms lose momentum, they require more time to reach the maximum of the light-shift–induced potential. Since the optical pumping rate P is proportional to the intensity, the condition for optimised pumping requires that the laser intensity be kept low to minimise the temperature. The time required for the atom to reach thermal equilibrium can then become arbitrarily long. The description of the cooling process presented here becomes invalid at low intensities and the temperature cannot be lowered any further simply by reducing the intensity. The next limit that becomes important is the single photon recoil limit, i.e., the temperature velocity spread due to the spontaneous emission of a single photon.

At this temperature, the de Broglie wavelength of the atoms becomes comparable to the optical wavelength, which defines the variations of the optical potential. Under these conditions, it is no longer possible to treat the motion of the atoms classically, but it becomes necessary to use a quantum mechanical description that takes the finite extension of the wave packet into account. The finite length of the de Broglie wavelength can become directly observable in magnetooptical traps, where the atoms are cooled to low enough velocities. The three-dimensional polarisation gradients generate a three-dimensional potential in whose minima the atoms can be bound in a crystallike arrangement (Grynberg et al. 1993; Hemmerich, Zimmermann and Hänsch 1993). The atoms then have stationary states best seen as vibrational states in an effective potential.

11.4.4 Stimulated magnetooptic force

Experimental arrangement

The comparison of spontaneous and stimulated forces acting on two-level atoms showed that stimulated forces are not subject to the spontaneous emission bottleneck and may therefore become arbitrarily large at high laser intensities. On the other hand, the example of the electric dipole force showed

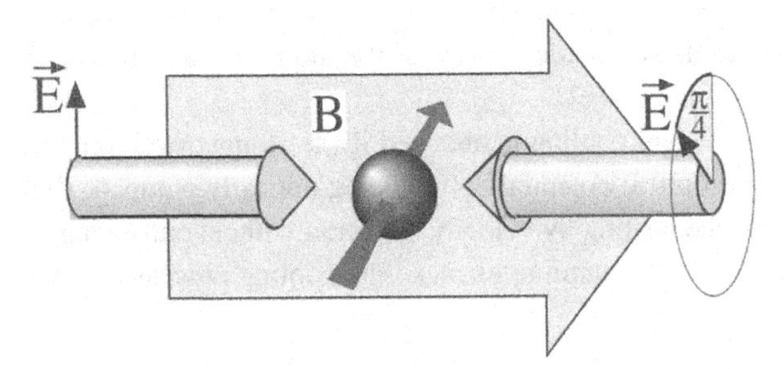

Figure 11.23. Arrangement for the stimulated magnetooptic force. The polarisation of the two counterpropagating laser beams is linear, with an angle of 45° between them. The magnetic field along the laser beams determines the direction of the resulting force.

that for stimulated forces, it is more difficult to obtain a nonvanishing spatial average. In the case of multilevel atoms, the level structure and the existence of a magnetic dipole moment add degrees of freedom that can be exploited for constructing useful effective forces. A good example for this possibility is the stimulated magnetooptic radiation force (Grimm et al. 1991a; Grimm et al. 1991b; Söding et al. 1992; Grimm et al. 1993; Söding and Grimm 1994), whose principle is outlined in Figure 11.23.

Two counterpropagating laser beams generate the force. The polarisations of the two beams are linear, with an angle $\pi/4$ between them. A magnetic field is oriented parallel to one of the two beams. The orientation of the magnetic field, together with the sign of the angle between the two optical polarisations, determines the direction of the resulting force.

Effective force

The stimulated magnetooptic force can again be understood as arising from the stimulated redistribution of photons between the two laser beams. As in the case of the rectified dipole force, the atom absorbs photons preferentially from one of the two laser beams and emits them in a stimulated process into the second beam. A complete absorption–emission cycle transfers two quanta of momentum between the laser beams and the atom. The task of the magnetic field in this process is to make the stimulated emission asymmetric. It suppresses the stimulated emission into the beam from which the photon was absorbed and enhances the emission into the other beam.

This magnetic-field–induced asymmetry is quite analogous to the Hanle effect: The interaction with the first laser beam induces an atomic dipole moment that couples not only to the laser field, but also to the magnetic field. The Larmor precession rotates the dipole moment in such a way that the interaction with the first laser beam decreases when the atom is in the excited state, simultaneously increasing the interaction with the counterpropagating beam. To optimise this effect, it is necessary to synchronise the Larmor precession with the optical Rabi frequency. For a given laser intensity, the force shows a dispersionlike dependence on the magnetic field strength, reaching a maximum when the Larmor frequency becomes $\Omega_L = \omega_x/\alpha$, where α is a numerical factor of order unity that depends on the angular momentum configuration of the transition. For a $J = 0 \leftrightarrow J' = 1$ transition, e.g., $\alpha = \sqrt{8}$. The stimulated magnetooptic force has found applications in a coherent beam splitter for atoms (Pfau et al. 1993) and appears to allow the construction of alternative magnetooptic traps (Emile et al. 1992).

11.4.5 Raman transitions

The "Doppler limit" for the temperature that can be reached with spontaneous force cooling of two-level atoms can be understood as originating from the uncertainty of the excited state energy. The temperature is directly proportional to the natural linewidth of the atomic resonance line. Cooling to lower temperatures, i.e., "measuring" the atomic velocity more precisely, requires longer interaction times. This is possible if both states involved in the transition are electronic ground states. The only limit to their lifetime is the time they spend inside the experimental apparatus. Laser radiation cannot excite transitions between electronic ground states directly. Instead, one uses a Raman scheme, where Λ-type multilevel atoms interact with two frequency components of a laser field.

This cooling scheme allows cooling not only below the Doppler limit, but even below the single photon recoil limit (Aspect et al. 1988; Aspect et al. 1989), reaching temperatures as low as 100 nK in one dimension (Kasevich and Chu 1992) or 650 nK in three dimensions (Kastberg et al. 1994). The long lifetime of the ground-state levels involved in the Raman transition leads to a very narrow linewidth. As the discussion of coherent population trapping in Section 3.4.1 showed, the trapping is extremely sensitive to the mismatch between the laser detunings of the two involved fields. Only if the frequency difference between the two fields exactly matches the energy separation of the two states that are coupled in the Raman transition can the trapping be effective. If the two fields propagate in opposite directions, this sensitivity to the frequency difference becomes a sensitivity for velocity. Any velocity component in the direction of the laser beams shifts their two frequencies in the

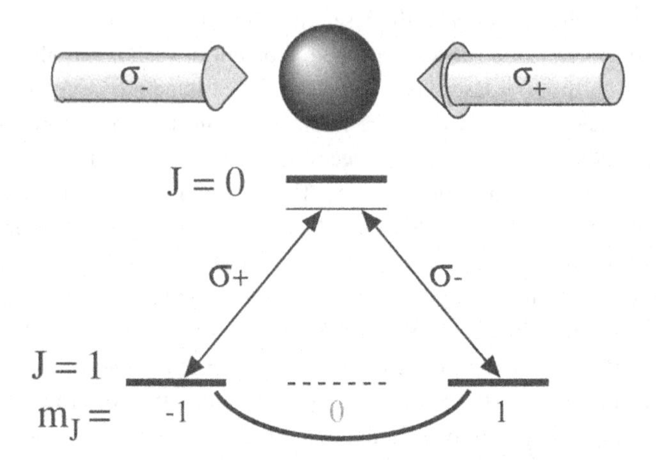

Figure 11.24. Velocity-selective coherent population trapping in a $J = 1$ ground state with two counterpropagating laser beams. The curved line between the two ground states with $m_J = \pm 1$ indicates the trapping state. The third level is not involved in the process.

atomic rest frame in opposite directions. Only the atoms at rest with respect to the laser experience zero frequency difference and fall into the trapping state.

Figure 11.24 summarises the basic experimental setup. Two counterpropagating laser beams with opposite circular polarisations drive the Raman transition. The two laser beams couple to different transitions between ground-state sublevels and the $J = 0$ excited state of the model atom. Together, they can induce Raman transitions between the $m_J = \pm 1$ sublevels, thereby exciting a sublevel coherence between the two states. In close analogy to coherent population trapping, Raman pulses can specifically transfer those atoms that have a low velocity into states that no longer couple to the field (Kasevich et al. 1991). Atoms that end up in this state remain trapped there, as they no longer absorb photons. The Raman transition that excites this state requires equal laser frequencies in both beams. If both beams originate from the same laser, they have identical frequencies in the rest frame of the atom only if its velocity component in the direction of the laser beams vanishes. This condition selects atoms with small velocities for the trapping state.

The Raman pulse by itself does not cool the atoms; it merely selects the coldest ones from the thermal ensemble. The two laser beams also form an optical molasses, however, and the velocity diffusion constantly thermalises the atoms not in the trapping state, thus creating more cold atoms that can be trapped. Once they are in the trapping state, they are no longer subject to the velocity diffusion process. The Raman process therefore "protects" the coldest part of the ensemble from the thermalisation effect. This scheme allows very low temperatures to be reached, which are limited only by the duration of the interaction time. At these low temperatures, the essentially classical picture that we use here for describing the atoms reaches its limits, as the de Broglie wavelength increases and may become larger than the optical wavelength (Cohen-Tannoudji 1992a).

References

Abella, I. D., N. A. Kurnit, and S. R. Hartmann (1966). Photon echoes. *Phys. Rev. 141*, 391–406.

Abragam, A. (1961). *The principles of nuclear magnetism.* Oxford University Press, Oxford.

Abramochkin, E., and V. Volostnikov (1991). Beam transformations and nontransformed beams. *Optics Commun. 83*, 123–135.

Adams, C. S., M. Sigel, and J. Mlynek (1994). Atom optics. *Phys. Rep. 240*, 143–210.

Aitchison, J. S., A. M. Weiner, Y. Silberberg, M. K. Oliver, J. L. Jackel, D. E. Leaird, E. M. Vogel, and P. W. E. Smith (1990). Observation of spatial optical solitons in a nonlinear glass waveguide. *Optics Lett. 15*, 471–473.

Aleksandrov, E. B. (1964). Luminescence beats induced by pulse excitation of coherent states. *Opt. Spectroscopy 17*, 522–523.

Allen, L., and J. H. Eberly (1987). *Optical resonance and two-level atoms.* Dover Publications, Mineola, NY.

Allen, L., M. W. Beijersbergen, R. J. C. Spreeuw, and J. P. Woerdman (1992). Orbital angular momentum of light and the transformation of Laguerre-Gaussian laser modes, *Phys. Rev. A 45*, 8185–8189.

Alzetta, G., A. Gozzini, L. Moi, and G. Orriols (1976). An experimental method for the observation of r.f. transitions and laser beat resonances in oriented Na vapour. *Nuovo Cimento 36B*, 5–20.

Ambrose, W. P., and W. E. Moerner (1991). Fluorescence spectroscopy and spectral diffusion of single impurity molecules in a crystal. *Nature 349*, 225–227.

Aminoff, C. G., J. Javanainen, and M. Kaivola (1983). Collision effects in velocity-selective optical pumping of sodium. *Phys. Rev. A 28*, 722–737.

Aminoff, C. G., A. M. Steane, P. Bouyer, P. Desbiolles, J. Dalibard, and C. Cohen-Tannoudji (1993). Cesium atoms bouncing in a stable gravitational cavity. *Phys. Rev. Lett. 71*, 3083–3086.

Anderson, L. W. (1979). Optically pumped electron spin polarized targets for use in the production of polarized ion beams. *Nucl. Instrum. Meth. 167*, 363–370.

Appelt, S., P. Scheufler, and M. Mehring (1989). Direct observation of single- and double-quantum sublevel coherence in Rubidium vapor by optical Raman beat detection. *Optics Commun. 74*, 110–114.

Arditi, M., and T. R. Carver (1961). Pressure, light and temperature shifts in optical detection of 0–0 hyperfine resonance in alkali metals. *Phys. Rev. 124*, 800–809.

Arimondo, E., and G. Orriols (1976). Nonabsorbing atomic coherences by coherent two-photon transitions in a three-level optical pumping. *Lett. Nuovo Cimento 17,* 333–338.

Ashkin, A. (1970a). Acceleration and trapping of particles by radiation pressure. *Phys. Rev. Lett. 24,* 156–159.

Ashkin, A. (1970b). Atomic-beam deflection by resonance-radiation pressure. *Phys. Rev. Lett. 25,* 1321–1324.

Ashkin, A. (1978). Trapping of atoms by resonance radiation pressure. *Phys. Rev. Lett. 40,* 729–732.

Ashkin, A., and J. M. Dziedzik (1971). Optical levitation by radiation pressure. *Appl. Phys. Lett. 19,* 283–285.

Ashkin, A., and J. M. Dziedziec (1987). Optical trapping and manipulation of viruses and bacteria. *Science 235,* 1517–1520.

Ashkin, A., and J. Gordon (1979). Cooling and trapping of atoms by resonance radiation pressure. *Optics Lett. 4,* 161–163.

Ashkin, A., J. M. Dziedzic, and T. Yamane (1987). Optical trapping and manipulation of single cells using infrared laser beams. *Nature 330,* 769–771.

Askar'yan, G. A. (1962). Effects of the gradient of a strong electromagnetic beam on electrons and atoms. *Sov. Phys. JETP 15,* 1088–1090.

Aspect, A. (1992). Laser cooling and trapping of neutral atoms. Experiments. *Phys. Rep. 219,* 141–152.

Aspect, A., E. Arimondo, R. Kaiser, N. Vansteenkiste, and C. Cohen-Tannoudji (1988). Laser cooling below the one-photon recoil energy by velocity-selective coherent population trapping. *Phys. Rev. Lett. 61,* 826–829.

Aspect, A., E. Arimondo, R. Kaiser, N. Vansteenkiste, and C. Cohen-Tannoudji (1989). Laser cooling below the one-photon recoil energy by velocity-selective coherent population trapping: theoretical analysis. *J. Opt. Soc. Am. B 6,* 2112–2124.

Bagnato, V. S., G. P. Lafyatis, A. G. Martin, E. L. Raab, and D. E. Pritchard (1987). Continuous stopping and trapping of neutral atoms. *Phys. Rev. Lett. 58,* 2194–2197.

Bai, Y. S., and R. Kachru (1991). High-resolution spectroscopy of hyperfine structure using phase-correlated four-wave mixing. *Phys. Rev. Lett. 67,* 1859–1862.

Bai, Y. S., W. R. Babbitt, and T. W. Mossberg (1986). Coherent transient optical pulse-shape storage/recall using frequency-swept excitation pulses. *Optics Lett. 11,* 724–726.

Balling, L. C. (1975). Optical pumping, in *Advances in quantum electronics*, ed. D. W. Goodwin. Academic Press, London.

Balykin, V. I., V. S. Letokhov, and V. I. Mishin (1980a). Cooling of sodium atoms by resonant laser emission. *Sov. Phys. JETP 51,* 692–696.

Balykin, V. I., V. S. Letokhov, and V. I. Mishin (1980b). Observation of the cooling of free sodium atoms in a resonance laser field with scanning frequency. *JETP Letters 29,* 650–654.

Balykin, V. I., V. S. Letokhov, and A. I. Sidorov (1986). Focusing atomic beams by the dissipative radiation-pressure force of laser light. *JETP Letters 43,* 217–220.

Balykin, V. I., V. S. Letokhov, Y. V. Ovchinnikov, and A. I. Sidorov (1987). Reflection of an atomic beam from a gradient of an optical field. *JETP Letters 45,* 353–356.

Balykin, V., V. S. Letokhov, V. G. Minogin, Y. V. Rozhdestvensky, and A. I. Sidorov (1985). Radiative collimation of atomic beams through two-dimensional cooling of atoms by laser-radiation pressure. *J. Opt. Soc. Am. B 2,* 1776–1783.

Balykin, V. I., V. S. Letokhov, Y. B. Ovchinnikov, and A. I. Sidorov (1988). Quantum-state-selective mirror reflection of atoms by laser light. *Phys. Rev. Lett. 60,* 2137–2140; erratum: *61*: 902.

Barkov, L. M., D. A. Melik-Pasheayev, and M. S. Zolotorev (1989). Nonlinear Faraday rotation in Samarium vapor. *Optics Commun. 70,* 467–472.

Barrat, J. P., and C. Cohen-Tannoudji (1961a). Elargissement et déplacement des raies de résonance magnétique causés par une excitation optique. *J. Phys. Rad. 22,* 443–450.

Barrat, J. P., and C. Cohen-Tannoudji (1961b). Etude du pompage optique dans le formalisme de la matrice densité. *J. Phys. Rad. 22,* 329–336.

Barrat, J. P., C. Cohen-Tannoudji, and M. G. Ribaud (1961). Elargissement et déplacement des raies de résonance magnétique causés par le pompage optique. *Compt. Rend. 252,* 255–256.

Basche, T., W. E. Moerner, M. Orrit, and H. Talon (1992). Photon antibunching in the fluorescence of a single dye molecule trapped in a solid. *Phys. Rev. Lett. 69,* 1516–1519.

Bassini, M., F. Biraben, B. Cagnac, and G. Grynberg (1977). Raman transients observed in Doppler-free two-photon excitation. *Optics Commun. 21,* 263–266.

Bell, W. E., and A. L. Bloom (1961a). Observation of forbidden resonances in optically driven spin systems. *Phys. Rev. Lett. 6,* 623–624.

Bell, W. E., and A. L. Bloom (1961b). Optically driven spin precession. *Phys. Rev. Lett. 6,* 280–281.

Bernheim, R. (1965). *Optical pumping.* Benjamin, New York.

Berns, M., W. H. Wright, B. J. Tromberg, G. A. Profeta, J. J. Andrews, and R. J. Walter (1989). Use of a laser-induced optical force trap to study chromosome movement on the mitotic spindle. *Proc. Natl. Acad. Sci. USA 86,* 4539–4543.

Berry, M. V. (1987). The adiabatic phase and Pancharatnam's phase for polarized light. *J. Mod. Optics 34,* 1401–1407.

Beth, R. A. (1936). Mechanical detection and measurement of the angular momentum of light. *Phys. Rev. 50,* 115–125.

Bhandari, R., and J. Samuel (1988). Observation of topological phase by use of a laser interferometer. *Phys. Rev. Lett. 60,* 1211–1213.

Bhaskar, N. D. (1993). Concentration of atomic population in any single-ground-state magnetic sublevel in alkali-metal vapors. *Phys. Rev. A 47,* R4559–R4562.

Bicchi, P., L. Moi, and B. Zambon (1979). Rotatory power of sodium vapour oriented by laser radiation. *Nuovo Cimento 49B,* 9–22.

Bicchi, P., C. Marinelli, E. Mariotti, M. Meucci, and L. Moi (1994). Radiation trapping and vapor density of indium confined in quartz cells. *Optics Commun. 106,* 197–201.

Bitter, F. (1949). The optical detection of radiofrequency resonance. *Phys. Rev. 76,* 833–835.

Bitto, H., and J. R. Huber (1990). Molecular quantum beat spectroscopy. *Optics Commun. 80,* 184–198.

Bitto, H., A. Levinger, and J. R. Huber (1993). Optical-rf double resonance spectroscopy with quantum beat detection. *Z. Phys. D 28,* 303–310.

Bjorkholm, J. E., and A. Ashkin (1974). Cw self-focusing and self-trapping of light in sodium vapor. *Phys. Rev. Lett. 32,* 129–132.

Bjorkholm, J. E., P. F. Liao, and A. Wokaun (1982). Distortion of on-resonance two-photon spectroscopic lineshapes caused by velocity-selective optical pumping. *Phys. Rev. A 26,* 2643–2655.

Bjorkholm, J. E., R. E. Freeman, A. Ashkin, and D. B. Pearson (1978). Observation of focusing of neutral atoms by the dipole forces of resonance-radiation pressure. *Phys. Rev. Lett. 41,* 1361–1364.

Bjorkholm, J. E., R. E. Freeman, A. Ashkin, and D. B. Pearson (1980). Experimental observation of the influence of the quantum fluctuations of resonance-radiation pressure. *Optics Lett. 5,* 111–113; erratum: *5*:210.

Blasberg, T., and D. Suter (1992). Displacement of a laser beam by a precessing magnetic dipole. *Phys. Rev. Lett. 69,* 2507–2510.

Blasberg, T., and D. Suter (1993). Determination of the absolute sign of nuclear quadrupole interactions by laser-radio frequency double resonance experiments. *Phys. Rev. B 48,* 9524–9527.

Blasberg, T., and D. Suter (1994). Excitation of coherent Raman beats in rare earth solids with a bichromatic laser field. *Optics Commun. 109,* 133–138.

Bloch, D., and M. Ducloy (1981). Polarisation selection rules and disorienting collision effects in resonant degenerate four-wave mixing. *J. Phys. B 14,* L471–L476.

Bloch, F. (1946). Nuclear induction. *Phys. Rev. 70,* 460–485.

Bloch, F., and A. Siegert (1940). Magnetic resonance for nonrotating fields. *Phys. Rev. 57,* 522–527.

Bloch, F., W. W. Hansen, and M. Packard (1946). Nuclear induction. *Phys. Rev. 69,* 127.

Bloembergen, N., P. S. Pershan, and L. R. Wilcox (1960). Microwave modulation of light in paramagnetic crystals. *Phys. Rev. 120,* 2014–2023.

Bloom, D. M., P. F. Liao, and N. P. Economou (1978). Observation of amplified reflection by degenerate four-wave mixing in atomic sodium vapor. *Optics Lett. 2,* 58–60.

Bloom, D. M., J. F. Young, and S. E. Harris (1975a). Mixed metal vapor phase matching for third-harmonic generation. *Appl. Phys. Lett. 27,* 390–392.

Bloom, D. M., G. W. Bekkers, J. F. Young, and S. E. Harris (1975b). Third harmonic generation in phase-matched alkali metal vapors. *Appl. Phys. Lett. 26,* 687–689.

Boden, C., M. Dämmig, and F. Mitschke (1992). Light-shift induced chaos in a nonlinear optical resonator. *Phys. Rev. A 45,* 6829–6837.

Born, M., and E. Wolf (1986). *Principles of optics.* Pergamon Press, Oxford.

Bos, F. C., W. J. Buma, and J. Schmidt (1985). The triplet state of pyridine: a magnetic resonance study using electron-spin-echo spectroscopy. *Chem. Phys. Lett. 117,* 203–207.

Bose, S. K., and B. Dutta-Roy (1991). Geometry of quantum evolution and the coherent state. *Phys. Rev. A 43,* 3217–3220.

Boshier, M. G., and W. J. Sandle (1982). Self-focusing in a vapour of two-state atoms. *Opt. Commun. 42,* 371–376.

Boyd, R. W., M. G. Raymer, P. Narum, and D. J. Harter (1981). Four-wave mixing parametric interactions in a strongly driven two-level system. *Phys. Rev. A 24,* 411–423.

Bradley, L. C. (1992). Pulse-train excitation of sodium for use as a synthetic beacon. *J. Opt. Soc. Am. B 9,* 1931–1944.

Breiland, W. G., C. B. Harris, and A. Pines (1973). Optically detected electron spin echoes and free precession in molecular excited states. *Phys. Rev. Lett. 30,* 158–161.

Brewer, R. G. (1977a). *Coherent optical spectroscopy.* Frontiers of Laser Spectroscopy, Les Houches, North Holland.

Brewer, R. G. (1977b). Coherent optical transients. *Physics Today 30* (May 1977). 50–59.

Brewer, R. G., and A. Z. Genack (1976). Optical coherent transients by laser frequency switching. *Phys. Rev. Lett. 36,* 959–962.

Brewer, R. G., and E. L. Hahn (1973). Coherent Raman beats. *Phys. Rev. A 8,* 464–472.

Brewer, R. G., and E. L. Hahn (1975). Coherent two-photon processes: transient and steady-state cases. *Phys. Rev. A 11,* 1641–1649.

Brewer, R. G., and R. L. Shoemaker (1971). Photo echo and optical nutation in molecules. *Phys. Rev. Lett. 27,* 631–634.

Brewer, R. G., and R. L. Shoemaker (1972). Optical free induction decay. *Phys. Rev. A 6,* 2001–2007.

Brink, D. M., and G. R. Satchler (1962). *Angular momentum.* Clarendon Press, Oxford.

Brossel, J., and F. Bitter (1952). A new "double resonance" method for investigating atomic energy levels. Application to Hg $^3P_1$. *Phys. Rev. 86,* 308–316.

Brossel, J., A. Kastler, and J. Winter (1952). Generation optique d'une inégalité de population entre les sous-niveaux Zeeman de l'état fondamental des atomes. *J. Phys. Radium 13,* 668–668.

Brun, E., B. Derighetti, D. Meier, R. Holzner, and M. Ravani (1985). Observation of order and chaos in a nuclear spin-flip laser. *J. Opt. Soc. Am. B 2,* 156–167.

Burggraf, H., M. Kuckartz, and H. Harde (1986). Observation of 517 GHz fine-structure quantum-beats in Na, in *Methods of laser spectroscopy*, ed. Y. Prior, A. Ben-Reuven, and M. Rosenbluh. Plenum Press, New York.

Burns, M., P. Pappas, M. S. Field, and D. E. Murnick (1977). Laser induced nuclear orientation effects. *Nucl. Instr. Meth. 141,* 429–432.

Candela, D., M. E. Hayden, and P. J. Nacher (1994). Steady-state production of high nuclear polarization in 3He-4He mixtures. *Phys. Rev. Lett. 73,* 2587.

Carlson, N. W., W. R. Babbitt, and T. W. Mossberg (1983). Storage and phase conjugation of light pulses using stimulated photon echoes. *Optics Lett. 8,* 623–625.

Carnal, O., and J. Mlynek (1991). Young's double slit experiment with atoms: a simple atom interferometer. *Phys. Rev. Lett. 21,* 2689–2692.

Carusotto, S., G. Fornaca, and E. Polacco (1968). Radiation beats and rotating systems. *Nuovo Cimento 53,* 87–97.

Cates, G. D., D. R. Benton, M. Gatzke, W. Happer, K. C. Hasson, and N. R. Newbury (1990). Laser production of large nuclear-spin polarization in frozen xenon. *Phys. Rev. Lett. 65,* 2591–2594.

Chang, S., and S. S. Lee (1985). Optical torque exerted on a homogeneous sphere levitated in the circularly polarized fundamental-mode laser beam. *J. Opt. Soc. Am. B 2,* 1853–1860.

Chebotayev, V. P., N. M. Dyuba, M. I. Skvortsov, and L. S. Vasilenko (1978). Coherent radiation in time separated fields. *Appl. Phys. 15,* 319–322.

Chen, X., V. L. Telegdi, and A. Weis (1987). Magneto-optical rotation near the Caesium D_2 line (Macaluso-Corbino effect) in intermediate fields: I. Linear regime. *J. Phys. B 20,* 5653–5662.

Chen, X., V. L. Telegdi, and A. Weis (1990a). Quantitative observation of the non-linear dichroic Voigt effect near the Cs D_2–line. *Optics Commun. 78,* 337–340.

Chen, X., V. L. Telegdi, and A. Weis (1990b). Quantitative study of the nonlinear Macaluso-Corbino (resonant Faraday) effect in Cs. *Optics Commun. 74,* 301–305.

Chen, Y. C., K. Chiang, and S. R. Hartmann (1979). Photon echo relaxation in LaF3:Pr3+. *Optics Commun. 29,* 181–185.

Chen, Y. C., K. Chiang, and S. R. Hartmann (1980). Spectroscopic and relaxation character of the $^3P_0$ - $^3H_4$ transition in $LaF_3:Pr^{3+}$ measured by photon echoes. *Phys. Rev. B 21,* 40–47.

Chiao, R. Y., and Y. S. Wu (1986). Manifestation of Berry's topological phase for the photon. *Phys. Rev. Lett. 57,* 933–936.

Chiao, R. Y., E. Garmire, and C. H. Townes (1964). Self-trapping of optical beams. *Phys. Rev. Lett. 13,* 479–482.

Chiao, R. Y., A. Antaramian, K. M. Ganga, H. Jiao, and S. R. Wilkinson (1988). Observation of a topological phase by means of a nonplanar Mach-Zehnder interferometer. *Phys. Rev. Lett. 60,* 1214–1217.

Chow, W. W., M. O. Scully, and J. O. Stoner (1975). Quantum-beat phenomena described by quantum electrodynamics and neoclassical theory. *Phys. Rev. A 11,* 1380–1388.

Chu, S. (1991). Laser manipulation of atoms and particles, *Science 253,* 861–866.

Chu, S., J. E. Bjorkholm, A. Ashkin, and A. Cable (1986). Experimental observation of optically trapped atoms. *Phys. Rev. Lett. 57,* 314–317.

Chu, S., L. Hollberg, J. E. Bjorkholm, A. Cable, and A. Ashkin (1985). Three-dimensional viscous confinement and cooling of atoms by resonance radiation pressure. *Phys. Rev. Lett. 55,* 48–51.

Cohen, M. H., and F. Reif (1957). Quadrupole effects in nuclear magnetic resonance studies of solids. *Solid State Physics 5,* 321–438.

Cohen-Tannoudji, C. (1961). Observation d'un déplacement de raie de résonance magnetique causé par l'excitation optique. *Comptes Rend. 252,* 394–396.

Cohen-Tannoudji, C. (1962). Théorie quantique du cycle de pompage optique. *Ann. Phys. 7,* 423–460.

Cohen-Tannoudji, C. (1992a). Atomic motion in laser light, in *Fundamental systems in quantum optics; proceedings of the Les Houches summer school,* ed. J. Dalibard, J. M. Raimond, and J. Zinn-Justin. North-Holland, Amsterdam.

Cohen-Tannoudji, C. (1992b). Laser cooling and trapping of neutral atoms: theory. *Phys. Rep. 219,* 153–164.

Cohen-Tannoudji, C., and J. Dupont-Roc (1972). Experimental study of Zeeman light shifts in weak magnetic fields. *Phys. Rev. A 5,* 968–984.

Cohen-Tannoudji, C., J. Dupont-Roc, and G. Grynberg (1992). *Atom-photon interactions / basic processes and application.* Wiley Interscience, New York.

Cohen-Tannoudji, C., J. Dupont-Roc, S. Haroche, and F. Laloe (1970a). Diverses résonances de croisement de niveaux sur des atomes pompés optiquement en champ nul I. Théorie. *Rev. Phys. Appl. 5,* 95–101.

Cohen-Tannoudji, C., J. Dupont-Roc, S. Haroche, and F. Laloe (1970b). Diverses résonances de croisement de niveaux sur des atomes pompés optiquement en champ nul II. applications a la mesure de champs faibles. *Rev. Phys. Appl. 5,* 102–108.

Colegrove, F. D., P. A. Franken, R. R. Lewis, and R. H. Sands (1959). Novel method of spectroscopy with applications to precision fine structure measurements. *Phys. Rev. Lett. 3,* 420–422.

Cook, R., and R. K. Hill (1982). An electromagnetic mirror for neutral atoms. *Optics Commun. 43,* 258–260.

Cornell, E. A., C. Monroe, and C. E. Wieman (1991). Multiply loaded, ac magnetic trap for neutral atoms. *Phys. Rev. Lett. 67,* 2439–2442.

Corney, A. (1977). *Atomic and laser spectroscopy.* Clarendon Press, Oxford.

Corney, A., and G. W. Series (1964). Double resonance excited by modulated light. *Proc. Phys. Soc. London 83,* 213–216.

Corney, A., B. P. Kibble, and G. W. Series (1966). The forward scattering of resonance radiation, with special reference to double resonance and level crossing experiments. *Proc. Roy. Soc. London A 293,* 70–93.

Courty, J. M., P. Grangier, L. Hilico, and S. Reynaud (1991). Quantum fluctuations in optical bistability: calculations from linear response theory. *Optics Commun. 83,* 251–256.

Dalibard, J., and C. Cohen-Tannoudji (1985). Dressed-atom approach to atomic motion in laser light: the dipole force revisited. *J. Opt. Soc. Am. B 2,* 1707–1720.

Dalibard, J., and C. Cohen-Tannoudji (1989). Laser cooling below the Doppler limit by polarization gradients: simple theoretical models. *J. Opt. Soc. Am. B 6,* 2023–2045.

Davies, P. B. (1981). Laser magnetic resonance spectroscopy. *J. Chem. Phys. 85,* 2599–2607.

Dehmelt, H. G. (1957a). Modulation of a light beam by precessing absorbing atoms. *Phys. Rev. 105,* 1924–1925.

Dehmelt, H. G. (1957b). Slow spin relaxation of optically polarized Sodium atoms. *Phys. Rev. 105,* 1487–1489.

Dehmelt, H. G. (1958). Spin resonance of free electrons polarized by exchange collisions. *Phys. Rev. 109,* 381–385.

Dehmelt, H. (1990). Experiments with an isolated subatomic particle at rest. *Rev. Mod. Phys. 62,* 525–530.

Delsart, C., J. C. Keller, and V. P. Kaftandjian (1980). Experimental evidence for light-shift induced zero-field level crossing ("optical Hahnle effect"). *Optics Commun. 32,* 406–410.

DeMartini, F., and G. R. Jacobovitz (1988). Anomalous spontaneous-stimulated-decay phase transition and zero-threshold laser action in a microscopic optical cavity. *Phys. Rev. Lett. 60,* 1711–1714.

Demtröder, W. (1991). *Laser Spektroskopie.* Springer, Berlin.

DeOliveira, F. A., B. J. Dalton, and P. L. Knight (1987). Squeezing and trapping in three-level atoms. *J. Opt. Soc. Am. B 4,* 1558–1564.

Derrick, G. H. (1964). Comments on nonlinear wave equations as models for elementary particles. *J. Math. Phys. 5,* 1252–1254.

DeVoe, R. G., and R. G. Brewer (1978). Optical coherent transients by laser-frequency switching: subnanosecond studies. *Phys. Rev. Lett. 40,* 862–865.

DeVoe, R. G., and R. G. Brewer (1979). Subnanosecond optical free-induction decay. *Phys. Rev. A 20,* 2449–2458.

DeVoe, R. G., and R. G. Brewer (1983). Experimental test of the optical Bloch equations for solids. *Phys. Rev. Lett. 50,* 1269–1272.

DeVoe, R. G., A. Szabo, S. C. Rand, and R. G. Brewer (1979). Ultraslow optical dephasing of LaF3:Pr3+. *Phys. Rev. Lett. 42,* 1560–1563.

Diedrich, F., and H. Walther (1987). Nonclassical radiation of a single stored ion. *Phys. Rev. Lett. 58,* 203–206.

Dodd, J. N., R. D. Kaul, and D. M. Warrington (1964). The modulation of resonance fluorescence excited by pulsed light. *Proc. Phys. Soc. London 84,* 176–178.

Dodd, J. N., and G. W. Series (1978). Time-resolved fluorescence spectroscopy. *Progr. Atomic Spectroscopy 14,* 639–677.

Dodd, J. N., W. J. Sandle, and D. Zissermann (1967). Study of resonance fluorescence in cadmium: modulation effects and lifetime measurement. *Proc. Phys. Soc. 92,* 497–504.

Dohnalik, T., J. Koperski, M. Stankiewicz, J. Zakrzewski, and K. Zyczkowski (1984). Quantum beats in forward scattering of resonance radiation. *Acta Phys. Pol. A66,* 493–501.

Dooghin, A. V., N. D. Kundikova, V. S. Liberman, and B. Y. Zel'dovich (1992). Optical magnus effect. *Phys. Rev. A 45,* 8204–8208.

Drake, K. H., W. Lange, and J. Mlynek (1988). Nonlinear Faraday and Voigt effect in a J = 1 to J' = 0 transition in atomic Samarium vapour. *Optics Commun. 66,* 315–320.

Drummond, P. D., R. M. Shelby, S. R. Friberg, and Y. Yamamoto (1993). Quantum solitons in optical fibres. *Nature 365,* 307–313.

Ducloy, M., M. P. Gorza, and B. Decomps (1973). Higher-order nonlinear effects in a gas laser: creation and detection of a hexadecupole moment in the Neon $^2p_4$ level. *Optics Commun. 8,* 21–25.

Ducloy, M., and D. Bloch (1994). Spectroscopy and phase conjugation by resonant four wave mixing, in *Optical phase conjugation*, ed. M. Gower and D. Proch. Springer, Berlin.

Dupont-Roc, J., N. Polonsky, C. Cohen-Tannoudji, and A. Kastler (1967). Lifting of a Zeeman degeneracy by interaction with a light beam. *Phys. Lett. 25A,* 87–88.

D'Yakonov, M. I. (1965). Theory of resonance scattering of light by a gas in the presence of a magnetic field. *Sov. Phys. JETP 20,* 1484–1489.

Dyson, F. J. (1949). The radiation theories of Tomonaga, Schwinger, and Feynman. *Phys. Rev. 75,* 486–502.

Eckart, C. (1930). The application of group theory to the quantum dynamics of monatomic systems. *Rev. Mod. Phys. 2,* 305–380.

Edmonds, A. R. (1974). Angular momentum in quantum mechanics. Princeton University Press, Princeton.

Einstein, A. (1917). Zur Quantentheorie der Strahlung. *Phys. Zeitschrift 18,* 121–128.

Emile, O., F. Bardou, C. Salomon, P. Laurent, A. Nadir, and A. Clairon (1992). Observation of a new magneto-optical trap. *Europhys. Lett. 20,* 687–691.

Endo, T., S. Nakanishi, T. Muramoto, and T. Hashi (1982). Coherent Raman beats associated with superhyperfine structure in Ruby. *Optics Commun. 43,* 359–363.

Enk, S. J. v., and G. Nienhuis (1992). Entropy production and kinetic effects of light. *Phys. Rev. A 46,* 1438.

Erickson, L. E. (1977). Optical measurement of the hyperfine splitting of the $^1D_2$ metastable states of Pr^{3+} in LaF_3 by enhanced and saturated absorption spectroscopy. *Phys. Rev. B 16,* 4731–4736.

Erickson, L. E. (1979). Hyperfine interaction in the lowest levels of the trivalent praseodymium in yttrium aluminum perovskite ($YAlO_3$). *Phys. Rev. B 19,* 4412–4420.

Erickson, L. E. (1985a). NMR measurement of an ion in an excited state by indirect optical detection: A $^1G_4$ level of $LiYF_4$:Pr^{3+}. *Phys. Rev. B 32,* 1–5.

Erickson, L. E. (1985b). Optically detected nuclear magnetic resonance of $LiYF_4$:Eu^{3+} in the ground electronic state $^7F_0$. *J. Phys. C 18,* 2935–2942.

Erickson, L. E. (1987). A low-magnetic-field Raman heterodyne study of the Van-Vleck paramagnet $YAlO_3$:Eu^{3+}. *J. Phys. C 20,* 291–298.

Erickson, L. E. (1990). Optical-pumping effects on Raman-heterodyne-detected multipulse rf nuclear-spin-echo decay. *Phys. Rev. B 42,* 3789–3797.

Erickson, L. E. (1991). Two-pulse and stimulated nuclear-quadrupole-resonance echoes in YAlO3:Pr3+. *Phys. Rev. B 43,* 12723–12728.

Erickson, L. E. (1993). Nuclear-quadrupole-resonance measurement of the ^{27}Al frozen core in $YAlO_3$:Pr^{3+}. *Phys. Rev. B 47,* 8734–8738.

Ernst, R. R. (1966). Sensitivity enhancement in magnetic resonance. *Adv. Magn. Reson. 2,* 1–135.

Ernst, R. R., and W. A. Anderson (1966). Application of Fourier transform spectroscopy to magnetic resonance. *Rev. Sci. Instrum. 37,* 93–102.

Ernst, R. R., G. Bodenhausen, and A. Wokaun (1987). *Principles of nuclear magnetic resonance in one and two dimensions.* Oxford University Press, Oxford.

Ertmer, W., R. Blatt, J. L. Hall, and M. Zhu (1985). Laser manipulation of atomic

beam velocities: demonstration of stopped atoms and velocity reversal. *Phys. Rev. Lett. 54,* 996–999.
Estermann, I., and O. Stern (1930). Beugung von Molekularstrahlen. *Z. Phys. 61,* 95–125.
Fano, U. (1957). Description of states in quantum mechanics by density matrix and operator techniques. *Rev. Mod. Phys. 29,* 74–93.
Feynman, R. P., F. L. Vernon, and R. W. Helwarth (1957). Geometrical representation of the Schrödinger equation for solving maser problems. *J. Appl. Phys. 28,* 49–52.
Fill, E. E., M. O. Scully, and S. Y. Zhu (1990). Lasing without inversion via the lambda quantum-beat laser in the collision-dominated regime. *Optics Commun. 77,* 36–40.
Fischer, E. (1959). Die dreidimensionale Stabilisierung von Ladungsträgern in einem Vierpolfeld. *Z. Phys. 156,* 1–26.
Franken, P. A. (1961). Interference effects in the resonance fluorescence of "crossed" excited atomic states. *Phys. Rev. 121,* 508–512.
Franken, P. A., A. E. Hill, C. W. Peters, and G. Weinreich (1961). Generation of optical harmonics. *Phys. Rev. Lett. 7,* 118–119.
Franz, F. A., and J. R. Franz (1966). Excited-state mixing in the optical pumping of alkali-metal vapors. *Phys. Rev. 148,* 82–89.
Franz, F. A., and C. Volk (1976). Spin relaxation of rubidium atoms in sudden and quasimolecular collisions with light-noble-gas atoms. *Phys. Rev. A 14,* 1711–1728.
Freyberger, M., and A. M. Herkommer (1994). Probing a quantum state via atomic deflection. *Phys. Rev. Lett. 72,* 1952–1955.
Frisch, R. (1933). Experimenteller Nachweis des Einsteinschen Strahlungsrückstosses. *Z. Phys. 86,* 42–48.
Fukuda, Y., J. Hayashi, K. Kondo, and T. Hashi (1981). Synchronized quantum beat spectroscopy using periodic impact excitations with cw mode-locked laser pulses. *Optics Commun. 38,* 357–360.
Fukuda, Y., K. Yamada, and T. Hashi (1983a). Sublevel echoes induced by resonant light pulses: quantum beat echoes. *Optics Commun. 44,* 297–300.
Fukuda, Y., M. Tanigawa, T. Hashi, and K. Kondo (1983b). High-resolution synchronized-quantum-beat spectroscopy using modulation sidebands: measurement of pressure shifts of the hyperfine splitting in sodium 3 $^2S_{1/2}$. *Optics Lett. 8,* 301–303.
Gabrielse, G., and H. Dehmelt (1985). Observation of inhibited spontaneous emission. *Phys. Rev. Lett. 55,* 67–70.
Gaeta, A. L., R. W. Boyd, J. R. Ackerhalt, and P. W. Milonni (1987). Instabilities and chaos in the polarizations of counterpropagating light fields. *Phys. Rev. Lett. 58,* 2432–2435.
Gauthier, D. J., M. S. Malcuit, and R. W. Boyd (1988). Polarization instabilities of counterpropagating laser beams in sodium vapor. *Phys. Rev. Lett. 61,* 1827–1830.
Gawlik, W., J. Kowalski, R. Neumann, and F. Träger (1974a). Observation of the electric hexadecupole moment of free Na atoms in a forward scattering experiment. *Optics Commun. 12,* 400–404.
Gawlik, W., J. Kowalski, R. Neumann, and F. Träger (1974b). Strong narrowing of the Na forward scattering signals due to the interaction with an intense dye laser field. *Phys. Lett.* A *48,* 283–284.
Gel'mukhanov, F. K., and A. M. Shalagin (1980). Theory of optically induced diffusion in gases. *Sov. Phys. JETP 51,* 839–846.

Genack, A. Z., R. Macfarlane, and R. G. Brewer (1976). Optical free-induction decay in LaF3:Pr3+. *Phys. Rev. Lett. 37,* 1078–1080.
Gerlach, W., and O. Stern (1924). Über die Richtungsquantisierung im Magnetfeld. *Ann. d. Phys. 74,* 673–699.
Geschwind, S., R. J. Collins, and A. L. Schawlow (1959). Optical detection of paramagnetic resonance in an excited state of $C^{r3}+$ in $A_{l}2_{O3}$. *Phys. Rev. Lett. 3,* 545–548.
Gheri, K. M., C. Saavedra, and D. F. Walls (1993). Intracavity second-harmonic generation using an electromagnetically induced transparency. *Phys. Rev. A 48,* 3344–3361.
Ghosh, A. P., C. D. Nabors, M. A. Attili, and J. E. Thomas (1985). $^3P_1$–orientation velocity-changing collision kernels studied by isolated multipole echoes. *Phys. Rev. Lett. 54,* 1794–1797.
Gibbs, H. M. (1985). *Optical bistability: controlling light with light.* Academic Press, Orlando.
Gibbs, H. M., S. L. McCall, and T. N. C. Venkatesan (1976). Differential gain and bistability using a sodium-filled Fabry-Perot interferometer. *Phys. Rev. Lett. 36,* 1135–1138.
Giordmaine, J. A., and W. Kaiser (1966). Light scattering by coherently driven lattice vibrations. *Phys. Rev. 144,* 676–688.
Giordmaine, J. A., and R. C. Miller (1965). Tunable coherent parametric oscillation in LiNbO3 at optical frequencies. *Phys. Rev. Lett. 14,* 973–976.
Glaser, S., G. Wäckerle, and K. P. Dinse (1985). High-resolution spectroscopy of Pr^{3+}:$YAlO_3$ by stimulated photon-echo envelope modulation. *Chem. Phys. Lett. 121,* 267–271.
Glauber, R. J. (1963a). Coherent and incoherent states of the radiation field. *Phys. Rev. 131,* 2766–2788.
Glauber, R. J. (1963b). The quantum theory of optical coherence. *Phys. Rev. 130,* 2529–2539.
Golub, J. E., Y. S. Bai, and T. W. Mossberg (1988). Radiative and dynamical properties of homogeneously prepared atomic samples. *Phys. Rev. A 37,* 119–124.
Gorcester, J., and J. H. Freed (1988). Two-dimensional Fourier transform ESR correlation spectroscopy. *J. Chem. Phys. 88,* 4678–4693.
Gordon, J. P., and A. Ashkin (1980). Motion of atoms in a radiation trap. *Phys. Rev. A 21,* 1606–1617.
Gould, P. L., G. A. Ruff, and D. E. Pritchard (1986). Diffraction of atoms by light: the near-resonant Kapitza-Dirac effect. *Phys. Rev. Lett. 56,* 827–830.
Goy, P., J. M. Raimond, M. Gross, and S. Haroche (1983). Observation of cavity-enhanced single-atom spontaneous emission. *Phys. Rev. Lett. 50,* 1903–1906.
Gozzini, S., J. H. Xu, C. Gabbanini, G. Paffuti, and L. Moi (1989). Light-induced drift dynamics in an optically thin regime: monochromatic and broadband laser excitation. *Phys. Rev. A 40,* 6349–6353.
Grangier, P., G. Roger, and A. Aspect (1986). Experimental evidence for a photon-anticorrelation effect on a beam splitter: a new light on single-photon interferences. *Europhys. Lett. 1,* 173–179.
Grantham, J. W., H. M. Gibbs, G. Khitrova, J. F. Valley, and X. Jiajin (1991). Kaleidoscopic spatial instability: bifurcations of optical transverse solitary waves. *Phys. Rev. Lett. 66,* 1422–1425.
Gray, H. R., R. M. Whitley, and C. R. Stroud (1978). Coherent trapping of atomic populations. *Optics Lett. 3,* 218–220.
Grimm, R., J. Söding, and Y. B. Ovchinnikov (1994). Coherent beam splitter for atoms based on a bichromatic standing light wave. *Optics Lett. 19,* 658–660.

Grimm, R., V. S. Letokhov, Y. B. Ovchinnikov, and A. I. Sidorov (1991). Observation of a magnetooptic radiation force. *JETP Letters 54,* 611–614.

Grimm, R., Y. B. Ovchinnikov, A. I. Sidorov, and V. S. Letokhov (1990). Observation of a strong rectified dipole force in a bichromatic standing light wave. *Phys. Rev. Lett. 65,* 1415–1418.

Grimm, R., Y. B. Ovchinnikov, A. I. Sidorov, and V. S. Letokhov (1991). Dipole force rectification in a monochromatic laser field. *Optics Commun. 84,* 18–22.

Grimm, R., J. Söding, Y. B. Ovchinnikov, and A. I. Sidorov (1993). Sub Doppler manifestation of the magneto-optical radiation force. *Optics Commun. 98,* 54–60.

Grischkowsky, D. (1970). Self-focusing of light by potassium vapor. *Phys. Rev. Lett. 24,* 866–869.

Grischkowsky, D., E. Courtens, and J. A. Armstrong (1973). Observation of self-steepening of optical pulses with possible shock formation. *Phys. Rev. Lett. 31,* 422–425.

Grison, D., B. Lounis, C. Salomon, J. Y. Courtois, and G. Grynberg (1991). Raman spectroscopy of cesium atoms in a laser trap. *Europhys. Lett. 15,* 149–154.

Grynberg, G., B. Lounis, P. Verkerk, J. Y. Courtois, and C. Salomon (1993). Quantized motion of cold cesium atoms in two- and three-dimensional optical potentials. *Phys. Rev. Lett. 70,* 2249–2252.

Gustafson, T. K., P. L. Kelley, R. Y. Chiao, and R. G. Brewer (1968). Self-trapping in media with saturation of the nonlinear index. *Appl. Phys. Lett. 12,* 165–168.

Gustafson, T. K., J. P. E. Taran, H. A. Haus, J. R. Lifsitz, and P. L. Kelley (1969). Self-modulation, self-steepening, and spectral development of light in small-scale trapped filaments. *Phys. Rev. 177,* 306–313.

Güttler, F., M. Pirotta, A. Renn, and U. P. Wild (1993). Single molecule spectroscopy: stark effect of pentacene in p-terphenyl, in *Spectral holeburning and luminescence line narrowing: science and applications.* Monte verità, Ascona, Switzerland.

Güttler, F., J. Sepiol, M. Pirotta, A. Renn, and U. P. Wild (1993). *Polarization measurements on single molecules.* QELS. Baltimore.

Hack, E., and J. R. Huber (1991). Quantum beat spectroscopy of molecules. *Int. Rev. Phys. Chem. 10,* 287–317.

Häger, C., and F. Kaiser (1992). Bifurcation structures into chaos of delay-differential equations for a passive optical ring resonator. *Appl. Phys. B 55,* 132–137.

Hahn, E. L. (1950). Spin echoes. *Phys. Rev. 80,* 580–594.

Hakuta, K., L. Marmet, and B. P. Stoicheff (1991). Electric-field-induced second-harmonic generation with reduced absorption in atomic hydrogen. *Phys. Rev. Lett. 66,* 596–599.

Hakuta, K., L. Marmet, and B. P. Stoicheff (1992). Nonlinear optical generation with reduced absorption using electric-field coupling in atomic hydrogen. *Phys. Rev. A 45,* 5152–5159.

Hall, J. L., M. Zhu, and P. Buch (1989). Prospects of using laser-prepared atomic fountains for optical frequency standards applications. *J. Opt. Soc. Am. B 6,* 2194–2205.

Hamilton, M. W., R. J. Ballagh, and W. J. Sandle (1982). Polarization switching in a ring cavity with resonantly driven J = 1/2 to J = 1/2 atoms. *Z. Phys. B 49,* 263–272.

Hanle, W. (1924). Über magnetische Beeinflussung der Polarisation der Resonanzfluoreszenz. *Z. Physik 30,* 93–105.

Hänsch, T. W., and A. L. Schawlow (1975). Cooling of gases by laser radiation. *Optics Commun. 13,* 68–69.

Happer, W. (1970). Light propagation and light shifts in optical pumping experiments, in *Progress in quantum electronics 1*, ed. K. W. H. Stevens and J. H. Sanders. Pergamon Press, Oxford.

Happer, W. (1972). Optical pumping. *Rev. Mod. Phys. 44*, 169–249.

Happer, W., and B. S. Mathur (1967a). Effective operator formalism in optical pumping. *Phys. Rev. 163*, 12–25.

Happer, W., and B. S. Mathur (1967b). Off-resonant light as a probe of optically pumped alkali vapors. *Phys. Rev. Lett. 18*, 577–580.

Happer, W., E. Miron, S. Schaefer, D. Schreiber, W. A. v. Wijngaarden, and X. Zeng (1984). Polarization of the nuclear spins of noble-gas atoms by spin exchange with optically pumped alkali-metal atoms. *Phys. Rev. A 29*, 3092–3110.

Harde, H., and H. Burggraf (1982). Ultrahigh-resolution spectroscopy by means of periodic excitation with picosecond pulses. *Optics Commun. 40*, 441–445.

Harde, H., H. Burggraf, J. Mlynek, and W. Lange (1981). Quantum beats in forward scattering: subnanosecond studies with a mode-locked dye laser. *Optics Lett. 6*, 290–292.

Haroche, S. (1976). Quantum beats and time-resolved fluorescence spectroscopy, in *High-resolution laser spectroscopy*, ed. K. Shimoda. Springer, Berlin.

Haroche, S. (1992). Cavity quantum electrodynamics, in *Fundamental systems in quantum optics; proceedings of the Les Houches summer school*, ed. J. Dalibard, J. M. Raimond, and J. Zinn-Justin. North-Holland, Amsterdam.

Haroche, S., and D. Kleppner (1989). Cavity quantum electrodynamics. *Physics Today* (January 1989), 24–30.

Haroche, S., and J. M. Raimond (1985). Radiative properties of Rydberg states in resonant cavities, in *Advanced atomic molecular physics, 20*, ed. D. Bates. Academic Press, Orlando.

Haroche, S., J. A. Paisner, and A. L. Schawlow (1973). Hyperfine quantum beats observed in Cs vapor under pulsed dye laser excitation. *Phys. Rev. Lett. 30*, 948–951.

Harris, S. E., J. E. Field, and A. Imamoglu (1990). Nonlinear optical processes using electromagnetically induced transparence. *Phys. Rev. Lett. 64*, 1107–1110.

Hartmann, S. R., and E. L. Hahn (1962). Nuclear double resonance in the rotating frame. *Phys. Rev. 128*, 2042–2053.

Hashi, T., Y. Fukuda, and M. Tanigawa (1992). Optical excitation and detection of spin precession, in *Pulsed magnetic resonance: NMR, ESR, and optics*, ed. D. M. S. Bagguley. Oxford University Press, Oxford.

Haus, H. A., and J. L. Pan (1993). Photon spin and the paraxial wave equation. *Am. J. Phys. 61*, 818–821.

Hawkins, W. B. (1955). Orientation and alignment of sodium atoms by means of polarized resonance radiation. *Phys. Rev. 98*, 478–486.

Heer, C. V. (1960). A low temperature atomic beam oscillator, in *Quantum electronics*, ed. C. H. Townes. Columbia University Press, New York.

Heer, C. V. (1963). Feasibility of containment of quantum magnetic dipoles. *Rev. Sci. Instrum. 34*, 532–537.

Heinzen, D. J., and M. S. Feld (1987). Vacuum radiative level shift and spontaneous emission linewidth of an atom in an optical resonator. *Phys. Rev. Lett. 59*, 2623–2626.

Heitler, W. (1953). *The quantum theory of radiation*. Dover Publications, New York.

Hellwarth, R. W. (1963). Theory of stimulated Raman scattering. *Phys. Rev. 130*, 1850–1852.

Hemmerich, A., C. Zimmermann, and T. W. Hänsch (1993). Sub-kHz Rayleigh resonance in a cubic atomic crystal. *Europhys. Lett. 22*, 89–94.

Hercher, M. (1964). Laser induced damage in transparent media. *J. Opt. Soc. Am. 54*, 563.

Heritage, J. P., T. K. Gustafson, and C. H. Lin (1975). Observation of coherent transient birefringence in CS_2 vapor. *Phys. Rev. Lett. 34*, 1299–1302.

Hilico, L., C. Fabre, S. Reynaud, and E. Giacobino (1992a). Linear input-output method for quantum fluctuations in optical bistability with two-level atoms. *Phys. Rev. A 46*, 4397–4405.

Hilico, J., J. M. Courty, C. Fabre, E. Giacobino, I. Abram, and J. L. Oudar (1992b). Squeezing with $\chi^{(3)}$ materials. *Appl. Phys. B 55*, 202–209.

Hinds, E. A. (1991). Cavity quantum electrodynamics, in *Advanced atomic molecular optic physics 28*, ed. D. Bates. Academic Press, Boston.

Hocker, G. B., and C. L. Tang (1968). Observation of the optical transient nutation effect. *Phys. Rev. Lett. 21*, 591–594.

Hoff, P. W., H. A. Haus, and T. J. Bridges (1970). Observation of optical nutation in an active medium. *Phys. Rev. Lett. 25*, 82–84.

Holliday, K., and U. P. Wild (1993). Spectral hole-burning, in *Molecular luminescence spectroscopy, part 3*, ed. S. G. Schulman. John Wiley,

Holliday, K., M. Croci, E. Vauthey, and U. P. Wild (1993). Spectral holeburning and holography in a Y_2SiO_5:Pr^{3+} crystal. *Phys. Rev. B 47*, 14741–14751.

Holliday, K., X. F. He, P. T. H. Fisk, and N. B. Manson (1990). Raman heterodyne detection of electron paramagnetic resonance. *Optics Lett. 15*, 983–985.

Holstein, T. (1947). Imprisonment of resonance radiation in gases. *Phys. Rev. 72*, 1212–1233.

Holstein, T. (1951). Imprisonment of resonance radiation in gases. II. *Phys. Rev. 83*, 1159–1168.

Holzner, R., P. Eschle, A. W. McCoord, and D. M. Warrington (1992). Transverse bouncing of polarized laser beams in sodium vapor. *Phys. Rev. Lett. 69*, 2192–2195.

Huber, G., F. Touchard, S. Büttgenbach, C. Thibault, R. Klapisch, H. T. Duong, S. Liberman, J. Pinard, J. L. Vialle, P. Juncar, and P. Jacquinot (1978). Spins, magnetic moments, and isotope shifts of $^{21\text{-}31}$Na by high resolution laser spectroscopy of the atomic D_1 line. *Phys. Rev. C 18*, 2342–2354.

Hulet, R. G., E. S. Hilfer, and D. Kleppner (1985). Inhibited spontaneous emission by a Rydberg atom. *Phys. Rev. Lett. 55*, 2137–2140.

Hurst, G. S., M. G. Payne, S. D. Kramer, and J. P. Young (1979). Resonance ionization spectroscopy and one-atom detection. *Rev. Mod. Phys. 51*, 767–819.

Ikeda, K. (1979). Multiple-valued staionary state and its instability of the transmitted light by a ring cavity system. *Optics Commun. 30*, 257–261.

Ikeda, K., H. Daido, and O. Akimoto (1980). Optical turbulence: chaotic behavior of transmitted light from a ring cavity. *Phys. Rev. Lett. 45*, 709–712.

Imamoglu, A., J. E. Field, and S. E. Harris (1991). Lasers without inversion: a closed lifetime broadened system. *Phys. Rev. Lett. 66*, 1154–1156.

Ingold, M., M. Duelli, P. Günter, and M. Schadt (1992). All-optical associative memory based on a nonresonant cavity with image-bearing beams. *J. Opt. Soc. Am. B 9*, 1327–1337.

Jaynes, E. T., and F. W. Cummings (1963). Comparison of quantum and semiclassical radiation theories with application to the beam maser. *Proc. IEEE 51*, 89–109.

Jeener, J. (1971). Ampère International Summer School II. Basko Polje, Yugoslavia, unpublished.

Jeener, J., B. H. Meier, P. Bachmann, and R. R. Ernst (1979). Investigation of exchange processes by two-dimensional NMR spectroscopy. *J. Chem. Phys. 71*, 4546–4553.

Jenkins, B. K., A. A. Sawchuk, T. C. Strand, R. Forchheimer, and B. H. Soffer (1984). Sequential optical logic implementation. *Appl. Opt. A 23,* 3455–3464.

Jhe, W., A. Anderson, E. A. Hinds, D. Meschede, L. Moi, and S. Haroche (1987). Suppression of spontaneous decay at optical frequencies: a test of vacuum-field anisotropy in confined space. *Phys. Rev. Lett. 58,* 666–669.

Jiao, H., S. R. Wilkinson, and R. Y. Chiao (1989). Two topological phases in optics by means of a nonplanar Mach-Zehnder interferometer. *Phys. Rev. A 39,* 3475–3486.

Kachru, R., T. W. Mossberg, and S. R. Hartmann (1980). Noble-gas broadening of the sodium D lines measured by photon echoes. *J. Phys. B 13,* L363–L368.

Kachru, R., T. W. Mossberg, E. Whittaker, and S. R. Hartmann (1979). Optical echoes generated by standing-wave fields: observation in atomic vapors. *Optics Commun. 31,* 223–227.

Kaftandjian, V. P., C. Delsart, and J. C. Keller (1981). Optical Hanle effect. *Phys. Rev. A 23,* 1365–1374.

Kanorsky, S. I., A. Weis, J. Wurster, and T. W. Hänsch (1993). Quantitative investigation of the resonant nonlinear Faraday effect under conditions of optical hyperfine pumping. *Phys. Rev. A 47,* 1220–1226.

Karlov, N. V., J. Margerie, and Y. Merle-D'Aubigne (1963). Pompage optique des centres F dans KBr. *J. Phys. 24,* 717–723.

Karlsson, M. (1992). Optical beams in saturable self-focusing media. *Phys. Rev. A 46,* 2726–2734.

Kasevich, M., and S. Chu (1991). Atomic interferometry using stimulated Raman transitions. *Phys. Rev. Lett. 67,* 181–184.

Kasevich, M., and S. Chu (1992). Laser cooling below a photon recoil with three level atoms. *Phys. Rev. Lett. 69,* 1741–1744.

Kasevich, M., D. S. Weiss, E. Riis, K. Moler, S. Kasapi, and S. Chu (1991). Atomic velocity selection using stimulated Raman transitions. *Phys. Rev. Lett. 66,* 2297–2300.

Kastberg, A., P. S. Jessen, W. D. Phillips, S. L. Rolston, and R. J. C. Spreeuw (1994). 3–D optical lattices for cesium atoms. *Phys. Rev. Lett.* submitted, .

Kastler, A. (1950). Quelques suggestions concernant la production optique et la détection optique d'une inégalité de population des niveaux de quantification spatiale des atomes. Application a l'éxperience de Stern et Gerlach et a la résonance magnétique. *J. Physique Rad. 11,* 255–265.

Kastler, A. (1967). Optical methods for studying Hertzian resonances. *Science 158,* 214–221.

Kazantsev, A. P., and I. V. Krasnov (1989). Rectification effect of a radiation force. *J. Opt. Soc. Am. B 6,* 2140–2148.

Keith, D., C. R. Ekstrom, Q. A. Turchette, and D. E. Pritchard (1991). An interferometer for atoms. *Phys. Rev. Lett. 66,* 2693–2696.

Kenty, C. (1932). On radiation diffusion and the rapidity of escape of resonance radiation from a gas. *Phys. Rev. 42,* 823–842.

Ketterle, W., K. B. Davis, M. A. Joffe, A. Martin, and D. E. Pritchard (1993). High densities of cold atoms in a dark spontaneous-force optical trap. *Phys. Rev. Lett. 70,* 2253–2256.

Khitrova, G., J. F. Valley, and H. M. Gibbs (1988). Gain-feedback approach to optical instabilities in sodium vapor. *Phys. Rev. Lett. 60,* 1126–1129.

Kimble, H. J. (1992). Squeezed states of light: an (incomplete) survey of experimental progress and prospects. *Phys. Rep. 219,* 227–234.

Kitano, M., T. Yabuzaki, and T. Ogawa (1981a). Optical tristability. *Phys. Rev. Lett. 46,* 926–929.

Kitano, M., T. Yabuzaki, and T. Ogawa (1981b). Self-sustained precession in an optical tristable system. *Phys. Rev. A 24,* 3156–3159.

Klepel, H., and D. Suter (1992). Transverse optical pumping with polarization-modulated light. *Optics Commun. 90,* 46–50.

Kleppner, D. (1981). Inhibited spontaneous emission. *Phys. Rev. Lett. 47,* 233–236.

Kocharovskaya, O. (1992). Amplification and lasing without inversion. *Phys. Rep. 219,* 175–190.

Kohler, B., S. Bernet, A. Renn, and W. P. Wild (1993). Storage of 2000 holograms in a photochemical hole-burning system. *Optics Lett. 18,* 2144–2146.

Köhler, J., J. A. J. M. Disselhorst, M. C. J. M. Donckers, E. J. J. Groenen, J. Schmidt, and W. E. Moerner (1993). Magnetic resonance detection of a single molecular spin. *Nature 363,* 242–243.

Kohmoto, T., Y. Fukuda, M. Tanigawa, T. Mishina, and T. Hashi (1983). Quantum-beat free-induction decay in Tm^{2+}:SrF_2: Fourier transform ESR spectroscopy by optical means. *Phys. Rev. B 28,* 2869–2872.

Köster, E., J. Mlynek, and W. Lange (1985). Zeeman coherences in degenerate backward four-wave mixing: saturation studies on coupled transitions. *Optics Commun. 53,* 53–58.

Kristensen, M., and J. P. Woerdman (1994a). Is photon angular momentum conserved in a dielectric medium? *Phys. Rev. Lett. 72,* 2171–2174.

Kristensen, M., M. A. v. Eijkelenborg, and J. P. Woerdman (1994b). Faraday effect as a probe of hindered electronic precession in atoms. *Phys. Rev. Lett. 72,* 2155–2158.

Kügler, K. J., W. Paul, and U. Trinks (1978). A magnetic storage ring for neutrons. *Phys. Lett. 72B,* 422–424.

Kulina, P., and R. H. Rinkleff (1982). Determination of the tensor polarizabilities in Yb and Sm by quantum beat spectroscopy. *Z. Phys. A 304,* 371–372.

Kupriyanov, D. V., and M. Sokolov (1988a). Quantum features in the noise spectrum in radiation that has passed through a nonequilibrium gaseous medium. *Sov. Phys. JETP 68,* 1145–1149.

Kupriyanov, D. V., I. M. Sokolov, and S. V. Subbotin (1988b). Effect of the polarization of atoms on the propagation of quasi-resonance emission in a gaseous medium. *Sov. Phys. JETP 66,* 71–78.

Kurnit, N. A., I. D. Abella, and S. R. Hartmann (1964). Observation of a photon echo. *Phys. Rev. Lett. 13,* 567–568.

Lamb, W. E., and R. C. Retherford (1947). Fine structure of the hydrogen atom by a microwave field. *Phys. Rev. 72,* 241–243.

Lamb, W. E., R. R. Schlicher, and M. O. Scully (1987). Matter-field interaction in atomic physics and quantum optics. *Phys. Rev. A 36,* 2763–2772.

Lange, W., and J. Mlynek (1978). Quantum beats in transmission by time-resolved polarization spectroscopy. *Phys. Rev. Lett. 40,* 1373–1375.

Lange, W., G. Ankerhold, M. Schiffer, D. Mutschall, and T. Scholz (1992). *Nonlinear radiation trapping in alkaline vapors.* 18 IQEC '92, Vienna.

Laubereau, A., G. Wochner, and W. Kaiser (1976). Ultrafast coherent excitation and probing of molecular vibrations with isotopic substructure. *Phys. Rev. A 13,* 2212–2225.

LeBerre, M., E. Ressayre, and A. Tallet (1982). Self-focusing and spatial ringing of intense cw light propagating through a strong absorbing medium. *Phys. Rev. A 25,* 1604–1618.

LeBerre, M., E. Ressayre, and A. Tallet (1984a). Resonant self-focusing of a cw intense light beam. *Phys. Rev. A 29,* 2669–2676.

LeBerre, M., E. Ressayre, and A. Tallet (1985). Quasi-trapping of Gaussian beams in two-level systems. *J. Opt. Soc. Am. B 2,* 956–967.

LeBerre, M., E. Ressayre, A. Tallet, K. Tai, H. M. Gibbs, M. C. Rushford, and N. Peyghambarian (1984b). Continuous-wave off-resonance rings and continuous-wave on-resonance enhancement. *J. Opt. Soc. Am. B 1,* 591–605.

Legowski, S. (1964). Relaxation of optically pumped cesium atoms by different buffer gases. *J. Chem. Phys. 41,* 1313–1317.

Lehmann, J. C. (1964). Etude de l'influence de la structure hyperfine du niveau excité sur l'obtention d'une orientation nucléaire par pompage optique. *J. Phys. 25,* 809–824.

Lehmitz, H., and H. Harde (1986). Polarization-selective detection of hyperfine quantum beats in Cs, in *Methods of laser spectroscopy,* ed. A. B.-R. Y. Prior and M. Rosenbluh. Plenum Press, New York.

Lehmitz, H., W. Kattau, and H. Harde (1986). Modulated pumping in Cs with picosecond pulse trains, in *Methods of laser spectroscopy,* ed. A. B.-R. Y. Prior and M. Rosenbluh. Plenum Press, New York.

Lett, P. D., R. N. Watts, C. I. Westbrook, W. D. Phillips, P. L. Gould, and H. J. Metcalf (1988). Observation of atoms laser cooled below the Doppler limit. *Phys. Rev. Lett. 61,* 169–172.

Levenson, M. D., and G. L. Eesley (1979). Polarization selective optical heterodyne detection for dramatically improved sensitivity in laser spectroscopy. *Appl. Phys. 19,* 1–17.

Lindquist, K., M. Stephens, and C. Wieman (1992). Experimental and theoretical study of the vapor-cell Zeeman optical trap. *Phys. Rev. A 46,* 4082–4090.

Lorentz, H. A. (1880). Über die Beziehung zwischen der Fortpflanzungs-geschwindigkeit des Lichtes und der Körperdichte. *Ann. Phys. Chem. 9,* 641–665.

Lorenz, L. (1881). Über die Refractionsconstante. *Wiedem. Ann. 11,* 70–103.

Loudon, R. (1983). *The quantum theory of light.* Clarendon Press, Oxford.

Lowe, I. J., and R. E. Norberg (1957). Free-induction decays in solids. *Phys. Rev. 107,* 46–61.

Lowe, R. M., D. S. Gough, R. J. McLean, and P. Hannaford (1987). Determination of ground-state depolarization rates by transmission Zeeman beat spectroscopy: Application to Sm I $4f^6 6s^2\ ^7F_1$. *Phys. Rev. A 36,* 5490–5493.

Lukac, M., and E. L. Hahn (1988). External reflection and transmission spectroscopy of Pr3+:LaF3 by Stark modulated optical pumping. *J. Luminesc. 42,* 257–265.

Macaluso, D., and O. M. Corbino (1898). Sopra una nuova azione che la luce subisce attraversando alcuni vapori metallici in un campo magnetico. *Nuovo Cimento Ser. 4 8,* 257–258.

Macaluso, D., and O. M. Corbino (1899). Sulle relazione tra il fenomeno di Zeeman e la rotazione magnetica anomala del piano di polarizzazione della luce. *Nuovo Cimento 9,* 384–389.

Macfarlane, R. M., R. M. Shelby, and R. L. Shoemaker (1979). Ultrahigh-resolution spectroscopy: Photon echoes in $YAlO_3:Pr^{3+}$ and $LaF_3:Pr^{3+}$. *Phys. Rev. Lett. 43,* 1726–1730.

Maeda, M. W., P. Kumar, and J. H. Shapiro (1987). Squeezing experiments in sodium vapor. *J. Opt. Soc. Am. B 4,* 1501–1513.

Maki, J. J., W. V. Davis, R. W. Boyd, and J. E. Sipe (1992). Phase conjugation using the surface nonlinearity of a dense potassium vapor. *Phys. Rev. A 46,* 7155–7161.

Maneuf, S., and F. Reynaud (1988a). Quasi-steady state self-trapping of first, second and third order subnanosecond soliton beams. *Optics Commun. 66,* 325–328.

Maneuf, S., R. Desailly, and C. Froehly (1988b). Stable self-trapping of laser beams: observation in a nonlinear planar waveguide. *Optics Commun. 65,* 193–198.

Manson, N. B., X. F. He, and P. T. H. Fisk (1990). Raman heterodyne detected electron-nuclear-double-resonance measurements of the nitrogen-vacancy center in diamond. *Optics Lett. 15,* 1094–1096.

Marburger, J. H., and E. Dawes (1968). Dynamical formation of a small-scale filament. *Phys. Rev. Lett. 21,* 556–558.

Mariotti, E., J. H. Xu, M. Allegrini, G. Alzetta, S. Gozzini, and L. Moi (1988). Light-induced drift stationary states. *Phys. Rev. A 38,* 1327–1334.

Martin, P. J., B. G. Oldaker, A. H. Miklich, and D. E. Pritchard (1988). Bragg scattering of atoms from a standing light wave. *Phys. Rev. Lett. 60,* 515–518.

Marxer, H., B. Derighetti, and E. Brun (1983). Superfluorescent transients of an inhomogeneously broadened Q-switched nuclear-magnetic-resonance system. *Phys. Rev. Lett. 50,* 958–961.

Mathur, B. S., H. Tang, and W. Happer (1968). Light shift in the alkali atoms. *Phys. Rev. 171,* 11–19.

Maudsley, A. A., A. Wokaun, and R. R. Ernst (1978). Coherence transfer echoes. *Chem. Phys. Lett. 55,* 9–14.

McClelland, D. E., R. J. Ballagh, and W. J. Sandle (1986). Simple analytical approximation to continuous-wave on-resonance beam reshaping. *J. Opt. Soc. Am. B 3,* 212–218.

McClelland, J. J., R. E. Scholten, E. C. Palm, and R. J. Celotta (1993). Laser-focused atomic deposition. *Science 262,* 877–880.

McCord, A. W., R. J. Ballagh, and J. Cooper (1988). Dispersive self-focusing in atomic media. *J. Opt. Soc. Am. B 5,* 1323–1334.

McLean, R. J., P. Hannaford, and R. M. Lowe (1990). Transmission Zeeman beat spectroscopy: application to the ^7F ground level term of Sm I. *Phys. Rev. A 42,* 6616–6628.

Meer, G. J. v. d., J. Smeets, E. R. Eliel, P. L. Chapovsky, and L. J. F. Hermans (1993). Velocity-dependent collision rates from light-induced drift experiments: C2H4–noble gas mixtures. *Phys. Rev. A 47,* 529–537.

Melsen, A. G. M. v. (1957). *Atom gestern und heute.* Verlag Karl Alber, München.

Meystre, P., and M. Sargent (1990). *Elements of quantum optics.* Springer, Berlin.

Migdall, A., J. V. Prodan, W. D. Phillips, T. H. Bergeman, and H. J. Metcalf (1985). First observation of magnetically trapped atoms. *Phys. Rev. Lett. 54,* 2596–2599.

Minogin, V. G., and O. T. Serimaa (1979). Resonant light pressure forces in a strong standing laser wave. *Optics Commun. 30,* 373–379.

Mishina, T., Y. Fukuda, and T. Hashi (1988). Optical generation and detection of $\Delta m = 2$ Zeeman coherence in the Cs ground state with a diode laser. *Optics Commun. 66,* 25–30.

Mishina, T., M. Tanigawa, Y. Fukuda, and T. Hashi (1987). Synchronized quantum beat echoes in Cs vapor with diode lasers. *Optics Commun. 62,* 166–170.

Mitschke, F., R. Deserno, W. Lange, and J. Mlynek (1986). Magnetically induced optical self-pulsing in a nonlinear resonator. *Phys. Rev. A 33,* 3219–3231.

Mitsunaga, M., E. S. Kintzer, and R. G. Brewer (1984). Raman heterodyne interference of inequivalent nuclear sites. *Phys. Rev. Lett. 52,* 1484–1487.

Mitsunaga, M., N. Uesugi, and K. Sugiyama (1993). Kilohertz-resolution pump-probe spectroscopy in Pr^{3+}:$YAlO_3$. *Optics Lett. 18,* 1256–1258.

Mitsunaga, M., R. Yano, and N. Uesugi (1991). Time- and frequency-domain hybrid optical memory: 1.6 kbit data storage in Eu^{3+}:Y_2SiO_5. *Optics Lett. 16,* 1890–1892.

Mlynek, J., and W. Lange (1979). A simple method for observing coherent ground state transients. *Optics Commun. 30,* 337–340.

Mlynek, J., R. Grimm, E. Buhr, and V. Jordan (1988). Raman heterodyne Ramsey spectroscopy in a Sm atomic beam. *Appl. Phys. B 45,* 77–82.

Mlynek, J., W. Lange, H. Harde, and H. Burggraf (1981a). High-resolution spectroscopy using pulse trains. *Phys. Rev. A 24,* 1099–1102.

Mlynek, J., C. Tamm, E. Buhr, and N. C. Wong (1984). Raman heterodyne detection of radio frequency resonances in Sm vapour: effects of velocity-changing collisions. *Phys. Rev. Lett. 53,* 1814–1817.

Mlynek, J., K. H. Drake, G. Kersten, D. Frölich, and W. Lange (1981b). Double-resonance polarization spectroscopy using modulation sidebands. *Optics Lett. 6,* 87–89.

Mlynek, J., N. C. Wong, R. G. DeVoe, E. S. Kintzer, and R. G. Brewer (1983). Raman heterodyne detection of nuclear magnetic resonance. *Phys. Rev. Lett. 50,* 993–996.

Moerner, W. E., and L. Kador (1989). Optical detection and spectroscopy of single molecules in a solid. *Phys. Rev. Lett. 62,* 2535–2538.

Möller, M., and W. Lange (1992). Measuring chaotic scenarios in a sodium-filled resonator. *Appl. Phys. B 55,* 126–131.

Morris, G. A., and R. Freeman (1978). Selective excitation in Fourier transform nuclear magnetic resonance. *J. Magn. Reson. 29,* 433–462.

Moskowitz, P. E., P. L. Gould, and D. E. Pritchard (1985). Deflection of atoms by standing-wave radiation. *J. Opt. Soc. Am. B 2,* 1784–1790.

Mossberg, T., A. Flusberg, R. Kachru, and S. R. Hartmann (1977). Tri-level echoes. *Phys. Rev. Lett. 39,* 1523–1526.

Narum, P., A. L. Gaeta, M. D. Skeldon, and R. W. Boyd (1988). Instabilities of laser beams counterpropagating through Brillouin-active medium. *J. Opt. Soc. Am. B 5,* 623–628.

Neuhauser, W., M. Hohenstatt, P. Toschek, and H. Dehmelt (1978). Optical-sideband cooling of visible atom cloud confined in parabolic well. *Phys. Rev. Lett. 41,* 233–236.

Noll, G., U. Siegner, S. G. Shevel, and E. O. Göbel (1990). Picosecond stimulated photon echo due to intrinsic excitations in semiconductor mixed crystals. *Phys. Rev. Lett. 64,* 792–795.

Olsen, M. K., K. M. Gheri, and D. F. Walls (1994). Bright squeezing from self-induced transparencies in dressed three-level atoms. *Phys. Rev. A 50,* 5289–5300.

Omont, A. (1977). Irreducible components of the density matrix. Application to optical pumping. *Progr. Quant. Electr. 5,* 69–138.

Oria, M., D. Bloch, M. Fichet, and M. Ducloy (1989). Efficient phase conjugation of a cw low-power laser diode in a short vapor cell at 852 nm. *Optics Lett. 14,* 1082–1084.

Ovchinnikov, Y. B., R. Grimm, A. I. Sidorov, and V. S. Letokhov (1994). Resonant light pressure on atoms in a bichromatic standing wave. *Optics and Spectroscopy 76,* 188–197.

Pancharatnam, S. (1956). Generalized theory of interference and its applications. *Proc. Ind. Acad. Sci. 44,* 247–262.

Pancharatnam, S. (1966). Light shift in semiclassical dispersion theory. *J. Opt. Soc. Am. B 56,* 1636–1636.

Pancharatnam, S. (1968). Modulated birefringence of a spin-assembly in magnetic resonance. *Phys. Lett. 27A,* 509–510.

Pappas, P. G., M. M. Burns, D. D. Hinshelwood, M. S. Feld, and D. E. Murnick (1980). Saturation spectroscopy with laser optical pumping in atomic barium. *Phys. Rev. A 21,* 1955–1968.

Partridge, R. B., and G. W. Series (1966). The modulated absorption of light in an optical pumping experiment on ^{4}He. *Proc. Phys. Soc. 88,* 969–982.

Paul, W. (1990). Electromagnetic traps for charged and neutral particles. *Rev. Mode. Phys. 62,* 531–540.

Pauliat, G., M. Ingold, and P. Günter (1989). Analysis of the buildup of oscillations in self-induced photorefractive ligh resonators. *IEEE J. Quantum Electron. 25,* 201–207.

Pearson, D. B., R. R. Freeman, J. E. Bjorkholm, and A. Ashkin (1980). Focusing and defocusing of neutral atomic beams using resonance-radiation pressure. *Appl. Phys. Lett. 36,* 99–101.

Penning, F. M. (1936). Die Glimmentladung bei niedrigem Druck zwischen koaxialen Zylindern in einem axialen Magnetifeld. *Physica 3,* 873–894.

Perkins, T. T., D. E. Smith, and S. Chu (1994a). Direct observation of tube-like motion of a single polymer chain. *Science 264,* 819–822.

Perkins, T. T., S. R. Quake, D. E. Smith, and S. Chu (1994b). Relaxation of a single DNA molecule observed by optical microscopy. *Science 264,* 822–826.

Pfau, T., C. Kurtsiefer, C. S. Adams, M. Sigel, and J. Mlynek (1993). Magneto-optical beam splitter for atoms. *Phys. Rev. Lett. 71,* 3427–3430.

Phillips, W. D. (1992). Laser cooling, optical traps and optical molasses, in *Fundamental systems in quantum optics; proceedings of the Les Houches summer school,* ed. J. Dalibard, J. M. Raimond and J. Zinn-Justin. North-Holland, Amsterdam.

Phillips, W., and H. Metcalf (1982). Laser deceleration of an atomic beam. *Phys. Rev. Lett. 48,* 596–599.

Phillips, W. D., J. V. Prodan, and H. J. Metcalf (1985). Laser cooling and electromagnetic trapping of neutral atoms. *J. Opt. Soc. Am. B 2,* 1751–1767.

Pinard, M., C. G. Aminoff, and F. Laloe (1979). Velocity-selective optical pumping and Doppler-free spectroscopy. *Phys. Rev. A 19,* 2366–2370.

Pinard, M., D. Grancelement, and G. Grynberg (1986). Continuous-wave self-oscillation using pair production of photons in four-wave mixing in sodium. *Europhys. Lett. 2,* 755–760.

Pines, A., M. G. Gibby, and J. S. Waugh (1973). Proton-enhanced NMR of dilute spins in solids. *J. Chem. Phys. 59,* 569–590.

Pirotta, M., F. Güttler, J. Sepiol, A. Renn, and U. P. Wild (1993). Single molecule spectroscopy: fluorescence lifetime of pentacene in p-terphenyl, in *Spectral holeburning and luminescence line narrowing: science and applications.* Monte verità, Ascona, Switzerland.

Ponti, A., and A. Schweiger (1994). Echo phenomena in electron paramagnetic resonance spectroscopy. *Appl. Magn. Reson. 7,* 363–403.

Prise, M. E., N. C. Craft, M. M. Downs, R. E. LaMarche, L. A. D'Asaro, L. M. F. Chirovsky, and M. J. Murdocca (1991). Optical digital processor using arrays of symmetric self-electrooptic effect devices. *Appl. Opt. 30,* 2287–2296.

Prodan, J., A. Migdall, W. D. Phillips, I. So, H. Metcalf, and J. Dalibard (1985). Stopping atoms with laser light. *Phys. Rev. Lett. 54,* 992–995.

Purcell, E. M., H. C. Torrey, and R. V. Pound (1946). Resonance absorption by nuclear magnetic moments in a solid. *Phys. Rev. 69,* 37–38.

Raab, E. L., M. Prentiss, A. Cable, S. Chu, and D. E. Pritchard (1987). Trapping of neutral sodium atoms with radiation pressure. *Phys. Rev. Lett. 59,* 2631–2634.

Rabi, I. I., J. R. Zacharias, S. Millman, and P. Kusch (1938). A new method of measuring nuclear magnetic moment. *Phys. Rev. 53,* 318–318.

Raftery, D., H. Long, L. Reven, P. Tang, and A. Pines (1992). NMR of optically pumped xenon thin films. *Chem. Phys. Lett. 191,* 385–390.

Raftery, D., H. Long, T. Meersmann, P. J. Grandinetti, L. Reven, and A. Pines (1991). High-field NMR of adsorbed xenon polarized by laser pumping. *Phys. Rev. Lett. 66,* 584–587.

Raizen, M. G., R. J. Thompson, R. J. Brecha, H. J. Kimble, and H. J. Carmichael (1989). Normal-mode splitting and linewidth averaging for two-state atoms in an optical cavity. *Phys. Rev. Lett. 63,* 240–243.

Ramsey, N. F. (1990). Experiments with separated oscillatory fields and hydrogen masers. *Rev. Mod. Phys. 62,* 541–552.

Reddy, B. R., and L. E. Erickson (1983). Nuclear-quadrupole-resonance study of the ground state of praseodymium in lanthanum trifluoride. *Phys. Rev. B 27,* 5217–5223.

Reintjes, J., C. Y. She, R. C. Eckardt, N. E. Karnangelen, R. C. Elton, and R. A. Andrews (1976). Generation of coherent radiation at 53.2 nm by fifth-harmonic conversion. *Phys. Rev. Lett. 37,* 1540–1543.

Reintjes, J., C. Y. She, R. C. Eckardt, N. E. Karnangelen, R. C. Elton, and R. A. Andrews (1977). Seventh harmonic conversion of mode-locked laser pulses to 38.0 nm. *Appl. Phys. Lett. 30,* 480–482.

Rempe, G., H. Walther, and N. Klein (1987). Observation of quantum collapse and revival in a one-atom maser. *Phys. Rev. Lett. 58,* 353–356.

Reynaud, S., A. Heidmann, E. Giacobino, and C. Fabre (1992). Quantum fluctuations in optical systems, in *Progress in optics 30,* ed. E. Wolf. Elsevier, Amsterdam.

Rosatzin, M., D. Suter, and J. Mlynek (1990). Light-shift-induced spin echoes in a J = 1/2 atomic ground state. *Phys. Rev. A 42,* 1839–1841.

Rosatzin, M., D. Suter, W. Lange, and J. Mlynek (1990). Phase and amplitude variations of optically induced spin transients. *J. Opt. Soc. Am. B 7,* 1231–1238.

Rosenfeld, L. (1929). Zur Theorie des Faradayeffektes. *Z. Phys. 57,* 835–854.

Rosker, M. J., F. W. Wise, and C. L. Tang (1986). Femtosecond relaxation dynamics of large molecules. *Phys. Rev. Lett. 57,* 321–324.

Salomon, C., J. Dalibard, A. Apsect, H. Metcalf, and C. Cohen-Tannoudji (1987). Channeling atoms in a laser standing wave. *Phys. Rev. Lett. 59,* 1659–1662.

Salomon, C., J. Dalibard, W. Phillips, A. Clarion, and S. Guellati (1990). Laser cooling of cesium atoms below 3μK. *Europhys. Lett. 12,* 683–688.

Schenzle, A., S. Grossman, and R. G. Brewer (1976). Theory of modulated photon echoes. *Phys. Rev. A 13,* 1891–1897.

Schenzle, A., N. C. Wong, and R. G. Brewer (1980). Theorem on coherent transients. *Phys. Rev. A 22,* 635–637.

Schenzle, A., N. C. Wong, and R. G. Brewer (1981). Theorem on coherent transients: response to a comment. *Phys. Rev. A 24,* 2250–2252.

Schiffer, M., G. Ankerhold, E. Cruse, and W. Lange (1994). Role of radiation trapping in degenerate four-wave mixing experiments. *Phys. Rev. A 49,* R1558–R1561.

Schlesser, R., and A. Weis (1992). Light-beam deflection by cesium vapor in a transverse-magnetic fields. *Optics Lett. 17,* 1015–1017.

Schuller, F., M. J. D. MacPherson, D. N. Stacey, R. B. Warrington, and K. P. Zetie (1991). The Voigt effect in a dilute atomic vapor. *Optics Commun. 86,* 123–127.

Schweiger, A., and H. H. Guenthard (1981). Electron nuclear double resonance with

circularly polarized radio frequency fields (CP-ENDOR); theory and applications. *Mol. Phys. 42,* 283–295.
Scully, M. O. (1991). Enhancement of the index of refraction via quantum coherence. *Phys. Rev. Lett. 67,* 1855–1858.
Scully, M. O. (1992). From lasers and masers to phaesonium and phasers. *Phys. Rep. 219,* 191–201.
Segert, J. (1987). Non-abelian Berry's phase effects and optical pumping of atoms. *Ann. Phys. 179,* 294–312.
Sepiol, J., F. Güttler, M. Pirotta, A. Renn, and U. P. Wild (1993). Single molecule spectroscopy: polarization measurements of pentacene in p-Terphenyl, *Spectral holeburning and luminescence line narrowing: science and applications.* Monte verità, Ascona, Switzerland.
Serber, R. (1932). The theory of the Faraday effect in molecules. *Phys. Rev. 41,* 489.
Series, G. W. (1966). Theory of the modulation of light in optical pumping experiments. *Proc. Phys. Soc. 88,* 957–968.
Sesko, D. W., T. G. Walker, and C. E. Wieman (1991). Behavior of neutral atoms in a spontaneous force trap. *J. Opt. Soc. Am. B 8,* 946–958.
Shang, S. Q., B. Sheehy, H. Metcalf, P. v. Straten, and G. Nienhuis (1991). Velocity-selective resonances and sub-Doppler laser cooling. *Phys. Rev. Lett. 67,* 1094–1097.
Shelby, R. M., and R. M. Macfarlane (1984). Hyperfine spectroscopy and spin dynamics by optical and sublevel coherence. *J. Luminesc. 31,* 839–844.
Shelby, R. M., A. C. Tropper, R. T. Harley, and R. M. Macfarlane (1983). Measurement of the hyperfine structure of Pr^{3+}:YAG by quantum-beat free-induction decay, hole-burning, and optically detected nuclear quadrupole resonance. *Optics Lett. 8,* 304–306.
Shen, Y. R. (1984). *The principles of nonlinear optics.* Wiley, New York.
Shoemaker, R. L., and R. G. Brewer (1972). Two-photon superradiance. *Phys. Rev. Lett. 28,* 1430–1433.
Silverman, M. P., S. Haroche, and M. Gross (1978a). General theory of laser-induced quantum beats. I. Saturation effects of single laser excitation. *Phys. Rev. A 18,* 1507–1516.
Silverman, M. P., S. Haroche, and M. Gross (1978b). General theory of laser-induced quantum beats: II. Sequential laser excitation; effects of external static fields. *Phys. Rev. A 18,* 1517–1528.
Simonyi, K. (1990). *Kulturgeschichte der Physik.* Verlag Harri Deutsch, Thun.
Sleator, T., T. Pfau, V. Balykin, and J. Mlynek (1992). Imaging and focusing of an atomic beam with a large period standing light wave. *Appl. Phys. B 54,* 375–379.
Slusher, R. E., L. W. Hollberg, B. Yurke, J. C. Mertz, and J. F. Valley (1985). Observation of squeezed states by four-wave mixing in an optical cavity. *Phys. Rev. Lett. 55,* 2409–2412.
Slusher, R. E., B. Yurke, P. Grangier, A. LaPorta, D. F. Walls, and M. Reid (1987). Squeezed-light generation by four-wave mixing near an atomic resonance. *J. Opt. Soc. Am. B 4,* 1453–1464.
Snyder, A. W., D. J. Mitchell, L. Poladian, and F. Ladouceur (1991). Self-induced optical fibers: spatial solitary waves. *Optics Lett. 16,* 21–23.
Söding, J., and R. Grimm (1994). Stimulated magneto-optical force in the dressed-atom picture. *Phys. Rev. A 50,* 2517–2527.
Söding, J., R. Grimm, J. Kowalski, Y. B. Ovchinnikov, and A. I. Sidorow (1992).

Observation of the magnetooptical radiation force by laser spectroscopy. *Europhys. Lett. 20,* 101–106.
Solomon, I. (1958). Multiple echoes in solids. *Phys. Rev. 110,* 61–65.
Steel, D. G., and R. C. Lind (1981). Multiresonant behavior in nearly degenerate four-wave mixing: the ac Stark effect. *Optics Lett. 6,* 587–589.
Steinberg, A. M., and R. Y. Chiao (1994a). Dispersionless, highly superluminal propagation in a medium with a gain doublet. *Phys. Rev. A 49,* 2071–2075.
Steinberg, A. M., and R. Y. Chiao (1994b). Tunneling delay times in one and two dimensions. *Phys. Rev. A 49,* 3283.
Stenholm, S. (1986). The semiclassical theory of laser cooling. *Rev. Mod. Phys. 58,* 699–739.
Suter, D. (1992a). Optically excited zeeman coherences in atomic groundstates: nuclear spin effects. *Phys. Rev. A 46,* 344–350.
Suter, D. (1992b). Sensitivity of optically excited and detected magnetic resonance. *J. Magn. Reson. 99,* 495–506.
Suter, D. (1993). Polarization oscillations of coupled laser beams in an optically pumped atomic vapour. *Optics Commun. 95,* 255–259.
Suter, D., and T. Blasberg (1993). Stabilisation of transverse solitary waves by a nonlocal response of the nonlinear medium. *Phys. Rev. A 48,* 4583–4587.
Suter, D., and H. Klepel (1992a). Indirect observation of 'forbidden' Raman transitions by laser-induced coherence transfer. *Europhys. Lett. 19,* 469–474.
Suter, D., and H. Klepel (1992b). Two-dimensional spectroscopy of 'forbidden' Raman transitions by laser-induced coherence transfer. *Helv. Phys. Acta 65,* 832–833.
Suter, D., and T. Marty (1993a). Coherent Raman beats: high order interference effects. *Optics Lett. 18,* 1663–1665.
Suter, D., and T. Marty (1993b). Laser induced dynamics of atomic sublevel coherences. *Optics Commun. 100,* 443–450.
Suter, D., and J. Mlynek (1991a). Dynamics of atomic sublevel coherences during modulated optical pumping. *Phys. Rev. A 43,* 6124–6134.
Suter, D., and J. Mlynek (1991b). Laser excitation and detection of magnetic resonance, in *Advanced Magnetic Optics Resonance*, ed. W. S. Warren. Academic Press, San Diego.
Suter, D., and J. G. Pearson (1988). Experimental classification of multi-spin coherence under the full rotation group. *Chem. Phys. Lett. 144,* 328–332.
Suter, D., J. Aebersold, and J. Mlynek (1991a). Evanescent wave spectroscopy of sublevel resonances near a glass/vapor interface. *Optics Commun. 84,* 269–274.
Suter, D., H. Klepel, and J. Mlynek (1991b). Time-resolved two-dimensional spectroscopy of optically driven atomic sublevel coherences. *Phys. Rev. Lett. 67,* 2001–2004.
Suter, D., M. Rosatzin, and J. Mlynek (1991c). Optically induced coherence transfer echoes between Zeeman substates. *Phys. Rev. Lett. 67,* 34–37.
Szabo, A. (1986). On-axis photon echo modulation in Ruby. *J. Opt. Soc. Am. B 3,* 514–522.
Szabo, A., and R. Kaarli (1991). Optical hole burning and spectral diffusion in ruby. *Phys. Rev. B 44,* 12307–12313.
Szabo, A., T. Muramoto, and R. Kaarli (1990). ^{27}Al nuclear-spin dephasing in the ruby frozen core and Cr^{3+} spin-flip-time measurements. *Phys. Rev. B 42,* 7769–7776.
Szoke, A., V. Daneu, J. Goldhar, and N. A. Kurnit (1969). Bistable optical element and its applications. *Appl. Phys. Lett. 15,* 376–379.

Tam, A. C., and W. Happer (1977). Long-range interaction between cw self-focused laser beams in an atomic vapour. *Phys. Rev. Lett. 38,* 278–282.
Tamm, C., E. Buhr, and J. Mlynek (1986). Raman heterodyne studies of velocity diffusion effects in radio-frequency-laser double resonance. *Phys. Rev. A 34,* 1977–1993.
Tanaka, M., T. Ohshima, K. Katori, M. Fujiwara, T. Itahashi, H. Ogata, and M. Kondo (1990). Depolarization of optically pumped sodium atoms by wall surfaces. *Phys. Rev. A 41,* 1496–1504.
Tanigawa, M., Y. Fukuda, T. Mishina, and T. Hashi (1991). Optical-pumping model of synchronized quantum beat echoes. *J. Opt. Soc. Am. B 9,* 313–320.
Tanigawa, M., Y. Fukuda, T. Kohmoto, K. Sakuno, and T. Hashi (1983). Sublevel echoes selectively excited by light-pulse trains: synchronized-quantum-beat echoes. *Optics Lett. 8,* 620–622.
Tanner, C. E., and C. E. Wieman (1988). Precision measurement of the Stark shift in the $6S_{1/2}$ -> $6P_{3/2}$ cesium transition using a frequency-stabilized laser diode. *Phys. Rev. A 38,* 162–165.
Thompson, R. J., G. Rempe, and H. J. Kimble (1992). Observation of normal-mode splitting for an atom in an optical cavity. *Phys. Rev. Lett. 68,* 1132–1135.
Tomita, A., and R. Y. Chiao (1986). Observation of Berry's topological phase by use of an optical fiber. *Phys. Rev. Lett. 57,* 937–940.
Torrey, H. C. (1949). Transient nutations in nuclear magnetic resonance. *Phys. Rev. 76,* 1059–1068.
Tremblay, P., A. Michaud, M. Levesque, S. Thériault, M. Breton, J. Beaubien, and N. Cyr (1990). Absorption profiles of alkali-metal D lines in the presence of a static magnetic field. *Phys. Rev. A 42,* 2766–2773.
Ungar, P. J., D. S. Weiss, E. Riis, and S. Chu (1989). Optical molasses and multilevel atoms: theory. *J. Opt. Soc. Am. B 6,* 2023–2071.
vanStryland, E. W., and R. L. Shoemaker (1979). Modulated coherent Raman beats. *Phys. Rev. A 20,* 1376–1384.
Vedenin, V. D., F. S. Ganikhanov, S. Dinev, N. I. Koroteev, V. N. Kuliasov, V. B. Morozov, and V. G. Tunkin (1989). Time-domain polarization coherent anti-Stokes Raman spectroscopy of Tl atoms: dephasing measurements of separate multipole moments. *Optics Lett. 14,* 113–115.
Veer, W. E. v. d., R. J. J. v. Diest, A. Doenszelmann, and H. B. v. L. v. d. Heuvell (1993). Experimental demonstration of light amplification without population inversion. *Phys. Rev. Lett. 70,* 3243–3246.
Voelker, S. (1989). Hole-burning spectroscopy. *Ann. Rev. Phys. Chem. 40,* 499–530.
Vogelsanger, B., and A. Bauder (1990). Two-dimensional microwave Fourier transform spectroscopy. *J. Chem. Phys. 92,* 4101–4114.
Vogelsanger, B., A. Bauder, and H. Mäder (1989). Two-dimensional experiments with collision-induced transfer of populations in microwave Fourier transform spectroscopy. *J. Chem. Phys. 91,* 2059–2068.
Voigt, M. (1908). *Magneto- und Elektrooptik.* Teubner, Leipzig.
Voitsekhovich, V. S., M. V. Danileiko, A. M. Negriiko, V. I. Romanenko, and L. P. Yatsenko (1989). Observation of a stimulated radiation pressure of amplitude-modulated light on atoms. *JETP Letters 49,* 161–164.
Voitsekhovich, V. S., M. V. Danileiko, A. M. Negriiko, V. I. Romanenko, and L. P. Yatsenko (1991). Stimulated light pressure on atoms in counterpropagating amplitude-modulated waves. *Sov. Phys. JETP 72,* 219–227.
Wagner, W. G., H. A. Haus, and J. H. Marburger (1968). Large-scale self-trapping of optical beams in the paraxial ray approximation. *Phys. Rev. 175,* 256–266.

Wald, L. L., E. L. Hahn, and M. Lukac (1992). Variation of the Pr^{3+} nuclear quadrupole resonance spectrum across the inhomogeneous optical line in Pr^{3+}:LaF_3. *J. Opt. Soc. Am. B 9,* 784–788.

Walker, T., D. Sesko, and C. Wieman (1990). Collective behaviour of optically trapped neutral atoms. *Phys. Rev. Lett. 64,* 408–411.

Walls, D. F. (1979). Evidence for the quantum nature of light. *Nature 280,* 451–454.

Walmsley, I. A., M. Mitsunaga, and C. L. Tang (1988). Theory of quantum beats in optical transmission-correlation and pump-probe experiments for a general Raman configuration. *Phys. Rev. A 38,* 4681–4689.

Walther, H. (1988a). The single atom maser and the quantum electrodynamics in a cavity. *Physica Scripta T23,* 165–169.

Walther, H. (1988b). Single-atom oscillators. *Europhysics News 19,* 105–108.

Walther, H. (1992). Experiments on cavity quantum electrodynamics. *Phys. Rep. 219,* 263–281.

Wei, C., and N. B. Manson (1994). Experimental investigations of the absorption and dispersion profiles of a strongly driven transition: two-level system with a weak probe. *Phys. Rev. B 49,* 4751–4764.

Weis, A., J. Wurster, and S. I. Kanorsky (1993). Quantitative interpretation of the nonlinear Faraday effect as a Hanle effect of light-induced birefringence. *J. Opt. Soc. Am. B 10,* 716–724.

Weissbluth, M. (1978). *Atoms and molecules.* Academic Press, San Diego.

Weisskopf, V., and E. Wigner (1930). Berechnung der natürlichen Linienbreiten auf Grund der Diracschen Lichttheorie. *Z. Phys. 63,* 54–73.

Werij, H. G. C., J. P. Woerdman, J. J. M. Beenakker, and L. Kuscer (1984). Demonstration of a semipermeable optical piston. *Phys. Rev. Lett. 52,* 2237–2240.

Whittaker, E. A., and S. R. Hartmann (1982). Hyperfine structure of the $^1D_2$–$^3H_4$ levels of Pr^{3+}:LaF_3 with the use of photon echo modulation spectroscopy. *Phys. Rev. B 26,* 3617–3621.

Wigner, E. P. (1931). *Gruppentheorie und ihre Anwendung auf die Quantenmechanik der Atomspektren.* Vieweg, Braunschweig.

Wild, U. P., S. Bernet, B. Kohler, and A. Renn (1992). From supramolecular photochemistry to the molecular computer. *Pure Appl. Chem. 64,* 1335–1342.

Wild, U. P., F. Guettler, M. Pirotta, and A. Renn (1992). Single molecule spectroscopy: Stark effect of pentacene in p-terphenyl. *Chem. Phys. Lett. 193,* 451–455.

Wineland, D., and H. Dehmelt (1975). Proposed 10^{-14} $\Delta\nu/\nu$ laser fluorescence spectroscopy on Tl^+ mono-ion oscillator III. *Bull. Am. Phys. Soc. 20,* 637–637.

Wineland, D. J., R. E. Drullinger, and F. L. Walls (1978). Radiation-pressure cooling of bound resonant absorbers. *Phys. Rev. Lett. 40,* 1639–1642.

Woerdman, J. P., G. Nienhuis, and I. Kuscer (1992). Is it possible to rotate an atom? *Optics Commun. 93,* 135–144.

Wokaun, A., and R. R. Ernst (1977). Selective excitation and detection in multilevel spin systems: application of single transition operators. *J. Chem. Phys. 67,* 1752–1758.

Wong, G. K. L., and Y. R. Shen (1972). Study of spectral broadening in a filament of light. *Appl. Phys. Lett. 21,* 163–165.

Wong, N. C., E. S. Kintzer, J. Mlynek, R. G. DeVoe, and R. G. Brewer (1983). Raman heterodyne detection of nuclear magnetic resonance. *Phys. Rev. B 28,* 4993–5010.

Wrachtrup, J., C. v. Borczyskowski, J. Bernard, M. Orrit, and R. Brown (1993). Optical detection of magnetic resonance in a single molecule. *Nature 363*, 244–245.

Wu, F. Y., S. Ezekiel, M. Ducloy, and B. R. Mollow (1977). Observation of amplification in a strongly driven two-level atomic system at optical frequencies. *Phys. Rev. Lett. 38*, 1077.

Wyborne, B. G. (1965). *Spectroscopic properties of rare earths.* Wiley Interscience, New York.

Yabuzaki, T., H. Hori, and M. Kitano (1982). Theoretical study on optical-pumping self-focusing. *Jap. J. Appl. Phys. 21*, 504–512.

Yano, R., M. Mitsunaga, and N. Uesugi (1991). Ultralong optical dephasing time in $Eu^{3+}:Y_2SiO_5$. *Optics Lett. 16*, 1884–1886.

Yodh, A. G., T. W. Mossberg, and J. E. Thomas (1986). Multipole-specific, model independent velocity-change spectra of collisionally perturbed $^3P_1$–state ^{174}Yb atoms. *Phys. Rev. A 34*, 5150–5153.

Yodh, A. G., J. Golub, N. W. Carlson, and T. W. Mossberg (1984). Optically inhibited collisional dephasing. *Phys. Rev. Lett. 53*, 659–662.

Zavoisky, E. (1945). Spin-magnetic resonance in paramagnetics. *J. Phys. 9*, 245.

Zemansky, M. W. (1932). Note in the equivalent absorption coefficient for diffused resonance radiation. *Phys. Rev. 42*, 843–847.

Zetie, K. P., R. B. Warrington, M. J. D. Macpherson, D. N. Stacey, and F. Schuller (1992). Interpretation of nonlinear Faraday rotation in Sm vapour. *Optics Commun. 91*, 210–214.

Zewail, A. H. (1980). Optical molecular dephasing: principles of and probings by coherent laser spectroscopy. *Acc. Chem. Res. 13*, 360–368.

Zewail, A. H. (1988). Laser femtochemistry. *Science 242*, 1645–1653.

Zhu, Y. (1992). Lasing without inversion in a closed three-level system. *Phys. Rev. A 45*, R6149–R6152.

Index

Λ-type systems, 92
Absorption, 42, 46, 62, 104–105, 203, 282
Absorption coefficient, 206, 208
Absorption/emission cycle, 396
Absorption losses, 372
Absorptive detection, 235–236
Adiabatic approximation, 113
Adjacent transitions, 92
Alignment, 187, 189
Alkali atom, 163
Amplitudes, 311
Angular momentum, 131–137; conservation, 151–152, 161, 229; multiple reservoirs, 153–154; selection rules, 152–153
Anisotropic collisions, 161
Anisotropic susceptibility, 241
Annihilation operators, 9
Anticrossing, 43, 59
Anti-Helmholz configuration, 413
Antiholes, 201, 258
Antiparticles, 360
Anti-Stokes, 106, 248, 250, 278–279
Apparent magnetic field, 306
Arbitrary polarisation, 223–225
Aristotle, 1
Asymmetry parameter, 252–253
Atomic diffusion, 364
Atomic fountain, 414
Atomic hypothesis, 1–2
Atomic velocity, 386
Atom–light interaction, 12–20
Atoms: early, 1–3; energy levels, 3–4; internal structure, 2–3; sublevels (substates), 4–6

Balanced detection, 233–234
Basis operators, 49
Basis transformation, 102
Beam diameter, 371
Beam displacement, 229, 230
Beam profile, 362–363
Beam splitters, 404–405, 419
Beat signal, 266
Bessel function, 380, 381
Bichromatic excitation, 97–98, 274–276
Bichromatic field, 82–83, 97, 268, 272–273
Birefringence, 25, 224–225; circular, 210, 215, 223, 225, 247; linear, 240–241
Bloch equations, 33, 48, 60, 150, 169
Bloch–Siegert shift, 56
Block-diagonal structure, 41
Block structure, 85
Bohr, Niels, 3
Bohr–Sommerfeld model, 134
Boltzmann's law, 18
Bottleneck, 402
Buffer gas, 23, 164, 281

Capture range, 394
Chaos, 374, 375, 382–383
Chirp cooling, 411
Circular birefringence, 210, 215, 223, 225, 247
Circular dichroism, 210, 215, 223, 224
Circular polarisation, 12
Clock transition, 147
Closed two-level system, 410
Coherence, 33, 75–79, 281
Coherences between sublevels, 209

Coherence transfer, 77, 85–87, 90, 324–336, 338, 348, 350, 352
Coherence transfer echoes, 321, 349–353
Coherent emmission, 67–69
Coherent population trapping, 100–105
Coherent Raman beats, 266–269, 272, 275
Coherent Raman processes, 248–279; frequency domain experiments, 256–263; overview, 248–256; time-resolved experiments, 263–279
Coherent Raman scattering, 105–109, 226, 239, 249–251, 258–260, 266, 267
Coherent state, 10, 45
Collapse, 367
Collisional broadening, 283
Collision-induced relaxation, 167, 313
Collisions, 60, 110, 161, 164
Commutation relations, 48, 78
Complete basis, 139
Complex Lorentzian, 214
Cone, 174
Connected transition, 89
Conservation laws, 160–161
Conservation of angular momentum, 120, 129–130, 229
Continuous-wave (cw) excitation, 270
Continuous-wave (cw) experiments, 30
Coordinate rotations, 121
Coordinate system, 120
Coulombs law, 140
Counterpropagating laser beams, 392, 394, 412–414
Counterrotating components, 180
Coupled absorption, 359–361
Coupled beams, 374–376
Coupled wave equations, 227
Coupling coefficients, 128
Coupling scheme, 154
Coupling strengths, 97
Creation operators, 9
Critical intensity, 366
Critical value, 378
Cross peaks, 321, 327, 328, 335
Crystal, 417
Crystal field, 251–252, 254
Cummings collapse and revival, 46
Curie point, 120
Cycling transition, 410

D_1 line, 159
Damping, 99–100, 115–117, 162, 171, 174, 196–198, 229
Dark resonance, 100, 102
Dark state, 101
deBroglie wavelength, 37, 417, 421
Decay rate, 108, 170
Decreased absorption, 201
Defocusing, 298
Democritus, 1, 2
Density operator, 49–52; components, 336–338
Dephasing, 71, 111, 353
Depopulation pumping, 193
Derrick's theorem, 369
Destructive interference, 298
Detection period, 316
Detection system, 283
Diagonal initial condition, 84–85
Diagonal peaks, 321
Diamagnetic ground states, 198–200
Dichroism, 25; circular, 210, 215, 223, 224
Dielectric susceptibility, 11
Difference frequency, 109
Difference frequency mismatch, 98, 276
Difference in dispersion, 236
Differential absorption, 236
Diffusion, 109, 161, 320, 363–364
Dipole force, 388, 399–402
Dipole operator, 47
Dirac, P. A. M., 6
Dispersion, 46, 62, 104–105, 203
Dispersive detection, 236–237, 296
Displacement, 219, 228–230
Doping, 256
Doppler broadening, 87, 90, 281, 283, 323
Doppler limit, 397–399, 415, 419
Doppler shift, 60, 71, 390, 391, 394, 412, 414
Double quantum coherence, 352
Double resonance experiments, 315
Dressed states, 43, 58

Echo, 321
Echo formation, 89–90, 265–266, 322
Echo modulation, 263–265, 269
Effective atomic density, 210
Effective coupling strength, 103
Effective evolution, 118

Effective field, 57, 59, 174, 290, 300
Effective frequency, 99
Effective interactions, 28
Effective observables, 333–334, 338–339
Effective optical path length, 370
Effective potential, 417
Effective quadrupole interaction, 253
Effective Rabi frequency, 96
Eigenmodes, 7–9
Eigenpolarisations, 221–222
Einstein, Albert, 386
Electric dipole approximation, 14
Electric dipole interaction, 40, 140–141
Electric dipole moment, 48, 400
Electric dipole moment operator, 213
Electric dipole transitions, 75, 151–159
Electric field gradient tensor, 143
Electrodynamics, 3
Electronic structure of rare earth ions, 251–252
Electron paramagnetic resonance (EPR), 29, 34
Electrooptic modulator (EOM), 182, 376
Emission, 42
Enhanced nuclear Zeeman interaction, 254–255
Ensemble of atoms, 49
Entropy, 161
Equilibrium, far from, 325
Equilibrium magnetisation, 170
Euler angles, 123, 125
Evanescent wave, 403–404
Evaporative cooling, 411
Even harmonics, 184
Evolution, 84, 86–87, 93–95, 315, 326
Exchange of coherence, 264
Exchange of order, 325
Excitation bandwidth, 96, 185
Excitation efficiency, 95, 98, 276–277
Excitation period, 273
Excited state reorientation, 193, 246
Expansion coefficients, 138
Experimental setup, 283
Exponential decay, 64
External degrees of freedom, 16, 119–120

Faraday, Michael, 3
Faraday effect, 215–217, 220
Faraday rotation, 13, 216–217
Feedback, 377, 381
Feynman, Richard, 6
Field gradient (EFG) tensor, 252–253
Field inhomogeneity, 320
Field states, 9–10
Fine structure, 136
Focusing atomic beam, 405–406
Forbidden multipoles, 336–353
Forbidden transition, 90
Formalism, 10–11, 212–213
Fourier transformation, 271, 272, 276, 281, 288, 308, 309, 311, 313, 315, 326, 334, 346
Fourier transform pair, 70, 71
Fractal dimension, 382
Fraunhofer, Joseph, 3
Free atoms, 61
Free induction decay (FID), 285, 286, 291–296, 315, 317–319, 352
Free precession, 63–64, 66–67, 264, 265, 309, 316, 317
Frequency chirp, 391, 399
Frequency dependence, 62–63
Frequency-domain, 231, 270
Frequency-domain experiments, 256–263, 269, 307–308
Frequency doubling, 358
Frequency mismatch, 98
Friction force, 394

Gaussian beam, 300, 362–363, 405, 406
Generalised Rabi frequency, 43
Gradient (dipole) force, 388, 399–402
Grating, 404
Ground-state coherences, 116, 244
Group theory, 120
Growth rate, 378

Hahn echo, 352
Hanle effect, 218, 219
Harmonic oscillator, 7–8, 51
Heisenberg, Werner Karl, 3, 6
Heisenberg's uncertainty relation, 36
Heterodyne detection, 69–70, 231–233, 250
Hexadecupole element, 337
Hexadecupole moment, 139
Higher harmonics, 239
High-order effects, 237–239
Hilbert space, 84, 85, 126
Hole, 201, 258
Hyperfine coupling, 136
Hyperfine coupling constant, 136

Hyperfine interaction, 144–145, 191–192
Hyperfine multiplets, 147–148, 247
Hyperfine pumping, 187–188
Hyperfine splitting, 186, 246, 312
Hyperfine structure, 137

Ideal probe, 355
Incoherent mechanism, 75
Increased absorption, 257
Increased transmission, 257
Independent subspaces, 85
Index of absorption, 222
Index of refraction, 222
Index profile, 364, 365
Induced dipoles, 205
Inertial reference frame, 53
Inhomogeneous broadening, 201–202, 256, 267–268, 277
Inhomogeneous decay, 167
Inhomogeneous Hamiltonian, 71, 88
Inhomogeneous laser beam, 336
Initial condition, 171, 325
Intensity gradient, 408
Intensity modulation, 178, 273
Intensity profile, 364
Interaction Hamilton, 40, 77, 81–83
Interaction representation, 54, 80–82, 86, 107
Interference, 90
Intermediate field, 148
Internal angular momentum, 132
Internal degrees of freedom, 16, 119–159; angular momentum, 131–137; electric dipole transitions, 151–159; interaction with external fields, 140–151; multipole moments, 137–140
Internal dynamics, 314
Irradiation: of single transition, 79–81, 84; of two transitions, 81–83
Irreducible tensor operators, 77, 328, 330, 340
Isotropic atoms, 203–209

Jahn–Teller distortion, 120
Jaynes–Cummings model, 14, 38–39, 41–45, 58

Kerr effect, 358, 362
Kerr media, 366
Kramers–Konig relations, 104

Ladder-type systems, 92
Landé factor, 144, 145
Larmor precession, 149–151, 170, 175, 238, 339–340, 419
Laser cooling, 403
Laser detuning, 290, 291
Laser frequency detuning, 98–99
Laser frequency jitter, 278
Laser frequency mismatch, 276
Laser spectroscopy, 21–22; compared to magnetic resonance spectroscopy, 33–35
Lateral displacement, 229, 230
Lensing effect, 356, 406
Level crossing, 218
Level schemes, 22
Level shifts, 162
Lie group, 122
Lifetime broadening, 167, 320
Light: classical description, 10–12; mechanical effects of (*see* Mechanical effects of light); quantum theory of, 6–10
Light-induced drift, 162, 385
Light-induced forces, 385–388
Light shift, 24–25, 115–117, 171–174, 196–198, 229, 291, 326, 407, 415, 416
Limit cycles, 375, 380–381
Linear birefringence, 240–241
Linearly polarised light, 189, 191
Linear polarisation, 12
Lineshape function, 214
Local oscillator, 69–70
Longitudinal optical pumping, 165–166
Long-term behavior, 380
Loop gain, 381
Lorentz–Lorenz theory of dispersion, 16, 17, 46, 63, 203–204, 206–208
Losses, 381
Low-power approximation, 19

Magnetic dipole interaction, 143
Magnetic dipole moment, 143, 410–411
Magnetic dipole transition, 75, 90
Magnetic field, 130
Magnetic resonance spectroscopy, 5–6, 270; compared to laser spectroscopy, 33–35; evolution of technique, 29–30; principle, 28–29; sensitivity, 31–32
Magnetisation components, 212, 287–288

Magnetisation vector, 292
Magnetooptic effects, 215–220, 281
Magnetooptic traps, 412–414
Material excitation, 106, 108
Matrix elements, 47, 156–157
Maxwell, James Clerk, 6, 203, 386
Maxwell demon, 72
Maxwell's equations, 10–12, 27, 203, 205, 220, 221
Mechanical effects of light, 385–421; light-induced forces, 385–388; multilevel atoms, 409–421; spontaneous forces, 389–399; stimulated forces, 399–408
Microscopic order, 231, 280, 281
Microscopic processes, 211
Microscopic theory, 12–14
Mirror for atoms, 404
Mismatch, 97
Mixed phase lines, 343
Mixing, 115, 117
Mixing angle, 59, 93, 96, 99
Mixing matrix, 317
Mixing of density operator elements, 330
Mixing pulse, 316, 317, 321, 348
Modulated excitation, 303–308
Modulated light, 303
Modulated pump pulses, 306
Modulation phase, 352
Modulation schemes, 177
Momentum conservation, 386
Momentum transfer, 405
Monochromatic laser field, 81–82
Multilevel atoms, 409–421
Multilevel ground states, 186–202; observables in, 239–242
Multilevel systems, 17–20, 325, 367
Multiple absorption processes, 337
Multiple echoes, 321, 322, 349
Multiple scattering, 106, 239
Multipole components, 334, 342, 346, 350
Multipole expansion, 139, 168
Multipole moments, 137–140, 324, 337, 338
Multistability, 375

Nonequilibrium populations, 209
Nonlinear dynamics, 354–384; experimental observation, 369–372; light-induced waveguides, 361–366; overview, 354–361; self-focusing, 365–371; temporal instabilities, 374–384
Nonlinear Faraday effect, 217
Nonlocal medium response, 368, 369
Nuclear magnetic resonance (NMR), 29, 31, 34, 90, 176
Nuclear quadrupole coupling, 143
Nuclear quadrupole moment, 143, 252
Nuclear spin, 136, 158–159, 252–255, 312
Nuclear Zeeman interaction, 149–150, 311, 312
Number operator, 9
Number state, 9, 10

Object rotations, 121
Observables, 325; in multilevel ground states, 239–242
Observation, 321
Observation time, 282
Octupole moment, 139
Odd harmonics, 183
Off-resonance effects, 95–97
One-sided interaction, 354, 374
One-way processes, 15
Optical bistability, 383
Optical coherence, 107, 244
Optical dephasing rate, 214
Optical detuning, 290
Optical free induction decay, 69–71, 87, 88
Optically anisotropic vapours, 203–248; anisotropic media, 209–220; isotropic atoms, 203–209; polarisation-selective detection, 230–247; propagation, 220–230
Optical molasses, 393–395, 397, 398, 403, 414
Optical nonlinear media, 20, 356
Optical polarisation, 52, 66–68, 72, 220, 242, 244
Optical properties, 25–28; of polarised atomic media, 211–215
Optical pumping, 23–25, 34–35, 75, 102–103, 113–116, 160–202, 330, 358, 359, 361–362, 373, 377, 409–410, 415; modulated pumping, 174–186; multilevel ground states, 186–202; principle and overview, 160–163; two-level ground states, 163–174
Optical pumping rate, 164, 192–194

Optical Rabi frequency, 96, 277, 390, 393
Optical spectroscopy, 5, 31
Optical susceptibility, 220, 246
Optical tweezers, 402
Optical wave, 68
Orbital angular momentum, 132, 134–135
Orbitals, 3–4
Orientation dependence, 209
Oscillation amplitudes, 51
Oscillations, 289
Oscillatory exchange, 87

Packet of magnetisation, 176
Parity, 90, 142
Pauli, Wolfgang, 6
Periodic system of the elements, 2
Period-three orbit, 383
Perturbation theory, 55–56
Phase, 83
Phase factors, 124, 343, 351
Phase-sensitive detection, 305–307
Phase shifts, 56–57
Phase transition, 355
Phenomenological theories, 12–14
Photon echo, 70–73, 263
Photon momentum, 386
Photon recoil, 386, 389, 396
Plane waves, 11; eigenpolarisations of, 221–222
Polarisation, 12, 205
Polarisation instabilities, 384
Polarisation modulation, 181–186
Polarisation of the fluorescence, 190, 218, 284
Polarisation-selective detection, 230–247, 283, 284, 336
Polarisation states, 132
Polarised atomic media, optical properties of, 211–215
Polar vectors, 142
Population difference, 79, 114
Position operator, 141
Power broadening, 62
Poynting vector, 228
Precession, 63, 150–151, 174, 289, 292
Preparation period, 315
Pressure broadening, 109–110
Principal axis system, 254
Product states, 43
Propagtion, 220–230
Pseudo-quadrupole interaction, 253
Pseudo-spin, 48, 50–53, 56, 63–67
"Pudding with raisins" analogy, 2
Pulsed magnetic field, 297
Pulse durations, 277, 300–301
Pulses, 65–66; π pulse, 66
Pulse separation, 321
Pump beam polarisation, 376
Pump probe experiments, 256–257, 282–283, 313

Quadratic detector, 109, 232
Quadratic Zeeman interaction, 146–147, 312, 329
Quadrupole magnetic field, 413
Quadrupole moment (alignment), 139
Quadrupole splitting, 252–253
Quality factors, 14
Quantisation axis, 57, 212
Quantisation direction, 324
Quantum beats, 90–91
Quantum electrodynamics (QED), 6, 91, 207
Quantum mechanics, 3–6
Quantum optics, 35
Quantum theory of light, 6–10
Quantum well structures, 39
Quasi-stationary state, 112

Rabi flopping, 63–67
Rabi oscillation, 45
Rabi's beam experiment, 30
Rabi signal, 278
Radiation field, 131–134
Radiation trapping, 365
Raman beats, 92, 250, 268; coherent, 266–269, 272, 275–276
Raman excitation, 91–97, 106, 249–250, 276
Raman field, 105, 108, 109, 266
Raman heterodyne spectroscopy, 258–263
Raman process, 77 (*see also* Coherent Raman processes)
Raman pulses, 420
Raman scattering, 105–109, 240
Raman transitions, 419–421
Rare earth ions, 248; electronic structure of, 251–252
Rate equations, 110
Recoil limit, 417, 419
Rectified dipole force, 406–408

Reduced density operator, 49
Reduced dynamics, 163
Reduced matrix element, 157
Refocusing, 72–73, 263, 265, 298–300
Refocusing efficiency, 301–302
Refocusing pulse, 73, 264, 265
Refractive index, 206, 208, 361, 363, 365, 366, 373
Regular behaviour, 382
Relative orientation, 135
Relaxation, 59–60, 109, 111, 166–168
Relaxation rate, 165, 172
Rephasing, 353
Repopulation, 193
Repumping, 410
Repumping laser, 410
Reservoirs, 153–154
Resonant enhancement, 175–176
Ring structure, 369, 372
Rotating coordinate system, 79, 179
Rotating frame, 53, 55–56, 68–69, 303, 331, 341
Rotating frame approximation, 56
Rotating wave approximation, 41, 180
Rotation, 299
Rotational symmetry, 119–131, 137, 324, 339; Hamiltonian, 128–129
Rotation angle, 66, 300
Rotation axis, 173
Rotation matrices, 124
Rotation properties, 121, 126
Rotations, types of, 120–121
Rutherford, Ernest, 2

Saturation, 208
Saturation parameter, 61
Scaling factor, 181
Scattering force, 389–392
Scattering rate, 62, 390
Schrödinger, Erwin, 3
Schrödinger equation, 44–45, 52, 54, 60, 84, 120, 128–131, 150
Schwinger, Julian, 6
Selection rules, 151–154, 314, 327, 338, 410
Selective population, 160
Self-defocusing, 372
Self-focusing, 27–28, 365–371
Self-organisation, 375
Self-phase modulation, 373–374
Self-pulsing, 374, 383
Self-steepening, 374
Self-trapping, 366, 367, 372
Semiclassical description, 15–16, 46–63
Semiclassical theory, 207–208
Signal phase, 296
Sign of quadrupole interaction, 261
Single photon recoil limit, 417
Single transition: excitation of, 83–85; irradiation of, 79–81, 84
Single-transition operators, 77–79
Sisyphus cooling, 415–417
Sodium ground state, 191–196, 242–247
Soliton pulses, 374
Source, 220
Spectral density, 167–168
Spectral hole burning, 200, 201, 256–258, 261
Spectroscopic resolution, 26
Spectroscopy, 3 (*see also* Magnetic resonance spectroscopy; Two-dimensional spectroscopy)
Spherical basis, 133
Spherical harmonics, 124–125
Spherical symmetry, 120
Spherical tensor operators, 126–127
Spin angular momentum, 135
Spin echoes, 296–302
Spin nutation, 285–291
Spin operators, 49
Spin–orbit interaction, 135–136, 144, 154–158
Splitting, 96
Spontaneous decay rate, 111
Spontaneous emission, 44, 59, 62, 265, 387, 395, 402, 416
Spontaneous Raman scattering, 105
Spontaneous scattering force, 387, 388
Spontaneous symmetry breaking, 355
Stability of atoms, 3
Standing light waves, 404
Standing wave field, 406
Stark effect, 142
Stationary magnetisation, 172, 174
Stationary solution, 61–62, 172, 204, 214
Stationary state, 99, 107–108
Stimulated emission, 387
Stimulated forces, 399–408
Stimulated magnetooptic radiation force, 418–419
Stimulated Raman scattering, 106
Stokes, 105, 106, 248, 250, 278–279
Stokes parameters, 235

Strange attractor, 382
Strongly coupled system, 154
Sublevel coherences, 209, 217, 241, 242, 244, 322
Sublevel dynamics, 22–25, 280–313; experimental arrangement, 280–286; free induction decay (FID), 285, 286, 291–296; modulated excitation, 303–308; spin echoes, 296–302; spin nutation, 285–291; time-domain experiments, 308–313
Sublevel populations, 194
Sublevels (substates), 4–6
Sublevel transition, 276
Subspaces, 84
Superposition, 90
Superposition principle, 360
Superposition states, 5
Susceptibility, 208
Susceptibility tensor, 211, 220, 244–247
Symmetry adapted basis, 329–330
Symmetry-adapted representation, 130–131
Symmetry axis, 331
Symmetry properties, 120

Tensorial order, 343, 346
Tensor operators, 137
Thermodynamics, 160
Third harmonic generation, 358
Third-order tensor, 344
Thomson, Sir Joseph, 2
3J symbol, 128
Three-level echoes, 87–90, 353
Three-level system, 22, 74–118; dynamics, 83–99; overdamped systems, 109–118; phenomenological introduction, 74–77; steady-state effects, 99–109; system and Hamiltonian, 77–83
Three-wave mixing, 108
Tilted coordinate system, 63
Time-domain experiments, 30, 231, 269–274, 308–313, 315
Time-domain signal, 271, 288, 308, 310–311
Time-resolved experiments, 263–279
Tomonaga, Shinichiro, 6
Torque, 149
Transfer matrix, 321
Transformation properties, 54, 126
Transition strengths, 159, 415
Translational energy, 396–397
Translational temperature, 397–398
Transverse component, 291; of pseudospin, 50–51, 66–67
Transverse cooling, 399
Transverse magnetic fields, 174–175
Transverse optical pumping, 168–171, 283, 286
Transverse solitons, 366
Trapping, 36, 402; magnetooptic, 412–414; of neutral atoms, 403
Trapping state, 101–102, 420
Tri-level echo, 87–90
Triple resonance, 261–263
Two-beam Raman heterodyne experiment, 262
Two-dimensional Fourier transformation, 318–319
Two-dimensional Lorentzian, 319–320
Two-dimensional spectroscopy, 314–353; coherence transfer, 324–336; forbidden multipoles, 336–353; fundamentals, 314–324
Two-level atoms, 38–73; dynamics, 63–73; quantum mechanical description, 38–46; semiclassical description, 46–63
Two-level system, 16–22, 207
Two-photon process, 94
Two-photon transition, 249
Two-quantum spectrum, 348
Two-quantum transitions, 347–349

Uncoupled transition, 94
Unitary transformation, 86, 93

Velocity capture range, 394
Velocity diffusion, 395–398
Velocity distribution, 391
Velocity-selective optical pumping, 160–163, 188, 189
Vibrational states, 427
Virtual field, 297
Virtual magnetic field, 172
Voigt effect, 219–220
V-type systems, 92

Wall coatings, 23
Wave equation, 206, 227, 367–369
Waveguides, light-induced, 361–366

Wave mixing, 26–27, 77, 226–227, 357–359
Waves and particles, 35–37
Wigner–Eckart theorem, 127–128, 155
Wigner rotation matrices, 125
Zeeman effect, 143–147, 177, 270
Zeeman tuning, 411–413
Zero-field splitting, 142, 270
Zero-point energy, 8